I ıea
fin
re
Boo'

Fiber Optic Smart Structures

WILEY SERIES IN PURE AND APPLIED OPTICS

Founded by Stanley S. Ballard, University of Florida

EDITOR: Joseph W. Goodman, Stanford University

Fiber Optic Smart Structures

Edited By
ERIC UDD
Blue Road Research

A Wiley-Interscience Publication

John Wiley & Sons, Inc.

New York / *Chichester* / *Brisbane* / *Toronto* / *Singapore*

Library of Congress Cataloging in Publication Data:

Fiber optic smart structures / edited by Eric Udd.
      p.   cm. -- (Wiley series in pure and applied optics)
   "A Wiley-Interscience publication."
   Includes bibliographical references and index.
   ISBN 0-471-55448-0 (cloth)
   1. Fiber optic detectors. 2. Smart structures. 3. Composite
materials. I. Udd, Eric. II. Series.
TA1815.F53   1994
681'.2--dc20   93-37681

Printed in the United States of America

10 9 8 7 6 5 4 3 2

Contributors

ANDREW S. BICOS, McDonnell Douglas Space Systems Company, 5301 Bolsa Avenue, M/S 13-3, Huntington Beach, CA 92647

ZAFFIR CHAUDHRY, Department of Mechanical Engineering, Virginia Tech, Blacksburg, VA 24061

TIM CLARK, Amphenol Fiber Optic Products, 1925A Ohio Street, Lisle, IL 60532

RICHARD O. CLAUS, Fiber Optic Research Center, Virginia Tech, Blacksburg, VA 24061

JOHN P. DAKIN, University of Southampton, Optical Fiber Group, Department of Electronics & Computer Science, Southampton S09 5NH, United Kingdom

ABHIGIT DASGUPTA, Department of Mechanical Engineering, University of Maryland, College Park, MD 20742

JIM R. DUNPHY, United Technology Research Center, Silver Lane, East Hartford, CT 06108

PETER L. FUHR, Computer Science-Electrical Engineering Department, University of Vermont, Burlington, VT 05405

BARRY G. GROSSMAN, Department of Electrical/Computer Engineering, Florida Institute of Technology, Melbourne, FL 32901-6988

JORN S. HANSEN, Institute for Aerospace Studies, University of Toronto, 4925 Dufferin Street, Downsview, Ontario M3H 5T6, Canada

DRYVER R. HUSTON, University of Vermont, Department of Civil & Mechanical Engineering, Burlington, VT 05405

DAVID W. JENSEN, Civil and Environmental Engineering, Brigham Young University, Provo, UT 84602

ALAN D. KERSEY, NRL Code 6574, Naval Research Lab, 4555 Overlook Avenue. SW, Washington, DC 20375

MICHEL LE BLANC, Institute of Aerospace Studies, University of Toronto, 4925 Dufferin Street, Downsview, Ontario M3H 5T6, Canada

CHUNG E. LEE, FFPI Industries, Inc., 2306 Sandy Lane, Bryan, Texas 77801

RAM L. LEVY, McDonnell Douglas Aerospace, Mail Code 111 1041, P.O. Box 516, St. Louis, MO 63166

JEFFERY R. LORD, Simmonds Precision, Instrument Systems Division, Vergennes, VT 05491

RUSSELL G. MAY, Fiber Optic Research Center, Virginia Tech, Blacksburg, VA, 24061

RAYMOND M. MEASURES, Institute for Aerospace Studies, University of Toronto, 4925 Dufferin Street, Downsview, Ontario M3H 5T6, Canada

GERALD MELTZ, United Technology Research Center, Silver Lane, East Hartford, CT, 06108

WILLIAM W. MOREY, United Technology Research Center, Silver Lane, East Hartford, CT 06108

KENT A. MURPHY, Fiber Optic Research Center, Virginia Tech, Blacksburg, VA 24061

CRAIG ROGERS, Department of Mechanical Engineering, Virginia Tech, Blacksburg, VA 24061

SCOTT D. SCHWAB, McDonnell Douglas Aerospace, Mail Code 111 1041, P.O. Box 516, St. Louis, MO 63166

JAMES S. SIRKIS, Department of Mechanical Engineering, University of Maryland, College Park, MD 20742

HERB SMITH, McDonnell Douglas Aerospace, Mail Code 102 1310, P.O. Box 516, St. Louis, MO 63166

WILLIAM B. SPILLMAN, Simmonds Precision, Instrument Systems Division, Vergennes, VT 05491

V. S. SUDARSHANAM, Fiber Optic Research Center, Virginia Tech, Blacksburg, VA 24061

HENRY F. TAYLOR, Department of Electrical Engineering, Texas A & M University, College Station, TX 77843

MICHAEL H. THURSBY, Department of Electrical/Computer Engineering, Florida Institute of Technology, Melbourne, FL 32901-6988

ERIC UDD, Blue Road Research, 2555 N.E. 205th Avenue, Troutdale, OR 97060

ASHISH M. VENGSARKER, Fiber Optic Research Center, Virginia Tech, Blacksburg, VA 24061

ANBO WANG, Fiber Optic Research Center, Virginia Tech, Blacksburg, VA 24061

Contents

Preface

All around us are living things that sense and react to their environment in sophisticated ways. As structures have become more complex and are being asked to perform ever more difficult missions, there has been an ever-increasing need to build "intelligence" into them so that they can sense and react to their environment. Examples include buildings that can sense, react, and survive earthquakes and spacecraft that can sense and repair damage autonomously. To perform these functions successfully a "nervous system" is required that performs in a manner analogous to those of living things sensing the environment, conveying the information to a central processing unit (the brain) and reacting appropriately. Fiber optic technology has enabled the nerves of the system to be realized. Using hair-thin glass fibers as information carriers and sensors that may be built directly into the fibers with no increase in overall size, it is possible to create long strands of fiber sensors capable of measuring strain, pressure, temperature, and other key parameters. These sensor "strings" may then be embedded into the structural materials with no degradation in overall strength, resulting in "smart structures" with built-in nervous systems.

The functions of fiber optic smart structures are fourfold. The first function is to monitor key manufacturing parameters, such as temperature, pressure, viscosity, and water vapor content, during the manufacturing process using fiber sensors that have been embedded in various parts of a device or structure. The key function of fiber optic smart structures here is to augment the manufacturing process, improving yield and quality. The second aspect of fiber optic smart structures is to perform nondestructive evaluation of parts once they have been fabricated. This may involve the use of fiber sensors buried deep within the parts in combination with external sources generating acoustic or electromagnetic waves. The third function is to integrate the fiber sensors in the various parts into a health monitoring system capable of assessing the structural integrity of the system. Examples here include buildings that could phone in their status after an earthquake, or a launch vehicle on the pad that could assess its readiness for launch automatically. The fourth function is active control, where a structure senses environmentally induced structural changes and reacts in real time. An example is cracks in an airframe, resulting in automatic changes in the flight envelope.

This book provides a comprehensive overview of the field of fiber optic smart structures by many of the people who have pioneered the field. It is divided into three main parts. The first part provides an overview of the field and issues associated with the implementation of fiber optic smart structures. The second part covers fiber sensors that are being used to support fiber optic smart structures as well as the actuators and neural network processing that are important companion technologies. The final part reviews some of the most important emerging areas, including aerospace and civil structure applications.

The first part of the book, consisting of Chapters 1 through 6, opens with Chapter 1 by Eric Udd, which introduces the field and provides an overview of the book. This is followed by Chapter 2, also by Eric Udd, which provides the reader with a framework for the technology associated with the field and its interrelationship with real systems. The remaining chapters of this part are directed toward exploring the technology associated with fiber optic smart structures in detail. This begins with Chapter 3 by J. S. Hansen. Because of the light weight and high strength of composite materials, their use in structures of the future is becoming increasingly important. They are also highly compatible with the use of embedded fiber optic sensors, and this chapter serves as an introduction to these types of materials. Chapter 4, by J. Sirkis and A. Dasgupta, considers the problem of embedding glass fibers into materials and the determination of what parameters are actually sensed. These considerations are critical to the implementation of effective systems. Chapter 5, by D. Jensen and J. Sirkis, covers issues associated with embedding optical fibers into composite materials, including orientation and size. Once the issues associated with embedding the optical fibers into the material have been resolved, the next issue is getting the fiber into and out of the parts in which they are manufactured. This problem is addressed by W. B. Spillman and J. R. Lord in Chapter 6.

The second part of the book, which consists of Chapters 7 through 17, is concerned with the technology associated with the field of fiber optic smart structures. It begins with Chapter 7, by Eric Udd, who overviews the sensor technology used to support fiber optic smart structures and sets the stage for the second portion of the book. Ray Measures follows with a comprehensive overview of fiber optic strain sensing in Chapter 8, which is one of the most important types of fiber optic sensors that is used to support systems measuring structural integrity. These introductory chapters are then followed by a series of chapters providing detailed description of fiber optic sensors that have demonstrated considerable utility and promise for applications. C. E. Lee and H. F. Taylor begin in Chapter 9 with an introduction to sensors for smart structures based on the Fabry–Perot interferometer. These sensors allow measurements to be made by determining environmentally induced changes in the position of mirrors built into the fiber. They have been used for sensing in natural gas engines and have recently become commercially available. J. R. Dunphy, G. Meltz, and W. W. Morey continue in Chapter 10 with a discussion of optical fiber Bragg grating sensors. These sensors can be used to support strain and temperature measurements and are particularly exciting because they have the prospect of being very low in cost, as it has recently been demonstrated that these sensors may be written into optical fiber while it is being drawn. In Chapter 11, A. M. Vengsarkar, K. A. Murphy, and R. O. Claus explore elliptical-core two-mode optical fiber sensors, interferometric sensors that have the potential for high sensitivity to support strain and temperature measurements while retaining the advantages of single-fiber construction. In Chapter 12, Tim Clark and Herb Smith introduce microbend fiber optic sensors, which have the potential of being very low cost, since they can use standard telecommunication grade optical fiber. Clark and Smith describe the wide-ranging methods that have been used to implement this type of fiber sensor. One of the most important fields of fiber optic smart structures is their implementation in process control. By using optical fiber to carry light to a material that can be stimulated to fluoresce and carry the resultant signal back, a wide range of parameters in organic materials can be sensed, including viscosity, water vapor content, and degree of cure. Ram Levi and Scott Schwab overview these fluorescence optrode sensors in Chapter 13. Chapters 14 and 15, by John Dakin and Alan Kersey, are concerned with providing broad area coverage by either stringing fiber sensors together along a single line or investigating fiber sensors that have widely dispersed coverage. In many cases these techniques are essential to making cost-effective systems. After all the information is brought back to a central location, the next major task is the processing of the information. Neural networks offer the prospect of processing large amounts of information rapidly and are a natural complement to large-scale fiber optic sensor systems. B. G. Grossman and M. H. Thursby explore this in Chapter 16. Once the information has been processed, the structure must react if necessary to the environmental effect. Zaffir Chaudhry and Craig Rogers discuss the wide range of available technologies in Chapter 17. The implementation of fiber optic sensors in combination with these actuators forms the basis for many smart materials and structures.

The last part of the book is concerned with the application of fiber optic smart structures. One of the most important application areas is for situations where the temperatures involved are beyond the capabilities of conventional electronic sensors. In Chapter 18, R. O. Claus, K. A. Murphy, A. Wang, and R. G. May overview this important application area and the type of fiber sensors that are employed. For nondestructive evaluation of parts once they have been manufactured, one of the important techniques is the application of ultrasonic waves. R. O. Claus, V. S. Sundarshanam, and K. A. Murphy review this area in Chapter 19. The next three chapters cover the major applications to aircraft and spacecraft. Michel Le Blanc and Ray Measures discuss fiber optic damage assessment in Chapter 20. While many of the examples are from aerospace, the same technology can be applied to other structures. Herb Smith reviews fiber optic smart structures for aircraft in Chapter 21 from the point of view of one of the first aircraft structural engineers to apply the technology. Chapter 22, by Andrew Bicos, reviews requirements for control of spacecraft and the various technologies that could be used to implement these control systems. Chapter 23, by D. R. Huston and P. L. Fuhr, describes perhaps the largest and most rapidly emerging application. This area includes earth-quake-resistant buildings, bridges that perform self-diagnostics, as well as smart dams and highways. Dryver Huston and Peter Fuhr were among the first to conduct widespread field tests of fiber optic sensors in civil structures, and this chapter provides an overview of some recent tests.

I would like to thank the contributors to this book for their diligent efforts to turn out a high-quality and timely product. I would also like to thank the many people at McDonnell Douglas Aerospace who have supported my efforts on fiber optic smart structures, most notably Richard Cahill and Jim Dorr, who recognized the merits of this technology early and have provided continuous encouragement and support. I would also like to acknowledge all the members of the McDonnell Douglas Smart Structure Working group, including Chris Gackscatter, John Belk, Diana Holdinghausen, Russ Farles, Kelli Corona (now at Production Products), Perry Wims, Dr. Wil Otaguro, John Paul Theriault, Jeff Eck, Dr. Don Edberg, Dr. John Tracy, and the McDonnell Douglas contributors to this book who have worked to bring fiber optic smart structures into reality. Finally, I would like to acknowledge Jim Kelly of ARPA and Bill Stange of WPAFB, who have acted as sponsors of our early work on fiber optic smart structures.

I would also like to acknowledge the support, encouragement, and understanding of my wife, Holly, and my daughters, Emelia and Ingrid. I would like to dedicate my efforts in bringing this book into being to my parents, John and Irene Udd.

ERIC UDD

Blue Road Research
Troutdale, Oregon

Fiber Optic Smart Structures

1

The Evolution of Fiber Optic Smart Structures

ERIC UDD
Blue Road Research,
Troutdale, Oregon

On the scale of time used to measure the existence of the earth, humans have been on this planet a very short time and the period of time for recorded history is but a small fraction of this time. For the past half-century revolutionary changes have been occurring at an amazing pace in many scientific and engineering fields. Traditionally, many fields of study have been separate and distinct. Recently, there has been considerable movement and convergence between these fields of endeavor and the results have been astonishing. Fiber optic smart structures is one of these fields.

Imagine for a moment structures that can assess their own health and perform self-repair, bridges and buildings that can phone in their status, aircraft and spacecraft that make critical adjustments in flight as conditions change, and gossamer webs of lightweight strands that float and change shape and structural properties as needed. All these things are becoming possible through the merger of material science, structural mechanics, sensor technology, advanced processing techniques, and actuators. This new field, termed *smart structures*, offers the prospect of adding effective nervous and responsive systems to robots, civil structures, and vehicles of all sorts.

One of the key enabling technologies for this field involves the emergence of fiber optic sensors and communication links that allow critical parameters of materials and structures to be sensed while offering light weight, immunity to electromagnetic interference, the ability to be embedded under hostile environments, and extremely high bandwidth capability. The net result is that designers were able to add fiber optic nervous systems with tremendous improvements in capability and flexibility relative to the prior art electronic sensors, and the field of fiber optic smart structures was born.

This book, which provides a survey of the field of fiber optic smart structures, is divided into three main parts. The first part, consisting of

Fiber Optic Smart Structures, Edited by Eric Udd.
ISBN 0-471-55448-0 © 1995 John Wiley & Sons, Inc.

Table 1.1   Fiber Optic Smart Structure Technology

Composite Materials	Structural Mechanics	Fiber Optic Sensors	Multiplexing
Chapter 3 · Fundamentals	Chapter 4 · Embedded fibers/ structural integrity	Chapters 7–15 · Gratings	Chapter 14 · Distributed
Chapters 4, 5 · Interactions with fibers	Chapter 5 · Glass–material interface Chapter 6 · Ingress/egress	· Fabry–Perot etalons · Dual mode · Microbend · Optrodes · Blackbody · Distributed · Interferometric	Chapter 15 · Multiplexing techniques

Signal Processing	Assessment Systems	Control Systems	
Chapter 16 · Neural networks	Chapter 19 · Ultrasonics Chapter 20 · Damage Chapters 21, 22 · Aerospace Chapter 23 · Civil	Chapter 17 · Actuators Chapter 22 · Space structures	

Chapters 1 through 6, provides an overview of the technology and the issues associated with placing optical fibers into composite materials. The second part, comprised of Chapters 7 through 15, provides an introduction to many of the fiber optic sensors that have been used successfully to support fiber optic smart structure systems. The last part, consisting of Chapters 16 through 23 surveys technologies associated with fiber optic smart structures, including neural networks and actuators, as well as how this technology is or will be applied. Table 1.1 outlines the technical areas associated with fiber optic smart structures, including composite materials, structural mechanics, chemistry, fiber sensors, multiplexing techniques, signal processing, assessment, and control systems. Each of these disciplines is an integral part of fiber optic smart structure technology, and the chapters that contain substantial discussions of these aspects of the field are indicated in the table.

Nature provides many examples of smart structures. The simple one-celled animal of Fig. 1.1 illustrates the fundamentals. Various environmental effects act on the outer layer of the one-celled animal. These effects may be in the form of radiant energy, chemical reactions, or electric and magnetic fields. Sensors in the outer layer detect these environmental effects, and the resulting information is conveyed for signal processing and interpretation, at which point the

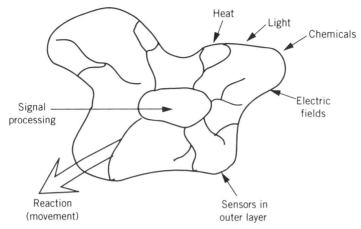

Figure 1.1. A single-celled animal is a sophisticated smart structure.

single-celled animal reacts to its environment in a number of ways, including movement, changing chemical composition, and reproductive actions dependent on the specific environmental effect. The ultimate goals of the field of smart structures are to bring many of these aspects of life into humanmade constructs. It is rather humbling that at this stage even nature's simplest examples of life possess a sophistication that is far beyond today's ablest humanmade smart structure. Yet nature has had billions of years and a vast laboratory to develop life, whereas humankind has just begun to create autonomous smart structures. Still the results of even the earliest attempts have been impressive. The *Voyager* spacecraft sent back spectacular photos. Automobiles using sensors and feedback control on engines have greatly improved performance, and robots are being widely used in manufacturing.

With the arrival of new technologies that offer significant enhancements to the designer's ability to create an effective smart structure, the number of applications will increase rapidly. Table 1.2 illustrates how signal processing, fiber optic sensors, and communication technology are evolving. Both signal-processing capability and fiber optic communications began to enter the market in the 1980 time period and by 1990 has established multibillion-dollar markets. Each area has exhibited improvements in performance of approximately two orders of magnitude, while prices for comparable components have dropped by similar amounts. Derivative markets will continue to expand in both of these areas by the year 2000. Fiber optic sensors were just beginning to enter the market in 1990 and will dominate many sensor markets by 2000. The field of fiber optic smart structures depends heavily on the availability of high-performance low-cost fiber optic sensors, and in the 1990 period the first limited fiber optic smart structure demonstrations were being performed on a level of sophistication comparable to fiber optic sensor demonstrations in 1980. By 2000, fiber optic smart structure systems will begin to enter the market and should have a strong presence by 2010.

Table 1.2   Evolution of Fiber Optic Smart Structure Technologies

1980	1990	2000
64-kB memories	4 MB memories	256-MB memories
8-bit computers	32-bit computers	128-bit computers
8080	80486	80 × 86
Entry of personal computer market	Neural networks entry	Widespread use of neural networks
	Parallel processing	
	Widespread pc	Intelligent services: telecom, cable
Laboratory fiber sensors	Fiber sensors start to enter market	Fiber sensors dominate many sensor markets
Fiber communication links enter market	Fiber communication links dominate land-based market	Fiber communication links dominate worldwide fixed-site telecom market, provide fiber to the home services
	Limited fiber optic smart structure demonstrations	Fiber optic smart structures begin to enter the market

The results of these efforts may provide planes and spacecraft that are able to diagnose their status prior to takeoff, greatly reducing personnel requirements while increasing safety. It would also allow onboard systems to provide maintenance schedule information and diagnostic support, thus reducing downtime and lowering costs. The field opens up the possibility of buildings, bridges, and dams with integral capabilities that alow the detection of fire, water damage, structural assessment, and intrusion alarms that could be routed via fiber optic or wireless communication links to a centralized location for fire, police, and maintenance. Other areas that promise significant enhancements are manufacturing process control, monitoring of natural structures for prediction of earthquakes and volcanic activity, and future medical systems that might include artificial limbs and nerves.

The possibilities offered by the field of fiber optic smart structures are widespread and often dramatic in scope. It is the hope of the editor and contributors of this book that it will help the designers of these future systems in their efforts to make these dreams a reality.

2

Fiber Optic Smart Structure Technology

ERIC UDD
Blue Road Research,
Troutdale, Oregon

2.1. INTRODUCTION

The revolutions in the fiber optic telecommunication and optoelectronic industries have enabled the development of fiber optic sensors that offer a series of advantages over conventional electrical sensors. This development, in combination with advances in composite material technology, has opened up the new field of fiber optic smart structures which offers mechanical and structural engineers the possibility of incorporating fiber optic nervous systems into their designs. Buildings and bridges that can call up central maintenance depots and report on their status after an earthquake, storms, or simply with the passage of time; aircraft that "know" and communicate if it is safe to take off and monitor and correct for structural changes in flight; and artificial limbs that can feel, react, and touch are all manifestations of the dreams of engineers and scientists working on the emerging field of smart structures (1–9). These dreams require materials that are lighter in weight, have superior strength, and have the ability to change shape, degree of stiffness, and mechanical and electrical properties as required. The new materials must be "smart," with the ability to sense environmental changes within or around the structure and have the ability to interpret and react to these changes.

The realization of these dreams has required a nervous system capable of sensing change in the material while being part of the structure itself. Fiber optic sensor technology has enabled the implementation of this nervous system by (1) providing sensors that are small and rugged enough that they can be integrated and consolidated directly into materials, (2) enabling sensors to be multiplexed in substantial numbers along a single line, allowing weight reduction and minimizing points of ingress and egress into parts, (3) providing electrical isolation and immunity to electromagnetic interference, and (4) supporting the high bandwidth necessary for large numbers of high-performance sensors. With the complementary revolutions taking place in the optoelec-

Fiber Optic Smart Structures, Edited by Eric Udd.
ISBN 0-471-55448-0 © 1995 John Wiley & Sons, Inc.

tronic and fiber optic telecommunication industry, the designer of the fiber optic nervous system is continually being offered higher-performance components at lower cost.

Development of full systems require the coordination of many disciplines, including experts in the fields of materials, structures, actuators, signal processing, sensors, and systems. In this chapter we provide an overview of how these disciplines interact to form the field of fiber optic smart structures. We conclude with a review of some of the emerging application areas.

2.2. ASPECTS OF FIBER OPTIC SMART STRUCTURES

There are four main aspects to fiber optic smart structures. The first is the use of optical fibers embedded or attached to materials to augment the manufacturing process by monitoring such parameters as temperature, pressure, strain, degree of cure, chemical content, and viscosity. Once a part has been made, the next aspect of fiber optic smart structures is to enhance nondestructive evaluation. This aspect allows the assessment of the integrity of parts prior to assembly for damage or defects that may have arisen during manufacture or handling. The parts can then be assembled and sensors integrated to form a health management system to assess overall structural integrity. The highest level of integration involves combining actuator systems with a health monitoring system and signal processing to form control systems. These systems could be used to augment flight control or enable a building to "react" to an earthquake in a manner that minimizes damages.

2.3. IMPLEMENTATION AND CHALLENGES OF FIBER OPTIC SMART STRUCTURES

The basic functions of a fiber optic smart structure system are to (1) sense environmental conditions in or around the structure, (2) convey the information back to an optical and or electronic signal processor, and (3) perform an action as a result of the information sensed. Figure 2.1 shows a basic system in block diagram form. An environmental effect acts on and or around a structure that has embedded and/or attached multiplexed fiber optic sensors. The information is then conveyed via a fiber optic link to an optical–electronic signal processor. After processing, the information is conveyed to a control system that may be tasked to do damage, performance, or health management functions. This system may also have the capability of taking corrective action via actuators that are directed to react to the environmental phenomenon.

The technologies associated with fiber optic smart structures and their interrelationship are illustrated by Fig. 2.2. The first major issue is embedding the optical fibers into the structure (10–15) in such a way that structural integrity is not compromised and ensuring that the interface between the

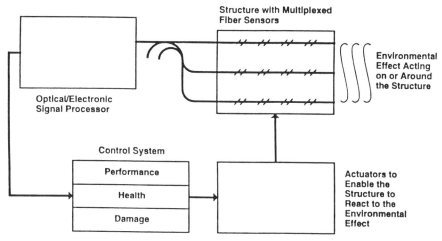

Figure 2.1. Basic block diagram of a fiber optic smart structure system.

optical fiber and the surrounding material allows accurate measurement of the environmental effects of interest. There are a wide range of materials of interest, including carbon epoxy, thermoplastics, and concrete on the low end of the processing temperature range, to titanium and carbon–carbon for very-high-temperature applications. For most low- to moderate-temperature applications, silica-based optical fibers with a melting temperature of about 1400°C are adequate. Generally, the practical limit is somewhat lower than this since at about 1000°C the migration of dopants from the core region of the fiber becomes a serious problem. Most designs using silica-based fiber have an upper operating range of about 700°C to allow for an adequate margin. For very high temperatures work has been done on using optical fibers based on sapphire (16, 17) that have the potential to operate up to about 2000°C. While most current work has been done using silica-based optical fibers because of their low cost and availability, sapphire fibers can be expected to play a significant

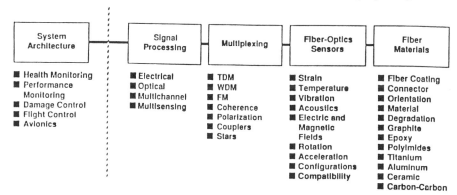

Figure 2.2. Technologies associated with fiber optic smart structures. (From Eric Udd, ed., *Fiber Optic Sensors: An Introduction for Engineers and Scientists*, Wiley, New York, copyright 1991; used with permission.)

role in the higher-temperature regimes, where very few sensors can survive.

To embed optical fibers into materials successfully, the coating is critical. One of the first roles of the coating is to protect the optical fiber from moisture. This is due to the significant strength degradation that may occur in the fiber when microcracks on the surface of the fiber are penetrated by moisture, causing the cracks to propagate.

Another important factor is that the coating must form an appropriate interface between the optical fiber and the host material, allowing accurate measurement of this environmental effect to be monitored. For organic composite materials such as carbon–epoxy or thermoplastics, it is particularly important that the fiber coating be chemically compatible with the resin of the host material. For example, to measure strain in a thermoplastic material a polyimide coating would be an appropriate choice whereas an epoxy–acrylate would not. The polyimide coating consolidates with the thermoplastic and provides an excellent interface to the glass, whereas the epoxy–acrylate does not consolidate well with the thermoplastic, allowing a potential reduction in the inherent strength of the part, and it does not provide a rigid interface to the glass, allowing slippage under strain conditions. For the case of embedding optical fibers in metals, the critical parameter is to match the metallurgy of the coating to that of the metal into which it is being embedded. Examples include using an aluminum-coated fiber for embedding into an aluminum part, or graphite–aluminum or gold coatings for placing fibers into titanium (18, 19). In the case of reinforced concrete, most concerns to date have focused on survival of the optical fiber and protection from moisture. It is less clear what coatings are most appropriate for nonhomogeneous materials, and further work is under way. Once an appropriate coating has been chosen, the next is to embed the fiber into the material itself. In the case of composite materials it is important to consider the orientation of the fiber in the part as well as the overall size of the optical fiber and its coating (11, 13, 14, 20, 21). Generally, running the optical fiber parallel to the strength-member fibers of the composite material minimizes potential strength degradation of the part. The orientation of the optical fiber also becomes less critical as the size of the fiber is decreased.

Using these two design considerations in combination with an appropriate fiber coating, it is possible to embed large numbers of optical fibers into a composite part without significant strength degradation. After the optical fibers are in place, the next major issue is ingress and egress out of the part and connectors (11, 22). To implement a connection successfully, it is extremely important to make adequate provision for strain relief and protection of the fiber at the point of ingress/egress, as this is its most vulnerable point.

After the details of appropriate coatings, fiber size, and placement into the part have been worked out, it becomes necessary to select a fiber sensor (23–25) that will be able to measure the environmental effect with sufficient sensitivity, dynamic range, and scaling to meet the performance requirements of the system designer. Table 2.1 shows some of the fiber sensors that have been

Table 2.1 Representative Fiber Optic Sensor Candidates for Smart Structures

	Multiplexing Capability	Single-Fiber Configurations	Single-Ended Configuration	Sensitivity	Suitability for Short Range Length	Embedibility IO Sensing Locations
Sagnac/Mach–Zehnder	Good	N	Y	High	Low	Poor
Michelson	Good	N	Y	High	Medium	Poor
Sagnac	Good	N	N	Medium	Low	Poor
Distributed sensors						
Raman	Good	Y	Y	Low	Low	Good
Rayleigh	Good	Y	Y	Low	Low	Good
Sagnac/Mach–Zehnder	Good	Y	N	Low	Low	Good
Fabry–Perot etalon	Good	Y	Y	High	High	Good
Fiber grating	Good	Y	Y	High	High	Good
Microbending	Good	Y	Y	Low–medium	Medium–high	Good
Dual-mode sensor	Poor	Y	Y	High	High	Poor
Polarization	Poor	Y	Y	High	High	Poor–medium
Break sensors	Poor	Y	Y	Low–medim	High	Good

used to support fiber optic smart structure systems. The classic interferometric fiber sensors—the Mach–Zehnder, Michelson, and Sagnac interferometers—have been used primarily to support characterization studies (26–28). Although these sensors have high sensitivity and can be made to have good dynamic range, it is difficult to embed several of these sensors multiplexed together, due to the large number of fiber optic leads and components. The distributed fiber sensors (29–31) offer the prospect of supplying environmental information at locations along a single fiber. Although the long-term prospects of these sensors look promising, the low sensitivity and spatial resolution have caused many researchers to look at alternative approaches. The two that are most promising in terms of multiplex potential, sensitivity, and spatial resolution and that are the focus of a good deal of activity are the Fabry–Perot etalon (32–39) and fiber grating (40–43). These sensors also have the advantages of allowing a single point of ingress and egress and have spectrally dependent signals that minimize problems associated with variable losses associated with connectors and fiber leads. Two relatively low-cost options—the microbending (44,45) and break (9) fiber sensors—are available for applications that have less stringent requirements as to accuracy. Two other sensors that are more accurate but are difficult to multiplex are the dual-mode (46–49) and polarimetric (50) fiber sensors. Most of the work done so far with these sensors has been for single-sensor configurations.

After the sensors are selected, the next issue facing the fiber optic smart structure designer is the choce of multiplexing techniques necessary to support the required number of sensors along a single fiber. There are five basic

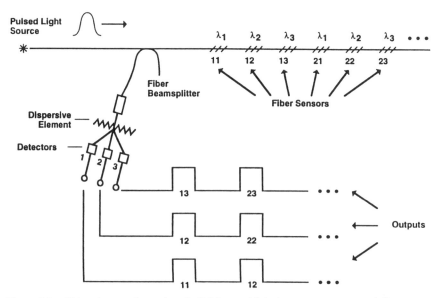

Figure 2.3. Using time- and wavelength-division multiplexing to support several fiber sensors along a single fiber.

techniques: wavelength, time division, frequency, polarization, and coherence multiplexing (29, 51). The two most commonly used techniques are wavelength-division multiplexing, where each sensor is color encoded, and time-division multiplexing, which can be used to separate sensors out spatially. Figure 2.3 illustrates a simple example of combining time- and wavelength-division multiplexing. In this case a spectrally broadband light source is pulsed and the resultant light beam propagates down an optical fiber reaching sensors 11, 12, and 13, which reflect light centered about the wavelengths λ_1, λ_2, and λ_3, which are within the spectral bandwidth of the source. The resultant sensor signals are reflected back toward the light source and are coupled via the fiber beamsplitter to the dispersive elements, which seperate the colors centered about λ_1, λ_2, and λ_3 onto the three output detectors 1, 2, and 3, respectively. Similarly, a second set of signals results from reflections from the second set of fiber sensors, 21, 22, and 23. The process can be repeated until optical loss budgets are exceeded with subsequent sets of sensors. It is also possible to use frequency and coherence multiplexing effectively. Polarization multiplexing has been used mainly for two sensors in line and has had limited utility. In general, the type of multiplexing technique selected will depend strongly on the fiber sensor utilized and its expected performance characteristics.

The next issue to be addressed involves the processing of the information from the multiplexed fiber sensors. Some of the processing may be handled optically via filters or active optical elements. Electronic signal processing would then be used to complete the transformation of the signals to a form usable by the control system. Since large numbers of sensors and correspond-ing data may be involved, there is considerable interest in using neural networks (52–56) to process the data.

The control system is defined by a system designer, who in turn imposes performance specifications on the sensors, multiplexing techniques, and pro-cessing. To make the fiber optic smart structure system work effectively, all these technologists must work together through the design and development process. Since many of the technologists come from disciplines that have traditionally had little interaction, this is one of the great challenges in this field. At the same time it is also one of its strong points. As so little has been done in the past, a good deal of progress remains to be made in the future.

2.4. APPLICATIONS OF FIBER OPTIC SMART STRUCTURES

One way to look at fiber optic smart structures is to us the analogy to living things. The basic idea is to develop a fiber optic nervous system capable of collecting information on the condition of the structure/organism. The infor-mation is then relayed back to a processor/brain, where the information is sorted and analyzed. Based on this information, actuators/muscles may be activated, processing/body temperature raised or lowered, or chemicals/hor-mones injected into the structure/organism. The contribution of fiber optics is primarily to allow the formation of nervous systems under conditions that were at best extremely difficult with prior art technology and on a scale that offers orders of magnitude more information with simultaneous size and weight savings due to the high bandwidth/short wavelength nature of light. To illustrate this potential, the remainder of this sector is devoted to a series of application areas that hold considerable promise for the near-term realization of the potential of fiber optic smart structures.

2.4.1. Aerospace Applications

As an example of the application of fiber optic smart structures to an aerospace application, consider the case of a launch vehicle such as a rocket. This type of platform consists of a series of subsystems that could potentially benefit by the integration of fiber optic smart structure systems. The first use of these systems will be in the manufacture of composite parts. In the case of a launch vehicle this could involve sensors that could be used to support the manufacture of cryogenic tanks, rocket nozzles, farings, solid rocket booster casings, and interstages. Figure 2.4 illustrates the case for a tank that is being filament wound. Fiber sensors could be wound in directly with the preimpregnated material and used to monitor the consolidation process. These sensors could then be used to augment nondestructive evaluation techniques prior to

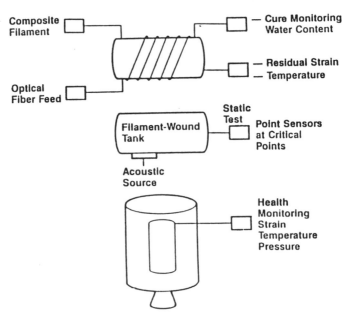

Figure 2.4. Usage of fiber sensors to support the manufacture of a cryogenic tank and a health monitoring system. (From Eric Udd, ed., *Fiber Optic Sensors: An Introduction for Engineers and Scientists*, Wiley, New York, copyright 1991; used with permission.)

installation on the rocket. Once the tank has been installed, these fiber sensors or others that might be placed at a different stage of the manufacturing process would be used as part of a vehicle health monitoring system to monitor structural changes of the tank and warn of leaks. Other potential application areas on launch vehicles would include monitoring rocket nozzles for areas of excessive burning, augmenting separation systems between stages, vibration and acoustic damping of the payload and other key components, and eventually, support of vehicle shape control systems that would allow adaptive guidance.

Space-based applications for fiber optic smart structures can also be expected to include large space platforms. In this case there are future needs for space-based platforms that are extremely light in weight and yet rigid. A section of this type of structure is illustrated by Fig. 2.5. Fiber optic smart structures for this type of platform can be expected to manifest themselves in the form of health monitoring systems that would assess changes in the structural integrity of the platform due to such events as docking, impacts, or on orbit aging. It can also be expected that fiber optic smart structures would be used in combination with actuator systems for vibration control in order to isolate critical areas requiring a zero-g environment and to damp out oscillations that would degrade pointing and tracking accuracy. In some cases these platforms could include habitats such as that shown in Fig. 2.6. Here the potential exists for building fiber optic damage assessment systems directly into

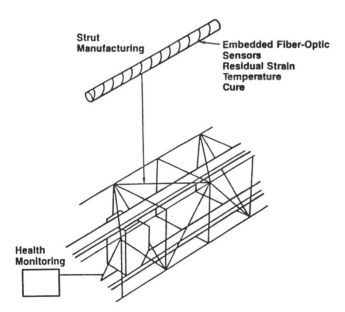

Figure 2.5. Fiber optic smart structure used to support a space-based platform. (From Eric Udd, ed., *Fiber Optic Sensors: An Introduction for Engineers and Scientists*, Wiley, New York, copyright 1991; used with permission.)

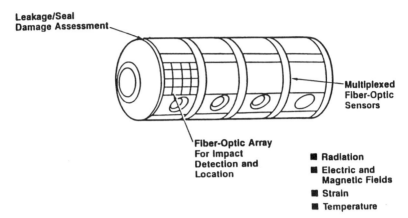

Figure 2.6. Fiber optic damage and health monitoring systems for a space habitat. (From Eric Udd, ed., *Fiber Optic Sensors: An Introduction for Engineers and Scientists*, Wiley, New York, copyright 1991; used with permission.)

the walls of the structure to monitor the location and severity of an impact. This type of system could potentially be designed to support a self-healing structure that detects and assesses damage and then automatically initiates repair. Other features that could be built into the habitat include distributed acoustic sensors to locate gas leaks and radiation, electromagnetic, pressure, and temperature sensors, used to measure the environment around the structure.

For the near future, commercial and military aircraft can be expected to be the first major aerospace users of fiber optic smart structures. Initial application areas would include the manufacturing and processing of parts and the introduction of simple health monitoring and structural control systems. Examples of these early systems would be icing indicators, vibration, and localized strain monitoring systems. Later efforts would include automatic maintenance systems that would perform preflight and postflight assessments of the aircraft. These systems would also be used to improve and simplify repair procedures and to provide in-flight structural integrity warning systems. Finally fiber optic smart structures would evolve into an integral part of smart aircraft. This could feature deformable structures that radically change shapes as required, self-healing systems performing damage assessment, flight corrections and repair in flight, and automatic flight control systems featuring environmental awareness around the aircraft and real-time corrective actions.

The first fiber optic smart structure systems are likely to consist of a small number of sensors. To access fully the power of this technology on large platforms, thousands and in some cases tens of thousands of sensors and/or discernible sensing points will be required. As an example, consider the case of a composite wing where it may be desirable to measure strain every 10 cm to map out the strain field fully. For a modest 20-m^2 wing, this would involve

2000 sensors. Because of processing requirements, it is likely that a large aerospace platform will have two classes of fiber sensors. The first would be distributed fiber sensors (29–31) that would be used to localize an event such as damage. The second set of fiber sensors would be discrete strings of fiber optic sensors that could be used to support detailed assessments. To be practical these sensors must meet a series of requirements. The first and most important is that the fiber sensors be low in cost and that the function performed by the fiber optic smart structure system they support be of significant value to the user to justify the increased cost. For example, the fiber optic smart structure system might be used to enhance maintainability, improve reliability, or enhance the performance of the platform. In each case these improvements have definable economic value and the fiber optic smart structure system must pass the test of improving the overall value of the platform to the end user. Additional customer-driven considerations that are important are the redundancy and survivability of the system, installation of the system onto the platform, and repairability. Requirements for the fiber optic sensors that are driven by structural considerations and manufacturing are that the fiber sensors have a single point of ingress/egress, that the fiber sensors be no larger or very close to the diameter of the fiber, and that a significant number of sensors be multiplexed along a single fiber line. Two candidates for meeting these requirements among existing fiber optic sensors are fiber optic grating (40–43) and etalon-based fiber sensors (32–39). Potentially, these sensors may be multiplexed into "strings" of fiber optic sensors. To hold costs down, these strings could be multiplexed using an optical switch that could be used to interrogate sensors as needed. A block diagram of a portion of the fiber optic smart structure architecture is shown in Fig. 2.7. A fiber sensor demodulator is used to extract the data from the sensors in the string being accessed. The data are then formatted and transmitted to a system signal

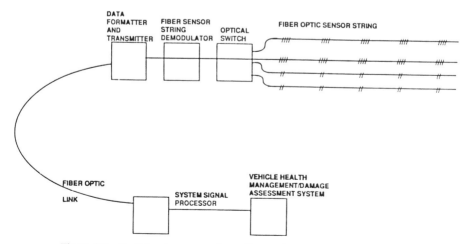

Figure 2.7. Partial system supporting a fiber optic smart structure architecture.

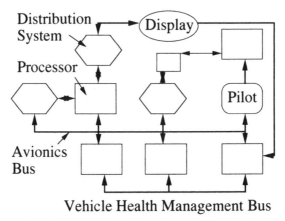

Figure 2.8. Integrated avionics system. (From Eric Udd, *Fiber Optic Sensor Workbook*, Blue Road Research, Troutdale, Oreg., copyright 1993 and 1994; used with permission.)

processor, which in turn transfers the data to the vehicle health management/damage assessment system. As an example of how these data would be handled on an aerospace platform, Fig. 2.8 illustrates an avionics system for a fighter aircraft in block diagram form. The data transmitted from the fiber optic sensor system signal processor would be transferred to the vehicle health management bus. This, in turn, would be transferred to the vehicle management system data processors and used to support the other elements of the avionics system.

2.4.2. Medical Applications and Biological Analogies

Some smart structure medical products already exist. One employs sensors to determine the amount of medication in the blood and readjusts the rate at which medication is supplied intravenously. Another would be closed-loop treadmills where a patient's exertion level is adjusted to correspond to his or her heart rate. However, possibilities abound for future fiber optic structure–based products, including artificial limbs that readjust themselves as required by the user's action, artificial hearts that include oxygen, pressure, and temperature sensors to regulate heartbeat, beds that sense the pressure distribution of patients lying on them and adjust themselves to avoid bedsores, and rehabilitation systems that help the patient enough to complete exercise successfully but act to challenge the patient. Consequently, virtually any biological system is a candidate for replacement by a fiber optic smart structure–based system. For medical applications, fiber optics hold the additional advantage of using passive dielectric devices that do not pose an electrical shock threat or involve radiating electromagnetic energy.

As an example of a biological analogy, consider the example of the tree of Fig. 2.9. Its leaves orient themselves toward sunlight by sensing light and

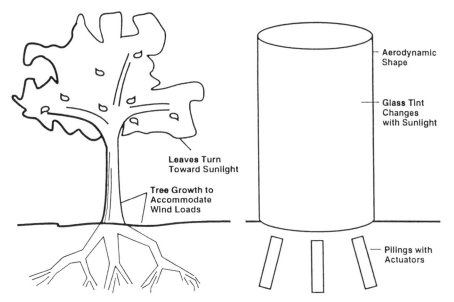

Figure 2.9. Biological analogies to smart structures.

changing hydraulic actuators. The tree's roots search out and grow toward water and the overall growth pattern of the tree adopts to wind loading. One can easily carry these analogies over when designing a building. The building can be shaped aerodynamically, the glass tint may be made to change as light intensity varies, and pilings with actuators or rollers could be used to help the building to adjust to windloads or an earthquake.

2.4.3. Civil Structure Applications

In addition to integrating fiber optic smart structures into buildings, efforts are under way to place fiber optics smart structures into highways, bridges, and dams (57–60). Many of these systems are being placed to monitor the long-term health of the structures, but they will also be used to augment damping and for control systems such as internal climate control for buildings. One can easily go further by linking systems together between buildings that could serve to provide communities with fire protection and emergency services, as well as monitoring such functions as gas and electric use.

2.4.4. Natural Structures

One recently proposed (61) application area of fiber optic smart structures is the use of long-gauge-length strain sensors based on the Sagnac interferometer to monitor large natural structures. Application areas here would include monitoring strain buildup between earth plates prior to slippage that results in earthquakes, strain due to magma flow in volcanoes, and movement of the earth around power utilities and tanks containing hazardous waste materials. The information gathered by these strain sensors could be routed into local or national fiber optic telecommunication links and processed at a convenient location.

2.5. SUMMARY

Fiber optic smart structures are an enabling technology that will allow engineers to add nervous systems to their designs, enabling damage assessment, vibration damping, and many other capabilities to structures that would be very difficult to achieve by other means. This technology can be expected to have dramatic future impact in the aerospace, medical, and civil structure fields.

REFERENCES

1. E. Udd, ed., *Fiber Optic Smart Structures and Skins*, Proc. SPIE, Vol. 986. SPIE, Bellingham, WA, 1989.

2. E. Udd, ed., *Fiber Optic Smart Structures and Skins II*, Proc. SPIE, Vol. 1170. SPIE, Bellingham, WA, 1990.

3. E. Udd and R. O. Claus, ed., *Fiber Optic Smart Structures Skins III*, Proc. SPIE, Vol. 1370. SPIE, Bellingham, WA, 1991.

4. R. O. Claus and E. Udd, eds., *Fiber Optic Smart Structures and Skins IV*, Proc. SPIE, Vol. 1588. SPIE, Bellingham, WA, 1992.

5. G. J. Knowles, ed., Active Materials and Adaptive Structures. *Proceedings of the ADPA/AIAA/ASME/SPIE Conference on Active Materials and Adaptive Structures.* IOP Publishing, Blacksbury, VA, 1991.

6. B. Culshaw, P. T. Gardiner, and A. McDonach, eds., *First European Conference on Smart Structures and Materials*, SPIE, Vol. 1777. SPIE, Bellingham, WA, May 1992.

7. R. O. Claus and R. S. Rogowski, eds., *Fiber Optic Smart Structures and Skins V*, Proc. SPIE, Vol. 1798. SPIE, Bellingham, WA, 1993.

8. E. Udd, Embedded Sensors Make Structures Smart. *Laser Focus*, May, p. 138 (1988).

9. R. M. Measures, Smart Structures in Nerves of Glass. *Prog. Aerosp. Sci.* **26**, 289 (1989).

10. E. H. Urruti, P. E. Blaszyk, and R. M. Hawk, Optical Fibers for Structural Sensing Applications. *Proc. SPIE—Int. Soc. Opt. Eng.* **986**, 158 (1989).

11. R. L. Wood, A. K. Tay, and D. A. Wilson, Design and Fabrication Considerations for Composite Structures with Embedded Fiber Optic Sensors. *Proc. SPIE—Int. Soc. Opt. Eng.* **1170**, 160 (1990).

12. W. Maslach, Jr. and J. S. Sirkis, Strain or Stress Component Separation in Surface Mounted Interferometric Optical Fiber Strain Sensors. *Proc. SPIE—Int. Soc. Opt. Eng.* **1170**, 452 (1990).

13. D. W. Jensen and J. Pascual, Degradation of Graphite/Bismaleimide Laminates with Multiple Embedded Fiber Optic Sensors. *Proc. SPIE—Int. Soc. Opt. Eng.* **1370**, 228 (1991).

14. J. S. Sirkis and A. Dasgupta, Optimal Coatings for Intelligent Fiber Optic Sensors. *Proc. SPIE—Int. Soc. Opt. Eng.* **1370**, 129 (1991).

15. C. DiFrancia, R. O. Claus, and T. C. Ward, Role of Adhesion in Optical-Fiber-Based Smart Composite Structures and Its Implementation in Strain Analysis for Modeling of an Embedded Optical Fiber. *Proc. SPIE—Int. Soc. Opt. Eng.* **1588**, 44 (1992).

16. A. R. Raheem-Kizchery, S. B. Desu, and R. O. Claus, High-Temperature Refractory Coating Materials for Sapphire Waveguides. *Proc. SPIE—Int. Soc. Opt. Eng.* **1170**, 513 (1990).

17. K. A. Murphy, B. R. Fogg, C. Z. Wang, A. M. Vengsarker, and R. C. Claus, Sapphire Fiber Interferometer for Microdisplacement Measurements at High Temperatures. *Proc. SPIE—Int. Soc. Opt. Eng.* **1588**, 117 (1992).

18. S. E. Baldini, E. Nowakowski, H. G. Smith, E. J. Freible, M. A. Putnam, R. Royowski, L. D. Melvin, R. O. Claus, T. Tran and M. S. Helben, Jr., Cooperative Implementation of a High-Temperature Acoustic Sensor. *Proc. SPIE—Int. Soc. Opt. Eng.* **1588**, 125 (1992).

19. S. E. Baldini, D. J. Tubbs, and W. A. Stange, Embedding Fiber Optic Sensors in Titanium Matrix Composites. *Proc. SPIE—Int. Soc. Opt. Eng.* **1370**, 162 (1991).

20. A. Dasgupta, Y. Wan, J. S. Sirkis, and H. Singh, Micromechanical Investigation of an Optical Fiber Embedded in a Laminated Composite. *Proc. SPIE—Int. Soc. Opt. Eng.* **1370**, 129 (1991).

21. A. M. Vengsarker, K. A. Murphy, M. F. Gunther, A. J. Plante, and R. O. Claus, Low Profile Fibers for Embedded Smart Structure Applications. *Proc. SPIE—Int. Soc. Opt. Eng.* **1588**, 2 (1992).

22. W. B. Spillman, Jr., Fiber Optic Sensors for Composite Monitoring. *Proc. SPIE—Int. Soc. Opt. Eng.* **986**, 6 (1989).

23. E. Udd, ed., *Fiber Optic Sensors: An Introduction for Engineers and Scientists.* Wiley (Interscience), New York, 1991.

24. J. Dakin and B. Culshaw, eds., *Optical Fiber Sensors: Principles and Components*, Vol. 1. Artech House, Boston, 1988.

25. B. Culshaw and J. Dakin, eds., *Optical Fiber Sensors: Systems and Applications*, Vol. 2. Artech House, Norwood, MA, 1989.

26. D. A. Brown, B. Tan, and S. L. Garret, Nondestructive Dynamic Complex Moduli Measurements Using a Michelson Interferometer and a Resonant Bar Technique. *Proc. SPIE—Int. Soc. Opt. Eng.* **1370**, 238 (1991).

27. J. A. Sirkis and H. W. Haslach, Jr., Complete Phase-Strain Model for Structurally Embedded Interferometric Optical Fiber Sensors. *Proc. SPIE—Int. Soc. Opt. Eng.* **1370**, 248 (1991).

28. E. Udd, R. J. Michal, S. E. Higley, J. P. Thériault, P. LeCong, D. A. Jolin, and A. M. Markus, Fiber Optic Sensor Systems for Aerospace Applications. *Proc. SPIE—Int. Soc. Opt. Eng.* **838**, 162 (1987).

29. A. D. Kersey, Distributed and Multiplexed Fiber Optic Sensors. In *Fiber Optic Sensors: An Introduction for Engineers and Scientists* (E. Udd, ed.). Wiley (Interscience), New York, p. 325, 1991.

30. J. P. Dakin, D. A. J. Pearce, A. P. Strong, and C. A. Wade, A Novel Distributed Optical Fiber Sensing System Enabling Location of Disturbances in a Sagnac Loop Interferometer. *Proc. SPIE—Int. Soc. Opt. Eng.* **838**, 325 (1987).

31. E. Udd, Sagnac Distributed Sensor Concepts. *Proc. SPIE—Int. Soc. Opt. Eng.* **1586**, 46 (1991).

32. C. E. Lee and H. F. Taylor, Interferometric Optical Fiber Sensors Using Internal Mirrors. *Electron. Lett.* **24**, 193 (1988).

33. C. E. Lee, R. A. Atkins, and H. F. Taylor, Performance of a Fiber-Optic Temperature Sensor from -200 to $1050°C$. *Opt. Lett.* **13**, 1038 (1988).

34. T. Valis, D. Hogg, and R. M. Measures, Fiber Optic Fabry–Perot Strain Sensor. *IEEE Photon. Technol. Lett.* **2**, 227 (1990).

35. C. E. Lee and H. F. Taylor, Fiber Optic Fabry–Perot Sensor Using a Low Coherence Source. *IEEE J. Lightwave Technol.* **LT-9**, 129 (1991).

36. C. E. Lee, H. F. Taylor, A. M. Markus, and E. Udd, Optical Fiber Fabry–Perot Embedded Sensor. *Opt. Lett.* **14**, 1225 (1989).

37. T. Valis, D. Hogg, and R. M. Measures, Composite Material Embedded Fiber Optic Fabry–Perot Strain Rosette. *Proc. SPIE—Int. Soc. Opt. Eng.* **1370**, 154 (1991).

38. K. A. Murphy, B. R. Fogg, G. Z. Wang, A. M. Vengsarkar, and R. O. Claus, Sapphire Fiber Interferometer for Microdisplacement Measurements at High Temperatures. *Proc. SPIE—Int. Soc. Opt. Eng.* **1588**, 117 (1992).

39. K. A. Murphy, M. F. Gunther, A. M. Vengsarkar, and R. O. Claus, Fabry–Perot Fiber Optics Sensors in Full Scale Testing on an F-15 Aircraft. *Proc. SPIE—Int. Soc. Opt. Eng.* **1588**, 134, (1992).

40. H. D. Simonsen, R. Paetsch, and J. R. Dunphy, Fiber Bragg Grating Sensor Demonstration in Glass Fiber Reinforced Polyester Composite. *Proc. SPIE—Int. Soc. Opt. Eng.* **1777**, 73 (1992).

41. G. Meltz, W. W. Morrey, and W. H. Glenn, Formation of Bragg Grating in Optical Fibers by a Transverse Holographic Method. *Opt. Lett.* **14**, 823 (1989).

42. W. W. Morey, Distributed Fiber Grating Sensors. *Proc. Opt. Fiber Sens. Conf., 7th, Sydney, 1990*, p. 285 (1990).

43. S. M. Melle, K. Liu, and R. M. Measures, Strain Sensing Using a Fiber Optics Bragg Grating. *Proc. SPIE—Int. Soc. Opt. Eng.* **1588**, 255 (1992).

44. H. Smith, Jr., A. Garrett, and C. R. Saff, Smart Structure Concept Study. *Proc. SPIE—Int. Soc. Opt. Eng.* **1170**, 224 (1990).

45. E. Udd, J. P. Theriault, A. Markus, and Y. Bar-Cohen, Microbending Fiber Optic Sensors for Smart Structures. *Proc. SPIE—Int. Soc. Opt. Eng.* **1170**, 478 (1990).

46. Z. J. Lu and F. A. Blaha, A Fiber Optic Strain and Impact Sensor System for Composite Materials. *Proc. SPIE—Int. Soc. Opt. Eng.* **1170**, 239 (1990).

47. D. A. Cox, D. Thomas, K. Reichard, D. Lindner, and R. O. Claus, Model Domain Fiber Optic Sensor for Closed Loop Vibration Control of a Flexible Beam. *Proc. SPIE—Int. Soc. Opt. Eng.* **1170**, 372 (1990).

48. B. Y. Kim, J. N. Blake, S. Y. Huang, and H. J. Shaw, Use of Highly Elliptical Core Fibers for Two-Mode Fiber Devices. *Opt. Lett.* **12**, 729 (1987).

49. J. N. Blake, S. Y. Huang, B. Y. Kim, and H. J. Shaw, Strain Effects on Highly Elliptical Core Two Mode Fibers. *Opt. Lett*, **12**, 732 (1987).

50. W. B. Spillman, Jr., L. B. Maurice, J. R. Lord, and D. M. Crowne, Quasi-distributed Polarimetric Strain Sensor for Smart Skins Applications. *Proc. SPIE—Int. Soc. Opt. Eng.* **1170**, 483 (1990).

51. A. D. Kersey and J. P. Dakin, eds., *Distributed and Multiplexed Fiber Optic Sensors*, Proc. SPIE, Vol. 1586. SPIE, Bellingham, WA, 1991.

52. B. G. Grossman, H. Hou, R. H. Nassar, A. Ren, and M. H. Thursby, Neural Network Processing of Fiber Optic Sensors and Arrays. *Proc. SPIE—Int. Soc. Opt. Eng.* **1370**, 205 (1991).

53. M. Thursby, K. Yoo, and B. Grossman, Neural Control of Smart Electromagnetic Structures. *Proc. SPIE—Int. Soc. Opt. Eng.* **1588**, 219 (1992).

54. B. Grossman, X. Gao, and M. Thursby, Composite Damage Assessment Employing an Optical Neural Network Processor and an Embedded Fiber Optic Sensor Array. *Proc. SPIE–Int. Soc. Opt. Eng.* **1588**, 64 (1992).

55. J. M. Mazzu, S. M. Allen, and A. K. Caglayen, Neural Network/Knowledge Based Systems for Smart Structures. *Proc. Act. Mater. Adap. Struct. Conf.*, Alexandria, VA, *1991*, p. 243 (1991).

56. M. R. Napolitano, C. I. Chen, and R. Nutter, Application of a Neural Network to the Active Control of Structural Vibration. *Proc. Act. Mater. Adapt. Struct. Conf.*, Alexandria, VA, *1991*, p. 247 (1991).

57. D. R. Huston, Smart Civil Structures—An Overview. *Proc. SPIE—Int. Soc. Opt. Eng.* **1588**, 182 (1992).

58. D. R. Huston, P. L. Fuhr, P. J. Kajenski, T. P. Ambrose, and W. B. Spillman, Installation and Preliminary Results from Fiber Optics Sensors Embedded in a Concrete Building. *Proc. SPIE—Int. Soc. Opt. Eng.* **1777**, 409 (1992).

59. H. D. Wright and R. M. Lloyd, Monitoring the Performance of Real Building Structures. *Proc. SPIE—Int. Soc. Opt. Eng.* **1777**, 219 (1992).

60. A. Holst and R. Lessing, Fiber-Optic Intensity-Modulated Sensors for Continuous Observation of Concrete and Rock-Filled Dams. *Proc. SPIE—Int. Soc. Opt. Eng.* **1777**, 223 (1992).

61. E. Udd, R. G. Blom, D. Tralli, E. Saaski, and R. Dokka, Application of the Sagnac Interferometer Based Strain Sensor to an Earth Movement Detection System. *Proc. SPIE—Int. Soc. Opt. Eng.* **2191**, p. 126 (1994).

3

Introduction to Advanced Composite Materials

JORN S. HANSEN
Institute for Aerospace Studies
University of Toronto
Downsview, Ontario, Canada

3.1. INTRODUCTION

This chapter is an introduction to advanced structural composite materials. The aim is to give standard terminology, to provide an introductory theoretical background, and to indicate the capabilities and problems arising from the use of these materials. The presentation emphasizes fiber-reinforced polymer-matrix laminated composites to the exclusion of other composite systems. This seems a logical choice because of widespread use and because these materials are prime candidates to receive embedded sensor systems.

Advanced composite technology represents the highly refined application of an idea used for centuries: combine stiff, strong reinforcement with material that will bind and protect the reinforcement. One of the first such applications was probably mud reinforced by straw for the construction of dwellings. Later, in civil engineering, reinforced concrete became and still is one of the most commonly used and flexible construction materials available. It combines stiff, strong fibers (reinforcing steel) with a protective matrix (concrete) and allows almost unlimited geometric and design flexibility.

In a similar manner, advanced composites combine reinforcement with a matrix. Now, however, exotic (expensive) fibers such as glass, graphite, or Kevlar are combined with exotic matrices such as epoxy, polyimide, or PEEK. Furthermore, the geometry of the reinforcement is specialized; the fibers are long (continuous) and are placed parallel to one another or are woven as in a piece of cloth. The sole objective is to obtain high performance; this is true whether the application is a racing bicycle or an aircraft. High performance is synonymous with high stiffness, high strength, and low weight. For example, a stiff lightweight bicycle allows better handling and better energy transfer from

Fiber Optic Smart Structures, Edited by Eric Udd.
ISBN 0-471-55448-0   © 1995 John Wiley & Sons, Inc.

the rider to the ground, and a lightweight aircraft allows increased payload or decreased operating costs. Therefore, the measures of merit are high specific stiffness (stiffness/density) and high specific strength (strength/density).

As can be seen from Fig. 3.1 significant advantages are available merely by replacing metal by composite. However, that is not the entire story because design flexibility and cleverness also play a role. To provide design and construction flexibility, advanced composites are usually laminated, that is, formed from many layers or plies in which the fibers can be oriented in prescribed directions. This allows tailoring to place stiffness and strength only where it is required. Much as in reinforced concrete, the ability to place the stiff, strong reinforcement in preferred directions and locations provides the possibility of optimal performance. Therefore, the role of fiber orientation and location in the structure is an overriding concern; structural characteristics are dominated by the fibers and their location. A considerable effort must be expended to keep track of the fibers and their resulting effects.

The analysis and physical understanding of composite structures is very different from that with metallic structures; intuition often fails to provide an accurate image of the response. On the other hand, unusual characteristics allow designs that are not possible for metal structures. Because of this flexibility, special requirements are placed on the use of composites; if a designer does not understand the full capabilities, potential gains are lost. Furthermore, nothing comes free. With advantage come disadvantages and a designer must also be aware of weakness. Composite laminates often exhibit low fracture toughness and are plagued by failure modes such as delamination which are unknown in metals. Therefore, it is necessary to understand what can and cannot be done. In the presentation, an effort has been made to discuss some of these special characteristics. This understanding is at the heart of using composites to the fullest advantage. In places, the presentation is a bit theoretical; it was felt that this was necessary in order that the reader be able to understand the subject matter and to appreciate the subtleties. This is not always easy but the subject of composites is not simple.

In Section 3.2 we consider a layer of composite material or the composite lamina. The presentation starts with an overview of material modeling, which leads to a description of the composite lamina. This is followed by stress, strain, and material property transformation and a discussion of thermal effects. In Section 3.3 we combine composite laminae to form a layered plate which introduces the composite laminate. The presentation is quite complete, starting with the essential assumptions of classical plate theory and ending with a formulation for composite plates. In addition, since composites are manufactured from materials with different coefficients of thermal expansion and manufacture takes place at elevated temperatures, the effect of thermal residual stresses is shown. In Section 3.4 we introduce a shorthand notation for laminates and then discuss commonly encountered special laminates. In Section 3.5 a brief discussion of composite failure theory is presented. Commonly used failure criteria are illustrated and discussed. The chapter closes

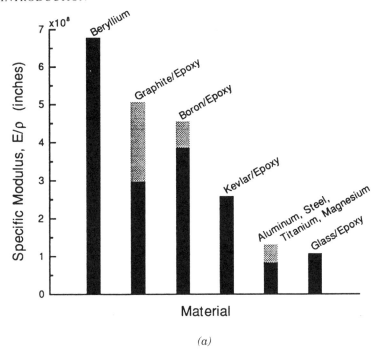

(a)

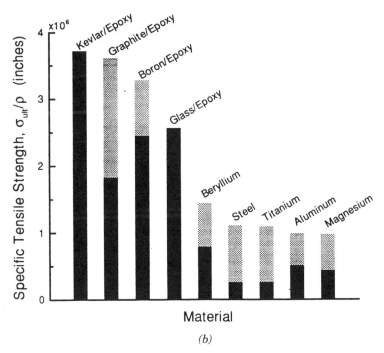

(b)

Figure 3.1. Comparison (a) of specific moduli and (b) specific tensile strengths for various materials.

with short discussion of fiber optic sensors as related to composite materials. In all sections an attempt has been made to provide physical insights into the composite characteristics being discussed.

3.2.   COMPOSITE LAMINA

A layer or ply of a composite structure is called a *lamina*. An understanding of the composite lamina is fundamental to an understanding of composite structures, and therefore a complete description is given. The presentation begins with a discussion of material modeling and some associated definitions. This leads to the specification of a lamina, a description of stresses and strains for a lamina, and appropriate constitutive relations for a lamina. Considered next is the transformation of stress and strain for an arbitrarily oriented lamina, which in turn leads to constitutive relations for arbitrarily oriented laminae. A description of thermal effects and the introduction of thermal coefficients of expansion for a lamina follows. The section closes with definitions of some special laminae.

3.2.1. Material Characteristics

Before any discussion concerning materials can proceed, it is necessary to present idealized conceptual or mathematical models of the material. Furthermore, if these models are to have value, they must be appropriate for the problem at hand. There are no universally applicable material models. For example, if the aluminum skin on an aircraft wing is being considered and the deflection of the wing is being calculated, the atomic structure of aluminum is not important but average properties giving material stiffness are important. On the other hand, if the fatigue characteristics of a metal depend on the crystal structure, it is wise not to ignore the crystal structure and deal only with "smeared-out" properties. In the current discussion it is assumed that the material microstructural scale is small compared to the response-of-interest scale. Therefore, it is safe to smear-out or average the microstructure and assume that the material can be represented as a *continuum*. Within the context of a continuum, there are two characteristics of importance. These describe directional or orientation effects on material properties and the uniformity of properties from location to location in a material. With respect to orientation, there are two extremes:

- *Isotropic Materials.* Material properties are invariant with regard to orientation. Thus the properties will not change if they are measured with respect to different coordinate systems. There are an infinity of material symmetry planes.

- *Anisotropic Materials.* These are the converse of isotropic material; that is, the material properties are different in all directions. There are no planes of material symmetry.

In a similar manner, there are two extremes when discussing spatial uniformity:

- *Homogeneous Materials.* The properties are the same from location to location.
- *Heterogeneous Materials.* The properties vary from location to location. Any material that is not homogeneous is heterogeneous.

It is clear from the definitions above that all materials are anisotropic and heterogeneous if they are examined at a fine-enough scale. Therefore, this material characterization is meaningful only if adopted at a macroscopic scale within the context of a continuum representation. Many materials, including metal, plastic, and wood, are treated as homogeneous even though it is not true on a fine scale. That is, for the intended purpose the material properties are satisfactorily defined by an averaged behavior of the individual constituents. It should be cautioned that if a fine-scale response of a material is of interest, it is inappropriate to assume that a material is homogeneous. It is to be noted that even when a material is assumed to be homogeneous, it does not necessarily follow that it is also isotropic. Wood is a perfect example, as it has very different properties along and perpendicular to the grain even though the properties are relatively uniform from location to location.

3.2.2. Lamina Description

A composite lamina is a thin layer of material composed of reinforcing fibers surrounded by a matrix. Typically, a lamina is about 5×10^{-3} in. thick and graphite fibers are about 0.2×10^{-3} in. in diameter. The fibers provide strength and stiffness, and the matrix binds the fibers together and provides protection from a hostile environment. For example, the strength of glass fibers is severely degraded by exposure to water.

The lamina is the building block of a laminated composite structure. A typical lamina with unidirectional reinforcement is illustrated in Fig. 3.2; a lamina for a woven fabric would have sets of fibers at right angles to each other. In this figure, particular attention should be paid to the coordinate denoted by $(1, 2, 3)$, which are referred to as the *principal material coordinates*, or more simply, the *principal* or *material coordinates*. For unidirectionally reinforced laminae, convention dictates that the 1-axis be oriented in the fiber direction, the 2-axis be perpendicular to the 1-axis in the plane of the material, and the 3-axis be normal to the 1–2 plane. The 1–2 plane lies at the geometric middle surface of the lamina. For the case of woven materials the 1- and 2-axes will be aligned with the fibers if they are orthogonal, whereas the 3-axis is always normal to the plane. The (x, y, z) axes are termed the *structural axes*. The choice of orientation of the structural axes is arbitrary but usually results from consideration of the geometry of the structure. For a rectangular plate the structural axes would normally be chosen parallel to the sides of the plate.

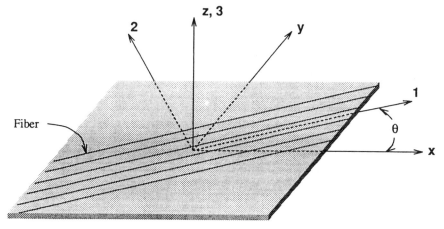

Figure 3.2. Typical composite lamina.

The angle θ measures the rotation of the 1-axis relative to the x-axis and is defined positive in a counterclockwise direction. The 3- and z-axes are always coincident and rotations always occur about this axis.

It is clear that a lamina has a relatively large microscale; the reinforcement diameter is relatively large compared to the thickness of the laminae. Fibers are definitely not microscopic compared to the lamina thickness. However, for most applications it has been found appropriate to assume that the reinforcing and matrix phases are smeared and to model a lamina as a continuum. Caution should be exercised; if fracture or other small-scale phenomena are important, such an assumption may not be appropriate. The ratio of fiber to matrix is assumed the same from location to location; therefore, the lamina is taken to be homogeneous. Furthermore, a fiber-reinforced composite lamina is not isotropic since the material properties must depend on orientation; the material stiffness will be different in the fiber direction (1-axis) than in the matrix direction (2-axis). It is clear, however, that this structure does exhibit some directional material symmetry. There are three such symmetry planes; planes parallel to the 1–2, 2–3, and 1–3 planes. Materials with such symmetries are said to be *orthotropic*.

3.2.3. Stress and Strain

Stress is defined as force divided by area; the orientation of the force and the area are important for calculating the stress. As a result, when the stress is evaluated in different coordinate systems, the transformation involved is not the standard vector transformation but the transformation of a second-rank tensor. Similarly, strain is a second-rank tensor. This is extremely important because it is always necessary to make transformations from material to structural coordinates, and vice versa, and the correct transformation must be used.

Stresses and strains are designated by σ_{ij}, ε_{ij}; $i, j = 1, 3$. In this notation the first subscript gives the direction of the stress or strain and the second subscript gives the direction of the normal to the reference area. Stress and strain may be thought of as two-dimensional arrays where the diagonal entries (repeated subscripts—that is, when the direction and the normal to the area are coincident) correspond to *normal stresses* or *strains* while the off-diagonal entries (different subscripts—that is, when the direction and the normal to the area are perpendicular) correspond to *shear stresses* or *strains*. In general, there are nine independent components of stress or strain. However, from equilibrium considerations it can be shown that $\sigma_{ij} = \sigma_{ji}$ and from the definition of the strains, $\varepsilon_{ij} = \varepsilon_{ji}$. Therefore, it is redundant to carry nine variables since only six are truly independent. This leads to the introduction of a contracted notation for stress and strain, which is summarized in Table 3.1.

There are a number of points that should be made about this contracted notation. Stress and strain now have the appearance of a vector and are usually written that way. However, they are not vectors and are not transformed using standard vector transformations. Also, it is not usual to use the contracted notation consistently; in particular the shear components are often written with a double subscript. For shear stresses the symbol σ_{ij} and τ_{ij} are freely interchanged and there should be no confusion. For the shear strains there is room for confusion, as the ε_{ij} and γ_{ij} representations differ by a factor of 2. The ε_{ij} is a tensor, whereas the γ_{ij} is not; γ_{ij} is referred to as the *engineering shear strain*. Why are there two different definitions of strain? This is a good question that can be answered, at least in part, by noting the following two points. First, engineering shear strain represents the change in angle of a right angle after the application of load; therefore, engineering shear strain represents a directly measurable quantity, in contrast to the tensor shear strain, which differs by a factor of 2. Second [see Eq. (2.8)], the constitutive relation is symmetric; this is possible only if the stress–strain relation is written in terms of engineering strain.

Table 3.1 Tensor and Contracted Notation

Stresses		Strains	
Tensor Notation	Contracted Notation	Tensor Notation	Contracted Notation
σ_{11}	σ_1	ε_{11}	ε_1
σ_{22}	σ_2	ε_{22}	ε_2
σ_{33}	σ_3	ε_{33}	ε_3
$\sigma_{23} = \tau_{23}$	σ_4	$2\varepsilon_{23} = \gamma_{23}$	ε_4
$\sigma_{31} = \tau_{31}$	σ_5	$2\varepsilon_{31} = \gamma_{31}$	ε_5
$\sigma_{12} = \tau_{12}$	σ_6	$2\varepsilon_{12} = \gamma_{12}$	ε_6

A few further notes. Caution should be exercised when dealing with the contracted representation of strain ε_i; these are not tensors but engineering strains, and because ε is used, there is a good chance of confusing them with the tensor strains ε_{ij}, which differ by a factor of 2 for the shear components. When a plane-stress assumption is made, only σ_{11}, σ_{22}, $\sigma_{12}(\tau_{12})$ and ε_{11}, ε_{22}, γ_{12} are retained in the analysis; this leads to an odd situation where the variables of interest are σ_1, σ_2, σ_6 and ε_1, ε_2, ε_6, which ruins the whole idea of representing the stresses and strains as vectors. Therefore, it is more common to write these quantities as σ_1, σ_2, $\sigma_{12}(\tau_{12})$ and ε_1, ε_2, γ_{12}. However, the remnant subscript "6" will crop up in subsequent sections in the subscripts for the constitutive relations.

3.2.4. Lamina Constitutive Relation

As noted above, the thickness of a composite lamina is small compared to its in-plane dimensions, and furthermore, when a plate is formed by stacking several laminae upon one another, the same relationship between the dimensions holds. Therefore, the assumption is made that the lamina is in a state of plane stress; that is, the out-of-plane stresses σ_3, τ_{13}, τ_{23} are small with respect to the in-plane stresses σ_1, σ_2, τ_{12} and may therefore be ignored. It is important to remember that the out-of-plane stresses are small and not zero; when failure analyses are undertaken, the exclusion of these stresses from the calculations may lead to serious difficulties. Based on the plane-stress assumption, the appropriate constitutive relation taken with respect to the principal material coordinates is

$$\begin{bmatrix} \sigma_1 \\ \sigma_2 \\ \tau_{12} \end{bmatrix} = \begin{bmatrix} Q_{11} & Q_{12} & 0 \\ Q_{12} & Q_{22} & 0 \\ 0 & 0 & Q_{66} \end{bmatrix} \begin{bmatrix} \varepsilon_1 \\ \varepsilon_2 \\ \gamma_{12} \end{bmatrix} \tag{3.1}$$

where Q_{ij} are material stiffness coefficients. This form of stress–strain relation is that of an orthotropic material, that is, a material which exhibits three planes of material symmetry, as noted earlier. The key characteristic of such a material is that the normal stresses and strains uncouple from the shear stresses and strains. A simple illustration of this uncoupling is the following. Consider a rectangular coupon made from an orthotropic material that has its sides cut parallel to the 1- and 2-axes; the fibers are parallel to the sides of the rectangle. If it is subjected to uniform normal stresses (some combination of σ_1 and σ_2), the coupon will deform into another rectangular shape and not a parallelogram; that is, there will be no shear deformation. In the above, notice the subscripts of Q_{66} even though it is the $(3, 3)$ entry in the constitutive relation; this is a carryover from the contracted notation.

The material stiffness coefficients in Eq. (3.17) are defined by

$$Q_{11} = \frac{E_{11}}{1 - v_{12}v_{21}} \qquad Q_{22} = \frac{E_{22}}{1 - v_{12}v_{21}}$$

$$Q_{12} = \frac{v_{12}E_{22}}{1 - v_{12}v_{21}} = \frac{v_{21}E_{11}}{1 - v_{12}v_{21}} \tag{3.2}$$

$$Q_{66} = G_{12}$$

The parameters E_{11}, E_{22} are the Young's modulus in the fiber direction and perpendicular to the fibers, respectively; v_{12}, v_{21} are the major and minor Poisson's ratios; and G_{12} is the in-plane shear modulus. Only three of E_{11}, E_{22}, v_{12}, v_{21} are independent, as can be seen from the definition of Q_{12}. It is usual to take v_{21} as the dependent variable because it is most difficult to determine accurately from experiments.

The evaluation of E_{11}, E_{22}, v_{12}, G_{12} is worth mentioning. These lamina properties depend directly on the fiber and matrix properties and it is possible to calculate the interrelationship; however, experimentally evaluated properties are more reliable. Typically, composite suppliers provide data sheets for their particular material and these data are usually adequate. However, a better approach is to determine E_{11}, E_{22}, v_{12}, G_{12} for the cure (consolidation) cycle and for the batch of material being used. Typically, the following experiments are conducted: four-ply, 0° flat specimens (fibers oriented along the length of the test coupon) tested in tension with a strain-gauge rosette mounted to determine E_{11}, v_{12}; eight-ply, 90° flat specimens (fibers oriented perpendicular to the length of the test coupon) with a strain-gauge rosette mounted and tested in tension to determine E_{22}, v_{21}; and four-ply, 90° tubular specimens (fibers oriented around the diameter of the tube) loaded torsionally and with a strain-gauge rosette mounted to determine G_{12}. In the first two experiments, tension tests are done because they are easier and cheaper than compression tests. In this regard, tension and compression experiments usually give slightly different results because the composite stiffness results from different mechanisms. In the tension case the fibers are stretched, whereas in the compression case the fibers have a tendency to bend. However, the tension results are used for both cases in an analyses because the use of different values of E_{11}, E_{22}, v_{12} in tension and compression would make an analysis exceedingly complicated. Also, the value obtained for v_{21} is only used as a check for compatibility with E_{11}, E_{22}, v_{12}. For G_{12} an average of the two strain-gauge results is taken since in theory both should give identical results and any discrepancy is due to experimental error. In all cases, results should be based on an average of at least three tests.

3.2.5. Transformation of Stress and Strain

When dealing with a composite structure, the orientation of the fibers and therefore the laminae are chosen in different directions to achieve desired structural properties. Therefore, it is necessary to introduce a reference system, the *structural axes*, which is common to all laminae. This makes it necessary to transform stresses, strains, and constitutive relations back and forth between the material and structural axes. The appropriate transformation for the stresses is given by

$$
\begin{bmatrix} \sigma_x \\ \sigma_y \\ \tau_{xy} \end{bmatrix} = T^{-1} \begin{bmatrix} \sigma_1 \\ \sigma_2 \\ \tau_{12} \end{bmatrix}
\tag{3.3}
$$

where

$$
T = \begin{bmatrix} \cos^2 \theta & \sin^2 \theta & 2\sin\theta\cos\theta \\ \sin^2 \theta & \cos^2 \theta & -2\sin\theta\cos\theta \\ -\sin\theta\cos\theta & \sin\theta\cos\theta & \cos^2\theta - \sin^2\theta \end{bmatrix}
\tag{3.4}
$$

The transformation matrix given in Eq. (3.4) is that which is appropriate for a second-rank tensor. As mentioned above, this transformation corresponds to a rotation about the z-axis. This transformation is precisely what is obtained graphically using a Mohr's circle approach. A similar relation exists for strain:

$$
\begin{bmatrix} \varepsilon_x \\ \varepsilon_y \\ \frac{1}{2}\gamma_{xy} \end{bmatrix} = T^{-1} \begin{bmatrix} \varepsilon_1 \\ \varepsilon_2 \\ \frac{1}{2}\gamma_{12} \end{bmatrix}
\tag{3.5}
$$

where the only feature of note is the factor $\frac{1}{2}$ in front of the engineering shear strain, which transforms the engineering strains to tensor strains. The inclusion of this factor is inconvenient, and as a result Eq. (3.5) is altered to

$$
\begin{bmatrix} \varepsilon_x \\ \varepsilon_y \\ \gamma_{xy} \end{bmatrix} = RT^{-1}R^{-1} \begin{bmatrix} \varepsilon_1 \\ \varepsilon_2 \\ \gamma_{12} \end{bmatrix}
\tag{3.6}
$$

where R is the Reuter matrix given by

$$
R = \begin{bmatrix} 1 & 0 & 0 \\ 0 & 1 & 0 \\ 0 & 0 & 2 \end{bmatrix}
\tag{3.7}
$$

It is extremely important not to transform engineering strains without the inclusion of the Reuter matrix; this is a common catastrophic error.

With the transformation of stress and strain established, it becomes a straightforward process to determine the constitutive relation of Eq. (3.1) in structural coordinates. This is done by transforming the stress and strain to structural coordinates and then writing the stress-strain relation in the form

$$\begin{bmatrix} \sigma_x \\ \sigma_y \\ \tau_{xy} \end{bmatrix} = \begin{bmatrix} \bar{Q}_{11} & \bar{Q}_{12} & \bar{Q}_{16} \\ \bar{Q}_{12} & \bar{Q}_{22} & \bar{Q}_{26} \\ \bar{Q}_{16} & \bar{Q}_{26} & \bar{Q}_{66} \end{bmatrix} \begin{bmatrix} \varepsilon_x \\ \varepsilon_y \\ \gamma_{xy} \end{bmatrix} \tag{3.8}$$

where the $\bar{Q}$ matrix gives the constitutive relation for an arbitrarily oriented lamina relative to the structural coordinates and is defined by

$$\bar{Q} = T^{-1}QRTR^{-1} = T^{-1}QT^{-T} \tag{3.9}$$

where T^{-T} is the inverse transpose of T. Notice that the character of an arbitrarily oriented lamina is quite different than when expressed in material coordinates. In general, $\bar{Q}_{16}, \bar{Q}_{26}$ are zero only when $\theta = 0°$ or $90°$, and these situations correspond to material coordinates. When $\bar{Q}_{16}, \bar{Q}_{26}$ are not zero, coupling exists between normal stresses and shear strains or shear stresses and normal strains. This does not happen for isotropic materials such as metals or for orthotropic materials in material coordinates. Physically, it means that if a rectangular piece of material is subjected to a normal stress, it will deform into a parallelogram; this is counterintuitive from a metal plate point of view. When structural coordinates do not correspond to lamina material coordinates, the three planes of material symmetry typical of an orthotropic material have apparently been lost and the material appears to be anisotropic. Of course, the lamina is still orthotropic and it is just a matter of expressing the $\bar{Q}$ matrix in material coordinates to recover the special orthotropic character.

When a lamina is rotated, the elastic properties vary significantly. These variations are what a competent designer can use to optimize composite structures. Typical results are shown in Fig. 3.3 for a graphite–epoxy lamina.

3.2.6. Thermal Effects

To this point, the development has ignored thermal effects. This is quite surprising since the fiber and matrix have very different thermal coefficients of expansion and composites are manufactured at elevated temperatures. For example, the thermal coefficient of expansion of epoxy is about 32.0×10^{-6} in./in./°F, while that of a graphite fiber is about 1.5×10^{-6} in./in./°F. Also, typically, thermosets (epoxy) are cured at about 350°F, while some thermoplastics (PEEK) are consolidated at about 700°F. Thermal effects arise from a difference in the thermal coefficient of expansion between the fiber and matrix

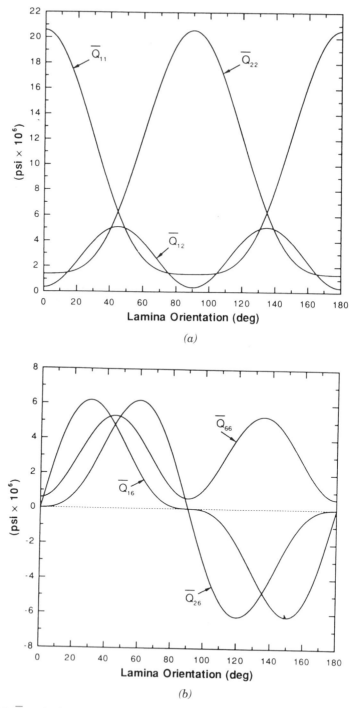

Figure 3.3. $\overline{Q}$ results for graphite–epoxy: (a) 3M T300/SP-288; (b) 3M T300/SP-288.

and because there are different thermal coefficients of expansion in the 1- and 2-directions. Therefore, when a component is used at room temperature, the possibility of large thermal stresses is quite real. Thermal stresses provide a constant shift in stress, which in theory does not affect elastic stiffness properties or the elastic response in a linear, elastic analysis. That is, thermal residual stresses will not affect deflection, natural frequency, or buckling calculations; of course, the stress state associated with such a calculation will be affected, as will failure calculations. Therefore, depending on the objective in an analysis, it may or may not be important to include thermal effects. It should be emphasized that for failure calculations the inclusion of thermal effects may be crucial. It is quite common for thermal residual stresses to represent a significant fraction of the failure allowable; in fact, test coupons have been known to fail when being cooled from the cure to room temperature. It is not uncommon to have thermal residual stresses that represent 40 to 60% of the ultimate strength of a laminate.

When thermal effects are included, the stress–strain relation for a lamina becomes

$$
\begin{bmatrix} \sigma_1 \\ \sigma_2 \\ \tau_{12} \end{bmatrix} = \begin{bmatrix} Q_{11} & Q_{12} & 0 \\ Q_{12} & Q_{22} & 0 \\ 0 & 0 & Q_{66} \end{bmatrix} \begin{bmatrix} \varepsilon_1 - \alpha_1 \Delta T \\ \varepsilon_2 - \alpha_2 \Delta T \\ \gamma_{12} \end{bmatrix} \tag{3.10}
$$

where α_1, α_2 are the lamina thermal coefficients of expansion in the 1- and 2-directions, and ΔT is the change in temperature.

This equation is really Eq. (3.1) rewritten with

$$
[\varepsilon_1 - \alpha_1 \Delta T \quad \varepsilon_2 - \alpha_2 \Delta T \quad \gamma_{12}]^T \tag{3.11}
$$

being the vector of elastic strains. The component

$$
[\varepsilon_1 \quad \varepsilon_2 \quad \gamma_{12}]^T \tag{3.12}
$$

is the *total strain*, the quantity a strain gauge would measure. The other component,

$$
[\alpha_1 \Delta T \quad \alpha_2 \Delta T \quad 0]^T \tag{3.13}
$$

is the thermal strain. When thermal effects are not considered the total strain and elastic strain are identical. This is precisely the situation in previous sections.

Two thought experiments nicely illustrate total and thermal strains. Consider the situation of a rectangular lamina with sides of the rectangle parallel to the 1- and 2-axes and subject the lamina to a uniform change in temperature ΔT. In the first experiment the lamina is unconstrained. This lamina will expand and from symmetry considerations (the sides of the rectangle are parallel to the 1- and 2-axes) remain rectangular. Since it is unconstrained, it will be stress free and from Eq. (3.10) this implies that the elastic strain will be zero. Therefore,

$$
\begin{bmatrix} \varepsilon_1 \\ \varepsilon_2 \\ \gamma_{12} \end{bmatrix} = \begin{bmatrix} \alpha_1 \Delta T \\ \alpha_2 \Delta T \\ 0 \end{bmatrix}
\tag{3.14}
$$

A strain gauge mounted on the specimen would record the total strain, which in this case is the thermal strain. The point to emphasize is there is no elastic distortion—only thermal distortion (at the macroscopic lamina scale, not the fiber–matrix scale).

In the second experiment the lamina is totally constrained on all four edges. The lamina cannot expand because of the constraints; the elastic strain must compensate for the thermal expansion since there can be no change in geometry. Therefore, the elastic strains are of equal magnitude but of opposite sign to the thermal strains. The important feature is that a strain gauge mounted on the specimen would record nothing. There is no total strain and the stress/elastic-strain state cannot be determined using strain gauges.

The first experiment illustrates thermal strains; the second illustrates thermal stresses. It is clear that the terms *thermal stress* and *thermal strain* are not synonymous and that the states do not coexist. Furthermore, the two experiments are at the extremes of reality and most situations will involve some degree of constraint and freedom. Therefore, the thermally induced stress and/or strain states will be some combination of the above.

The stresses expressed in structural coordinates are

$$
\begin{bmatrix} \sigma_x \\ \sigma_y \\ \tau_{xy} \end{bmatrix} = T^{-1} \begin{bmatrix} \sigma_1 \\ \sigma_2 \\ \tau_{12} \end{bmatrix} = T^{-1} Q R T R^{-1} \begin{bmatrix} \varepsilon_x \\ \varepsilon_y \\ \gamma_{xy} \end{bmatrix} - T^{-1} Q \begin{bmatrix} \alpha_1 \Delta T \\ \alpha_2 \Delta T \\ 0 \end{bmatrix}
\tag{3.15}
$$

As before, for an unconstrained lamina the total strains become

$$
T^{-1} Q R T R^{-1} \begin{bmatrix} \varepsilon_x \\ \varepsilon_y \\ \gamma_{xy} \end{bmatrix} = T^{-1} Q \begin{bmatrix} \alpha_1 \Delta T \\ \alpha_2 \Delta T \\ 0 \end{bmatrix}
\tag{3.16}
$$

and further, these observed strains are the thermal coefficients of expansion times ΔT. Thus

$$
T^{-1}QRTR^{-1}\begin{bmatrix} \alpha_x \\ \alpha_y \\ \alpha_{xy} \end{bmatrix} = T^{-1}Q\begin{bmatrix} \alpha_1 \\ \alpha_2 \\ 0 \end{bmatrix} \tag{3.17}
$$

which in turn yields

$$
\begin{bmatrix} \alpha_x \\ \alpha_y \\ \alpha_{xy}/2 \end{bmatrix} = T^{-1}\begin{bmatrix} \alpha_1 \\ \alpha_2 \\ 0 \end{bmatrix} \tag{3.18}
$$

This provides the relation between the thermal coefficient of expansion of a lamina in structural and material coordinates. This result is illustrated in Fig. 3.4.

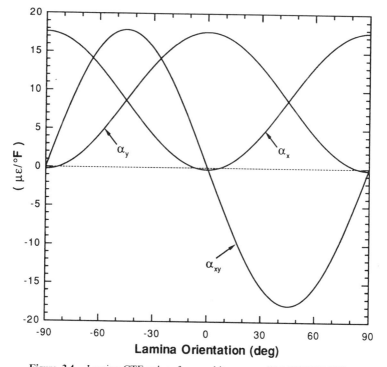

Figure 3.4. Lamina CTE values for graphite–epoxy (3M T300/SP-288).

3.2.7. Special Laminae

A *specially orthotropic composite lamina* is one for which the principal material axes are aligned with the structural axes. For example,

$$\begin{bmatrix} \bar{Q}_{11} & \bar{Q}_{12} & \bar{Q}_{16} \\ \bar{Q}_{12} & \bar{Q}_{22} & \bar{Q}_{26} \\ \bar{Q}_{16} & \bar{Q}_{26} & \bar{Q}_{66} \end{bmatrix} = \begin{bmatrix} \bar{Q}_{11} & \bar{Q}_{12} & 0 \\ \bar{Q}_{12} & \bar{Q}_{22} & 0 \\ 0 & 0 & \bar{Q}_{66} \end{bmatrix} \tag{3.19}$$

It is clear that this occurs only when $\theta = 0°$ or $90°$.

A *generally orthotropic composite lamina* is an orthotropic lamina in which the principal material axes are not aligned with the structural axes. Thus the $\bar{Q}$ matrix is uniquely defined by four material properties even though the matrix is apparently that of an anisotropic material. That is,

$$\begin{bmatrix} \bar{Q}_{11} & \bar{Q}_{12} & \bar{Q}_{16} \\ \bar{Q}_{12} & \bar{Q}_{22} & \bar{Q}_{26} \\ \bar{Q}_{16} & \bar{Q}_{26} & \bar{Q}_{66} \end{bmatrix} = T^{-1} \begin{bmatrix} Q_{11} & Q_{12} & 0 \\ Q_{12} & Q_{22} & 0 \\ 0 & 0 & Q_{66} \end{bmatrix} T^{-T} \tag{3.20}$$

3.3. LAMINATED COMPOSITE PLATE THEORY

The material presented in this section provides the basis upon which a group of laminae are combined to form a structural entity called a *laminate*. Plywood is a laminate. If layers of steel, brass, and aluminum are glued together, a laminated structure is obtained; this laminate would exhibit the combined response of the steel, brass, and aluminum layers. The objective of laminate theory is to calculate the combined structural properties; laminate properties are "average" properties of the included lamina. How are these averages determined? That depends very much on what is desired. For example, if a laminate is only subjected to in-plane loads, the situation is quite different than if bending loads will be applied. If the temperature varies, the effect of different coefficients of thermal expansion in the layers (e.g., steel, brass, aluminum) of the laminate must be considered. Various theoretical approaches can be adopted; the present is a straightforward extension of classical plate theory for isotropic, homogeneous plates to one for laminated plates. This theory has shortcomings and these will be discussed; the extension to more complex plate theories is only mentioned. The inclusion of thermally induced stresses and strains is considered, which is followed by an evaluation of lamina stresses and strains.

The goal of all plate theories is to reduce a three-dimensional problem to a two-dimensional problem, thereby simplifying calculations. For the case of composite plates this results, essentially in a "rule of mixtures" by which laminae properties are combined to obtain laminate properties.

3.3.1. Assumptions and Implications

A plate is a structure with two parallel surfaces or planes referred to as the *faces* of the plate. The plane midway between the faces is called the *middle plane* or *middle surface* of the plate. Rectangular structural coordinates (x, y, z) are adopted with (x, y) lying in the middle plane and z being normal to that plane. The displacements of a point (x, y, z) are denoted $\bar{u}(x, y, z)$, $\bar{v}(x, y, z)$, $\bar{w}(x, y, z)$, respectively. Furthermore, the displacements of the middle plane $(x, y, 0)$ are denoted $u(x, y)$, $v(x, y)$, $w(x, y)$.

The essential assumptions of this theory are:

1. The plate is thin. That is, the thickness h is small compared to other physical dimensions.
2. The displacements $\bar{u}$, $\bar{v}$, and $\bar{w}$ are small compared to the plate thickness.
3. The in-plane strains ε_x, ε_y, and γ_{xy} are small compared to unity.
4. The transverse normal strain ε_z and the transverse shear strains γ_{xz}, γ_{yz} are negligible.

The development of classical plate theory stems from the foregoing assumptions and will now be outlined. The linear strain–displacement relations are

$$\varepsilon_x = \frac{\partial \bar{u}}{\partial x} \qquad \varepsilon_y = \frac{\partial \bar{v}}{\partial y} \qquad \varepsilon_z = \frac{\partial \bar{w}}{\partial z}$$

$$\gamma_{xy} = \frac{\partial \bar{u}}{\partial y} + \frac{\partial \bar{v}}{\partial x} \qquad \gamma_{xz} = \frac{\partial \bar{u}}{\partial z} + \frac{\partial \bar{w}}{\partial x} \qquad \gamma_{yz} = \frac{\partial \bar{v}}{\partial z} + \frac{\partial \bar{w}}{\partial y} \tag{3.21}$$

Equation (3.21) combined with assumption 4 implies that

$$\varepsilon_z = \frac{\partial \bar{w}}{\partial z} = 0 \tag{3.22}$$

from which it follows that $\bar{w}(x, y, z)$ is independent of z. Therefore, $\bar{w}$ is equivalent to the middle-surface displacement, or

$$\bar{w}(x, y, z) = \bar{w}(x, y, 0) = w(x, y) \tag{3.23}$$

Assumption 4 also implies that

$$\gamma_{xz} = \frac{\partial \bar{u}}{\partial z} + \frac{\partial \bar{w}}{\partial x} = 0 \tag{3.24}$$

Noting $\bar{w} = w$ and rearranging yields

$$\frac{\partial \bar{u}}{\partial z} = -\frac{\partial w}{\partial x} \tag{3.25}$$

Since w is independent of z, both sides of Eq. (3.25) can be integrated with respect to z to yield

$$\bar{u}(x, y, z) = g(x, y) - z\frac{\partial w}{\partial x} \tag{3.26}$$

were $g(x, y)$ is a factor of integration that is independent of z. If $z = 0$ is substituted into Eq. (3.26) it may be seen that $g(x, y) = u(x, y)$, which then gives

$$\bar{u}(x, y, z) = u(x, y) - z\frac{\partial w}{\partial x} \tag{3.27}$$

In a similar manner,

$$\bar{v}(x, y, z) = v(x, y) - z\frac{\partial w}{\partial y} \tag{3.28}$$

Using these results, it is now possible to write the strain–displacement relations in terms of the middle-surface displacements. That is,

$$\begin{aligned}
\varepsilon_x &= \frac{\partial u}{\partial x} - z\frac{\partial^2 w}{\partial x^2} \\
\varepsilon_y &= \frac{\partial v}{\partial y} - z\frac{\partial^2 w}{\partial y^2} \\
\gamma_{xy} &= \frac{\partial u}{\partial y} + \frac{\partial v}{\partial x} - 2z\frac{\partial^2 w}{\partial x\, \partial y}
\end{aligned} \tag{3.29}$$

which may be expressed as

$$\varepsilon_x = \varepsilon_x^0 + z\kappa_x \qquad \varepsilon_y = \varepsilon_x^0 + z\kappa_y \qquad \gamma_{xy} = \gamma_x^0 + z\kappa_{xy} \tag{3.30}$$

where ε_x^0, ε_x^0, γ_x^0 are referred to as the middle-surface strains and κ_x, κ_y, κ_{xy} as the middle-surface curvatures. Equation (3.30) can be written concisely in vector form as

$$\varepsilon = \varepsilon^\circ + z\kappa \tag{3.31}$$

3.3.2. Stress and Moment Resultants

Stress resultants are defined as

$$N_x = \int_{-h/2}^{h/2} \sigma_x \, dz \qquad N_y = \int_{-h/2}^{h/2} \sigma_y \, dz \qquad N_{xy} = \int_{-h/2}^{h/2} \tau_{xy} \, dz \qquad (3.32)$$

while moment resultants are given by

$$M_x = \int_{-h/2}^{h/2} \sigma_x z \, dz \qquad M_y = \int_{-h/2}^{h/2} \sigma_y z \, dz \qquad M_{xy} = \int_{-h/2}^{h/2} \tau_{xy} z \, dz \qquad (3.33)$$

A more compact notation is

$$\left[\frac{N}{M} \right] = \int_{-h/2}^{h/2} \sigma \left[\frac{1}{z} \right] dz \qquad (3.34)$$

where N, M, σ are the vectors of stress resultants, moment resultants, and in-plane stresses, respectively. Stress and moment resultants N, M are interesting quantities; N is the average stress times the thickness, while M is the first moment of the stresses about the middle surface. Therefore, N and M are weighted averages of the stresses.

Since plate material properties change from lamina to lamina, the through-thickness integration must be completed stepwise over each ply. That is,

$$\left[\frac{N}{M} \right] = \sum_{k=1}^{N} \int_{h_{k-1}}^{h_k} (\sigma)_k \left[\frac{1}{z} \right] dz \qquad (3.35)$$

The index k is used to number the plies. The usual convention is that $k = 1$ corresponds to the bottom ply and $k = N$ to the top ply. Also, h_k and h_{k-1} are the upper and lower z coordinate of the kth ply. A schematic representation of a laminate and the conventional notation are given in Fig. 3.5. The subscript k on the stress σ means that the stress is evaluated in the domain of the kth ply. Substituting for stresses in terms of strains from Eq. (3.8) and using Eq. (3.31) yields

$$\left[\frac{N}{M} \right] = \sum_{k=1}^{N} \int_{h_{k-1}}^{h_k} \bar{Q}_k (\varepsilon^0 + z\kappa) \left[\frac{1}{z} \right] dz \qquad (3.36)$$

where $\bar{Q}_k$ are the material stiffness for the kth ply. The important feature of this result is that the z dependence of the expression within the integral is known; $\bar{Q}_k$, ε^0, κ are independent of z. The integral can be evaluated explicitly, thereby eliminating the dependence on z. This is precisely the objective in developing

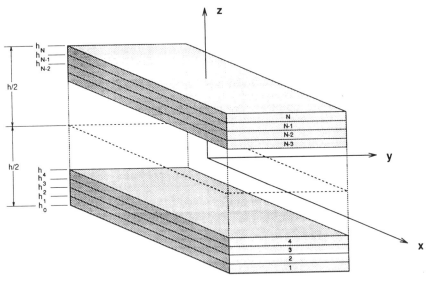

Figure 3.5. Composite laminate with notational convention.

plate theory, as the original three-dimensional equations in (x, y, z) can now be reduced to two-dimensional equations in (x, y).

Completing the integration gives the final result

$$\left[\frac{N}{M}\right] = \left[\begin{array}{c|c} A & B \\ \hline B & D \end{array}\right]\left[\frac{\varepsilon^0}{\kappa}\right] \tag{3.37}$$

where

$$A = \sum_{k=1}^{N} \bar{Q}_k(h_k - h_{k-1})$$

$$B = \tfrac{1}{2} \sum_{k=1}^{N} \bar{Q}_k(h_k^2 - h_{k-1}^2) \tag{3.38}$$

$$D = \tfrac{1}{3} \sum_{k=1}^{N} \bar{Q}_k(h_k^3 - h_{k-1}^3)$$

The A, B, D matrices are the heart of classical-composite-plate theory. These matrices define the average effect of laminae properties and stacking sequence on the plate response. The A matrix gives the in-plane stiffness of the laminate; that is, it relates the in-plane stress resultants N to the in-plane strains ε^0. In a similar manner the the D matrix gives the flexural stiffness of the laminate; it relates the moment resultants M to the curvatures κ. Both of these quantities have analogs in classical plate theory for isotropic, homogeneous plates. The matrix B is unique, as it acts to couple the stress resultants N to the curvatures κ or the moment resultants M to the in-plane strains ε^0. This means, for example, that if B is present, the application of an in-plane load will cause bending of a plate, a situation that is counter intuitive and does not occur for plates made from isotropic, homogeneous materials such as metals. It is worth noting that A is totally independent of the ply stacking sequence since $(h_k - h_{k-1})$ is merely the thickness of the kth ply. On the other hand, B and D are strongly dependent on sequence. This may be seen by examining the form of these matrices as given in Eq. (3.38).

3.3.3 Thermal Effects

As in the case of a lamina, the inclusion of thermal effects involves the introduction of the appropriate stress–strain equation [Eq. 3.10] in the calculations. Proceeding as before and interpreting ε^0, κ as total middle-surface strains and total curvatures yields

$$\left[\frac{N}{M} \right] = \left[\begin{array}{c|c} A & B \\ \hline B & D \end{array} \right] \left[\frac{\varepsilon^0}{\kappa} \right] - \left[\frac{J^*}{H^*} \right] \tag{3.39}$$

where

$$J^* = \int_{-h/2}^{h/2} T^{-1} Q \alpha_k \, \Delta T(z) \, dz$$

$$H^* = \int_{-h/2}^{h/2} z T^{-1} Q \alpha_k \, \Delta T(z) \, dz$$

and

$$\alpha_k = \begin{bmatrix} \alpha_1 \\ \alpha_2 \\ 0 \end{bmatrix}_k$$

Here the change in temperature for the laminate $\Delta T(z)$ is a function of z; however, to be consistent with current plate theory it can at most be a linear function of z.

If all the laminae are the same material and if the laminate is subjected to a uniform change in temperature ΔT, Eq. (3.39) simplifies to become

$$\left[\frac{N}{M} \right] = \left[\begin{array}{c|c} A & B \\ \hline B & D \end{array} \right] \left[\frac{\varepsilon^0}{\kappa} \right] - \Delta T \left[\frac{J}{H} \right] \alpha_m \tag{3.40}$$

where

$$J = \int_{-h/2}^{h/2} T^{-1}Q\,dz = \sum_{k=1}^{N} T_k^{-1}Q_k(h_k - h_{k-1})$$

$$H = \int_{-h/2}^{h/2} zT^{-1}Q\,dz = \tfrac{1}{2}\sum_{k=1}^{N} T_k^{-1}Q_k(h_k^2 - h_{k-1}^2)$$

$$\alpha_m = \begin{bmatrix} \alpha_1 \\ \alpha_2 \\ 0 \end{bmatrix}$$

To determine the laminate thermal coefficients of expansion, a laminate is taken under unconstrained conditions and is subjected to a change in temperature ΔT. Since there are no constraints, $N = M = 0$. Also, under these conditions the total middle-surface strain and the total curvature are the thermal coefficient of expansion and the thermally induced curvature per unit temperature times ΔT, respectively. That is,

$$\varepsilon^0 = \alpha_s\,\Delta T = \begin{bmatrix} \alpha_x \\ \alpha_y \\ \alpha_{xy} \end{bmatrix} \Delta T \qquad \kappa = \kappa_T\,\Delta T = \begin{bmatrix} \kappa_x^T \\ \kappa_y^T \\ \kappa_{xy}^T \end{bmatrix} \Delta T \qquad (3.41)$$

where α_s are the thermal coefficients of expansion and κ_T are the thermally induced curvature per unit change in temperature.

Substituting for N, M, ε^0, κ in Eq. (3.40) and manipulating yields

$$\begin{bmatrix} \alpha_s \\ \hline \kappa_T \end{bmatrix} = \begin{bmatrix} A & B \\ \hline B & D \end{bmatrix}^{-1} \begin{bmatrix} J \\ \hline H \end{bmatrix} \alpha_m \qquad (3.42)$$

which defines α_s and κ_T for the laminate.

3.3.4. Evaluation of Stress Fields

Once the response of a composite has been determined, the evaluation of the stresses or strains in a laminate can proceed. This is necessary to determine if design allowables have been exceeded or if failure has occurred. It is assumed that $u(x, y)$, $v(x, y)$, $w(x, y)$ can be determined; for actual situations such solutions are usually the result of a finite-element analysis. Strains follow directly from Eq. (3.29):

$$\varepsilon_x = \frac{\partial u}{\partial x} - z\frac{\partial^2 w}{\partial x^2}$$

$$\varepsilon_y = \frac{\partial v}{\partial y} - z\frac{\partial^2 w}{\partial y^2} \qquad (3.43)$$

$$\gamma_{xy} = \frac{\partial u}{\partial y} + \frac{\partial v}{\partial x} - 2z\frac{\partial^2 w}{\partial x\,\partial y}$$

Stress evaluation requires consideration at a lamina level since the stresses depend on material properties. Everything is given in structural coordinates, as can be seen from Eq. (3.8):

$$\begin{bmatrix} \sigma_x \\ \sigma_y \\ \tau_{xy} \end{bmatrix}_k = \begin{bmatrix} \bar{Q}_{11} & \bar{Q}_{12} & \bar{Q}_{16} \\ \bar{Q}_{12} & \bar{Q}_{22} & \bar{Q}_{26} \\ \bar{Q}_{16} & \bar{Q}_{26} & \bar{Q}_{66} \end{bmatrix}_k \begin{bmatrix} \varepsilon_x \\ \varepsilon_y \\ \gamma_{xy} \end{bmatrix}_k \qquad (3.44)$$

Note that the strains are linear functions of z, and therefore the specification of stress or strain in a lamina requires the specification of an appropriate z for that kth lamina subject to $h_{k-1} \leqslant z \leqslant h_k$. Another feature is that the stresses are inherently discontinuous across ply interfaces. This can be seen by referring to Eq. (3.44) and observing that the strains are continuous across ply boundaries while the $\bar{Q}$ matrices are different for each ply. Therefore, the stresses are discontinuous. This provides no problem for σ_x, σ_y, τ_{xy} since there is no requirement that they be continuous. However, if the transverse shear stresses τ_{xz}, τ_{yz} are of interest, there is a problem because equilibrium consider- ations demand continuity.

An additional step is required, since failure analyses are undertaken in material coordinates; the stresses must be rewritten as

$$\begin{bmatrix} \sigma_1 \\ \sigma_2 \\ \tau_{12} \end{bmatrix}_k = T_k \begin{bmatrix} \sigma_x \\ \sigma_y \\ \tau_{xy} \end{bmatrix}_k \qquad (3.45)$$

Also, since the stresses are a function of z, it is usual to evaluate the stress at the ply middle surface when investigating failure.

Only the in-plane stresses are determined from the current plate theory. This is a limitation since no information is gained regarding transverse stresses which are responsible for delamination effects. The difficulty can be overcome either by using more sophisticated plate theories or by using the current theory in conjunction with a slightly different approach. That is, rather than calculat- ing the transverse shear stresses based on the dispacement solution (which says that they are zero), they will be calculated based on equilibrium considerations while accepting the in-place stresses based on the displacement solution.

This second approach starts with the stress equilibrium equations in the absence of body forces:

$$\frac{\partial \sigma_x}{\partial x} + \frac{\partial \tau_{xy}}{\partial y} + \frac{\partial \tau_{xz}}{\partial z} = 0$$

$$\frac{\partial \tau_{xy}}{\partial x} + \frac{\partial \sigma_y}{\partial y} + \frac{\partial \tau_{yz}}{\partial z} = 0 \qquad (3.46)$$

$$\frac{\partial \tau_{xz}}{\partial x} + \frac{\partial \tau_{yz}}{\partial y} + \frac{\partial \sigma_z}{\partial z} = 0$$

The point to notice in the first two of these expressions is that when σ_x, σ_y, τ_{xy} are known, τ_{xz}, τ_{yz} can be calculated by an integration with respect to z. That is,

$$\tau_{xz} = -\int_{-h/2}^{z*} \frac{\partial \sigma_x}{\partial x} + \frac{\partial \tau_{xy}}{\partial y} dz$$

$$\tau_{yz} = -\int_{-h/2}^{z*} \frac{\partial \tau_{xy}}{\partial x} + \frac{\partial \sigma_y}{\partial y} dz \qquad (3.47)$$

where $z*$ is the through-thickness location at which the stress is to be evaluated. Although it is not shown explicitly, the integrals above must be determined in a piecewise fashion over each ply until the appropriate location corresponding to $z*$ is reached. The third part of Eq. (3.46) is not used.

Does this manipulation make sense? Perhaps. First, if τ_{xz}, τ_{yz} are calculated in this manner, they will be zero at the plate upper and lower surfaces $z = h/2$, $-h/2$, and they will be continuous at ply interfaces. These are equilibrium constraints that should be satisfied. Second, the in-plane stresses σ_x, σ_y, τ_{xy} are usually determined quite accurately in classical plate theory, and therefore it follows that τ_{xz}, τ_{yz} should be accurate. It is particularly comforting that this manipulation involves an integration with respect to z, since this also has a tendency to smooth the results. On the other hand, σ_x, σ_y, τ_{xy} have been differentiated in the equilibrium equations; differentiation of numerical results can sometimes lead to difficulties. A disconcerting aspect is that there are now two quite different stress predictions coming from the same piece of analysis, and the second prediction is based on equilibrium equations that were essentially ignored for the calculation of the displacement field. Comparisons to three-dimensional results seem to indicate that the second approach is sound.

3.4. COMPOSITE LAMINATES

In Section 3 a mathematical plate model was developed. Now the plate characteristics are described by the A, B, D matrices, which represent average or smeared properties of the laminate within the context of the plate approximations. That is, structural characteristics resulting from fiber stiffness properties and orientation in each lamina are contained in A, B, D. Can special classes of plates be identified by the form of these matrices? The answer is, yes. In this section we introduce a commonly used shorthand description for laminates. This is followed by a description of some special laminates and the relation between physical response characteristics of special laminates and the form of the A, B, D matrices.

3.4.1. Laminate Shorthand

A shorthand form for describing composite laminates has evolved, based on the following guidelines: All plies in a laminate are assumed to be of equal thickness and material properties unless stated otherwise, the laminate is described by the angle of each ply relative to structural coordinates, and the plies are numbered starting at the bottom of the plate. If there is any chance of ambiguity, the "bottom" should be identified explicitly; for example, in the case of cylindrical shells the numbering usually starts with the innermost ply. Thus a four-ply laminate where all plies are of equal thickness and have identical mechanical properties is described by

$$(\theta_1, \theta_2, \theta_3, \theta_4) \tag{3.48}$$

where θ_1, θ_2,... identify the angle of the fiber orientation or material 1-axis relative to the structural x-axis for ply number 1, 2,.... For the special situation that the laminate has a symmetric layup of plies relative to the plate middle surface, this symmetry is recognized. That is,

$$(\theta_1, \theta_2, \theta_2, \theta_1) \Rightarrow (\theta_1, \theta_2)_s \tag{3.49}$$

In a similar manner, if a laminate involves repetition, the description is shortened as follows:

$$(\theta_1, \theta_2, \theta_1, \theta_2) \Rightarrow (\theta_1, \theta_2)_2 \tag{3.50}$$

The number of possibilities is almost endless; following are some of the situations commonly encountered:

$$(\theta_1, -\theta_1, \theta_2, -\theta_1) \Rightarrow (\pm\theta_1, \pm\theta_2)$$

$$(\pm\theta_1, \pm\theta_2, \pm\theta_1, \pm\theta_2) \Rightarrow (\pm\theta_1, \pm\theta_2)_2 \tag{3.51}$$

$$(\pm\theta_1, \pm\theta_2, \pm\theta_1, \pm\theta_2, \mp\theta_2, \mp\theta_1, \mp\theta_2, \mp\theta_1) \Rightarrow (\pm\theta_1, \pm\theta_2)_{2s}$$

When a laminate is composed of unequal thickness plies, this must be stated explicitly: for example,

$$(\theta_1 @ h_1, \theta_2 @ h_2, \theta_3 @ h_3) \tag{3.52}$$

In a similar manner, if the laminate is composed of layers of different materials, the materials must be identified; that is,

$$(\theta_1 @ mat_1, \theta_2 @ mat_2, \theta_3 @ mat_3) \tag{3.53}$$

3.4.2. Special Laminates

The simplest laminate is a *single ply*. For this situation the relationship between lamina and laminate properties is straightforward but of interest: that is,

$$A = \bar{Q}h \qquad B = 0 \qquad D = \tfrac{1}{12}\bar{Q}h^3 \tag{3.54}$$

A class of laminates having considerable practical significance are the *symmetric laminates*. These laminates are such that all lamina properties are symmetric with regard to the laminate middle surface. Symmetry applies to both geometric and material properties. The single-ply laminate is the simplest symmetric laminate, but there are other possibilities. The following illustrate some of them:

$$(\theta_1, \theta_2, \theta_2, \theta_1)$$

$$(\theta_1 @ h_1, \theta_2 @ h_2, \theta_2 @ h_2, \theta_1 @ h_1) \tag{3.55}$$

$$(\theta_1 @ mat_1, \theta_2 @ mat_2, \theta_1 @ mat_1)$$

The important characteristic of symmetric laminates is that the B matrix vanishes. Why this happens can be seen by considering Eq. (3.38) in which B is defined. It vanishes because when two identical laminae are placed symmetrically about the plate middle surface, the $\bar{Q}_k$ matrices are equal while the factors $(h_k^2 - h_{k-1}^2)$ are of equal magnitude but opposite sign. Therefore, the contribution of each ply to B sums to zero since every ply has a partner located symmetrically about the middle surface. The implications of B vanishing are extremely significant; there will be no coupling between N and κ or M and ε. This means that for a symmetric laminate, in-plane loads will not cause the plate to bend, and bending loads will not cause stretching of the plate middle surface. By contrast, if the laminate is not symmetric, the application of an in-plane load will cause the plate to bend. This situation seems counterintuitive and is probably so because plates constructed from isotropic, homogeneous materials (metal) are symmetric and most of our experience is with this type of plate. Even more important, flat symmetric laminates remain flat when subjected to a change in temperature. In a symmetric laminate, H^*, H vanish as well as B; therefore, as can be seen from Eqs. (3.39) and (4.40), a change in temperature will not result in a thermally induced curvature. This has very significant implications in the manufacture of laminated composites since nonsymmetric laminates will always bend because of high-temperature cures. A bimetallic strip, which is a laminate formed from two different metals, is an example of a nonsymmetric laminate.

Another commonly encountered class are the *balanced laminates*. These are laminates in which for every $+\theta$ ply there is a $-\theta$ ply with identical material and geometric properties. A typical example is

$$(\theta_1, -\theta_2, -\theta_1, \theta_2, \theta_3, -\theta_3) \tag{3.56}$$

Note that the location of the $+\theta$ and $-\theta$ plies relative to one another and relative to the plate middle surface is not important. Balanced laminates exhibit the characteristic that the A matrix is *orthotropic* when the plies are orthotropic; that is,

$$A = \begin{bmatrix} A_{11} & A_{12} & 0 \\ A_{12} & A_{22} & 0 \\ 0 & 0 & A_{66} \end{bmatrix} \tag{3.57}$$

The A_{16}, A_{26} terms vanish because the $\bar{Q}_{16}$, $\bar{Q}_{26}$ are of opposite signs for $+\theta$ and $-\theta$ plies; therefore, the contribution from these plies sums to zero in Eq. (3.38). The B and D do not in general adopt this form because ply contributions are influenced by their location through the thickness. The significant physical result of a balanced laminate is that since $A_{16} = A_{26} = 0$ there is no coupling between normal stress resultants N_x, N_y and shear strains γ_{xy} nor between the shear resultant N_{xy} and the normal strains ε_x, ε_y. That is, if a rectangular specimen is subjected to an axial load, the specimen will extend axially and contract laterally but will remain rectangular.

Cross-ply laminates are composed of alternating $0°$ and $90°$ plies. Typical examples of cross-plies are

$$(0, 90, 0)$$

$$(0 @ h_1, 90 @ h_2, 0 @ h_3, 90 @ h_4) \tag{3.58}$$

The principal characteristic of cross-ply laminates composed of orthotropic laminae is that the A, B, and D matrices are orthotropic:

$$A = \begin{bmatrix} A_{11} & A_{12} & 0 \\ A_{12} & A_{22} & 0 \\ 0 & 0 & A_{66} \end{bmatrix} \quad B = \begin{bmatrix} B_{11} & B_{12} & 0 \\ B_{12} & B_{22} & 0 \\ 0 & 0 & B_{66} \end{bmatrix} \quad D = \begin{bmatrix} D_{12} & D_{12} & 0 \\ D_{12} & D_{22} & 0 \\ 0 & 0 & D_{66} \end{bmatrix}$$

$$\tag{3.59}$$

This orthotropic character results since $\bar{Q}_{16}$, $\bar{Q}_{26}$ are zero when an orthotropic lamina is oriented at $0°$ or $90°$ to the structural axes; that is, every ply is specially orthotropic. Therefore, all the $(1,6)$, $(2,6)$ entries in the A, B, D matrices vanish since they are a summation of $\bar{Q}_{16}$, $\bar{Q}_{26}$ through the thickness. The character leads to a laminate in which there is no coupling between normal stresses and shear strains or twisting curvature as well as normal bending moments and shear strains or twisting curvature. The laminate behaves very much like a metal except that in general the B matrix is present and therefore there is bending-membrane coupling.

Within the context of cross-ply laminates, there are two special cases of interest. *Symmetric cross-plies* are the simplest since then $B = 0$. On the other hand, *antisymmetric cross-plies* are antisymmetric about the middle surface. Illustrative of this situation is

$$(0 @ h_1, 90 @ h_2, 0 @ h_2, 90 @ h_1) \qquad (3.60)$$

A more restrictive situation is a *regular antisymmetric cross-ply*, given by

$$(0, 90)_n \qquad (3.61)$$

A laminate in which the laminae are stacked with alternating $+\theta$, then $-\theta$ angles is referred to as an *angle ply*. Typical examples are

$$(\pm\theta)_n, (+\theta, -\theta, +\theta), (\pm\theta)_s \qquad (3.62)$$

Perhaps the most important special laminates are the *quasi-isotropic laminates*. These laminates have the astonishing characteristic that they exhibit isotropic elastic in-plane properties. That is, the laminate in-plane elastic properties (the A matrix) are independent of the structural reference-axis orientation. In addition, quasi-isotropic laminates cannot be distinguished from isotropic materials using any external in-plane measurement system. (It is, of course, possible to determine that it is a composite if the laminate cross section is examined.) It should be emphasized that these laminates are not isotropic with respect to bending properties (D matrix) and that is why they are referred to as quasi-isotropic. Furthermore, if the laminate is not symmetric, the B matrix will be nonzero, which cannot occur for isotropic materials. The reason these laminates are so important is that they are often used as a direct substitute for metal (black aluminum) in order to save weight or to enhance structural properties. Such a substitution is not an optimal use of composites but may be considered a logical and straightforward first step in introducing composites into an existing metal design.

It requires quite a leap of faith to believe that it is possible to combine a finite number of orthotropic laminae and thereby create a quasi-isotropic laminate. As it turns out, the construction is quite straightforward, although not obvious. The laminates are formed from identical laminae stacked with equal angles between successive laminae. That is, for a laminate with N plies the angle separating each lamina is $180/N$, with the restriction that $N \geqslant 3$. Therefore, the orientation of the laminae are

$$\theta_1 = 180/N, \theta_2 = 2 \times 180/N, \ldots, \theta_k = k \times 180/N, \ldots, N \geqslant 3 \quad (3.63)$$

The details showing that this leads to a quasi-isotropic laminate will not be included here; reference may be made to any number of books on composite materials. However, it is useful to present the final form for A, which is

$$\begin{bmatrix} A_{11} & A_{12} & A_{16} \\ A_{12} & A_{22} & A_{26} \\ A_{16} & A_{26} & A_{66} \end{bmatrix} = h \begin{bmatrix} U_1 & U_4 & 0 \\ U_4 & U_1 & 0 \\ 0 & 0 & U_5 \end{bmatrix} \quad (3.64)$$

The first item to note is that the A_{ij} depend on only three parameters, which are material invariants U_1, U_4, U_5 and are characteristics of the individual lamina. Furthermore, $U_1 - U_4 = 2U_5$ which is precisely the relationship that exists between the shear modulus, Young's modulus, and Poisson's ratio for isotropic, homogeneous materials. (This relationship must exist; otherwise, the material properties cannot be invariant with respect to rotation.) The second feature, which is even more astounding, is that A does not have an explicit dependence on the number of lamina but only on the laminate thickness h. This means that all quasi-isotropic laminates manufactured from the same material have the same in-plane stiffness per unit thickness and are therefore identical from an elastic of view.

Does this means that all quasi-isotropic laminates are the same? The answer is, no. Because even though they have identical elastic properties, they do not have identical failure properties. Perhaps more important, none of these laminates are quasi-isotropic with respect to failure. For example, if coupons are cut at different angles to the structural axes and tested in tension to failure, two results become apparent. First, the stiffness of all tested coupons would be identical: that is, elastic isotropy. Second, the failure loads would be different: in other words, failure anisotropy. This means that in a particular design situation it is quite important which quasi-isotropic laminate is used and what orientation of that laminate is chosen.

3.5. LAMINA FAILURE ANALYSIS

Perhaps the most difficult and simultaneously most important problem for the design of composite structures is the prediction of failure. Failure may result from a broad range of loading conditions, such as static overload, fatigue, impact, or fracture. There are numerous modes of failure; fiber failure, matrix cracking, or delamination, to mention a few. Tensile failure and compression failure involve different mechanisms. For example, if a lamina is subjected to a tensile load in the 1-direction, the lamina strength is dominated by tensile strength of the fiber. On the other hand, if the lamina is subjected to the equivalent compressive load, the fibers may have a tendency to bend and are supported by the matrix. Therefore, there are dramatic differences in the tensile and compressive failure stresses in the 1-direction; tensile failure stresses are typically 1.5 times compressive failure stresses. Failure stresses in the 1- and 2-directions are dominated by the fibers and matrix, respectively. It is not unusual to see tensile failure stresses in the 1-direction which are 30 times those in the 2-direction. This occurs because the fibers provide the strength in the 1-direction, whereas the weaker matrix is the dominant factor in the 2-direction. This leads to great difficulties in developing failure models for composites since many factors must be considered.

When dealing with composite failure there are two different points of view that may be adopted: these are *mechanistic* and *phenomenological*. A *mechanistic approach* relates the failure properties of the lamina to constituent properties such as fiber strength, matrix strength, fiber–matrix interface strength, and so on. This approach is aimed at the microstructure of a composite and provides insight into the physics of failure. A mechanistic understanding is essential if improved composite systems are to be developed. However, this approach is limited since it is essentially intractable for a real composite laminate; application is restricted to simple fiber–matrix modeling situations. If failure were to be calculated using a mechanistic approach for a real composite using, for example, a finite element approach, the elements would have to be small enough to capture the fine details of the stress field around every fiber. This is not realistic with today's most powerful computers and will probably not be so in the near future. A *phenomenological approach* regards the material as a continuum, and a mathematical model is used to relate the stress or strain state of the continuum to the failure state. Failure is considered to be a material characteristic at the lamina or continuum scale and not at the fiber–matrix scale. This clearly involves the same type of thinking that was adopted to obtain the stress–strain continuum model of a lamina. The advantage of this approach is that it is relatively simple to use and, as will be seen, requires relatively few experimental data for calibration purposes. The disadvantages are that no understanding of the mechanisms of failure are obtained and no knowledge of failure types or material response is gained. Because of inherent simplicity and ease of application, phenomenological models are used almost exclusively for design.

In what follows, a description of some of the more commonly used phenomenological models will be presented. This will be followed by a discussion concerning the merits of these models, stress analysis requirements for application of these models, and some other possiblities.

3.5.1. Phenomenological Failure Models

Phenomenological failure models are mathematical models that define a failure surface. Failure is defined to occur when the stress or strain state in the material is altered such that the state passes through the failure surface. A schematic representation of a phenomenological failure surface in the σ_1, σ_2 plane is shown in Fig. 3.6; as noted, failure occurs when a particular combination of σ_1, σ_2 passes through the failure surface. In general, a failure surface would be six-dimensional, involving σ_1, σ_2, σ_3, τ_{23}, τ_{13}, τ_{12}; however, in composites since a plane-stress assumption is usually invoked, the failure surface is three-dimensional and involves σ_1, σ_2, τ_{12}. In metals, the commonly used von Mises yield criterion is representative of a phenomenological model. These models cannot be chosen arbitrarily but must satisfy a number of simple physical constraints in order that they make sense:

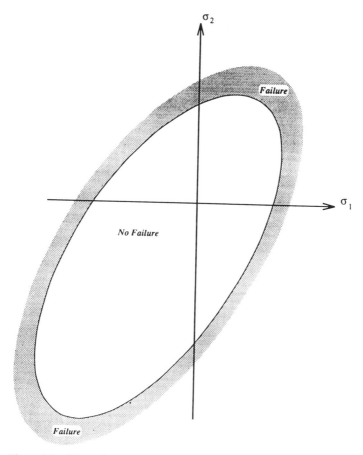

Figure 3.6. Schematic representation of a failure surface in the σ_1–σ_2 plane.

1. The failure surface must be closed. That is, if a particular load path is followed with increasing magnitude, failure must occur eventually.
2. There should be only one failure corresponding to a particular loading path.
3. The failure surface should be invariant with respect to the definition of reference axes. That is, a redefinition of the stress or strain should not affect the failure.
4. There should be unique conversions between stress and strain space.

Failure models are usually represented as algebraic relations expressed in terms of stress or strain. Therefore, they may be represented symbolically as being either

$$\text{stress based} \Rightarrow f(\sigma_{ij}, F) = 0$$

$$\text{strain based} \Rightarrow g(\varepsilon_{ij}, G) = 0$$

$$\text{energy based} \Rightarrow h(\sigma_{ij}\varepsilon_{ij}, H) = 0$$

where F, G, H are experimentally determined parameters. There are many failure criteria in use; some of the most common are maximum stress, maximum strain, tensor polynomial, Tsai–Hill, Tsai–Wu, and experimental. The choice of a particular criterion depends on a number of factors. The first and most important is that the criterion should predict accurately the failure of a lamina. On the other hand, there are also some very pragmatic constraints, not the least of which are the ease with which a criterion can be applied and the number of experimental data (cost) required to calibrate the model.

For illustration purposes the maximum stress, maximum strain, and tensor polynomial criteria are presented. These criteria are of interest because of their simplicity and widespread use.

The *maximum stress* failure criterion takes the form of a box (rectangular parallelepiped) in σ_1, σ_2, σ_6 space. Algebraically, it is expressed as

$$-X' < \sigma_1 < X$$

$$-Y' < \sigma_2 < Y \qquad\qquad (3.65)$$

$$-S' < \sigma_6 < S$$

Here X, X' are the tensile and compressive failure stresses in the fiber direction; Y, Y' are the tensile and compressive failure stresses perpendicular to the fiber direction; and S, S' are the positive and negative shear failure stresses in the (1, 2) plane. The shear failure stresses S, S' are equal; therefore, the evaluation of this failure model requires five independent experiments. This is unlike isotropic materials, which require only one experimental result. Also of note is the fact that the tensile and compressive failure stresses are different. This failure criterion is one of the most commonly used even though there are apparent shortcomings. The most important is the lack of interaction effects between the failure loads; for example, for a combined, tension stress state involving σ_1 and σ_2, the lamina is predicted to fail when $\sigma_1 = X$ (or $\sigma_2 = Y$), totally independent of the value of σ_2 (or σ_1). This is true even if $\sigma_2 = 99.9\% \, Y$ (or $\sigma_1 = 99.9\%_o X$). Such a situation is consistent with neither physical expectation nor experimental observation.

The *maximum strain* failure criterion is similar to the above, where strain replaces stress; that is,

$$-\varepsilon_{1_c} < \varepsilon_1 < \varepsilon_{1_t}$$

$$-\varepsilon_{2_c} < \varepsilon_2 < \varepsilon_{2_t} \qquad (3.66)$$

$$-\gamma_{12_c} < \gamma_{12} < \gamma_{12_t}$$

where $\varepsilon_{1_c}, \varepsilon_{1_t}, \varepsilon_{2_c}, \varepsilon_{2_t}, \gamma_{12_c}, \gamma_{12_t}$ are the respective compressive and tensile failure strains in the absence of all other strains.

This criterion suffers from the same lack of interaction in strain space that the maximum stress criterion suffered in stress space. In addition, it should be noted that the maximum stress and strain criteria are quite different; the one cannot be obtained from the other simply by transforming from stress to strain, or vice versa. Also, the experimental evaluation of the failure strains $\varepsilon_{1_c}, \varepsilon_{1_t}, \varepsilon_{2_c}, \varepsilon_{2_t}$ is more difficult than the determination of uniaxial failure stresses.

There are a number of failure criteria that include interaction effects; perhaps the most flexible of these is the *tensor polynomial* failure criterion. In the most general case this criterion takes the form

$$F_i \sigma_i + F_{ij} \sigma_i \sigma_j + F_{ijk} \sigma_i \sigma_j \sigma_k + \cdots < 1 \qquad i, j, k, \ldots = 1, 6 \qquad (3.67)$$

where F_i, F_{ij}, F_{ijk} are *principal material strength tensors*, which are determined from experiment. The first thing to notice about this criterion is that an unlimited number of strength tensors may be included, and therefore it should be possible to obtain a very accurate representation for the failure surface. The only problem is that every term requires experimental evaluation. Furthermore, most of these terms require experiments involving biaxial or triaxial stress states which are inherently more difficult and expensive to conduct. The important point to emphasize is that unlike the maximum stress and strain criteria, there is now interaction between the stresses. The interaction comes from all terms of the form $F_{ij} \sigma_i \sigma_j$, $F_{ijk} \sigma_i \sigma_j \sigma_k, \ldots$ when the $i, j, k, \ldots$ are different. How many terms are enough? This is difficult to answer because there is a double-edged sword involved. More terms will give better results; but, more terms requires greater experimental effort. Based on these considerations the most commonly used model is the *quadratic tensor polynomial*

$$F_i\sigma_i + F_{ij}\sigma_i\sigma_j < 1 \qquad i, j = 1, 6 \tag{3.68}$$

which for a plane-stress situation reduces to

$$F_1\sigma_1 + F_2\sigma_2 + F_{11}\sigma_1^2 + F_{22}\sigma_2^2 + F_{66}\sigma_6^2 + F_{12}\sigma_1\sigma_2 < 1 \tag{3.69}$$

This special case is also referred to as the *Tsai–Wu failure criterion*. In the above, F_6, F_{16}, F_{26} are absent because they are zero. This results because positive and negative shears give equal failure stresses.

The evaluation of the principal strength tensors is most easily accomplished with a specific set of experiments. To determine F_1 and F_{11}, a $0°$ coupon is used with the load applied in the fiber direction; thus $\sigma_2 = \sigma_6 = 0$. When a tensile load is applied, the failure stress state $\sigma_1 = X$, $\sigma_2 = \sigma_6 = 0$ is substituted into Eq. (3.69) to yield

$$F_{11}X^2 + F_1X = 1 \tag{3.70}$$

Doing the same for a compressive load gives

$$F_{11}(X')^2 - F_1X' = 1 \tag{3.71}$$

Solving for F_1 and F_{11} provides the final result:

$$F_1 = \frac{1}{X} - \frac{1}{X'} \qquad F_{11} = \frac{1}{XX'} \tag{3.72}$$

Following a similar procedure using a $90°$ coupon subjected to a load perpendicular to the fibers yields

$$F_2 = \frac{1}{Y} - \frac{1}{Y'} \qquad F_{22} = \frac{1}{YY'} \tag{3.73}$$

To determine F_{66} it is necessary to conduct a pure shear experiment $\sigma_1 = \sigma_2 = 0$. This is most easily accomplished by torsionally loading a 90° tubular specimen. It is clear on physical grounds that positive and negative torques give identical results; this is why $F_6 = 0$. Substituting the failure stress into Eq. (3.69) therefore gives

$$F_{66} = \frac{1}{S^2} \tag{3.74}$$

The experiments so far have been straightforward and are exactly the experiments required to calibrate the maximum stress failure criterion. The final parameter is F_{12}, which is determined from the application of a biaxial load state. Conceptually, this is most easily accomplished by assuming a prescribed load ratio between σ_1 and σ_2; that is, $\sigma_1/\sigma_2 = \lambda_{12}$. Measuring the failure stresses $\sigma_2 = Y_b$, $\sigma_6 = 0$ in this load state provides the data to determinate F_{12} from Eq. (3.69); that is,

$$F_{12} = \frac{1 - [F_1 \lambda_{12} Y_b + F_{11}(\lambda_{12} Y_b)^2 + {}^F_{22}(Y_b)^2]}{2\lambda_{12}(Y_b)^2} \tag{3.75}$$

Since the experimental application of a biaxial load state is difficult, other options have been proposed that avoid experimentation all together. The simplest of these is

$$F_{12} = 0 \tag{3.76}$$

Another option is to recognize that the failure criterion is similar in form to the von Mises yield criterion; if that model is extrapolated to orthotropic materials,

$$F_{12} = -\tfrac{1}{2}\sqrt{F_{11}F_{22}} \tag{3.77}$$

The latter two conditions have no direct physical basis whatsoever, but they do guarantee that the failure surfaces are closed, which is necessary physically. The expression in Eq. (3.75) appears to be the best alternative, but there are problems here as well. The fundamental question is: What load ratio λ_{12} should be chosen? There has been considerable attention paid to this problem in the literature and guidelines have been proposed. Some authors recommend that the data be obtained in the compression–compression quadrant (σ_1 and

σ_2 in compression). However, the fundamental difficulty is more of a philosophical problem. How can it be that a parameter evaluated in one load quadrant can influence failure in another quadrant? Does a failure surface evaluated when both σ_1 and σ_2 are negative mean anything when they are both positive? The answer must be "no" because totally different failure mechanisms are at work in the compression–compression quadrant as in the tension–tension quadrant. If this point of view is adopted, the idea of a smooth failure surface as given by the tensor polynomial suddenly loses some of its attractiveness. Notwithstanding this objection, it should be stated that in many cases the quadratic tensor polynomial does an excellent job of predicting lamina failure.

3.5.2. Observations

In the analysis and design of composite material components, perhaps the weakest link is the prediction of failure. This is not surprising because the same difficulty exists for metallic structures. What needs to be done? First, existing stress analysis capabilities are inadequate. The plate theory presented herein and various extensions all serve their purpose but are not totally satisfactory for the prediction of failure. To predict delamination failures three-dimensional effects must be included in the analysis, and even the best plate theories do only an adequate job. Three-dimensional finite element analyses might be better but face the difficult trade-off between resolution of the stres field and computational effort. A great deal of work has been done on improving stress prediction in composite plates; a great deal more work remains.

If it is assumed for the moment that the stress field can be determined as accurately as possible, is the problem solved? The answer is again, no. Almost all failure criteria currently in use are lamina failure criteria and not laminate failure criteria. None of the criteria described can predict delamination failure or the effect of stacking sequence. The development of laminate failure criteria is continuing and in some respects involves a more fundamental question than choice of criteria. That is, what is the best way of modeling composite failure? Should failure be based on a failure criterion or a fracture mechanics approach? The answer to this question is probably "it depends."

3.6. CONCLUDING REMARKS

In this chapter we have provided an introduction to composite materials; however, the role of fiber optic sensing systems for composite systems has not been addressed. The comments that follow should be interpreted from the point of view of what a composites researcher would like to find in a sensing system. To put the comments in perspective, it will help to step back a little. Composite materials are attractive for design for a number of reasons. For high-performance applications it is the specific stiffness and specific strength (stiffness and strength per unit mass) that make composite systems particularly advantageous. In addition, composite materials are prime candidates for structural optimization. Optimal designs in which lamina type and fiber orientation are selected usually involve little or no additional material or construction expense compared to nonoptimal designs. Therefore, the optimization process has considerable practical implications.

What are some of the problems? As noted above, at the present time full utilization of composites is inhibited by an inability to predict laminate failure with confidence. This situation leads to overdesign and subsequent reduction of the advantages. The solution to this dilemma would be to improve stress analysis and failure prediction capabilities, as noted above. However, there is another problem; there are no sensors that may be used to verify that stress and failure predictions are in fact accurate or which can be used to detect incipient failure. What are the requirements for these sensors? They must have the capability of measuring all strain components ε_x, ε_y, ε_z, γ_{yz}, γ_{xz}, γ_{xy}. In addition, it is necessary to make the measurements at ply interfaces and at free edges. Furthermore, in the regions of interest there are very severe gradients in stresses and strains, and therefore the sensors must be small to avoid averaging effects; the stresses at free edges have a singular-like character which exists over a domain with dimensions on the order of a few ply thicknesses.

The challenge is therefore to develop a very special sensor. It must be capable of measuring strain in all directions, it must survive being embedded within a composite laminate, and it must be very small. Fiber optic sensors may very well be the answer.

ACKNOWLEDGMENTS

The author has received invaluable assistance from discussions with his colleagues and students. Mario Kataoka-Filho and Michael Montemurro were kind enough to review the manuscript and provide valuable comments. Graham Elliott deserves special thanks for his interest in this project, his constructive criticism, and for preparation of the figures.

BIBLIOGRAPHY

R. M. Christensen, *Mechanics of Composite Materials.* Wiley, New York, 1979.

R. M. Jones, *Mechanics of Composite Materials.* Scripta, Washington, DC, 1975.

4

Optical Fiber/Composite Interaction Mechanics

JAMES S. SIRKIS AND ABHIJIT DASGUPTA
University of Maryland
College Park, Maryland

4.1. INTRODUCTION

One of the reasons that fiber optic smart composite structures have become an active technical field is because the small diameter of optical fibers has led to a perceived similarity between optical fibers and the fibers commonly used to reinforce many advanced composite materials. One must recognize that optical fibers are foreign entities to the host structure and therefore will *always* alter the stress and strain state in the vicinity of the embedded sensor. This inescapable truth of continuum mechanics is a result of the material and geometric discontinuity introduced by the embedded optical fiber sensor, and occurs irrespective of the small size of the fiber. The questions that the fiber optic smart structures community must then consider are: (1) Does the embedded optical fiber sensor, by its very presence, somehow harm the performance of the structure whose health is to be monitored? and (2) What is the embedded optical fiber sensor really measuring? In this chapter we address the issue of "harming" the host structure by examining the local stress and strain states induced by the embedded optical fiber that would otherwise not be present. In fact, it is not altogether clear whether the altered stresses states created by embedded optical fibers do indeed "harm" the host structural system (at least on the local scale). In Chapter 5, structural degradation on a global scale is discussed. The issue of what embedded sensors actually measure is particularly significant in light of the fact that the embedded fiber inevitably perturbs the strain state it is intended to measure. Optical behavior of the sensor, mechanical response of the host–sensor systems, and the state of the interfaces and interphases between the optical fiber and host material all contribute to the complexity of this question.

The interaction continuum mechanics between the constituents that make

Fiber Optic Smart Structures, Edited by Eric Udd.
ISBN 0-471-55448-0 ⓒ 1995 John Wiley & Sons, Inc.

up a fiber optic smart structure hold the answers to many of the mechanical reliability and sensoral interpretation concerns confronting the technology. The tools available to explore the interaction mechanics include numerical, analytical, and experimental stress analysis techniques that are not unique to fiber optic smart structures. In fact, many of the results presented in this chapter are found using techniques that have long existed and are simply being applied to the problems presented by fiber optic smart structure technology. Nowhere is this more evident than in the analytical stress analysis techniques discussed in Section 4.2.1, where solutions developed for reinforcing fibers in fiber-reinforced laminated composites are adapted continually to structurally embedded optial fiber sensors.

The majority of this chapter is devoted to reviewing the state of the knowledge concerning local interaction between embedded optical fiber sensors and their host materials. In Section 4.2 we review analytical, numerical, and experimental stress and strain analysis techniques that have been used in an effort to quantify stress and strain concentrations caused by embedded optical fiber sensors. In Section 4.3 we review optical fiber microcracking in composite structures with embedded optical fiber sensors, including interfacial debonding mechanisms. In Section 4.4 we present a formalism by which embedded optical fiber sensor signals can be interpreted and by which the stress and strain concentrations and microcracking mechanisms can at least be taken into account. The reader will notice that many of the results presented in this chapter are illustrative in nature and that they do not lead to definitive conclusions, nor do they lead to fiber optic smart structure design guidelines. This fact simply attests to the relative infancy of the mechanisms-related aspects of the fiber optic smart structures technology. As a result, in the final section we outline many mechanics-related issues that should be addressed to take full advantage of the fiber optic smart structures technology. Finally, since surface-mounted optical fiber sensors do not perturb the host strain field as much as embedded sensors do, the discussion in this chapter is restricted to the embedded case only.

4.2. STRESS AND STRAIN CONCENTRATIONS

An abrupt change in any system is met with an associated abrupt systematic response. Eisenberg (22) puts it best when he points out that traffic snarls where highways converge from four to two lanes, turbulence erupts where large rivers are forced into narrow channels, and severe disturbances in stress and strain state are developed in the vicinity of geometric or material discontinuities. Consider the flat plate in Fig. 4.1, which has a circular hole located at its center and is loaded with an in-plane tensile force F. The average stress (force divided by area) applied to the plate is $\sigma = F/wt$, while the average stress in cross section passing through the center of the hole (A–A' in Fig. 4.1) is $\sigma_0 = F/(w - d)t$. The problem is that the stress is not distributed uniformly

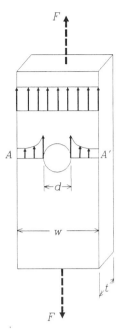

Figure 4.1. Plate with a hole loaded under uniaxial tension.

along this midplane cross section. In fact, the stress just adjacent to the hole is three times that of the average stress σ_0, so the hole is said to have a stress concentration factor of 3. The factor (σ/σ_0) of 3 is derived from the theory of elasticity and is independent of the diameter of the hole as long as w is much greater than d (14). This "plate with a hole" scenario is the most often cited example of stress concentrations (and rightfully so), since it illustrates that simple alterations in the geometric configurations can lead to significant ramifications in terms of the stress and strain distributions. This stress concentration behavior is common to holes, cracks, grooves, reentrent corners, and so on, and commonly is the driving consideration in mechanical design.

Consider Fig. 4.1 once more, but in this case imagine that the hole is filled with some material other than the plate material. The stress and strain distributions along A–A′ will still be nonuniform, although different from the plate with a hole. The filled material is now referred to as an *elastic inclusion* because it is a geometric entity included in (or surrounded by) the plate material and because its mechanical behavior is assumed to follow the theory of elasticity.

As in the case of the hole, this circular elastic inclusion induces stress and strain concentrations that otherwise would not be present. This particular example is very important to fiber optic smart structures because the optical fibers are themselves cylindrical elastic inclusions and therefore behave in a manner very similar to that shown in Fig. 4.1. In fact, this section is devoted to understanding, via modeling and experiments, the stress and strain concentrations caused by embedded optical fibers. The material surrounding the optical fiber will be referred to as the "host" throughout the discussion, and the host will either be monolithic materials (for ease of modeling) or laminated fiber-reinforced composites. Laminated composites are common host materials in fiber optic smart structures, owing to the previously noted geometric similarities between optical fibers and reinforcing fibers, and owing to the relative ease of embedding optical fibers in these material systems.

4.2.1. Analytical Investigations of Stress and Strain Concentrations

The strain concentrations caused by interactions between embedded optical fibers and the surrounding host can be analyzed by closed-form techniques for simple geometries such as circular fibers embedded in isotropic host materials or in transversely isotropic host materials (where the fiber axis is a material principal direction). Quantitative knowledge of this strain concentration is important because it determines both calibration and obtrusivity of fiber optic sensors. The term *calibration* is intended to represent the quantitative relationship between the magnitude of the state variable (strain, in this case) in the host and the magnitude in the sensor. *Obtrusivity* is a related concept and indicates the extent of perturbation in the average value of the state variable in the host, due to the presence of the sensor. Excessive obtrusivity not only affects the calibration of the sensor but poses potential reliability hazards due to large local strain concentrations. Closed-form analysis solutions for the strain fields can also be used to design coatings capable of tailoring the local stresses and strain states to the advantage of the structural designer. The purpose of the coating may be to maximize the load transfer capability and/or to minimize the stress and strain concentrations (obtrusivity) due to the

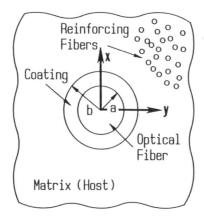

Figure 4.2. Schematic representation of a coated optical fiber embedded in an infinitely large transversely isotropic composite host. [From Dasgupta and Sirkis (17).] Copyright © 1991 AIAA - Reprinted with permission.

embedded fiber optic sensor. The optical fiber (including core and cladding) and any coatings are typically modeled as concentric cylinders. This assembly is then embedded in a host of infinite extent, as shown in Fig. 4.2. This geometry is applicable when the optical fiber is embedded parallel to the reinforcing fibers in the surrounding ply of the host composite. In the following discussion we first present some of the general solution techniques prevelent in the mechanics literature for relevant geometries, and eventually, their application to optical fiber analysis is demonstrated. The fundamental solution technique consists of assuming series solutions for displacement fields which satisfy the Navier equations in cylindrical polar coordinates, for different load conditions. The series solutions are assumed in terms of unknown constants which are evaluated by satisfying interfacial continuity conditions and the far-field boundary conditions which depend on the loading conditions as r (the radial polar coordinate) approaches infinity. This solution technique has been well developed by traditional analysts of mechanics of composite materials for perfectly bonded interfaces and uncoated fibers (11). More recently, the scheme has been extended to the generalized case of reinforced composites with multiple phases and interphases (2, 15, 46, 71). Interphases of very small thickness, with different properties, are used as a means for approximate modeling of interfaces between different phases. An alternative method for modeling interfaces is by using jump functions (25, 28, 39). In the case of fibers of finite length, eigenfunction methods can be utilized effectively to obtain the approximate stress and strain fields at the fiber ends. Eigenfunction methods appropriate in this situation are either of the shear-lag type to predict the strain concentration at the fiber end (57), or of an axisymmetric asymptotic type to capture the singular fields at the fiber end (34).

The displacement methods discussed above work reasonably well, either when the distance between neighboring fibers is large enough to ignore the interactions between neighboring fibers, or when used in conjunction with a self-consistent scheme (11) for random fiber distributions. In the event of periodic fiber distributions, the interactions between neighboring fibers can be modeled by assuming cubic packing or hexagonal packing, and by imposing periodicity boundary conditions on a representative unit volume surrounding the fiber (4, 30, 37). In the event of nonperiodic fiber distributions (e.g., when two isolated fibers are in close proximity), the analysis is much more complicated and can be accomplished either by stress function techniques (68) or by methods of pseudo-tractions (29). Alternatively, numerical methods can be used, as discussed in Section 4.2.2.

It is worth mentioning that the stress analysis can also be achieved using eigenstrain techniques, as in Eshelby's equivalent inclusion concept (24, 49) by treating the optical fiber as a cylindrical elastic inclusion in the case of inifinitely long fibers, or as a prolate spheroid in the case of fibers of finite length. The interactions between neighboring fibers can also be modeled readily by this technique (48). However, the coatings and interphases are difficult to model by the eigenstrain method (50, 72).

The axisymmetric displacement-based analysis scheme mentioned above is ideal for analyzing optical fibers because of the axisymmetric multiphase nature of the embedded optical fiber. The phase in this case consist of the core, cladding, and coating. Following the techniques used by composites analysts, the interphases between these phases are modeled as additional phases of small thickness (6, 7, 16, 41, 42, 53). The only limitation is that the properties of the interphase must be uniform in the circumferential direction. The complete solution for the displacement field then leads to exact strain and stress solutions in each phase and at the interphases under linear elastic, small-displacement assumptions (40). Lu (41) provides examples of the radial distributions of hoop, axial, and radial stresses for a sample optical fiber, coated with polyimide and embedded in a graphite–epoxy unidirectional composite host, for radial applied stress in the far field. Similar examples of such stress distributions are available in the literature (6, 7, 16, 53). All other load states produce stresses that vary along the circumferential direction, and examples may be found in the literature (6, 7, 53).

All of the above mentioned (and subsequent) stress analysis techniques yield the same important finding: the state of strain in the optical fiber core is quite different from the far-field value in the host. This relationship between the host strain and the sensor strain must be determined to make accurate measurements with the fiber optic sensor (63). One approach to quantify this relationship is to define a strain transfer function as the ratio of the fiber strain to the far-field strain. Two examples of analytical determination of such strain transfer functions are shown in Figure 4.3a and b, for longitudinal shear load and uniaxial tensile load (perpendicular to the optical fiber), respectively, for several coating moduli (53). Figure 4.3a illustrates that under longitudinal shear, when the shear modulus of the coating is less than that of the surrounding host matrix, greater shear strain is induced in the optical fiber by a thinner coating. Conversely, for a coating that is stiffer than the surrounding host matrix, the strain transfer increases with the coating thickness. The shear strain transferred to the fiber was also shown to be maximized by choosing coating shear modulus to be the Euclidean norm of the fiber and host shear modulii (52). On the other hand, Fig. 4.3b illustrates that the maximum strain transfer for uniaxial tension (perpendicular to the sensor axis) occurs when there is no coating (53).

It is important to recognize that stress concentrations in the host and at the interfaes are sometimes tensile in nature and can therefore nucleate overstress damage under large excursions of load or nucleate fatigue damage under cyclic loading. This type of stress concentration has also been investigated numerically, as discussed in Section 4.2.2, and its influence on the strength of the host structure has been evaluated experimentally, as discussed in Section 4.2.3. Residual stresses caused by thermal expansion mismatches (during fabrication and fiber embedding) affect the tensile or compressive nature of the final stress state after loading and are extremely important for damage analysis. However, residual stresses are difficult to evaluate exactly, due to the transition of the

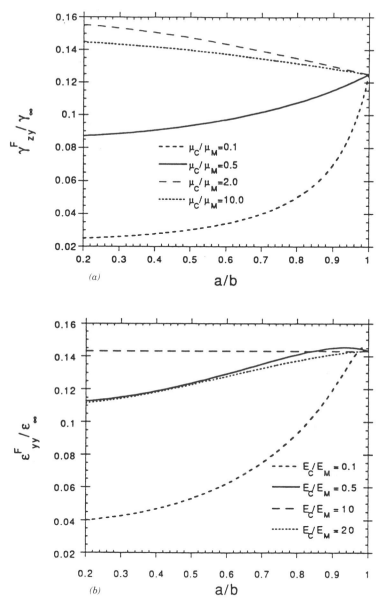

Figure 4.3. Strain transfer function for (a) longitudinal shear load and (b) transverse uniaxial tensile load. [From Pak et al. (54).]

material from viscous to viscoplastic to viscoelastic, and finally, to a fully elastic state. Approximate elastic analysis can be performed, as an alternative, by observing the thermoelastic residual stresses during cooldown of the assembly from the glass transition temperature of the polymeric host to room temperature. Such an approximation assumes that the polymeric host is not a load-bearing material above glass transition temperature and is a fully load-bearing elastic material at temperatures below glass transition. Any of the analytical methods discussed above can address residual stresses under these assumptions.

There is very little information in the literature at present regarding interactions between adjacent fiber sensors when there are multiple fibers in close proximity. Such analysis can be important for optical fiber networks as in fiber optic strain rosettes, and can best be performed using eigenstrain techniques (49). The eigenstrain method is also very well suited for analyzing the response of a sensor fiber whose cross section is experiencing highly nonuniform strains. Such a situation may be typical when strain sensors are placed in regions of large stress gradients: for example, near a cutout, reentrant corner, or other sources of stress concentration.

As discussed above, the displacement method is most appropriate for infinitely long fibers. In Fabry–Perot sensors, the strain state at a fiber end becomes important. The shear-lag eigenfunction analysis has been used by some investigators to analyze the strain concentration at the fiber end of intrinsic Fabry–Perot sensors (38). The asymptotic singular eigenfunction method is tedious to implement and is not popular among the fiber optic mechanics community.

In the vicinity of fiber optic sensors, the stress and strain concentrations discussed above can be tailored by the sensor designer by judicious use of coatings on the optical fiber sensor. The coating in the analysis provides an additional degree of freedom in the design process, and the coating parameters can be tailored to minimize detrimental stress concentrations reported in the sensor assembly and in the surrounding host (6, 7, 16, 35). Evidence is available that supports the existence of mechanically "optimal" coatings for the embedded fiber optic sensor. Optimization criteria for coating design are chosen both from damage mechanics rationale and from the perspective of improving the accuracy of the strain sensor (16). Coatings that produce a zero or fully compressive stress state in or near the embedded optical fiber are desirable. Further, since the optical fiber is typically calibrated under a state of zero transverse stresses, sensing accuracy is highest if the in situ transverse normal stresses in the fiber, under external loads, can be eliminated by proper coating design. It is not always possible to satisfy all of the optimization criteria simultaneously, and design trade-offs may be necessary.

The coated optical fiber that is embedded in a laminated composite structure parallel to the reinforcing fibers of the host composite is shown in Fig. 4.2. The optical fiber and coating are assumed isotropic, while the host is modeled as transversely isotropic. Stresses in this assembly are evaluated from closed-form elasticity solutions under axial loading parallel to the optical fiber. Without compromising the generality of the analytical technique, it is assumed as a first-order approximation that the residual stress state in the neighborhood of the sensor assembly due to fabrication processes is negligible. It is further assumed that perfect bonding exists at the interfaces between the optical fiber, coating, and the host and that no chemical interactions occur. The limitations of such assumptions are recognized, but the analysis is nevertheless worthwhile because the aim here is to illustrate that the coating design does have a significant influence on the stress state in the sensor assembly, and that its investigation is beneficial.

As an example, consider the stress state due to uniaxial external loading parallel to the optical fiber for varying combinations of coating Young's modulus, Poisson's ratio, and coating thickness. The existence of optimal combinations of coating parameters is observed when different transverse normal stress components (hoop or radial) in different phases (fiber, coating, or host) of the sensor assembly are targeted for minimization (or, in the absence of residual stresses, eliminated). For brevity, only results for minimizing transverse stresses in the host are presented in this section. For convenience, the optimal coating results are presented in a coating "design space" which is defined in terms of the coating stiffness and Poisson's ratio. The optimal thickness (b/a ratio) of the coating which will eliminate the host transverse stresses can be expressed in terms of the coating, host, and fiber properties as

$$\frac{(2G_c + C_f)(v_c - v_{31}^h)}{2(1 - v_c)} = \frac{C_f(v_f - v_{31}^h)}{1 - (b/a)^2} \tag{4.1}$$

where subscripts and superscripts f, c, and h signify the optical fiber, coating, and host, respectively. C is a function of Young's modulus and Poisson's ratio, v, is the Poisson's ratio of the fiber or coating and v_{32} is the longitudinal Poisson's ratio of the transversely isotropic composite host. A significant feature of this result is that the optimum combination of coating properties and geometry is independent of the host stiffness. As an example, Fig. 4.4 shows contours of optimal b/a ratios in the coating design space for a host with a longitudinal Poisson's ratio of 0.3. Similar plots for other optimization criteria are presented in the literature (16).

The optimal coating concept has also been explored by other investigators (6, 7), both for optical fiber sensors and for the reinforcing fibers of composite materials, by considering (1) hexagonal and diamond packing models, (2) loading transverse to the axis of the fiber, and (3) different optimization criteria formulated from failure models based on maximum shear stress, maximum principal stress, and maximum distortional energy. These results were based on graphite–epoxy hosts and concentrated on reducing the obtrusivity of either the reinforcing fibers of composite systems or of optical fiber sensors in smart structures. Figure 4.5 shows the variation of the maximum principal stress in the matrix material as a function of angular position around the embedded fiber–host interface for three different ratios of coating to matrix stiffness when the host is subjected to transverse tensile loading. The dashed line represents an optimum because the coating stiffness leads to an overall minimum in maximum principal stress.

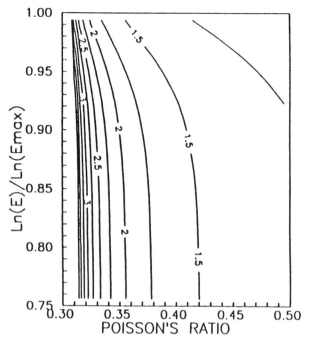

Figure 4.4. Optimal coating thickness (b/a ratio) in coating design space for graphite–epoxy host [From Dasgupta and Sirkis (17).]

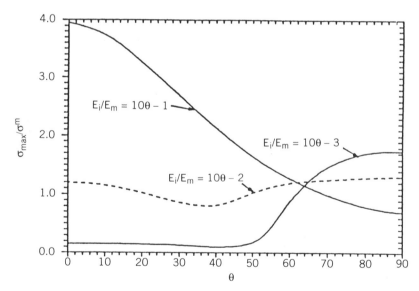

Figure 4.5. Circumferential distribution of stress concentration factors for different coating properties. [From Carman and Averall (5).]

The conclusions from these studies is that it is not possible to satisfy all desirable design criteria simultaneously, due to the mutually competing nature of the in-plane stress concentrations. However, experimental evidence suggests that it is the transverse tensile strength of the host composite system that is most sensitive to the presence of embedded fiber sensors (18). This sensitivity suggests that if all the potential optimization criteria cannot be satisfied, coatings that minimize transverse stress states may be preferable.

4.2.2. Numerical Investigations of Stress and Strain Concentrations

Many of the concerns about obtrusivity have centered on the interlaminar lenticular resin-rich pocket which forms when optical fibers are embedded in laminated composite hosts along any orientation other than parallel to the indigenous reinforcing fibers [see, e.g., Davidson *et al.* (19, 20) or Dasgupta *et al.* (18)]. Figures 4.6 and 4.7 shows an example of this resin-rich pocket for two composite systems. Notice that the resin pocket becomes smaller as the angle between the optical fiber and the reinforcing fibers directly adjacent to the optical fiber becomes smaller. The resin pocket also becomes smaller as the thickness of the laminate is reduced. This large resin pocket acts as an interlaminar discontinuity and poses a potential reliability hazard to both sensor and host under applied static, cyclic, or shock loads. The complexity of this lenticular geometry and the disparity in thermomechanical properties of

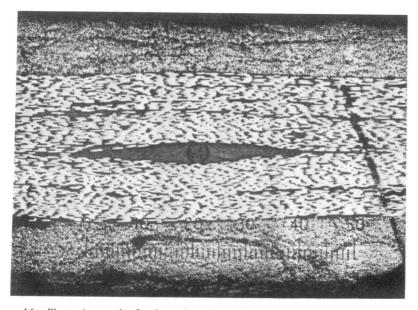

Figure 4.6. Photomicrograph of resin pocket observed around optical fiber in three laminate configurations $[90_4/O_4/OF(90)/SYM]$. [From Dasgupta et al. (17).]

the various constituents do not lend themselves to simple closed-form analysis. As a result, numerous studies using numerical techniques have been used to understand the induced stress and strain states. Several investigators have also attempted numerical analyses of the simple axisymmetric configurations described in Section 4.2.1, which occurs when the optical fiber is parallel to the reinforcing fibers of the composite in the surrounding ply. The purpose of these studies was primarily to verify the assumptions in the closed-form results. The numerical studies of the axisymmetric case include homogenization models (23), boundary element methods (53), and finite element methods (18). The findings of these studies agree very closely with the closed-form results discussed in Section 4.2.1.

The finite element method has been a popular method to investigate the stress concentration and the relationship between the fiber strain and the far-field host strain for complex geometries, as shown in Fig. 4.7. Investigations of the effect of the resin pocket geometry on the calculated strain concentrations under tensile loading of certain laminate sequences reveals that the strain concentration can be altered by nearly a factor of 2, depending on the choice of resin pocket geometry (59, 69). The geometry of the resin pocket in these studies was extracted from observations of enlarged micrographs. This approach required laborious cutting, polishing, and photographing of the resin pocket geometry. Salehi et al. (59) found the strain concentrations to range from 2.23 to 16.5, depending on the stacking sequence. Although these strain

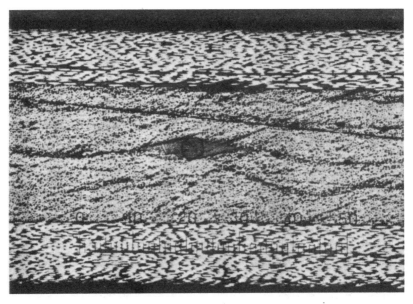

Figure 4.7. Photomicrograph of resin pocket observed around optical fiber in three laminate configurations $[0_4/+45_4/OF(90)/SYM]$.

concentrations did not agree with their own moiré interferometry experiments, they did agree with moiré interferometry experiments of other investigators (13, 62) (see Section 2.3). When residual stresses were included in the analysis, finite element analysis of resin pockets formed in APC-2 unidirectional and cross-ply composite laminates revealed overall stress concentrations ranging from 1.35 to 2.1 (18). This study is significant in that the effect of increasing ply disturbance was investigated by considering both 80–93-μm and 125–160-μm coated optical fibers in similar laminates. They found a 36% increase in stress concentration resulting from the larger optical fiber. Finite element studies of transverse compressive loading in graphite–epoxy cross-ply laminates (77) produced results that are qualitatively similar to experimental observations (62) with regard to the size effects of structurally embedded optical fiber sensors. Wan's (77) study is described in more detail below.

The tedious efforts required to obtain the resin pocket geometry in all the previous models clearly indicate that to make finite element analysis a feasible tool for understanding local interaction mechanics of embedded optical fiber sensors, one must first be able to predict the resin pocket geometry based on macroscale observables. Using energy methods in conjunction with polynomial trial functions from beam bending theory (17) or with trigonometric trial functions (8), it is possible to predict the precise geometry of this resin-rich region based on (1) the mechanical properties of the composite host and the optical fiber, (2) the volume fraction of reinforcing fibers in the composite, (3) the laminate stacking sequence, and (4) the radii of the optical fiber and of the indigenous reinforcing fibers. Close agreement has been obtained between these models and observed resin pocket geometries in graphite–epoxy specimens, as shown in Fig. 4.8. This figure shows the agreement between predicted and measured resin pocket aspect ratios for six different laminate stacking sequences. The uncertainty in the experimental values stems from the uncertainty in clearly identifying the tip of the resin pocket.

The crimp caused by the optical fiber in the reinforcing fibers tends to act as an initial imperfection or waviness and reduces the buckling strength of the host for compressive loading transverse to the optical fiber axis. The predictive model for the resin pocket geometry can also be used to provide predictions of the reduction of the transverse buckling strength due to the fiber crimp. Approximate closed-form predictions (8), and nonlinear large-deformation finite element predictions of buckling and postbuckling behavior (77) are all possible when the resin pocket is predictable. Figure 4.9 shows predictions of the buckling strength of a unidirectional laminate, as a function of the orientation of the optical fiber, relative to the loading direction (8). The reinforcing fibers of the unidirectional composite are oriented parallel to the loading direction. The buckling strength is seen to drop by more than a factor of 2 as the optical fiber diameter is increased from 100 μm to 300 μm. Results of nonlinear finite element postbuckling analysis (77) are shown in Figure 4.10, where deformation (v) in the thickness direction is plotted against the compressive loading (P) parallel to the major axis of the resin pocket. For convenience,

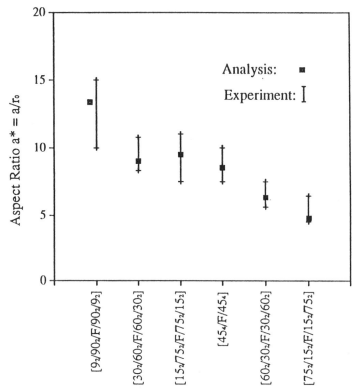

Figure 4.8. Comparison of predicted and measured aspect ratios of resin pocket (ratio of half-span a to radius r_f of optical fiber) for various laminate configurations. (From Dasgupta et al. (17).]

these plots show deformations of the top surface of the laminate as well as the interface between the host and the optical fiber. The increase in the fiber crimp due to increase in the optical fiber radius from 40 μm to 97.5 μm is seen to cause lamost twice as much postbuckling deformation in the thickness direction for the same magnitude of applied load (approximately 100 mN).

Using the predicted resin pocket geometry in an automatic meshing algorithm, Wan (77) conducted extensive finite element analyses to illustrate in symmetric graphite–epoxy laminates of the form [0/90̄/OF(O)/90/0] that significant stress concentrations can be expected at the interface between the optical fiber and the composite host under extensional loads parallel to the major axis of the resin pocket and at the tip of the resin pocket under transverse shear loads. The importance of these two types of loads stems from their dominance under flexural deformations in laminated beams and plates. As an example, Figs. 4.11 and 4.12 show the undeformed and deformed finite element meshes for the two load types. Figure 4.13 shows the stress distribution in the assembly for compressive loads parallel to the major axis of the resin

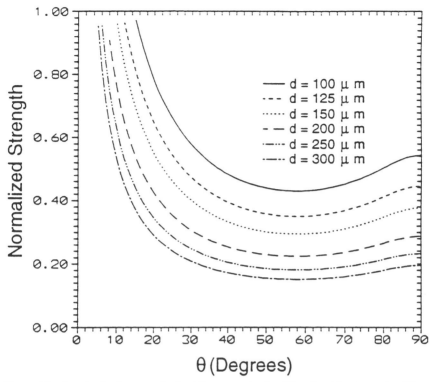

Figure 4.9. Longitudinal compresive buckling strength of unidirectional laminates for different play orientations and different diameters of optical fibers. [From Case and Carman (9).].

pocket. Clearly, there is a large stress concentration at the fiber–host interface along the vertical y-axis. Figure 4.14 shows a line plot of the von Mises stress distribution along the vertical midplane (parallel to the minor axis of the resin pocket) and provides further evidence of this stress concentration. The stress concentration factor is about 22, and the dependence of this stress concentration factor on the radius of the optical fiber is shown in Fig. 4.15. These results agree qualitatively with similar findings by other investigators (18, 59). Figure 4.16 illustrates similar stress distribution plots for the σ_{yy} extensional stresses under transverse shear loading. The corresponding transverse shear stress distribution is shown in Fig. 4.17. The stress concentrations for this loading mode are clearly at the tip of the resin-rich pocket along the x-axis. Figure 4.18 ilustrates the dependence of this stress concentration factor on the radius of the optical fiber. It is important to note that this stress concentration information must be normalized with respect to the strength of the individual phases (host and device), to draw conclusions regarding the influence of obtrusivity on structural integrity/reliability. The influence of the stress concentrations due to the obtrusivity of the optical fiber on the macroscale strength of the composite has been observed by several investigators and is addressed in Section 3.

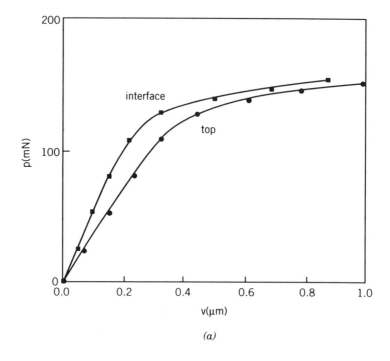

(a)

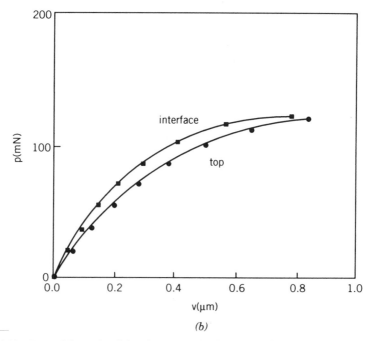

(b)

Figure 4.10. Large-deformation finite element result of transverse in-plane compressive load P versus transverse out-of-plane deformation v, indicating postbuckling behavior for two different diameters of optical fiber: (a) radius = 40 μm; (b) radius = 97.5 μm. [From (77).]

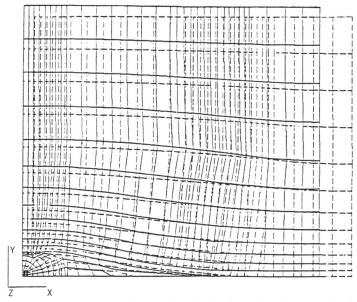

Figure 4.11. Finite-element meshes of symmetric half of resin pocket and surrounding host, used for analyzing stress distributions. Undeformed and deformed geometries have been superimposed for illustrative purposes. The loads modeled are uniaxial transverse in-plane compression.

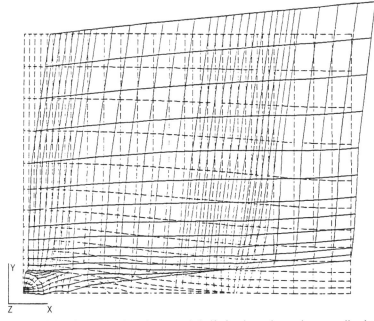

Figure 4.12. Finite-element meshes of symmetric half of resin pocket and surrounding host, used for analyzing stress distributions. Undeformed and deformed geomet'ries have been superimposed for illustrative purposes. The loads modeled are transverse sheer. [From (77).]

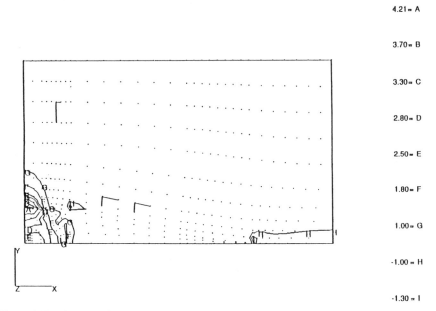

4.21 = A

3.70 = B

3.30 = C

2.80 = D

2.50 = E

1.80 = F

1.00 = G

-1.00 = H

-1.30 = I

Figure 4.13. Contour plot of transverse normal stress (σ_{yy}) distribution under uniaxial transverse in-plane compression, showing stress concentration at interface of host and optical fiber along the Y-axis [From (77).]

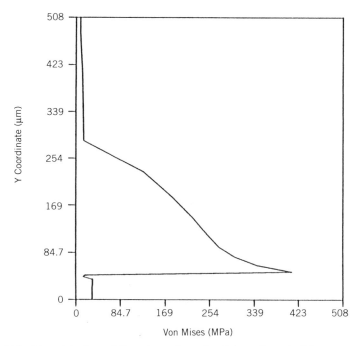

Figure 4.14. Line plot of von Mises stress along the Y-xis, under uniaxial transverse in-plane compression, showing stress concentration at interface between host and optical fiber. [From (77).]

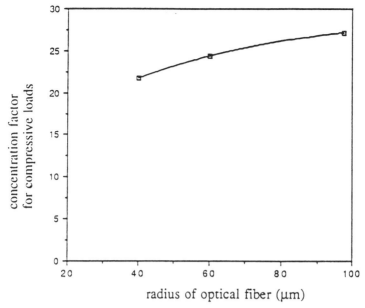

Figure 4.15. Dependence of stres concentration factor on radius of optical fiber for uniaxial transverse in-plane compression [From (77).]

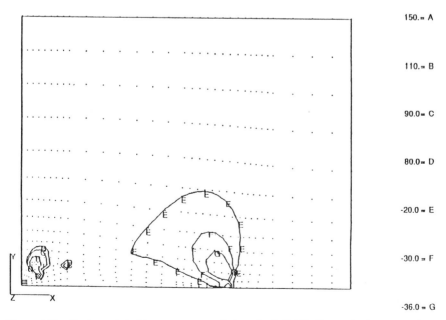

Figure 4.16. Contour plot of transverse normal stress (σ_{yy}) distribution under transverse shear load, showing stress concentration at tip of resin pocket, on the X-axis. [From (77).]

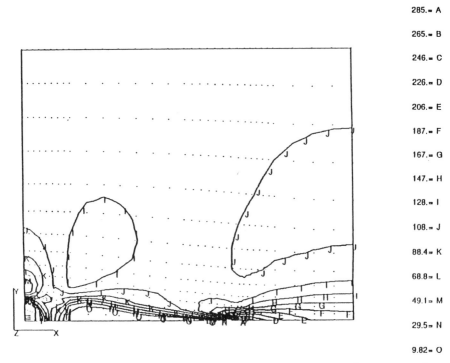

285.= A
265.= B
246.= C
226.= D
206.= E
187.= F
167.= G
147.= H
128.= I
108.= J
88.4= K
68.8= L
49.1= M
29.5= N
9.82= O

Figure 4.17. Contour plot of transverse shear stress (τ_{xy}) distribution under transverse shear load, showing stress concentration at tip of resin pocket, on the X-axis [From (77).]

Modeling of the damage behavior is very difficult, and preliminary attempts at failure prediction have typically used brittle fracture mechanics failure models for the optical fiber, ductile distortional-energy-based failure models for the coating and interphases, and a tensor polynomial Tsai–Wu criterion for the host composite (26, 40).

Preliminary results indicate that the static strengths, except for the longitudinal compressive buckling and transverse tensile strengths discussed above, may not be significantly degraded by the obtrusivity of the optical fiber sensor. However, fatigue behavior is much more difficult to predict because of the large number of possible failure modes and failure mechanisms. Accurate failure prediction is important and still remains an open research issue.

In closing, it is important to note that many issues in optical fiber mechanics are yet to be addressed and many of these issues will require numerical methods, due to the complexity of the geometry. These include interactions between neighboring fibers, response of fibers in nonuniform loading fields, and design of rosettes for multiparameter sensing.

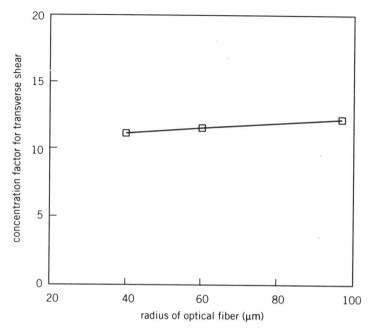

Figure 4.18. Dependence of stress concentration factor on radius of optical fiber for transverse shear loading. [From (77).]

4.2.3. Experimental Observations of Strain Concentrations

The experimental observations of strain concentrations in and around optical fiber sensors embedded in laminated composite materials are primarily responsible for the interest in interaction mechanisms issues associated with fiber optic smart structures. The most telling experimental results regarding induced strain concentrations has been produced via moiré interferometric studies. Although these studies are somewhat limited, they do provide a vivid picture of how optical fibers can act as cylindrical elastic inclusions. Virtually all moiré inferferometric results have confirmed the existence of strain concentrations resulting from embedded fiber sensors, but no consensus has yet been reached about how these strain concentrations alter the performance of the fiber–host system or if the systems performance is altered at all. This section serves as a pictorial review of some of the more revealing moiré interferometric results reported thus far. No attempt will be made to explain moiré interferometry or how to interpret the resulting displacement-encoded coherent optical fringe patterns. The interested reader can refer to Post (54) and Morimoto et al. (48) for a detailed discussion of this topic.

The first documented questions about potential mechanical implications of placing optical fibers within load-bearing structures were raised by Udd et al. (73) who were concerned about potential strength degradation of laminated composite materials with embedded optical fibers. These concerns seemed to motivate Czarnek et al. (13) to perform the first moiré interferometric study of the strain fields induced by optical fiber sensors embedded in graphite–epoxy-laminated host systems. This effort was significant because it clearly illustrated the existence of the lenticular resin-rich regions that formed around embedded optical fibers oriented in any direction other than parallel to the indigenous reinforcing fibers (as shown in Figure 4.7). Further, this work showed that local details such as the size and shape of the resin-rich zone and the stacking sequence of adjacent laminates contributed significantly to the induced strain concentrations. Figure 4.19 shows the vertical displacement fringe patterns found using moiré interferometry applied to $[0_2/90_2/OF(90)/90_2/0_2]$ and $[90_2/0_2/OF(90)/0_2/90_2]$ uniaxial tension specimens. This figure illustrates that the optical fiber causes a much larger perturbation in the local displacement (and strain) field when the fiber is not parallel to the reinforcing fibers. This figure also illustrates the utility of moiré interferometry for local scale displacement and strain investigations. Even without interpreting the optical fringe patterns, one can see that the optical fiber has a pronounced effect on the local displacement and strain field. Similar studies were conducted by Salehi et al. (59), who reproduced some of the original tests of Czarnek et al. (13) and added specimens of greater thickness. While their results also show perturbations in the local displacement fields, the calculated strain concentration levels were much more modest. One interesting fringe pattern recorded by Salehi et al. (59) is shown in Fig. 4.20, where a transverse matrix crack has propagated directly through the optical fiber–host composite interface in a $[90_4/O_4/OF(90)/O_4/90_4]$ graphite–epoxy laminate. This type of matrix crack is symptomatic of the low transverse strength of most laminated composites and is not generally indicative of an optical fiber/laminated composite system performance. There have been several other moiré interferometric studies of strain fields around optical fibers embedded in composite systems that have yielded interesting results. Moiré interferometry has been used, for example, to assess the size effects of embedded optical fibers (62), to determine the effects that optical fibers have on the interlaminar strain in thick composites (Fig. 4.21), and on sensor/sensor interaction of closely spaced fibers embedded in a thermally loaded composite laminate (Fig. 4.22).

Note that all of the investigations cited above have added to the overall understanding of the microinteraction puzzle. The optical fiber size-effect studies showed that the strain perturbations due to the embedded optical fiber became insignificant as the fiber diameter decreased to the order of 100 μm. These experiments were confirmed by the theoretical predictions of Case and Carman (8). The investigation into the thick composite systems indicates that

Figure 4.19. Moiré interferometric fringe patterns showing vertical displacement fields in (a) $.[O_2/90_2/OF(90)/90_2 0_2]$ and (b) $[90f_2/0_2/OF(90)/0_2/90_2]$ graphite–epoxy uniaxial tension specimens. [From Czarnek et al. (13).]

the interlaminar optical fiber has no measurable effect on the indigenous strain state. This conclusion is strengthened further by the thick composite studies of Guo et al. (27). On the other hand, Fig. 4.22 shows very clearly that the thermally induced strains are heavily concentrated between the two adjacent optical fibers.

In this section we have shown several visual examples of optical fibers acting as strain concentrators, but in actuality there has been little experimental work that provides values for strain concentrations near embedded optical fibers. So while the moiré interferometric studies have been quite striking, they only influence fiber optics smart structure design inasmuch as embedded optical fibers are generally embedded parallel to reinforcing fibers whenever possible. Since this optical fiber orientation will not always be possible or always desirable (44), studies of this type continue to add to the basic understanding of fiber optic smart structures.

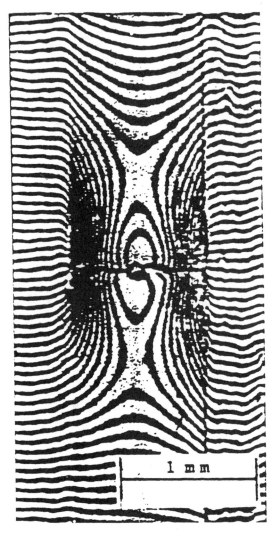

Figure 4.20. Moiré interferometric fringe pattern showing the vertical displacement field in a $[0_4/90_4/OF(90)/90_4/0_4]$ graphite–epoxy uniaxial tension specimen with a microcrack passing through the optical fiber. [From Salehi et al. (60).]

4.2.4. Summary

This section has provided an overview of the analytical, numerical, and experimental techniques that have been used to investigate the stress and strain concentrations that result from embedding optical fibers in load-bearing structures. Table 4.1 provides a convenient reference list of stress or strain concentrations that have thus far been reported in the literature, along with the host material system, optical fiber diameter, composite layup, and all

Figure 4.21. Moiré interferometric fringe pattern showing the vertical displacement field (with carriers of extension) in a graphite–epoxy uniaxial compression specimen with a multiple interlaminar embedded optical fiber.

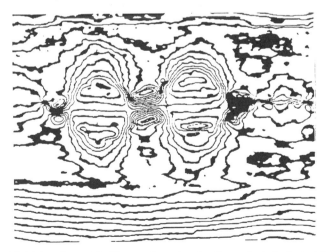

Figure 4.22. Moiré interferometric fringe pattern showing the vertical displacement field (with carriers of extension) around two closely spaced optical fibers embedded in graphite–epoxy and loaded with uniform temperature. (Courtesy of B. T. Han, IBM.)

other important descriptive information required to interpret the data. Note, however, that the data in this table should be met with a healthy degree of skepticism since they exhibit a large degree of scatter, are based on a remarkably small database, and the optical fibers and host material architectures are not uniform in all investigations. Nevertheless, the qualitative nature of the results is indisputable. The optical fiber sensor does indeed cause local stress and strain concentrations, and as a result, perturbs the very field variables it was designed to measure. The degree to which the stress concentrations affect the local integrity of the host material depends solely on the strength and toughness of the constituent materials. In fact, it has not yet been confirmed that the stress concentrations have a detrimental effect on the overall performance on the structural level, although there is strong evidence to suggest that transverse tensile and longitudinal compression strength of some composite systems are sensitive to alterations in the microarchitectural changes caused by optical fibers (5). From the sensoral perspective, however, the data presented in this section warrant careful examination of exactly what an embedded optical fiber sensor is actually measuring.

4.3. CRACKS

Composite material systems are generally considered brittle materials. Microcrack formation is an internal mechanism used to resist catastrophic failure by the redistribution of internal loads. Microcrack formation can be a precursor to total failure of the part or can act in a positive way by reducing residual stresses. One question that is commonly raised in the fiber optic smart structure community is whether or not induced stress concentrations caused by the structurally embedded optical fibers act as crack initiators or arrestors. Figure 4.20 showed that cracks do sometimes intersect embedded optical fibers, but limited experimental evidence is available that indicates the conditions under which embedded optical fibers initiate or terminate cracks. In this section we review microcrack initiation and propagation in laminated composites and document the available experimental results germane to crack initiation, propagation, and termination in composite systems with embedded optical fibers. We also discuss surface preparation and adhesion of the fiber–composite interface.

4.3.1. Microcracks in Optical Fiber Embedded Composites

It is evident from the discussion in the finite element and analytical stress analysis sections that significant effort has been expended to understand and modify the stress states near structurally embedded optical fibers. Much of this effort was aimed at preventing microcrack formation that might result from

Table 4.1 Documented Stress and Strain Concentrations Caused by Embedded Optical Fibers

Strain Conc. x	Strain Conc. y	Stress Conc. x	Stress Conc. y	Stress Conc. V.M.	Fiber Dia. (μm)	Coating Dia. (μm)	Coating Material	Host Material	Layup	Method of Calculation	Source
4.04	3.46	—	—	—	117	142	—	Gr/Ep	$[0_{12}/OF\{90\}/0_{12}]$	M.I.	Salehi et al. (59)
12.3	3.17	—	—	—	117	142	—	Gr/Ep	$[0_{12}/OF\{90\}/0_{12}]$	F.E.M.	Salehi et al. (59)
1.92	3.34	—	—	—	117	142	—	Gr/Ep	$[0_4/90_4/OF\{90\}/s]$	M.I.	Salehi et al. (59)
2.06	4.70	—	—	—	117	142	—	Gr/Ep	$[90_4/OF\{90\}/s]$	F.E.M.	Salehi et al. (59)
3.05	2.43	—	—	—	117	142	—	Gr/Ep	$[90_4/0_4/OF\{90\}/s]$	M.I.	Salehi et al. (59)
17.8	2.23	—	—	—	117	142	—	Gr/Ep	$[90_4/0_4/OF\{90\}/s]$	F.E.M.	Salehi et al. (59)
—	4.18	—	—	—	117	142	—	Gr/Ep	$[0_2/90_2/OF\{90\}/s]$	M.I.	Salehi et al. (59)
4.06	3.16	—	—	—	117	142	—	Gr/Ep	$[0_2/90_2/OF\{90\}/s]$	F.E.M.	Salehi et al. (59)
16.5	3.55	—	—	—	117	142	—	Gr/Ep	$[90_2/0_2/OF\{90\}/s]$	F.E.M.	Salehi et al. (59)
4.6	4.0	—	—	—	—	—	—	Gr/Ep	$[0_2/90_2/OF\{90\}/s]$	M.I.	Czarnek et al. (13)
14.2	3.6	—	—	—	—	—	—	Gr/Ep	$[0_2/90_2/OF\{90\}/s]$	M.I.	Czarnek et al. (13)
0.98	2.0	—	—	—	80	—	—	Gr/Ep	$[0_6/OF\{90\}/s]$	M.I.	Singh et al. (62)
—	2.04	—	—	—	80	125	Acrylate	Gr/Ep	$[0_6/OF\{90\}/s]$	M.I.	Singh et al. (62)
8.80	2.01	—	—	—	125	—	—	Gr/Ep	$[0_6/OF\{90\}/s]$	M.I.	Singh et al. (62)
—	3.50	—	—	—	125	250	Acrylate	Gr/Ep	$[0_6/OF\{90\}/s]$	M.I.	Singh et al. (62)
1.78	3.18	—	—	—	250	400	Acrylate	Gr/Ep	$[0_6/OF\{90\}/s]$	M.I.	Singh et al. (62)
4.50	3.05	—	—	—	250	600	Acrylate/ Tefzel	Gr/Ep	$[0_6/OF\{90\}/s]$	M.I.	Singh et al. (62)
—	—	—	—	1.34	80	93	Polyimide	CFR/PEEK	$[0_4/OF\{90\}/s]$	F.E.M.	Davidson and Roberts (19)
—	—	—	—	1.35	80	93	Polyimide	CFR/PEEK	$[90_4/0_4/OF\{90\}/s]$	F.E.M.	Davidson and Roberts (19)
—	—	—	—	2.1	—	140	Polyimide	CFR/PEEK	$[90_4/0_4/OF\{90\}/s]$	F.E.M.	Davidson and Roberts (19)
—	—	—	—	2.25	40–100	—	—	Gr/Ep	$[0_2/90_2/OF\{0\}/s]$	F.E.M.	Wan (77)
—	—	—	—	2.13	40–100	—	—	Fr/Ep	$[0_2/90_2/OF\{0\}/s]$	F.E.M.	Wan (77)

[a]V.M., Von Mises; F.E.M., finite-element method; M.I., Moiré interferometry.

local stress concentrations. It is interesting that very little effort has been
expended to verify experimentally if premature microcracking does indeed
result from embedded optical fibers. It seems that the effort in this regard is
strongest in the low-velocity-impact arena, with some recent effort in fa-
tigue-induced crack growth. Probably the most complete documentation of
microcracking near embedded optical fiber sensors is provided by Chang (9),
who was concerned with the local failure mechanics of optical fibers embedded
in low-velocity-impact specimens. This study included three different laminate
stacking sequences and six different optical fiber diameters. The general
conclusions reached by Chang's study were that the microcracks associated
with embedded optical fibers exhibit definite dependencies on fiber diameter,
coating compliance, and laminate stacking sequence. These results agree
with similar conclusions found for composite reinforcing fibers (1). As an
example, consider Fig. 4.23 and 4.24, where postimpact micrographs are shown
for an uncoated 200-μm optical fiber embedded in $[90_2/0_4/OF(90)/0_4/90_4]$ and
$[(+/-)45_3/OF(90)/(-/+)45^3]$ graphite–epoxy laminated composites. These
cross sections are taken 1.28 cm from the impact zone. Figure 4.23 shows a
microcrack emanating from a delamination between the 0 and 90° plies, and
propagating directly through the fiber–composite interface. Figure 4.24 does

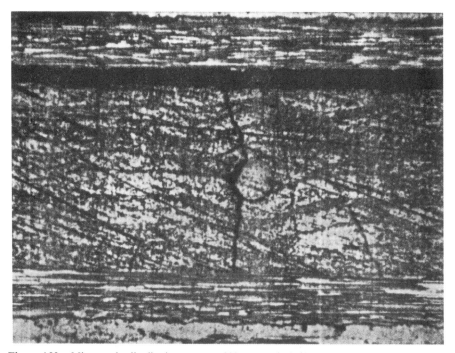

Figure 4.23. Microcrack distribution near a 200-μm optical fiber embedded in a $[0_4/90_4/$
$OF(90)/90_4/0_4]$ graphite–epoxy low-velocity-impact specimen. [From Chang (9).]

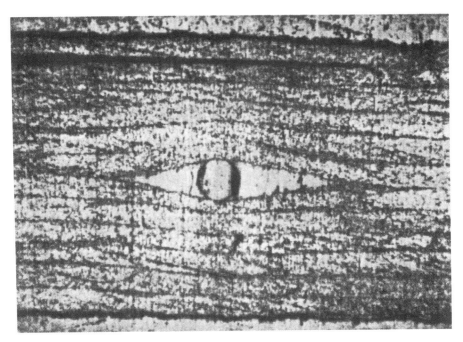

Figure 4.24. Microcrack distribution near a 200-μm optial fiber embedded in a $[(+/-)45_3/OF(90)/(-/+)45_3]$ graphite–epoxy low-velocity-impact specimen. [From Chang (9).]

not show any microcracking associated with the embedded optical fiber; instead, back-plane delamination typical of this stacking sequence exists. The fact that microcracking occurred in the $0°$ lamina is attributable to its low transverse tensile strength. Tay et al. (70) considered more complex laminates with several coated optical fibers (diameters not specified) embedded between plies. These specimens were also subjected to low-velocity impact but did not exhibit microcracking that can be associated with the optical fiber. To the contrary, the extensive microcracking and delamination evident in these specimens propagated seemingly unaware of the presence of the optical fibers.

The limited fatigue data that have thus far been reported in the literature sugests similar conclusions regarding microcrack formation. Melvin et al. (45) reported no fatigue crack formation in longitudinal tension–tension fatigue of unidirectional graphite–epoxy specimens with optical fibers embedded parallel to the reinforcing fibers. They did see interface failures between ultraviolet-curing acrylate coatings and the host material, but these were probably caused by coating mass loss during the composite cure cycle (60). Roberts and Davidson (56), on the other hand, report that microcracks can form near optical fibers under fatigue loading if the fibers are placed near the interface between cross-plies. Their specimens were cross-ply laminates produced from a toughened graphite–epoxy system and were loaded in low-cycle tension–

tension fatigue. The experimental results were accompanied by a finite element analysis that showed that the optical fiber interacted sufficiently with the cross-ply interface to produce stress concentrations. However, the highest stresses were near the optical fiber but not coincident with it. As a result, the cracks generally did not pass through the optical fiber.

4.3.2. Crack Suppression

It is not evident whether the microcracks that have been observed initiate from the optical fibers, simply pass through the fibers, or indeed terminate at the fibers. It is well known in the composite materials community that microcracks intersecting a stiff fiber will propagate through that fiber, whereas a crack intersecting a stiff fiber that is surrounded by a compliant coating will be restricted to the coating–composite interface (Fig. 4.25). This crack is therefore arrested. There is evidence, such as the micrograph in Fig. 4.26 obtained from a low-velocity-impact specimen (9, 10), that a similar phenomenon can occur in laminated composites with embedded optical fibers. Further, Blagojevic et al. (3) reported that the interlaminar fracture toughness of certain composites increases due to the interlaminar placement of optical fibers. Although no effort has been made to explore the possible crack blunting and crack deflection effects of embedded optical fibers, this does seem to be an unexpected benefit that should be investigated.

4.3.3. Interface Preparation and Adhesion

There is much concern that microcracking can propagate to the interface between the fiber and its coating or between the coating the host composite materials. As mentioned above, this can indeed occur if the coating stiffness is

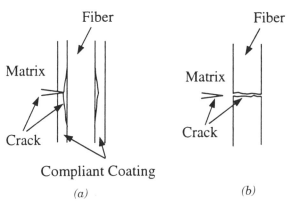

Figure 4.25. Crack propagation through fiber–matrix systems with (a) a compliant coating and (b) no coating.

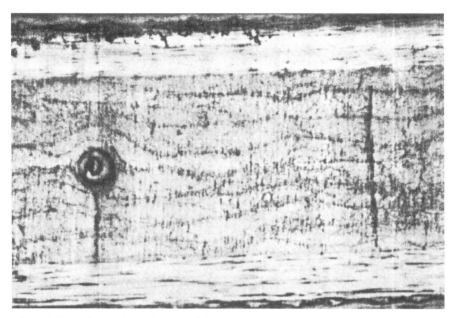

Figure 4.26. Microcrack terminated or initiated near an acrylate-coated optical fiber embedded in a $[O_4/90_4/OF(90)/90_4/0_4]$ graphite–epoxy low-velocity-impact specimen. [From Chang (9).]

much less than that of the fiber and host or if the adhesion between the fiber and the various constituents is poor. Since off-the-shelf optical fiber coatings are generally polymers, the former case is generally true. Ultraviolet-curable acrylate coatings are particularly susceptible to interfacial microcracking because their stiffness is a factor of 20 less than silica, and their already poor adhesion properties are exasperated by outgassing and shrinkage that occurs during the composite curing cycle (55). Figure 4.27 shows a complete interfacial debond between the coating and optical fiber. This micrograph is typical of acrylate-coated optical fibers embedded in laminated composites. Figure 4.28, on the other hand, shows a polyimide-coated optical fiber that was placed 1 cm away from the fiber in Fig. 4.27 in the same laminate so that the two fiber coating systems experienced identical curing cycles. No interfacial cracks form in the polyimide-coated fiber systems, mainly because polyimide possesses good thermal stability at thermoset curing temperatures (51). Fiber pullout tests have been used to characterize the adhesion between polyimide-coated optical fibers and various host polymers (21). These tests confirmed that the choice of coating–host combination determines the fiber's interfacial shear strength, and that polyimides again seem to have superior performance. This evidence is in agreement with the analytical results of Carman et al. (7), who argue based on composite strength predictions that polyimide is the superior choice for embedded optical fiber coatings. In addition, the pullout tests

Figure 4.27. Interfacial debond between an optical fiber and its acrylate coating. The coated fiber is embedded in an AS4/3501-6 graphite–epoxy laminate. (Courtesy of L. Melvin, NASA Langley Research Center.)

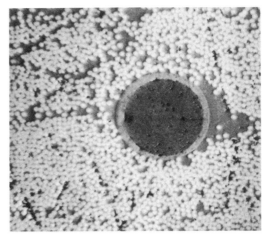

Figure 4.28. Polymide-coated optical fiber embedded in the same laminate as in Fig. 4.6. No interfacial debonding is visible. (Courtesy of L. Melvin, NASA Langley Research Center.)

showed no consistent tendency regarding pullout failure mode. The fracture surface generally reveals that the failure is a complex combination of cohesive and adhesive debonding.

There have been several attempts to discover what effect poor adhesion might have on the strain transfer from the host material to the optical fiber sensor (76, 12, 31, 32, 45, 75). These tests generally use optical fiber sensor response to infer adhesion quality. The most revealing tests of this kind were provided by Waite et al. (75), who treated similar optical fibers with mold release (release agent) and silane (interface enhancing agent) and embedded them in glass–epoxy three-point-bend specimens. These experimental extremes resulted in entirely different phase sensitivities for the embedded sensor, thereby confirming that interface adhesion can corrupt optical fiber sensor response. Jensen and Koharchik (31, 32) performed a similar study where they embedded untreated acrylate-coated optical fiber parallel to the load direction in un-idirectional graphite/bismaleimide quasi-static uniaxial tension and tension–tension low-cycle fatigue specimens. Figure 4.29 shows a comparison between a surface-mounted resistance strain gauge and the embedded optical fiber strain gauge for a cyclic load of steadily increasing magnitude. Except for a minor calibration error, the two signals compared quite favorably. More important, the cross-plot of the resistance strain gauge data against fiber strain gauge data (Fig. 4.30) exhibits no hysteresis, which indicates that the optical fiber sensor adhesion is adequate and no frictional behavior is present even though acrylate is known to develop interfacial debonds (see Fig. 4.27). One can attribute the linear behavior in Fig. 4.30 to good mechanical clamping of the fiber by the host composite due to residual thermal streses (40) and because the combination of material properties (v_{host} greater than v_{fiber}) and mechanical loading (uniaxial tension) leads to compressive interfacial stresses. Melvin et al. (45) considered adhesion characteristics of both acrylate and polyim-ide-coated optical fibers embedded in similar graphite–epoxy laminates. In this case, however, the specimens were subjected to high-cycle fatigue. Nonetheless, the embedded optical fiber sensor showed no behavior indicative of adhesion

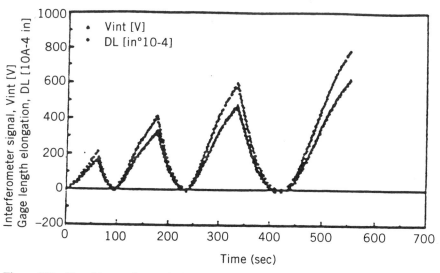

Figure 4.29. Time history of an embedded fiber optic sensor and a surface-mounted resistance strain gauge for a tension–tension graphite–bismilamide low-cycle-fatigue specimen. [From Jensen and Koharchik (32).]

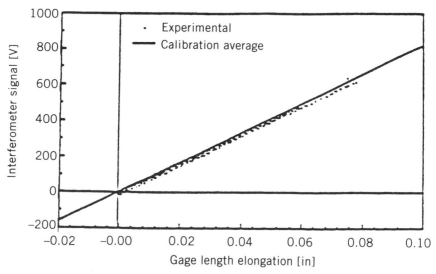

Figure 4.30. Cross-plot of the optical fiber and resistance strain gauge data presented in Fig. 4.29. [From Jensen and Koharchik (32).]

degradation. While micrographs do show fully detached interfaces in acrylate-coated sensors, the loading conditions were such that these microcracks did not limit the sensor's utility.

Perhaps the most comprehensive study of fiber–coating–host adhesion for optical fibers embedded in laminated composite systems is provided by Roberts and Davidson (55). They embedded optical fibers with 14 different fiber–coating combinations in graphite–epoxy host composites. The coatings they tested included polyimide, acrylate, carbon, aluminum, and bare fibers. The evaluation method consisted of standard mechanical strength and stiffness tests followed by electron microscopy of the optical fiber–coating–host interfaces. Poor adhesion is indicated by clean separation of the fiber from the coating, or the coating from the fiber.

Poor adhesion between an acrylate-coated (125- to 250-μm) optical fiber and its coatings are indicated in Fig. 4.31 by the clean separation of the two entities. Figure 4.32, on the other hand, shows good adhesion between a (80- to 93-μm) polyimide-coated optical fiber, its coating, and the host composite. Of all the coating materials that were tested, carbon showed the worst adhesion, and aluminum and polyimide showed the best adhesion. Based on these results, Carman et al. (7) used transverse tensile strength as the metric for assessing the importance of interface adhesion between coated optical fibers and graphite–epoxy host materials. Although their specimen geometry is not standard and there is a large degree of scatter in the data, the results indicate that intentionally reduced interfacial adhesion reduces the transverse tensile strength of the laminate.

Figure 4.31. Adhesive failure between a graphite–epoxy transverse tensile specimen and a DeSoto 131 acrylate-coated 125- to 250-μm fiber, indicating poor adhesion. [From Roberts and Davidson (56).]

4.4. SENSOR PERFORMANCE

The previous sections have described the mechanisms by which optical fiber sensors can interact mechanically with host structural components. These sections showed that complex stress states are produced in the host structure in the vicinity of the optical fiber. What the previous sections did not emphasize is that complex stress states are developed in the optical fiber as well. The stresses in the optical fiber sensor can potentially influence sensor performance in two ways. They may cause the optical fiber sensor to fail, thereby rendering it useless for further transduction tasks, or they may alter the interpretation of the embedded sensor signal. However, the size of surface flaws that result from optical fiber sensor manufacture are in the nanometer range, meaning that stress levels approaching gigapascal in the fiber will be required to induce brittle fracture in the fiber. It is therefore reasonable to

Figure 4.32. Cohesive failure between a graphite–epoxy transverse tensile specimen and poly-mide-coated 80- to 93-μm fiber, indicating strong adhesion. [From Roberts and Davidson (56).]

conclude that optical fiber fracture will not be a common phenomenon. This heuristic argument is borne out by the results of Waite et al. (76) but breaks down at sensor egress points from the host material or if flaws of more appreciable size are produced on the fiber surface as a result of the fiber optic smart structure manufacturing process. It is clear from the literature that with the possible exception of extrinsic Fabry–Perot sensors (38), premature frac-ture of completely embedded optical fiber sensors is not common and that optical fiber sensors generally survive loadings that are far in excess of the loads needed to cause failure in the host material system (9, 76). As a result, the stresses that are developed in the optical fiber mainly influence the interpretation of the sensor signal. The interpretation issues are manifested in transverse strain sensitivity and thermal apparent strain sensitivity of optical fiber sensors, in much the same way that they are manifested in surface-mounted resistance strain gauges (63).

A complete discussion of thermal and mechanical apparent strain is beyond the scope of this chapter, owing to the many popular sensor configurations.

Instead, a more generic presentation will be provided that imparts the important general concepts and some specific results. This section is further restricted to optical fiber sensors, whose performance can be traced to some form of coherent interference so that the measurand (strain, temperature, etc.) is related to an optical retardation or *phase change*. The phase change in any interferometrically based fiber sensor (including polarimetric and Bragg grating sensors) is the quantity that encodes all information about the measurand. It is hoped that one can uniquely relate the phase change to a specific measurand field such as axial strain. In its most revealing form, the optical retardation in any interferometric sensor can be expressed in terms of the strain- and temperature-dependent optical path length (64, 67)

$$\Delta\phi = \frac{2\pi}{\lambda} \int [(1 + S_1^f)n(S_i^f, T))] \, ds - \phi_0 \qquad (4.2)$$

where λ is the wavelength of light, n the effective refractive index, S_i^f the strain tensor describing the strain state that exists in the optical fiber, and S_1^f is the normal strain component everywhere tangent to the longitudinal axis of the optical fiber. The simplified expression in Eq. (4.1) is very useful in illustrating the concepts of transverse strain sensitivity and thermal apparent strain. Consider the embedded fiber sensor illustrated in Fig. 4.33. The system is subjected to radial compression but is prevented from axial expansion. This system is said to be in a state of plane strain since the axial normal strain components in the fiber and host are zero. If the axial strain in the fiber is zero (the fiber does not change length), should we expect the optical fiber sensor to experience a phase change? The answer is, yes. Even though the optical fiber sensor does not change in length, its optical path changes because the refractive index in Eq. (4.2) is independently a function of the strain state. This type of scenario was confirmed experimentally by Sirkis (63). The degree to which this transverse strain sensitivity causes errors in the sensor mesurements depends solely on the interpretation of the sensor signal. If, for example, the sensor is considered to contain axial strain information only (such as in the sur-

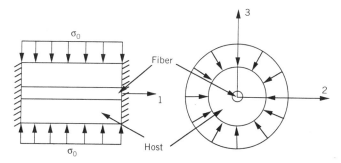

Figure 4.33. Optical fiber embedded in a host material subjected to plane-strain radial compression.

face-mounted case), the entire sensor response from the loading in Fig. 4.33 would be erroneous.

In a similar example, consider the embedded optical fiber sensor depicted in Fig. 4.34, in which the entire system is subjected to uniform thermal loading. For convenience the coefficient of thermal expansion of the host is taken to be larger than that of the optical fiber (this is generally true since $\alpha_{\text{fiber}} = 0.3 \times 10^{-6}$ per °C). The scenario in Fig. 4.34 illustrates the thermomechanical nature of the sensor response. It is clear that the applied temperature field will induce an optical phase change since the strain state is developed in the fiber, and the refractive index changes since it is an independent function of temperature and strain. Whether or not this thermally induced phase change introduces error into the sensor measurements again depends on the interpretation of the sensor signal. If the sensor is believed to depend only on mechanical strain, the entire sensor signal would be erroneous since it would indicate mechanical strain, when only thermal loading is present. This effect is known as *thermal apparent strain* since the fiber sensor apparently responds to a mechanical strain field that is in reality a thermally produced strain and thermooptic response. Regardless of the sensor interpretation, it is important to understand that the optical fiber strain state is not simply free thermal expansion. Under the temperature field, the optical fiber tries to expand in the radial and axial directions, only to be inhibited by the surrounding host medium. The strains in the fiber therefore contain contributions from the transverse mechanical restraining forces applied by the host. This point is made to emphasize that any attempt to design self-temperature-compensated (STC) embedded optical fiber sensors must account for the thermomechanical loading of the host on the optical fiber. This means that STC sensors will have to be designed with specific host material systems in mind.

It is possible to follow resistance strain gauge concepts to devise meaningful quantifications of errors due to transverse strain sensitivity and thermal apparent strain sensitivity (14). However, this approach requires certain preconceived notions regarding how an embedded optical fiber sensor should respond to general thermomechanical loading. The response of embedded phase-based optical fiber sensors can be generalized from Eq. (4.2) to

$$\delta = \frac{\Delta\phi}{\phi_0} = F_a S_1^f + F_{t2} S_2^f + F_{t3} S_3^f + F_T T^f \tag{4.3}$$

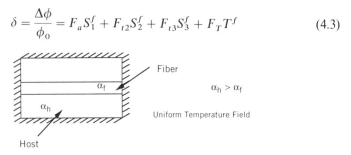

Figure 4.34. Fully constrained optical fiber–host system subject to uniform temperature loading.

where δ = normalized phase change of the sensor

ϕ_0 = absolute phase retardation of light propagating in the fiber sensor when it is free of thermomechanical loading

F_a = sensitivity of the sensor to axial strain

F_{t2} = sensitivity of the sensor to transverse strains in the 2-direction

F_{t3} = sensitivity of the sensor to transverse strains in the 3-direction

F_T = sensitivity of the sensor to temperature

Expressions for these sensitivities in terms of optical properties are provided in (65) for the seven popular phase-based sensor types. For convenience, all sensitivities are normalized by F_a, so that the sensor response becomes

$$\delta = F_a(S_1^f + K_{t2}S_2^f + K_{t3}S_3^f + \gamma T^f) \tag{4.4}$$

where K_{t2}, K_{t3}, and K_T are, respectively, the two transverse strain and the thermal apparent strain sensitivity fctors. Equation (4.4) again emphasizes the fact that the actual sensor response is dependent on more than just the axial strain. If one follows the precepts used by the resistance strain gauge community, embedded optical fibers are assumed (or more precisely, desired) to respond to axial strain only. Using this approach it is possible to define the normalized phase of the "ideal" optical fiber sensor as

$$\delta' = FS_1^f \tag{4.5}$$

where the superscript prime signifies the ideal sensor and F is the ideal axial sensitivity. The value of F will depend on the sensor type and calibration procedure. Errors due to transverse strains or thermal loading can then be quantified simply by finding the relative error between the actual and ideal sensor response

$$E = 100\,\frac{\delta - \delta'}{\delta'} = 100\left[(F_a - F) + K_{t2}\frac{S_2^f}{S_1^f} + K_{t3}\frac{S_3^f}{S_1^f} + \gamma\,\frac{T^f}{S_1^f}\right] \tag{4.6}$$

This expression reveals that the sensor error is a function of (1) the difference between the actual and calibrated axial sensitivity, (2) the transverse strain sensitivity factors, (3) the thermal apparent strain sensitivity factor, and (4) the loading ratios S_2^f/S_1^f, S_3^f/S_1^f, and T^f/S_1^f. This form is similar to the error equations developed for resistance strain gauges and quantifies the amount of error that is attributable to phase change associated with any loading other than axial strain. Equation (4.6) would predict, for example, an infinite error for the plane strain loading in Fig. 4.1 since S_1^f is zero by definition. This means simply that even though the entire sensor signal is assumed to be directly related to the axial strain [Eq. (4.5)], no portion of the phase change is actually caused by axial strain (in this case).

It is clear from the definition in Eq. (4.6) that errors can be quantified only after (1) the sensor type is specified so that F_a, K_{t2}, K_{t3}, and γ can be calculated, (2) the calibration procedure is specified so that F can be calculated, and (3) the load regime of operation is specified so that limits can be placed on the loading ratios S_2^f/S_1^f, S_3^f/S_1^f, and T^f/S_1^f. Further, the strains in Eq. (4.6) have contributions from both the thermal and mechanical loads, so that individual definitions for thermal apparent strain error and transverse strain sensitivity error must be derived (65). For example, consider an intrinsic Fabry–Perot optical fiber embedded parallel to the reinforcing fibers in a transversely isotropic laminated composite host material. Let the calibration consist of isothermal uniaxial loading in the fiber direction. The transverse strain sensitivity error computed for this sensor and calibration are provided in Fig. 4.35 for arbitrary limits on the mechanical loading ratios. This graph shows that an embedded optical fiber sensor of this type can operate in a specific band of loading ratios without incurring significant transverse strain sensitivity. The zero-error contour corresponds to the strain ratios produced by the calibration loads. This graph also shows that the transverse strain sensitivity error becomes dominant for off-diagonal combinations of mechanical loading ratios. While the Mach–Zehnder, Michelson, Bragg grating, and dual-mode sensors have similar transverse strain sensitivity error, as shown in Fig. 4.35, polarimetric sensors possess three-orders-of-magnitude greater sensitivity to transverse strains, and an extrinsic Fabry–Perot sensor possesses no

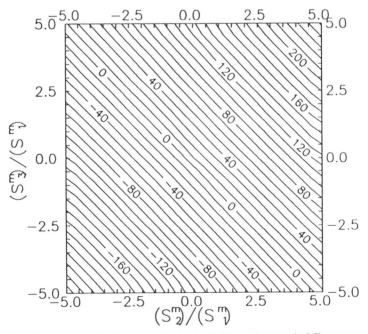

Figure 4.35. Transverse sensitivity error of an intrinsic Fabry–Perot optical fiber sensor embedded in a graphite–epoxy unidirectional laminate. [From Sirkis (65).]
Reprinted with permission.

transverse strain sensitivity at all. The polarimetric transverse strain sensitivity error is enormous because the temporary birefringence that governs this sensor's response is dominated by the transverse strains. The nonexistent transverse strain sensitivity in extrinsic Fabry–Perot sensors is attributable to its sensing region being composed of air whose refractive index is independent of strain (65).

Thermal apparent strain sensitivity error is provided in Fig. 4.36 for Mach–Zehnder, polarimetric, and extrinsic Fabry–Perot sensors assuming the same host structure and calibration used to produce Fig. 4.35. For ease of presentation, the polarimetric and extrinsic Fabry–Perot errors are, respectively, divided and multiplied by 10. This figure shows that not only is the extrinsic Fabry–Perot sensor insensitive to transverse strains, it is also least sensitive to thermal apparent strains. However, a certain level of caution is required when viewing Figs. 4.35 and 4.36. They have been developed under very specific assumptions — that the sensor ideally is sensitive to axial strain only, and that the sensor is embedded in a certain laminated composite system. These conditions may change depending on the sensor design. For example, it may be beneficial to design a polarimetric sensor that is ideally sensitive only to transverse strains. Axial strain sensitivity error would then have to be addressed.

4.5. SUMMARY

In this chapter we have provided an overview of local interaction mechanics research in the fiber optic smart structures field. The research in this area of smart structures is immature, as attested to by the dates of the literature cited. Although many of the interaction mechanics issues are generic to circular or cylindrical elastic inclusions, it was not until 1988 that these ideas were applied to structurally embedded optical fiber sensors. It is clear from the research

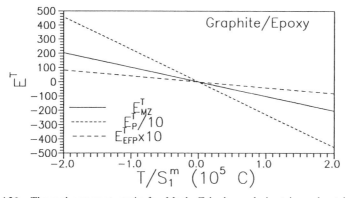

Figure 4.36. Thermal apparent strain for Mach–Zehnder, polarimetric, and extrinsic Fabry–Perot sensors embedded in a unidirectional graphite–epoxy laminate. [From Sirkis (65).] Reprinted with permission.

reviewed in this chapter that contradictory conclusions abound in the literature. However, two common themes seem to reoccur in the results.

1. There seems to exist a minimum acceptable optical fiber diameter that
   produces stress concentrations on the order of those produced by the
   indigenous reinforcing fibers, and this diameter seems to be roughly
   $100 \mu m$ (8, 9, 55, 62, 74).
2. There is both experimental and theoretical evidence that suggests that
   coating choice can alter the performance of the optical fiber embedded
   system either by reducing obtrusivity or by enhancing sensor performance (6, 7, 15, 52, 55, 62, 75). There is not yet a consensus on which
   coatings have the best overall characteristics, but acrylate coatings are
   losing favor at the expense of polyimide coatings, due to acrylate's
   thermal instability at typical composite curing temperatures.

This review shows a wealth of information about the local interaction
mechanics of optical fibers embedded in laminated graphite–epoxy host that
are subjected to quasi-static loading. However, there exist large gaps in the
state of the knowledge about optical fibers embedded in other host systems and
under time-dependent thermomechanical loading. Specific areas that need
concentrated effort are (1) experimental studies of the impact and fatigue
behavior of optical fiber embedded structures, (2) experimental and theoretical
studies into the effect of volume fraction of optical fiber sensors on the optical
and mechanical performance of the smart structures, (3) theoretical and
experimental studies into how nonlinear mechanical response of polymeric
matrix materials affect the local interaction mechanics of the fiber–host system,
(4) experimental studies to verify some of the optimal coating design theories
that have been presented, (5) studies into host microarchitectures other than
laminated composites, (6) studies to identify the conditions under which the
embedded optical fiber initiates or arrests cracking, (7) studies into the
importance of chemical sizing in the optical fiber embedding process, (8)
determination of whether or not fiber–host interface failure is indeed a critical
problem or if the manufacturing-induced residual stress automatically compen-
sates for "bad" interfaces, and (9) studies to determine if the local stress
concentrations induced by the embedded optical fiber is indeed a limitation to
the advancement of the technology. Many of these topics are interrelated, and
some have seen preliminary effort as seen by the material presented in the
chapter. For example, moiré interferometric studies have shown how viscoplastic response of epoxy can have a pronounced effect on sensor responses (66).
Jensen and Pascual (33) have studied optical fiber volume fraction effects on
the macrolevel mechanical properties of the host system. However, all of these
studies are preliminary and have not yet yielded definitive conclusions.

While the research into the various topics in local interaction mechanics
between embedded optical fiber sensors and their host materials has begun to
accelerate in the past year, it is still disproportionately limited. The state of the

knowledge has not yet progressed to the point where design guidelines can emerge, nor where in-service reliability and survivability can be addressed. However, the state of research has provided enough data to emphasize that a "system-level" approach to the analysis of fiber optic smart structures is warranted. This concept is made clear by the strong connection between calibration and obtrusivity.

ACKNOWLEDGMENT

The authors gratefully acknowledge the partial support of the Army Research Office under Grant DAAL03-92-G-0121 (Gary Anderson, Technical Monitor) during the preparation of this manuscript.

REFERENCES

1. B. D. Agarwal and L. J. Broutman. *Analysis and Performance of Fiber Composites.* Wiley (Interscience), New York, 1980.

2. Y. Benveniste, G. J. Dvorak, and T. Chen. Stress Fields in Composites with Coated Inclusions. *Mech. Mater.* 7, 305–317 (1989).

3. B. Blagojevic, W. Tsaw, K. McEwen, and R. M. Measures. The Influence of Embedded Optical Fibers on Interlaminar Fracture Toughness of Composite Materials. *Rev. Prog. Quant. Nondestr. Eval.*, Brunswick, ME, pp. 1213–1218.

4. G. P. Carman and R. C. Averill. (1992). Analytical Modeling of Micromechanical Stress Variations in Continuous Fiber-Reinforced Composites. *Proc. IUTAM Conf.* pp. 27–61.

5. G. P. Carman and G. Sendeckyj. A Review of Mechanics of Embedded Optical Fiber Sensors. *ASTM J. Compos. Technol. Res.* (to be published).

6. G. P. Carman, R. C. Averill, K. L. Reifsnider, and J. L. Reddy. Optimization of Coatings to Minimize Micromechanical Stress Concentrations in Composites. *J. Compos. Mater.* 27(6), 589–613 (1993a).

7. G. P. Carman, C. Paul, and G. Sendeckyj. Transverse Strength of Composites Containing Optical Fibers. *Proc. North Am. Conf. Smart Struct. Mater., 1st, 1993b.*

8. S. W. Case and G. P. Carman. Compressive Strength of Composites Containing Embedded Sensors or Actuators, *J Intell. Mater. Syst. Struct.* Vol. 5, No. 1, pp. 4–11.

9. C. C. Chang. Low Velocity Impact of Laminated Graphite/Epoxy Panels with Embedded Optical Fibers. M.S. Thesis, University of Maryland, College Park (1991).

10. C. C. Chang, J. S. Sirkis, and B. S. Smith. Low Velocity Impact of Optical Fiber Embedded Laminated Graphite/Epoxy Panels. In *Fiber Optic Smart Materials and Structures*, R. O. Claus, ed., pp. 131–136. IOP Publishing, Bristol, 1992.

11. R. M. Christensen, *Mechanics of Composite Materials*. Wiley, New York, 1979.

12. R. O. Claus and J. H. Cantrell. Detection of Ultrasonic Waves in Solids by An Optical Fiber Interferometer. *Proc. IEEE Ultrason. Symp.* **2**, 719 (1980).

13. R. Czarnek, Y. F. Guo, K. D. Bennett, and R. O. Claus. Interferometric Measurements of Strain Concentrations Induced by an Optical Fiber Embedded in a Fiber Reinforced Composite. *Proc. SPIE — Int. Soc. Opt. Eng.*, **986**, 120–129 (1989).

14. J. W. Dally, and W. F. Riley. *Experimental Stress Analysis*, 3rd ed. McGraw-Hill, New York, 1991.

15. A. Dasgupta and S. Bhandarkar. A Generalized Self-Consistent Mori-Tanaka Scheme for Fiber Composites with Multiple Interphases. *Mech. Mater.* **14**, 67–82 (1992).

16. A. Dasgupta and J. S. Sirkis. (1991). The Importance of Optimal Coatings to Structurally Embedded Optical Fiber Sensors in Smart Structures. *AIAA J.*, **30**(5), 1337–1343 (1991).

17. A. Dasgupta, Y. Wan, and J. S. Sirkis. Prediction of Resin Pocket Geometry for Stress Analysis of Optical Fibers Embedded in Laminated Composites. *Smart Mater. Struct.*, Vol. 1, No. 3, pp. 101–107, (1992).

18. R. Davidson and S. S. J. Roberts. Do Embedded Sensors Degrade Mechanical Performance of Host Composites? *Proc. Act. Mater. Adapt. Struct. Conf.*, Alexandria, VA, 1991, pp. 109–114.

19. R. Davidson and S. S. J. Roberts. Finite Element Analysis of Composite Laminates Containing Transversely Embedded Optical Fiber Sensors. *Proc. SPIE — Int. Soc. Opt. Eng.* **1777**, 115–122 (1992).

20. R. Davidson, D. H. Bowen, and S. S. J. Roberts. Composite Materials Monitoring Through Embedded Fiber Optics. *Int. J. Optoelectron.* **5**(5), 397–404 (1990).

21. C. DiFrancia, R. O. Claus, J. W. Hellgeth, and T. C. Ward. Discussion of Pull-out Tests of Polyimide Coated Optical Fibers Embedded in Neat Resin. *Fiber Optic Smart Materials and Structures*, R. O. Claus, ed., pp. 70–82. Technomic, Lancaster, PA, 1991.

22. M. A. Eisenberg, *Introduction to the Mechanics of Solids*. Addison-Wesley, Reading, MA, 1980.

23. D. Engrand, F. Lecuyer, and P. Sansonetti. *Non-destructive Testing of Composite Structures Using Embedded Optical Fibers: A Mechanical Interaction Model*, Dociete et Cie Report. Plaisis, France, 1990.

24. J. D. Eshelby, (1961). Elastic Inclusions and Inhomogeneities. *Prog. Solid Mech.* **2**, 89–140 (1961).

25. M. Gosz, B. Moran, and J. D. Achenbach. Effect of Viscoelastic Interface on the Transverse Behavior of Fiber-Reinforced Composites. *Int. J. Solids Struct.* **27**, 1757 (1991).

26. R. Grande. Nonlinear Thermomechanical Effects in Embedded Optical Fiber S. Thesis, Mechanical Engineering Dept., University of Maryland, (1993).

    ost, and B. Han. Thick Composites in Compression: An Experimental icromechanical Behavior and Smeared Engineering Properties. *J. ter.* **26**(1), 1930–1944 (1991).

28. Z. Hashin. Thermoelastic Properties of Particulate Composites with Imperfect Interface. *J. Mech. Phys. Solids* **39**(6), 745–762 (1991).

29. H. Horii and S. Nemat-Nasser, Elastic Fields of Interacting Inhomogeneities. *Int. J. Solids Struct.* **21**(7), 731–745 (1985).

30. T. Ishikawa, K. Koyama, and S. Kobayashi. Thermal Expansion Coefficients of Unidirectional Composites. *J. Compos. Mater.* **12**, 153–168 (1978).

31. D. W. Jensen and M. J. Koharchik. Calibration of Composite Embedded Fiber-Optic Strain Sensors. *Proc. 1990 Spring Conf. Soc. Exp. Mech.*, Albuquerque, NM, pp. 234–240 (1990).

32. D. W. Jensen and M. J. Koharchik. Cyclic Loading of Composite-Embedded Fiber-Optic Strain Sensors. *Proc. 1991 Spring Conf. Soc. Exp. Mech.*, Milwaukee, Wis., pp. 233–238 (1991).

33. D. Jensen and J. Pascual. Degradation of Graphite/Bismilamide Laminates with Multiple Embedded Fiber-Optic Sensors. *Proc. SPIE — Int. Soc. Opt. Eng.* **1370**, 228–237 (1991).

34. P. V. Kishore, A. C. W. Lau, and A. S. D. Wang. The Fiber Pull-out Problem: Matching of Singular and Complete Stress Fields. *Proc. Conf. Am. Soc. Compos.*, *6th*, pp. 1054–1063 (1991).

35. C. R. Kurkjian, and D. Inniss. Understanding Mechanical Properties of Lightguides: A Commentary. *Opt. Eng.* **30**(6), 681–689 (1991).

36. M. Leblanc and R. M. Measures. Micromechanical Considerations for Embedded Single-ended Sensors. *Proc. SPIE — Int. Soc. Opt. Eng.* **918** pp. 215–227 (1993).

37. A. W. Leissa, W. E. Clausen, and G. K. Agarwal. Stress and Deformation Analysis of Fibrous Composite Materials by Point Matching. *Int. J. Numer. Methods Eng.* **3**, 89–101 (1971).

38. J. J. Lesko, S. W. Case, B. R. Fogg, and G. P. Carman. Embedded Fabry–Perrot Fiber Optic Strain Rosette Sensor for Internal Stress State Assessment. *Proc. 7th Annu. Conf. Am. Soc. Compos.*, pp. 909–913 (1992).

39. A. Levy. The Debonding of Elastic Inclusions and Inhomogeneities. *J. Mech. Phys. Solids* **39**(4), 477–505 (1991).

40. I.-P. Lu. On Interphase Modeling for Optical Fiber Sensors Embedded in Laminated Composite Systems. M.S. Thesis, University of Maryland, College Park (1993).

41. I.-P. Lu and J. S. Sirkis. On Interphase Modeling for Optical Fiber Sensors Embedded in Laminated Composite Systems. ASME AD-Vol. 35, Adaptive Structures and Material Systems (Ed. G. P. Gorman and E. Garcia) pp. 419–426, (1993).

42. J. S. Madsen, A. P. Jardine, R. J. Meilunas, A. Tobin, and Y. E. Pak. Effect of Coating Characteristics on Strain Transfer in Embedded Fiber-Optic Sensors. *Proc. North Am. Conf. Smart Struct. Mater., 1st, 1993*.

43. C. T. Mathews and J. S. Sirkis. Interaction Mechanics of Interferometric Optical Fiber Sensors Embedded in a Monolithic Structure. *Proc. SPIE — Int. Soc. Opt. Eng.* **1370**, 142–153 (1991).

44. R. M. Measures, D. Hogg, R. D. Glossop, J. Lymer, M. LeBlanc, J. West, S. DuBois, W. Tsaw, and R. C. Tennyson. Structurally Integrated Fiber Optic Damage Assessment System for Composite Materials. *Appl. Opt.* **28**, 2626–2633 (1989).

45. L. D. Melvin, R. S. Rogowski, M. S. Holben, J. S. Namkung, K. Kahl, and J. S. Sirkis. Evaluation of Acrylate and Polyimide Coated Optical Fibers as Strain Sensors in Polymer Composites. *Proc. Act. Mater. Adapt. Struct. Conf.*, Alexandria, VA, pp. 801–804 (1991).

46. Y. Mikata and M. Taya. Stress Field in a Coated Continuous Fiber Composite Subjected to Thermomechanical Loadings. *J. Compos. Mater.* **19**, 554–578 (1985).

47. Y. Morimotto, Y. Seguchi, and T. Higashi. Application of Moiré Analysis of Strain Using Fourier Transform. *Opt. Eng.* **27**, 650–656 (1988).

48. Z. A. Moschovidis and T. Mura. Two-Ellipsoidal Inhomogeneities by the Equivalent Inclusion Method. *J. Appl. Mech.* **42**, 847–852 (1975).

49. T. Mura. *Micromechanics of Defects in Solids*, 2nd ed. Martinus Nijhoff, The Hague, The Netherlands, 1987.

50. T. Mura and R. Furahashi. The Elastic Inclusion with a Sliding Interface. *J. Appl. Mech.* **51**, 308–310 (1984).

51. D. K. Nath, G. W. Delson, S. E. Griffin, C. T. Harrington, Yi He, C. J. Rienhart, D. C. Paine, and T. F. Morse. Polyimide Coated Embedded Optical Fiber Sensors. *Proc. SPIE — Int. Soc. Opt. Eng.* **1489**, 17–32 (1991).

52. Y. E. Pak. Longitudinal Shear Transfer in Fiber Optic Sensors. *J. Smart Mater. Struct.* **1**(1), pp. 57–62 (1992).

53. Y. E. Pak, V. DyReyes and E. S. Schmuter. Micromechanics of Fiber Optic Sensors. *Proc. Act. Mater. Adapt. Struct. Conf.*, Alexandria, VA, **1991**, pp. 121–128 (1991).

54. D. Post. Moiré Interferometry. In *Handbook on Experimental Mechanics* A. S. Kobayashi, ed. Prentice-Hall, Englewood Cliffs, NJ, 1988.

55. S. S. J. Roberts and R. Davidson. Mechanical Properties of Composite Materials Containing Embedded Fibre Optic Sensors. *Proc. SPIE — Int. Soc. Opt. Eng.* **1588**, 326–341 (1992).

56. S. S. J. Roberts and R. Davidson. Short Term Fatigue Behavior of Composite Materials Containing Embedded Fiber Optic Sensors and Actuators. *Proc. SPIE — Int. Soc. Opt. Eng.* **1777**, 255–262 (1992).

57. W. Rosen. *Fiber Composite Materials*. Am. Soc. Mat., Metals Park, OH, 1964.

58. B. W. Rosen. Composite Materials Analysis and Design. *Engineered Materials Handbook*, Vol. 1. ASM International, Metals Park, OH, 1987.

59. A. Salehi, A. Tay, D. Wilson, and D. Smith. Strain Concentrations Around Embedded Optical Fibers by FEM and Moiré Interferometry. *5th Annu. ASM/ESD Adv. Compos. Conf./Exspos.*, Dearborn, MI, pp. 11–19 (1989).

60. P. Sansonetti, M. Lequiem, D. Engrand, J. J. Guering, R. Davidson, S. Roberts, B. Fornari, M. Martinelli, P. Escobar, V. Gusmeroli, P. Ferdinand, J. Plantey, M. Crowther, B. Culshaw, and C. Michie. Intelligent Composites Containing Measuring Fibre Optic Networks for Continuous Self-Diagnosis. *Proc. SPIE — Int. Soc. Opt. Eng.* 198–209 (1992).

61. J. S. Sirkis and H. S. Singh. Moiré Analysis of Thick Composites with Embedded Optical Fibers. To appear in Experimental Mechanics.

62. H. S. Singh, J. S. Sirkis, and A. Dasgupta. Micro-interaction of Optical Fibers Embedded in Laminated Composites. *Proc. SPIE — Int. Soc. Opt. Eng.* **1588**, 76–85 (1992).

63. J. S. Sirkis. Interpretation of Embedded Optical Fiber Sensor Signals. In *Applications of Optical Fiber Sensors in Engineering Mechanics*, F. Ansari, ed. Am. Soc. Civ. Eng., New York, pp. 85–99 (1993).

64. J. S. Sirkis. A Unified Approach to Phase-Strain-Temperature Models for Smart Structure Interferometric Optical Fiber Sensors. Part I. Development. *Opt. Eng.* **32**(4), 752–763 (1993).

65. J. S. Sirkis. A Unified Approach to Phase-Strain-Temperature Models for Smart Structure Interferometric Optical Fiber Sensors. Part II. Applications. *Opt. Eng.* **32**(4), 763–773 (1993).

66. J. S. Sirkis and A. Dasgupta. The Role of Local Interaction Mechanics in Fiber Optics Smart Structures. *J. Intell. Mater. Syst. Struct.* **3**(1), 101–107 (1993).

67. J. S. Sirkis and H. W. Haslach. Complete Phase-Strain Model for Structurally Embedded Interferometric Optical Fiber Sensors. *J. Intell. Mater. Syst. Struct.* **2**(1), 3–24 (1991).

68. E. Sternberg and M. A. Sadowsky. On the Axisymmetric Problem of the Theory of Elasticity for an Infinite Region Containing Two Spherical Cavities. *J. Appl. Mech.* **19**, 19–27 (1952).

69. A. Tay, D. A. Wilson, and L. Wood. Strain Analysis of Optical Fibers Embedded in Composite Materials Using Finite Element Modeling. *Proc. SPIE — Int. Soc. Opt. Eng.* **1170**, 521–533 (1990).

70. A. Tay, D. A. Wilson, A. C. Demirdogen, J. R. Houghton, and R. L. Wood. Microdamage and Optical Signal Analysis of Impact Induced Fracture in Smart Structures. *Proc. SPIE — Int. Soc. Opt. Eng.* **1370**, 328–343 (1991).

71. Y Tong and I. Jasiuk. Transverse Elastic Moduli of Composites Reinforced with Cylindrical Coated Fibers: Successive Iteration Method, *Proc. 5th Tech. Conference,* American Soc. for Composites, pp. 117–126, (1990).

72. E. Tsuchida, T. Mura and J. Dundios. The Elastic Field of an Elliptic Inclusion with a Slipping Interface, *J. of Applied Mechanics*, **53**, pp. 103–107 (1986).

73. E. Udd, R. J. Michal, S. E. Higlry, J. P. Theriault, P. LeChong and D. A. Holin. Fiber Optic Sensor System for Aerospace Applications, *SPIE*, **838**, 162–168 (1987).

74. A. Vengsarkar, K. A. Murphy, M. F. Gunther, A. J. Plante and R. O. Claus. Low Profile Fibers for Embedded Smart Structures Applications, *SPIE*, **1588** 2–13 (1991).

75. S. R. Waite, R. P. Tatam and A. Jackson. Use of Optical Fiber for Damage and Strain Detection in Composite Materials, **19**(6), 435–442 (1988).

76. S. R. Waite and G. N. Sage. The Failure of Optical Fibres Embedded in Composite Materials. *Composites.* Vol. 19, No. 4, July, pp. 288–294 (1988).

77. Y. Wan, Analytical Micromechanics of Optical Fiber Sensors Embedded in Laminated Composites, M.S. Thesis, Mechanical Engng. Dept., Univ. of Maryland, College Park, MD. (1990).

5

Integrity of Composite Structures with Embedded Optical Fibers

DAVID W. JENSEN
Brigham Young University
Provo, Utah

AND

JAMES S. SIRKIS
University of Maryland
College Park, Maryland

5.1. INTRODUCTION

In this chapter we address the effects that embedded optical fibers have on the global structural integrity of host structural systems produced from laminated advanced composite materials. The presentation here is strictly from a structural engineering perspective in the sense that optical fibers are treated simply as cylindrical elastic inclusions, irrespective of any sensing function for which they may be designed. *Structural integrity* implies that a structural system can satisfy its design performance without failure. Therefore, the "failure" or "loss of integrity" of the system must be defined carefully prior to any substantive discussion about the global performance of a host structure with embedded optical fibers. The most commonly used metrics for failure in composite material systems are loss of stiffness and/or loss of strength. *Stiffness* is a measure of the proportionality between stress (load per unit area) and strain (deformation per unit length), while *strength* is defined by the ultimate load-carrying capability of the structure. It is important to recognize that stiffness and strength of composite material systems are not scalar quantities because these systems are intrinsically anisotropic. For example, full characterization of strength of an orthotropic laminate requires the determination of five strength descriptors: (1) longitudinal tensile strength, (2) longitudinal compres-

Fiber Optic Smart Structures, Edited by Eric Udd.
ISBN 0-471-55448-0 © 1995 John Wiley & Sons, Inc.

sive strength, (3) transverse tensile strength, (4) transverse compressive strength, and (5) in-plane shear strength. To complicate the issue further, laminated composites are typically produced from a combination of unidirectional and/or woven layers so that the stiffness and strength vary on the lamina scale, depending on the relative orientation of adjacent plies. Laminated plate theories exist that can be used to predict the laminate properties, such as strength and stiffness, from those of the individual layers, provided that inclusions such as embedded optical fibers are either not present or do not alter the global behavior of the composite laminate (1).

Two approaches are generally used to characterize laminated composite structures. The first is to characterize the behavior of the unidirectional layers that form the basic building blocks of the entire laminated systems. One can infer the intrinsic behavior of the entire laminate from these measurements. The second approach to charcterization is simply to build the laminated structure and test it as a unit. Both of these approaches to testing are time consuming and expensive to perform since each of the material properties of these anisotropic materials must be determined by independent testing programs. This is one of the reasons why results on intrinsic strength and stiffness of optical fiber embedded composite materials are slow in being reported. Furthermore, the response is a function of the particular material system (i.e., fiber–matrix combination). Even components manufactured using the same composite material system, but with different reinforcing fiber volume fractions, can behave quite differently. For these reasons, caution must be exercised when attempting to make broad-ranging generalizations based on the information presented in this chapter.

While many of the results presented in this chapter represent classical loading cases that might be experienced by realistic structures, as pointed out by Sendeckyj and Paul (29), there is a real need to investigate thoroughly the effects of fiber optics embedded in composite structures under complex, expected service loading, including environmental effects. Clearly, much more testing is required before the full range of impact that embedded fiber optic sensors have on the structural behavior of smart structures can be understood fully.

Ultimately, the global intrinsic behavior of fiber optic smart structures must somehow be related to the detailed local behavior, such as that documented elsewhere in this book. Microcracking due to residual thermal stresses, mechanical overload, fatigue, and/or low-velocity impact, the influence of coatings and interphases on the local stress state, or any other local-scale stress raising mechanisms can translate to reduced global performance, and, therefore, loss of structural integrity. The relationships that bridge micromechanics and global continuum mechanics, however, are beyond the scope of this book. The discussion in this chapter is restricted to a survey of the structural-level integrity issues associated with fiber optic smart structures, although heuristic relationships to the microscale behavior are provided whenever appropriate.

5.2. FIBER OPTIC INCLUSIONS IN COMPOSITE MATERIALS

The possible presence of any inclusion in an advanced composite material is of serious concern to the structural engineer. The introduction of continuous optical fibers poses a potential threat to the integrity of the structural component. The load in an advanced composite material is carried primarily by relatively small diameter high-strength high-stiffness lightweight fibers such as graphite (6 to 11 μm diameter, 1.5 to 3.0 GPa tensile strength, 200 to 550 GPa tensile modulus, 1.7 to 2.0 g/cm^3), Kevlar (12 μm diameter, 2.8 to 3.6 GPa tensile strength, 62 to 124 GPa tensile modulus, 1.44 g/cm^3), or glass (9 to 13 μm diameter, 3.4 to 4.6 GPa tensile strength, 72 to 86 GPa tensile modulus, 2.5 g/cm^3). Since the diameter of the current generation of silica-based optical fibers is relatively large (80 to 500 μm in diameter) by comparison to the reinforcing fibers (on the order of 5 to 10 μm in diameter), serious concerns have been raised about the structural integrity of advanced composites — especially graphite-reinforced composites — which contain embedded optical fibers (14, 19). Jensen and Griffiths (14), for example, defined the various durability issues that need to be addressed, including both the static and dynamic response, such as the tensile and compressive strength and stiffness, and the damage tolerance considerations brought about by impact and fatigue loading. Mazur et al. (19) identified several structural integrity issues, including concerns about the orientation of the optical fibers relative to the reinforcing fibers, the coating on the optical fibers, and the interfacial bond strength between the optical fiber coating and the host material in which it is embedded.

Photomicrographs of graphite–epoxy laminate cross sections with optical fibers embedded in orientations ranging from parallel to angular to perpendicular relative to the reinforcing fibers reveal the geometric disturbances caused to load-carrying fibers by the fiber optic inclusions (2, 4, 15, 28, 30, 36). Sansonetti et al. (28) present photomicrographs of bare optical fibers parallel to the reinforcing fibers and acrylate-coated optical fibers perpendicular to the reinforcing fibers. In the latter case, the jacket is grossly distorted but still in contact with the embedded optical fiber. Bonniau et al. (4) present photomicrographs of bowtie fibers with polyimide and acrylate coatings, and side-hole fibers with acrylate and Teflon coatings. Vengsarkar et al. (36) show photomicrographs of various-diameter inclusions, ranging from 40 to 140 μm in diameter at angles of 0, 45, 60, and 90° relative to the reinforcing fibers. They also describe the production of low-profile fibers (i.e., fibers with thinner claddings, claddings only 35 to 40 μm thick). The possibility of using these smaller-diameter fiber optic sensors is naturally quite appealing.

Singh et al. (30) employed Moiré interferometry to examine the strain concentrations induced by the fiber optic inclusion in graphite–epoxy laminates. Their investigation was based on the edge view of a unidirectional laminate with the optical fibers embedded perpendicular to the reinforcing fibers (i.e., a $[90_6/\text{OF}/90_6]_T$ layup). Several configurations of optical fibers were examined, including an 80-μm-diameter uncoated optical fiber (bare and with a 125-μm-diameter jacket), a 125-μm-diameter fiber (bare and with a 250-μm diameter jacket), and a 250-μm-diameter fiber (bare and with both a 400- and a 600-μm-diameter jacket). This experiment provides both qualitative and quantitative information about the stress concentrations around the vicinity of the embedded optical fiber. In fact, the results indicate that *residual* (permanent) strain fields develop around the embedded optical fibers after repeated loading. Unfortunately, the experiment cannot correct, however, for the fact that the strain field that exists inside a real laminate is significantly altered by exposing a free surface for examination.

Various mathematical models have been developed to identify the critical parameters governing the behavior of embedded fiber optic sensors and to quantify their influence on the integrity of the host structure. A mathematical model of the inclusion developed by Sansonetti et al. (28) was used to explore the relationship between the properties of the coating on the fiber and the induced strain concentrations. The results indicate that the optimal fiber coating is a function of the ratios of both coating to composite modulus and that of coating thickness to optical fiber diameter. In general, a higher-modulus coating has been shown to be most effective for thicker coatings.

Significant research on the micromechanics of embedded fiber optic sensors has also been conducted by others (8, 31, 32, 34). Most of this is summarized in Chapter 4 of this book. Tay et al., for example, employed the finite element method to study the effects of the fiber optic sensor coating, the resin material properties, and the size of the resin-rich regions on the uniaxial response of 0, 45, and 90° layups. Their findings indicate that the strain induced by the inclusion is inversely proportional to the size of the resin-rich region. These strain concentration factors range from 2 to 6 in the applied loading direction and from 6 to 10 in the transverse direction. They also concluded that in most applications, the coating should be softer than the surrounding resin material.

Finally, it is important to note that the research results described in this chapter have been developed based on the application of intrinsic sensors. The deployment of localized (extrinsic) fiber optic sensors presents additional concerns that have not yet been addressed by the smart structures research community.

5.3. QUASI-STATIC LOADING

Some of the first published experimental data on the structural integrity of advanced composite materials with embedded optical fibers was reported by Measures et al. (20, 22, 23). This research was based on Kevlar–epoxy laminates tested in uniaxial tension and compression. The four-ply (tension) and eight-ply (compression) unidirectional laminates contained arrays of 125-μm-diameter bare optical fibers embedded in the midplane of the tension specimens and between every other ply (2:3, 4:5, 6:7 ply interfaces) in the compression specimens. The optical fibers were configured in a way that probably represents the worst case (i.e., embedded perpendicular to the adjacent reinforcing fibers and oriented perpendicular to the applied load). The results, however, suggested that embedded optical fibers slightly increase the ultimate tensile strength of Kevlar–epoxy laminates and have only a negligible adverse effect on the average compressive strength. The damage tolerance was also reported to be unaffected by the presence of the embedded optical fibers.

In a similar study, Blagojevic et al. (3) reported that optical fibers embedded in Kevlar–epoxy composites slightly improve the interlamina fracture toughness. Both Blagojevic's and Measure's results are predictable, however, given the relatively poor performance of nearly all Kevlar–epoxy laminates in compression. In particular, the extremely low interlaminar shear strength of Kevlar–epoxy introduces significant delamination which results in poor compressive properties (10, 27, 33). Since the adhesion between silica and epoxy is better than that between Kevlar and epoxy (33), the optical fibers may actually function as crack arrestors in Kevlar–epoxy laminates, eliminating any noticeable degradation in the compression performance.

Parallel experimental studies of the uniaxial tensile and compressive performance of relatively thin (approximately 1.5 mm) graphite–bismaleimide laminates with embedded optical fibers were conducted by Jensen and Pascual (16), Pascual (25), and Jensen et al. (17, 18). These experiments were based on NARMCO G40-600/5245C graphite–bismaleimide cross-ply laminates with a $[0_3/90_2/0]_s$ orthotropic layup. The optical fibers were embedded in configurations that were assumed to represent the extreme cases (i.e., embedded parallel or perpendicular to the adjacent reinforcing fibers and oriented parallel or perpendicular to the applied loading direction). The results indicate that embedded optical fibers have a relatively benign influence (less than 10% degradation in the worst case) on the tensile performance of thin composite structures but can have a rather significant adverse impact (up to 70%) on the compressive performance, depending on the orientation of the embedded optical fibers with respect to the adjacent reinforcing fibers and the direction of the applied loading. In particular, optical fibers embedded perpendicular to the applied uniaxial loading direction experience a 35 to 40% reduction in compressive strength, which increases to 70% if the optical fibers are also perpendicular to the adjacent reinforcing (graphite) fibers. If the fibers are aligned with the applied loading direction and are also parallel to the adjacent reinforcing fibers, the compressive strength degradation remains under 24%, even for very high optical fiber concentrations by volume (15% for low optical fiber concentrations by volume).

Sansonetti et al. (28) and Roberts and Davidson (26) performed a wide variety of mechanical performance tests on graphite–epoxy laminates with embedded optical fibers. The 1-mm-thick plates employed in their investigation consisted of eight plies of T300/F263 (Hexcel) graphite–epoxy with optical fibers embedded in the midplane of the laminates. Various coatings and optical fiber diameters were evaluated in this study. Among the various fiber configurations examined were assorted acrylate coatings, polyimide coatings, an aluminum coating, and bare fiber. The optical fiber claddings ranged in diameter from 80 to 300 μm, while the coatings ranged from 93 to 500 μm in diameter.

In the latter study, specimens loaded in longitudinal tension exhibited only minor strength degradation, while specimens loaded in longitudinal compression exhibited fairly significant strength degradation (up to 26% for both acrylate and polyimide-coated optical fibers). The results for in-plane shear were inconclusive, while interlaminar shear test results indicate a slight strength reduction (as high as 8% for acrylate-coated fibers; other configurations were not reported). The most interesting results, however, are from the specimens loaded in transverse tension, which were found to be highly dependent on the diameter of the inclusion, the elastic properties of the coating, and the adhesive strength of the cladding–coating and coating–matrix interfaces.

For the case of transverse tension, several interesting observations can be made. First, the degradation in performance was 5% or less for the configurations where the diameter of the coated optical fiber was less than one ply thickness (ca. 125 μm). The degradation increased, however, to a range of 13 to 22% for acrylate-coated fibers which were two to four plies thick. Moreover, the polyimide-coated fibers caused a 28% reduction in the transverse tensile strength even though the coating diameter was just over one ply thick. Similarly, the bare fibers just over two plies thick (300 μm) resulted in a 44% degradation of transverse tensile strength. Thus the bare fibers exhibited almost three times as much degradation in performance (from the original material without embedded optical fibers) as optical fibers with an acrylate coating of a similar diameter. Furthermore, some interesting qualitative observations include the fact that the acrylate coating debonds from the optical fiber under transverse tensile load, while the polyimide coating introduces significant stress concentrations. Nevertheless, despite these limitations, the 80:93 (cladding: coating, diameters in microns) polyimide-coated and 80:103 acrylate-coated fibers were found to be acceptable configurations inasmuch as these combinations exhibited almost no degradation in transverse tensile strength.

To explain the transverse compression strength degradation, Davidson and Roberts (9) used the finite element method to predict the stress concentrations introduced at the optical fiber interface due to transverse compression loading and residual thermal stresses. This investigation was based on optical fibers embedded parallel to the reinforcement fibers in a unidirectional graphite–epoxy composite. Although this configuration induces only a minimal perturbation to the load-bearing constituents in the composite, the research indicates that the stress distribution around the embedded optical fiber is related to the combined stiffness of the coating and optical fiber. Experimental verification of these results supports the conclusion that small-diameter optical fibers embedded in the direction of the reinforcing fibers do not degrade the transverse compression performance of the composite material. On the other hand, large-diameter fibers and/or thick coatings can cause severe degradation of material properties, particularly for bare and polyimide-coated fibers. This information can be utilized in practical implementations by specifying the coating thickness and coating material (modulus).

These conclusions were substantiated further by Pak et al. (24), who developed an isotropic elasticity solution to evaluate the influence of the coating properties on the strain transfer between the coating and the host material. The boundary element method was employed to confirm the results. The findings indicate that thinner coatings permit greater shear transfer from the host, provided that the shear modulus of the coating is softer than that of the host. On the other hand, the shear transfer increases with coating thickness when the coating is stiffer than the host. The optimal condition is for the shear modulus of the coating to be approximately equal to the geometric mean of the shear moduli of the optical fiber and the host material. Similarly, softer coatings were found to induce higher longitudinal stresses in the surrounding host material. These findings, however, are based on an isotropic host with stiffness equivalent to that of a typical matrix. A host composite material, however, is generally anisotropic and usually much stiffer than the typical matrix material.

In summary, the quasi-static response of advanced composite laminates with embedded optical fibers will depend on the specific configuration. Obviously, the stress concentrations are minimized if the optical fibers are embedded parallel to the adjacent reinforcing fibers. Similarly, the structural integrity is least affected when the optical fibers have an appropriate coating and are embedded parallel to the primary (most critical) loading direction.

5.4. LOW-VELOCITY-IMPACT LOADING

One of the first applications of optical fiber sensors was that of impact damage detection via fiber breakage sensors (7, 12, 23). Most of these investigations included some form of postimpact nondestructive evaluation of the optical fiber embedded composite systems. This type of nondestructive evaluation forms the cornerstone in addressing the germane structural integrity issues associated with low-velocity-impact damage. This type of information provides only qualitative information regarding the postimpact response of these systems; however, no postimpact strength or stiffness studies have yet been performed on laminate composite systems with embedded optical fibers. The most detailed investigations of impact damage zones in fiber optic smart structures have been performed by Measures and co-workers (11, 22, 23) on Kevlar–epoxy and by Chang (6) on graphite–epoxy. Both of these studies came to the same conclusion—that the damage induced by low-velocity impact is so traumatic that the presence of the optical fiber has no discernable effect on the distribution of damage in the composite. This result seems to hold for embedded fiber sensor networks of standard diameter and for isolated optical fibers of various diameters. Consider the damage zone shown in

Measures (21), where multiple optical fibers have been embedded in a complex Kevlar–epoxy laminate and subjected to a 14.6-J impact. Some of the embedded fibers were etched to induce fiber fracture (20), and some of the fibers still have the natural coatings. The important feature in this figure is that the damage zone shows no tendency to wick near the optical fibers. In fact, the damage zone is precisely as would be expected from a laminate of this type with no embedded optical fibers. Chang (6), on the other hand, considered low-velocity impact of graphite–epoxy plates with six different types of optical fibers embedded in three different layups. The optical fibers differed in diameter and coating material. Chang shows the results of x-ray nondestructive inspections of uncoated 200-μm-diameter fibers embedded in $[90_6/OF(0)]_s$, $[(+/-)45_3/OF(0)/(-/+)45_3]$, and $[90_2/0_4/OF(0)]_s$ laminates after impact at 3.9, 2.7, and 2.7 J, respectively. Not one of these figures indicates that the optical fiber interacted with the macroscale damage event. This behavior was consistently observed in all 72 of Chang's tests, as well as in C-scans presented by Tay et al. (35).

5.5. FATIGUE LOADING

Typically, any realistic structural component must be able to withstand various loads applied over the lifetime of the structure. These repeated loadings can accelerate the propagation of cracks to a critical stage. The presence of inclusions that cause stress concentrations will obviously degrade the fatigue performance of advanced composite materials.

Jensen and Brzenchek (13) reported the results from an experimental investigation of the tension–tension fatigue behavior of graphite–bismaleimide cross-ply laminates with embedded optical fibers. In their research the optical fibers were embedded either parallel or perpendicular to the adjacent reinforcing fibers and parallel or perpendicular to the direction of the applied tensile load. Four different configurations, representing these extreme combinations of optical fiber orientations, were evaluated and compared to results from a control group without any embedded optical fibers. All of the specimens were fabricated using NARMCO G40-600/5245C graphite–bismaleimide (Gr/BMI) unidirectional preimpregnated tape with a 12-ply, $[0_3/90_2/0]_s$, orthotropic layup. The optical fibers employed in this research were Lightwave Technologies F-1506C acrylate-buffered single-mode 125-μm optical fibers with a 250-μm outer diameter.

None of the specimens failed when subjected to 1 million cycles of tension–tension fatigue at a minimum-to-maximum load ratio of 0.10 and a maximum applied stress equal to 50% of the static ultimate tensile strength of the control laminate. After completion of the cyclic loading (to 1 million cycles), all of the specimens were loaded in quasi-static tension to failure (5). The fatigued specimens with embedded optical fibers exhibited only a slightly reduced residual tensile strength, except for one configuration: the specimens with optical fibers oriented perpendicular to the direction of the applied load and embedded perpendicular to the adjacent reinforcing fibers displayed a 25% reduction in residual tensile strength. In any real design, this situation should obviously be avoided whenever possible.

5.6. SUMMARY

To create smart structures applications, fiber optic sensors can either be attached to the surface or embedded within the structural material. Attaching fibers to the surface is usually the only option available on existing structures. This generally poses no threat to the integrity of the host structure. However, fiber optic sensors can easily be embedded in advanced composite materials during the fabrication of structural components. Embedded sensors present many advantages over surface-attached sensors, including protection from the environment and closer monitoring of the state of the particular parameter of interest, especially if internal.

Embedded fiber optic sensors, unfortunately, generally introduce stress concentrations into the material which degrade the performance and/or durability of the structure. The extent of degradation is a function of the optical fiber properties, the coating parameters, the host material properties, the relative orientation of the reinforcing and sensing fibers, the relative strength and stiffness of the same, and the type of loading experienced by the structural component. The results in this chapter provide the insight necessary to avoid the most critical situations.

As a general rule of thumb, to minimize any adverse influence of an embedded fiber optic sensor on the structural integrity of the host material, the following four considerations should be taken into account: (1) the fiber optic sensors should be as small as possible; (2) the coating on embedded optical fibers should be soft in comparison to the surrounding matrix material; (3) where possible, the optical fibers should be embedded parallel to the adjacent reinforcing fibers; and (4) if controllable, the loading should be applied in the direction that is parallel to the embedded fiber optic sensors. Finally, as with any design, the structural integrity of the smart system should be verified both analytically and experimentally.

REFERENCES

1. B. D. Agarwal and L. J. Broutman, *Analysis and Performance of Fiber Composites*, 2nd ed. Wiley (Interscience), New York, 1990.

2. A. S. Bicos and J. J. Tracy, Structural Considerations for Sensor Selection and Placement. *Proc. SPIE — Int. Soc. Opt. Eng.* **1170**, 70–76 (1990).

3. B. Blagojevic, W. Tsaw, K. McEwen, and R. M. Measures, The Influence of Embedded Optical Fibers on the Interlamina Fracture Toughness of Composite Materials. *Rev. Prog. Quant. Nondestr. Eval.*, Brunswick, ME, **1989** (1989).

4. P. Bonniau, J. Chazelas, J. Lecuellet, F. Gendre, M. Turpin, J.-P. LePesant, and M. Brevignon, Damage Detection in Woven-Composite Materials Using Embedded Fiber Optic Sensors. *Proc. SPIE — Int. Soc. Opt. Eng.* **1588**, 52–63 (1992).

5. D. Brzenchek, Tensile Fatigue of a Composite Laminate with Embedded Optical Fibers. M.S. Thesis, The Pennsylvania State University, University Park (1992).

6. C. C. Chang, Low Velocity Impact of Laminated Graphite/Epoxy Panels with Embedded Optical Fibers. M.S. Thesis, University of Maryland, College Park. (1991).

7. R. M. Crane, A. B. Mecander, and J. Gagoik, Fiber Optics for a Damage Assessment System for Fiber Reinforced Plastic Composite Structures. *Proc. Quant. Nondestr. Eval.* **2B**, 1419–1430 (1982).

8. A. Dasgupta, Y. Wan, J. S. Sirkis, and H. Singh, Micro-mechanical Investigation of an Optical Fiber Embedded in a Laminated Composite. *Proc. SPIE — Int. Soc. Opt. Eng.* **1370**, 119–128 (1991).

9. R. Davidson and S. S. J. Roberts, Do Embedded Sensor Systems Degrade Mechanical Performance of Host Composites? In *Active Materials and Adaptive Structures* (G. J. Knowles, ed.), pp. 109–114. Institute of Physics, London, 1991.

10. Du Pont, Characteristics and Uses of Kevlar 49 Aramid High Modulus Organic Fiber. *Du Pont Fibers Tech. Bull.* **K-5**, 6–7 (1981).

11. N. D. W. Glossop, *An Embedded Fiber Optic Sensor for Impact Damage Detection in Composite Materials*, Rep. No. 332. University of Toronto, Institute for Aerospace Studies, 1989.

12. B. Hofer, Fiber Optic Damage Detection in Composite Structures. *Composites* **18**(4), 309–316 (1987).

13. D. W. Jensen and D. Brzenchek, Fatigue of a Composite Laminate with Embedded Optical Fibers. *Proc. Int. Congr. Exp. Mech. Soc. Exp. Mech. 7th, 1992*, Vol. 2, pp. 1319–1325 (1992).

14. D. W. Jensen and R. W. Griffiths, Optical Fiber Sensing Considerations for a Smart Aerospace Structure. *Proc. SPIE — Int. Soc. Opt. Eng.* **986**, 70–76 (1989).

15. D. W. Jensen and M. J. Koharchik, Calibration of Composite-Embedded Fiber-optic Strain Sensors. *Proc. 1990 Soc. Exp. Mech. Spring Conf. Exper. Mech.* pp. 234–240 (1990).

16. D. W. Jensen and J. Pascual, Degradation of Graphite/Bismaleimide Laminates with Multiple Embedded Fiber-Optic Sensors. *Proc. SPIE — Int. Soc. Opt. Eng.* **1370**, 228–237 (1991).

17. D. W. Jensen, J. Pascual, and J. A. August, Performance of Graphite/Bismaleimide Laminates with Embedded Optical Fibers. Part I. Uniaxial Tension, *J. Smart Mater. Struct.* **1**(1) 24–30 (1992).

18. D. W. Jensen, J. Pascual, and J. A. August, Performance of Graphite/Bismaleimide Laminates with Embedded Optical Fibers. Part II. Uniaxial Compression. *J. Smart Mater. Struct.* **1**(1), 31–35 (1992).

19. C. J. Mazur, G. P. Sendeckyj, and D. M. Stevens, Air Force Smart Structures/Skins Program Overview. *Proc. SPIE — Int. Soc. Opt. Eng.* **986**, 19–29 (1989).

20. R. M. Measures, Fiber Optics Smart Structures Program at UTIAS. *Proc. SPIE — Int. Soc. Opt. Eng.* **1170**, 92–108 (1990).

21. R. M. Measures, Progress Towards Fiber Optic Smart Structures at UTIAS. *Proc. SPIE — Int. Soc. Opt. Eng.* **1370**, 46–68 (1991).

22. R. M. Measures, N. D. W. Glossop, J. Lymer, M. Leblanc, J. West, S. Dubois, W. Tsaw, and R. C. Tennyson, Structurally Integrated Fiber Optic Damage Assessment System for Composite Materials. *Proc. SPIE — Int. Soc. Opt. Eng.* **986**, 120–129 (1989).

23. R. M. Measures, N. D. W. Glossop, J. Lymer, M. LeBlanc, J. West, S. Dubois, W. Tsaw, and R. C. Tennyson, Structurally Integrated Fiber Optic Damage Assessment System for Composite Materials. *Appl. Opt.* **28**, 2626–2633 (1989).

24. Y. E. Pak, V. DyReyes, and E. S. Schmuter, Micromechanics of Fiber Optic Sensors. In *Active Materials and Adaptive Structures* (G. J. Knowles, ed.), pp. 121–128. Institute of Physics, London, 1991.

25. J. Pascual, Uniaxial Mechanical Performance of Smart Composite Structures with Embedded Optical Fibers. M.S. Thesis, The Pennsylvania State University, University Park (1991).

26. S. S. J. Roberts and R. Davidson, Mechanical Properties of Composite Materials Containing Embedded Fibre Optic Sensors. *Proc. SPIE — Int. Soc. Opt. Eng.* **1588**, 326–341 (1992).

27. B. W. Rosen, *Engineered Materials Handbook*, Vol. 1, p. 197. ASM International, Metals Park, OH, 1988.

28. P. Sansonetti, M. Lequime, D. Engrand, P. Ferdinand, J. Plantey, D. H Bowen, R. Davidson, S. S. J. Roberts, M. Crowther, M. Pleydell, B. Culshaw, C. Michie, M. Martinelli, P. Escobar, and B. Fornari, Intelligent Composites Containing Measuring Fibre Optic Networks for Continuous Self Diagnosis. *Proc. SPIE — Int. Soc. Opt. Eng.* **1170**, 211–223 (1990).

29. G. P. Sendeckyj and C. A. Paul, Some Smart Structures Concepts. *Proc. SPIE — Int. Soc. Opt. Eng.* **1170**, 2–10 (1990).

30. H. Singh, J. S. Sirkis, and A. Dasgupta, Micro-Interaction of Optical Fibers Embedded in Laminated Composites. *Proc. SPIE — Int. Soc. Opt. Eng.* **1588**, 76–85 (1992).

31. J. S. Sirkis and A. Dasgupta, Optimal Coatings for Intelligent Structure Fiber Optical Sensors. *Proc. SPIE — Int. Soc. Opt. Eng.* **1370**, 129–140 (1991).

32. J. S. Sirkis, A. Dasgupta, H. W. Haslach, Jr., I. Chopra, H. Singh, Y. Wan, C. Mathews, and K. Whipple, Smart Structures Mechanics Research at the University of Maryland. *Proc. SPIE — Int. Soc. Opt. Eng.* **1370**, 88–102 (1991).

33. A. B. Strong, *Fundamentals of Composites Manufacturing: Materials, Methods, and Applications.* Society of Manufacturing Engineers, Dearborn, MI, 1989.

34. A. Tay, D. A. Wilson, and L. Wood, Strain Analysis of Optical Fibers Embedded in Composite Materials Using Finite Element Modeling. *Proc. SPIE — Int. Soc. Opt. Eng.* **1170**, 521–533 (1990).

35. A. Tay, D. A. Wilson, A. C. Demirdogen, J. R. Houghton, and R. L. Wood, Microdamage and Optical Signal Analysis of Impact Induced Fracture in Smart Structures. *Proc. SPIE—Int. Soc. Opt. Eng.* **1370**, 328–343 (1990).

36. A. M. Vengsarkar, K. A. Murphy, M. F. Gunther, A. J. Plante, and R. O. Claus, Low Profile Fibers for Embedded Smart Structure Applications. *Proc. SPIE — Int. Soc. Opt. Eng.* **1588**, 2–13 (1992).

6

Methods of Fiber Optic Ingress/Egress for Smart Structures

WILLIAM B. SPILLMAN, JR., AND JEFFERY R. LORD
Intelligent Structural Systems Research Institute
College of Engineering
University of Vermont
Burlington, Vermont

6.1. INTRODUCTION

Development of systems for real-time health monitoring of composite structures is a continuing goal with application in both the fabrication and in-service periods. Much of the sensor research is aimed at developing sensing schemes that would ultimately behave as the structures' internal nervous system. This so-called nervous system would provide both point sensing and integrated structural sensing to assess internal and external effects on the structure. Actuators are also being developed to allow these structures to adapt to certain external stimuli. Incorporating both sensors and actuators into a composite structure would create the basis of a truly smart structure since it would be able to sense changes in its environment and then adapt to the environment based on these changes.

Fiber optics has been found to be a promising technology for use in the development of smart structures, for several reasons. Small size allows optical fibers to be integrated with various composite systems without compromising the structural integrity of the part. Integration can be either internal or external to the part. Internal integration is accomplished during the actual part fabrication, while external integration (i.e., surface mounting of the fiber) can be accomplished either during manufacture or as a separate operation after part fabrication. Optical fibers are insensitive to electromagnetic interference (EMI) and can be configured to sense a variety of physical effects, including temperature, pressure, electric field strength, magnetic field strength, stress, and strain (1). These sensors can function as point sensors or measure integrated

Fiber Optic Smart Structures, Edited by Eric Udd.
ISBN 0-471-55448-0 © 1995 John Wiley & Sons, Inc.

Figure 6.1. Composite test panel with integral optical fibers.

effects and can be multiplexed together in a variety of ways. Finally, optical fibers can be used strictly as data conduits with very high bandwidth. The combination of these attributes makes fiber optic technology an attractive candidate for integration into the smart structures now being developed.

With the large body of sensor activity seen in the literature, it is surprising that little attention has been paid to methods of practically accessing the enbedded optical fibers (2). These experiments have generally been carried out using only single-composite panels or beam elements and have not addressed the significant questions of how to efficiently access the embedded optical fibers at points of the fiber's ingress/egress from the composite or how to configure the individual structural elements to maintain the smart structure's data and sensing capabilities when elements are replaced from wear or damage. If large smart structures are to become a reality, these questions must be answered.

The laboratory demonstrations of composite sensor technology generally contain some variation of the fiber configuration shown in Fig. 6.1. Short lengths of fiber are embedded into the composite and later joined to longer fiber segments using any of the many splicing technologies available. Clearly, this configuration is not robust enough for field use. It is illustrative of the issues that are relevant for optical fiber interconnection. Whatever method is used to access the optical fiber must create an efficient optical coupling, provide strain relief for the optical fiber at the composite exterior interfaces, and allow for the replacement of worn or damaged sections.

This chapter will focus on the design issues relating to interconnection technology for embedded or attached optical fibers integrated with a composite smart structure. Whether the optical fibers are used for sensing, data communication or both, the development of these interfaces will be critical. We begin by discussing composite fabrication techniques and how these techniques drive the fiber optic connectorization issues. A discussion of the various methods of interconnection that have been demonstrated and the advantages and disadvantages of each follows. A detailed discussion of an actual fiber optic connector demonstrated with graphite–epoxy composites will then be presented. Finally, a discussion of possible improvements to advance the state of the art of methods of optical fiber ingress/egress for smart structures concludes the chapter.

6.2. COMPOSITE FABRICATION ISSUES

Composites are composed of a wide range of materials and are fabricated in a variety of ways. Each composite material system and fabrication technique poses unique challenges to both integrating optical fibers into the composite and in developing sites for optical fiber ingress/egress. Integration of optical fibers with the composite material is the first step toward creating a smart structure. The fact that optical fiber technology has the ability to carry large amounts of data, be multiplexed into system nets, and can also be utilized simultaneously for its sensing capability makes it a prime candidate for the nervous system that would be the core of any type of smart structure. In the smart structure, the optical fiber will transfer information between sensors and a central processor. This information will be interpreted and control data will be transmitted via these fibers to actuators that will adapt the smart structure in an appropriate manner to the environmental conditions sensed. For smart structures containing optical fibers to be practical, some general criteria must be met during the fabrication process of the composite material.

6.2.1. Optical Fiber/Composite Interaction

The optical fiber must not interact chemically with the composite system or be affected by the chemistry occurring during the curing process. This includes the bare optical fiber and protective coatings. Bare optical fibers are not generally suitable for direct embedding in composites. Removal of the protective coatings before embedment can cause serious fiber damage, as can the exposure to moisture in the atmosphere while the fiber is unprotected. The act of stripping the optical fiber requires sharp mechanical blades that can nick the glass, leaving a site for crack formation and propagation. Chemicals used to strip various coatings are quite aggressive and can potentially react and weaken the glass at the sites of microdefects that exist on all optical fibers. These defects are responsible for the lower tensile strengths seen in optical fibers over that of pristine glass. The tensile strength is generally in the range of 38.6 MPa (50,000 psi) for optical fibers. In addition, stripping the protective buffer coating provides opportunities for fiber-mishandling damage and exposes the glass to chemical attack from absorbed water. Water absorbed by glass is well known to weaken the glass rapidly at the sites of microdefects, eventually resulting in fiber failure (3). Handling long lengths of bare optical fiber without breakage is a near impossibility. Practical smart structures of useful sizes (i.e., wing sections or beams) will require that some sort of coating be left on the fiber to ensure its integrity during construction.

Figure 6.2. Optical fiber with acrylic buffer coating embedded within a composite structure.

6.2.1.1. Optical Fiber Coatings. Standard communication fibers with acrylic buffers are not optimum for protection of the optical fiber during embedding, particularly in the case of graphite–epoxy composites, as shown in Fig. 6.2. As can be seen, an "eye" of resin surrounds the buffered optical fiber and is roughly twice the thickness of a ply. The strength of the composite along the optical fiber could be reduced and would have to be compensated for through careful design. The cure temperature of the composite exceeds the maximum use temperature of an acrylic coating. Chemical changes to both the resin and the glass are possible, depending on the nature of the chemical changes occurring in the coating from the cure cycle. As can be seen, as a result of the cure, the acrylic coating on the fiber shown in Fig. 6.2 is no longer round. It has assumed an oval shape, indicating a permanent change in the coating due to the cure temperatures.

To minimize the size of the inclusion created by any embedded optical fiber, a thinner protective buffer on the fiber would clearly be desirable. Polyamide and various metals (gold, silver, aluminum) have been applied successfully to optical fibers in thicknesses of tens of microns, and these coatings are also more

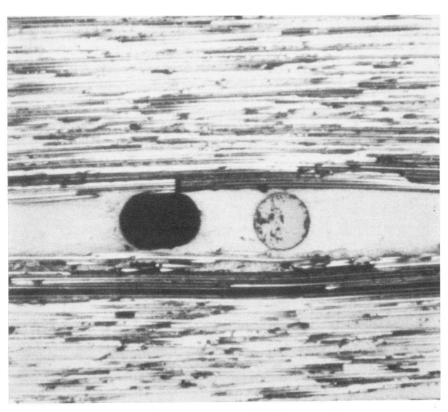

Figure 6.3. Stripped optical fiber embedded within a composite structure.

tolerant of the cure process than is an acrylic coating. These fibers have a number of potential advantages over acrylic-coated fibers. First, as seen in Fig. 6.3, they affect the composite structure less, since they create a smaller resin inclusion. The fiber shown here is stripped 125-μm fiber and appears similar in profile to metal- or polyamide-coated fibers. These would have diameters of approximately 150 μm compared to the 225-μm inclusion thickness shown in Fig. 6.2. The operating temperature range of these fibers is much higher than for acrylic-coated fibers. Temperatures in excess of 200°C are possible depending on the coating material selected. These coating materials can also exhibit better chemical resistance than is seen for the acrylic-coated fibers and can generally be applied more uniformly than acrylics. These coatings are available from specialty fiber manufacturers.

6.2.2.   Interconnecting Smart Parts

Composites for smart structures applications with integral optical fibers must be configured so that adjacent parts can be linked together, forming a larger structure. Ideally, two adjacent parts would be designed to locate the fibers embedded within them and form optical connections on assembly. This is shown schematically in Fig. 6.4. This situation is not practical, for many reasons. The tolerances required for successful optical connections are on the order of plus or minus a few microns for multimode fiber and $\pm 1\,\mu$m (ca. 0.00004 in.) for single-mode optical fiber. Although these tolerances are achievable in the highly precise ceramic parts used in optical connectors, it is unreasonable to think that a single-coated fiber could be placed inside a composite part even 10 cm on a side to this sort of tolerance. Another difficulty stems from the fact that many composite parts are trimmed to size after manufacture. Since optimum fiber optic interconnection requires smooth optical surfaces without defects, polishing of the optical fibers after the trimming in localized areas would be necessary to achieve the optical finish necessary. The difficulties of polishing the fiber after embedding can be seen in Fig. 6.3. The relatively soft resin inclusion allows chips of the polishing media

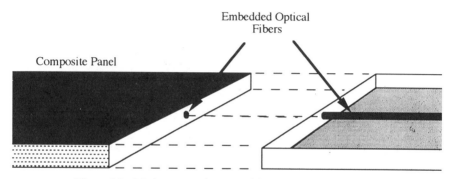

Figure 6.4.   Mating composite parts with embedded optical fibers.

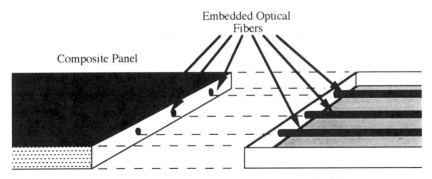

Figure 6.5. Mating composite parts with multiple optical fibers.

and glass to embed and subsequently release, thereby causing large scratches on the optical fiber surface. This effect can be seen dramatically in Fig. 6.2, where the very soft acrylic buffer allowed debris from polishing to chip away at the fiber edges. A final problem involves relative panel positioning as shown in Fig. 6.5. To achieve optimum optical coupling, the finished dimensions of the adjacent panels would need to be such that mating optical fibers would have to be located within a few microns of axial separation.

6.2.3. Composite/Optical Fiber Fabrication

To affect composite fabrication minimally, the incorporation of optical fibers must be compatible with conventional equipment and methods used in composite manufacturing. Optical fibers have been incorporated into the prepreg tape used in some composite fabrication. Alternatively, optical fiber spools can be used to feed the tape laying and graphite fiber winding equipment used commonly in composite manufacture. Utilizing this equipment, fibers can be incorporated at specific times and in specific places within the composite structure. Optical fibers can and have been incorporated successfully into composite structures during manufacture by a number of different organizations. The problems of accessing the fiber ends for connnectorization, however, have not been addressed so generally. Postcure composite part trimming and the difficulty of developing a new optical finish to fiber end faces after embedding define the choices necessary to instrument a composite with optical fibers. If the fiber is to exit the composite edge, no trimming can be done after fabrication of the part. If trimming is required, the optical fiber cannot exit the edge of the part. If the fiber does not exit the edge, it must rise through successive plies of the composite material to exit the part face (as shown in Fig. 6.6). Exiting the composite part in this manner adds manual labor due to the fact that each ply must be slit individually and the optical fiber fed through to the next layer until it finally exits the part as an optical fiber pigtail. In the cured composite, the point of exit for the optical fiber pigtail

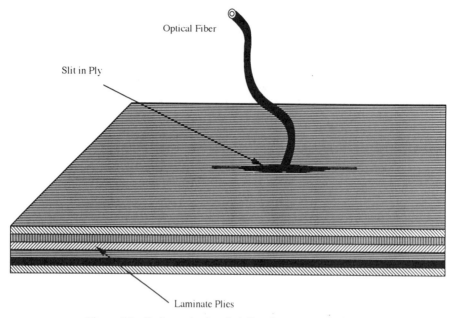

Optical Fiber

Slit in Ply

Laminate Plies

Figure 6.6.   Surface exit of optical fiber from composite layup.

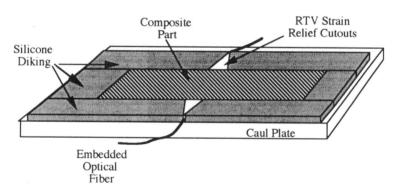

Composite
Part

RTV Strain
Relief Cutouts

Silicone
Diking

Caul Plate

Embedded
Optical
Fiber

Figure 6.7.   Composite panel with embedded optical fiber and RTV strain relief.

is a stress point for the fiber, and strain relief must be provided to prevent the optical fiber from bending over the sharp edge of the interface. Fig. 6.7 shows how this can be accomplished using room-temperature-vulcanized (RTV) silicone to surround the fiber exit. RTV silicone has the additional advantage of locating the fiber while the composite is cured.

6.2.3.1.   Optical Fiber Integration with Hand-Layup Composites. Whether the optical fiber exists the edge or the surface of the composite part, there are still significant manufacturing challenges to overcome. All composites use some type of form or mold to define the part shape during cure. Most composites contain a matrix material that if not liquid at the beginning of the fabrication process, passes through a liquid phase sometime during the cure cycle. The optical fiber must be incorporated into the mold design in some manner either to contain the fiber safely or to give safe passage for the fiber through the mold. The following procedure outlines how this can be carried in the fabrication of a typical laboratory test panel such as the one shown in Fig. 6.7.

1. The bottom caul plate is covered with mold release or Teflon sheeting.
2. The prepreg is cut and laid up to the layer at which the optical fiber will be incorporated.
3. The optical fiber is cleaned in the area to be embedded and placed on the composite.
4. Silicone diking material, similar to silicone self-adhesive weatherstripping, is placed around the part to maintain the part dimensions. Cutouts are made where the optical fiber exits the composite part (see Fig. 6.7).
5. The remainder of the composite part layup is completed.
6. RTV silicone is placed in the cutouts surrounding the optical fiber and allowed to cure.
7. A top sheet of Teflon is added to cover the composite part followed by the top caul plate.
8. The part is bagged, vacuum compacted, and cured.

From this procedure it can be seen that a number of measures are necessary to protect the optical fiber as it exits the composite part. Similar precautions are necessary when the fiber exits the face of the part, with the exception that the caul plate must allow passage of the fiber and also provide a means of securing the fiber in place while sealing the exit. Sealing is required due to the presence of the liquid phase and elevated pressure seen by the matrix material during the cure cycle. If the fiber is inadequately secured and its exit location poorly sealed, matrix material will be lost and the embedded position of the fiber may shift relative to the exit location, causing fiber breakage. Clearly, these issues of fiber positioning and mold sealing will be important no matter how we choose to exit the composite, whether through some sort of optical connection or by utilizing a cable pigtail.

6.2.3.2. Integration into Composites Without Plies. It it difficult to integrate optical fibers precisely into composites that do not have a ply structure. Examples of these composite are chopped fiber fiberglass, graphite–fiber epoxy, and concrete. Since these types of composites lack a ply structure, the optical fibers embedded within them prior to cure have the opportunity to migrate through the composite material during the cure cycle. Concrete provides the easiest method for avoidance of this problem since most concrete contains reinforcing rods and the optical fiber may be attached to the rod.

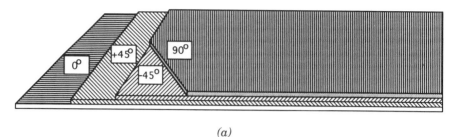

(a)

(b)

Figure 6.8. Quasi-isotropic (*a*) and pseudo-isotropic (*b*) fiber orientations within composite structures.

Attaching the fiber in this way, however, limits its sensitivities to structural regions in close proximity to the rods.

For fiber–matrix composites, a structure without plies is usually made by spraying the resin and the chopped fibers into a mold (4). The fibers of the composite can vary in length but are generally less than 5 cm (2 in.) in length. The orientation of the fibers is random and forms a "pseudo-isotope" configuration (Fig. 6.8). The optical fiber can be added at any point in the layup process but must be fixed after it exits the mold. The problem with optical fiber integration in this type of composite construction occurs when compaction is done. The short composite fiber lengths can allow shifting of the optical fiber, resulting in displacement of the optical fiber, which may or may not affect its performance. It is also possible that fiber shifts may produce so much strain in the fiber that it fails during cure. This would occur most often at the point where the optical fiber exits from the composite part and penetrates the mold. It is imperative for the survival of the optical fiber that once it is positioned within the composite, it remains, fixed in location, particularly at the part–mold interface.

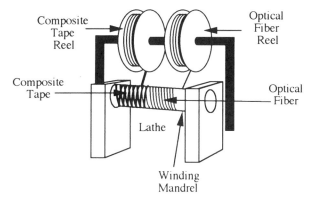

Figure 6.9.   Winding lathe for composite tape with an optical fiber winding.

6.2.3.3.   Filament-Wound Configurations. The issues of integration of optical fibers into filament-wound composite structures are similar to those that are important for hand-layup panel construction. The difference in this case is that the composite is fed from reels onto a mandrel, and at some point a spool of optical fibers is substituted for the composite tape reel (Fig. 6.9). In this type of construction it is difficult to access the optical fiber through the surface of the part, due primarily to the problems associated with controlling the optical fiber on a rotating mandrel. In addition, only the mandrel side of the composite is rigidly supported, requiring special fixtures to locate the surface exit of the optical fiber and to eliminate any possibility of optical fiber position shift during the curing process.

6.2.3.4.   Automated Composite Fabrication. In this type of composite construction, integration of the optical fiber into the structure requires that at the time at which the optical fiber is introduced, the automatic equipment used must be shut down while the fiber is added and fixed in place. This is true for both tape laying and filament winding machines. This requirement reduces the efficiency of the automated operations, but at present the technology for automatically integrating optical fibers and providing for their exit from a structure has not been demonstrated. Once the optical fiber is positioned and its point of egress from the part created, automated construction can resume.

6.3.   METHODS OF INGRESS/EGRESS

As shown above, optical fiber can be integrated and embedded into a composite part in a variety of configurations, as a number of workers in the field have demonstrated. Multiple fibers can be incorporated without much additional effort and may enter and exit the composite singly or at multiple fiber access points. The problem remains, however, that methods of optical fiber–composite part ingress/egress based on flexible strain relief coatings at the part surface will not generally survive field use. The coating provides the necessary protection for optical fibers to survive composite construction and cure under carefully controlled circumstances but will fail under operational conditions. For field use, full cabling complete with strength members is necessary to ensure the integrity of the optical fiber external to the composite part.

6.3.1. Connector Design Issues

The need for cabled fiber external to the composite for robustness in the field can be accomplished in a number of ways. Each provides a method of ingress/egress to the composite part and would include an optical connection at some point, at least to join together the various portions of the system creating the smart structure. No matter where in the system the optical connection is formed or by what technique, common conditions must be met to create a successful, low-loss optical connection.

6.3.1.1. Optical Fiber Parameters.

The heart of any optical fiber connection is the optical fiber itself. The positioning tolerances required for optical connections can be traced back to the tolerances arrived at from the optical fiber manufacture. Important aspects of optical fiber design for efficient light coupling are the core diameter, the cladding diameter, the core-to-cladding eccentricity, the overall fiber ellipticity, and the numerical aperture. The core and cladding diameters define the relative area of the fiber used for transmittance of light from one fiber to another. The fiber eccentricity provides a measure of how well the core is centered to the actual central axis of the fiber. The overall ellipticity measures the roundness of the outside diameter of the fiber. The numerical aperture is a measure of the half angle of the output light cone (Fig. 6.10) and defines how quickly the light beam exiting a fiber expands, or alternatively, the magnitude of the acceptance angle for

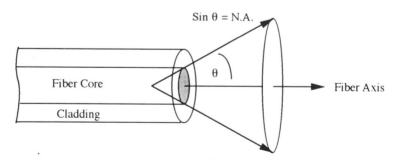

Figure 6.10.   Numerical aperture of an optical fiber.

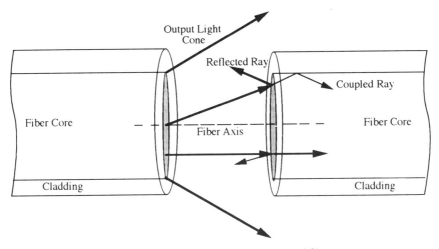

Figure 6.11. Coupling light between optical fibers.

coupling light into the fiber. The tolerancing of each of these parameters combines to define how well two fibers can match up and couple light from one to the other.

Communication-grade optical fibers are available in a wide range of dimensions and core–cladding diameters. Single-mode optical fibers typically have core diameters ranging from about 5 to 10 μm, with outside dimensions (OD) of 125 $\pm$ 1 μm. Multimode fiber can also be obtained with a 125-μm OD but will generally have core diameters of 50 or 62.5 μm. Another popular multimode fiber is 100-μm core, 140 μm OD. Multimode fiber is available in either graded- or step-index profile. Each of the fibers mentioned has slightly different requirements to achieve optimum coupling.

Coupling of light from one optical fiber to another is achieved by placing one optical fiber end face in close proximity to a second fiber end face, with fiber axes aligned accurately (Fig. 6.11). It can be seen from the figure that the fraction of light coupled between the two fibers will drop off very quickly as the fiber cores are misaligned. Allowing one of the fibers to tilt with respect to the other causes loss from both the effective core misalignment and from increased reflection from the second fiber end face. The amount of displacement or tilt can be accommodated depends on the amount of loss that can be tolerated in the system and is determined in large part by the core diameter and the numerical aperture of the fiber being used. Even with perfect alignment and end-face preparation (i.e., perpendicular to the fiber axis and a smooth optical finish) 4% of the optical radiation will be lost to Fresnel reflections at each fiber interface.

6.3.1.2. Multimode Fibers. The critical parameters for efficient coupling between two fibers are the core diameter cladding diameter, the ellipicity (out of roundness), core–cladding eccentricity, the fiber numerical aperture, and index

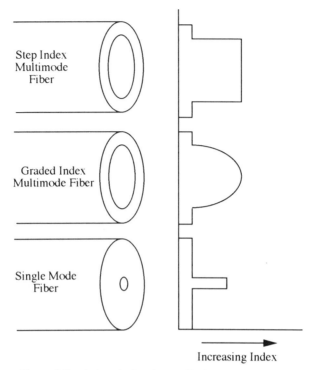

Figure 6.12.   Index of refraction profile for optical fibers.

profile. The accumulated tolerances allow for offsets between fibers of a few microns for acceptable connections but require that virtually no axial separation occur. The angular distribution profile of the optical power exiting a multimode fiber depends on the fiber's index profile (i.e., step index or graded index). The graded-index fiber generally has a narrower angular power distribution and a lower numerical aperture. Figure 6.12 illustrates the differences in the angular power distributions seen for step- and graded-index fibers. Typical fiber end displacements that can be tolerated are shown in Fig. 6.13.

6.3.1.3.   Single-Mode Fibers. Single-mode optical fibers have core diameters much smaller than those of multimode fibers (in the range of 5 to 10 μm) but have very similar outer diameters. The fiber parameters considered for the multimode-fiber case are still the critical ones for determining how efficiently light may be coupled from one fiber to another. Accurately aligning the small-diameter cores of single-mode fibers, however, requires a higher degree of precision than the alignment process for multimode fibers.

6.3.1.4.   End-Face Preparation. To couple light from one fiber to another, preparation of the end face is extremely important. Chips, scratches, the presence of particle contamination, and surface alignment to the fiber axis are

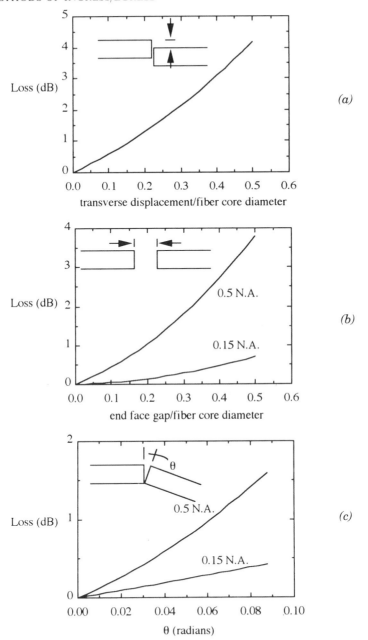

Figure 6.13.   Optical fiber coupling losses.

all important in determining the quality and efficiency of an optical connection. Chips can occur either during a cleaving process or as a result of mishandling the fiber. Small chips can be accommodated if they do not impinge on the fiber core area, but their presence provides a site for particle accumulation which

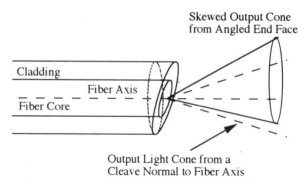

Figure 6.14. Angled fiber end face output cone of light.

later could create a problem on the fiber end face. A properly fabricated optical fiber termination should not have any chips on the end face. On the other hand, scratches can result from poor-quality polishing of the fiber or from the presence of grit between two optical fibers that are held in close proximity. Such scratches will not only result in slightly degraded coupling in the near term but will also serve as a site from which a crack could form later. A cracked fiber end face will significantly increase the fiber coupling loss.

6.3.1.5. Angled Fiber End Face and Fiber Tilt. The end-face alignment to the fiber axis is also critical to the fiber coupling efficiency. If the end face is formed by a cleaving process, the surface normal will not be parallel to the fiber axis if the fiber is twisted or the fiber is clamped along too small a radius during cleaving. The resultant angled end face refracts the optical energy, which effectively changes the numerical aperture of the fiber (Fig. 6.14). In addition, the fiber end face will be tilted (i.e., the two fiber axes will not be parallel) (Fig. 6.15), and this will reduce the coupling efficiency since additional light will be reflected from the angled end face and the tilted fiber collecting the light presents a smaller cross-sectional target to the incident light core.

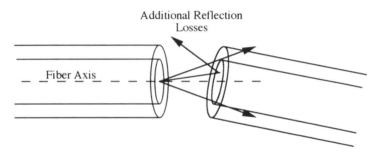

Figure 6.15. Optical coupling between tilted fibers.

6.3.2. Optical Connector Design Requirements

As we have seen, the process of coupling light from one fiber involves matching very precise alignment conditions. The simplest way to eliminate many of the alignment problems is to encase the optical fiber in a highly precise ceramic ferrule and then mate the opposing ferrules in an alignment sleeve. The fiber is then bonded in place, cleaved, and polished flat. This eliminates potential problems from poor cleaves or chipping of the end face that may have resulted from the cleaving process. Sizing outside diameters can be carried out with high precision since the center fiber bore of the ferrule can be located very accurately. Supporting the mating optical fibers within ceramic ferrules and use of the alignment sleeve during mating virtually eliminate any possibility of fiber tilt. In addition, the use of round ferrules and alignment sleeves allows the ferrule assemblies to be rotated to reduce losses due to slight fiber and ferrule eccentricities. A fiber optic connection is formed by the mating of two fiber end faces. This is true whether the connection is formed by connector pairs, as shown in Fig. 6.16, or in an optical splice (Fig. 6.17). In essence, the principal differences between optical connectors and optical splices lie in how quickly the mating is completed and whether or not the resultant optical interface is demateable. Both methods provide a means of aligning the two optical fibers

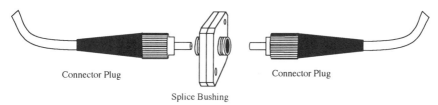

Connector Plug

Connector Plug

Splice Bushing

Figure 6.16. Mating optical connectors.

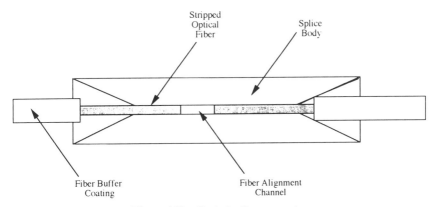

Stripped
Optical
Fiber

Splice
Body

Fiber Buffer
Coating

Fiber Alignment
Channel

Figure 6.17. Optical splice connection.

axially and a means of securing the fibers once aligned. To create an efficient connection, both methods must meet the same physical requirements.

As shown in Fig. 6.16, an optical connection is formed by three basic parts: two identical connectors, which house and protect the optical fiber, and a splice bushing (also called a connector adapter), which provides the means of aligning the two fiber end faces and the means of locking them in place once alignment has been accomplished.

6.3.3. Optical Cable Pigtail

One of the simplest ways to improve on the laboratory embedded fiber shown in Fig. 6.16 is to utilize the optical fiber cabling as an external pigtail, as shown in Fig. 6.18. With a suitable length of cable extending from the composite part, connectors of all types can be mounted with the same ease that field mounting of connectors is done by communications industry personnel. Adjacent composite parts of a smart structure can then be joined using standard components which are relatively inexpensive and are quite reliable. Provisions must be provided, however, for mounting of the splice bushings to some structural member so that they will remain immobile during their service life.

A cable prepared for embedding is shown schematically in Fig. 6.19. The center section of the external cabling along with the strength-member fibers, must be removed carefully, leaving only the buffer-coated fiber for embedding.

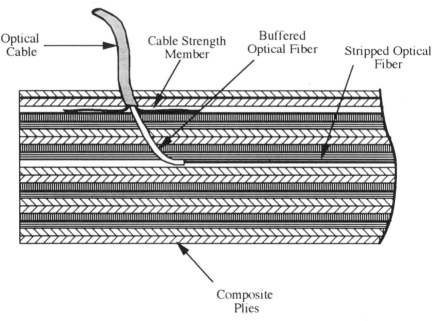

Figure 6.18.   Embedded optical cable.

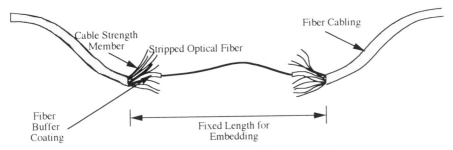

Figure 6.19. Center stripped cable for embedding.

Short lengths of the strength-member fibers are used to anchor the cable in the composite and provide strain relief, as seen in the inset of Fig. 6.18. As with the fiber buffer coating, the outer jacketing of the optical cable must be unaffected by the conditions of the composite cure. Standard communication-grade cables will not survive a typical cure cycle without damage. There are, however, high-temperature cables available that will survive the high temperature required. These cables use Teflon derivatives as an external jacket. If a strong bond to the matrix material is desired, the jacketing must be etched with a strong solvent which contains, among other substances, hydrofluoric acid. Hydrofluoric acid readily attacks glass, dissolving it rapidly. Therefore, it is extremely important to remove all traces of the solvent prior to integrating the cable into the composite part. Embedding the cable in the configuration shown requires that the fiber pass through successive plies until the cabled portion is reached. The strength members are then captured by placing them between plies several plies below the part surface. It is considerably easier

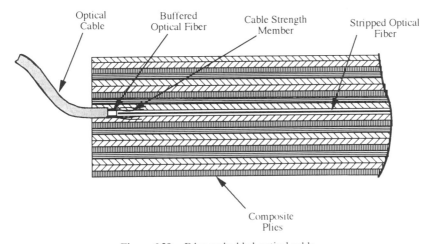

Figure 6.20. Edge embedded optical cable.

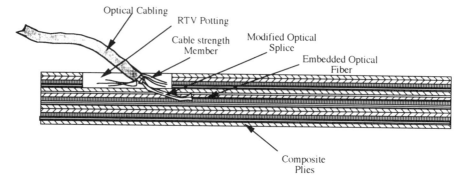

Figure 6.21.   Embedded optical splice connection.

to exit the composite part through its edge as shown in Fig. 6.20. Adjacent parts are likely to butt together, however, and a slot of some sort in the adjacent part would be required to accommodate the proper bending limits of the cabled fiber.

6.3.4.   Optical Splice Interconnection

A true demateable optical connection can be made by utilizing an optical splice of some sort. There are many types available, and most could be adapted for use with smart structures. We will consider only two of those available. The first makes use of a modified TRW Optasplice. As shown in Fig. 6.21, the embedded fiber is cleaved and bonded into the splice with epoxy. A minimum of epoxy is used so that none wicks through the microcapillary to foul the mating end of the fiber. A cutout in the composite is made to accommodate the splice as the composite is built up. The entire splice, except for the receptacle end, is then potted in RTV silicone.

The level of potting must rise slightly above the vertical dimensions of the part defined by the diking, which sets the separation of the panel caul plates. The caul plate then will press slightly on the potting, allowing the RTV silicone to seal against the caul plate so that resin is prevented from seeping into the splice during cure. As with other optical fiber/composite part integration, it is important that the optical fiber not shift a great deal during cure, or breakage at the splice will result.

Once cured, the splice allows for a demateable connection as long as a robust cabling configuration is not required. The mating optical fiber is stripped and cleaved prior to insertion into the splice. To make the connection more robust, a means of attaching the external cabling must be devised, requiring the use of additional potting material (RTV) to capture and bond the cable strain relief elements. The connection would then become permanent and nondemateable.

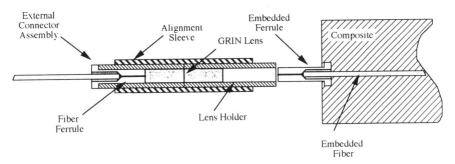

Figure 6.22. Embedded expanded beam demateable splice.

The second demateable splice type is somewhat of a hybrid between a splice and a true connector. It is composed of mated expanded beam connectors (Selfoc M-opal type). This type of connection is shown in Fig. 6.22. The optical fibers are terminated with the ferrules as shown prior to embedding. Fixturing of the ferrules is required during composite cure, as described previously, to prevent shifting of the ferrules during cure. The collimating lens is then bonded in place and an alignment sleeve is added to accommodate the mating external expanded beam assembly.

This type of connection was demonstrated several years ago (W. J. Rowe, Lockheed Georgia, private communication, 1987) and seen in an improved variation more recently (2). Some drawbacks exist in this type of connection, but they may be minimized through proper design. One problem involves the length of the ferrule available for embedding. Of the ferrule's total length (10.2 mm), 8 mm is required for insertion into the collimating lens assembly. This leaves roughly 2.2 mm of ferrule available for embedding. The diameter of the back portion of the ferrule is 2.1 mm. For many structural applications, an unacceptably large inclusion would be created by embedding a component of that size. This problem has been overcome through the use of a plastic sleeve to accommodate the embedded optical fiber (2). However, most plastics exhibit considerable thermal expansion, which can result in fiber pistoning over relatively low temperature ranges. Pistoning can cause the fiber to withdraw back into the ferrule or push forward into the grin lens, damaging either the fiber end face or the lens or both. In addition, plastics generally bond poorly with composite resin systems unless chemically etched prior to embedding. The etchant contains hydrofluoric acid along with some strong solvents, and residue from the etch could potentially cause damage to the optical fiber. One way to overcome these problems is to modify the ferrule. If the length dimension of the ferrule is increased by a factor of 2, approximately 10 mm of the ferrule would be available for embedding. If the outside diameter of the ferrule were machined to 1.5 mm (Fig. 6.23), this would provide the necessary ferrule area for embedding without changing the coupling of the type of demateable splice. Cutting threads (M 1.6, for example) into the collimating

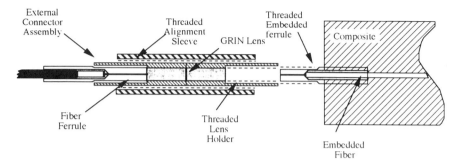

Figure 6.23. Modified expanded beam connection.

lens assembly, modified ferrule, mating ferrule, and alignment sleeve would also improve the design by allowing a more robust connection after assembly. Finally, drilling ports into the alignment sleeve and the collimating lens holder could allow adhesive to be introduced after assembly to stabilize the optical output seen through the connection.

6.4. CONNECTOR DESIGN CASE STUDY

To truly configure separate composite parts to form a smart structure, the method of ingress/egress must allow for the replacement of worn or damaged fiber optic interconnection elements. With optical connections that are basically "hardwired" by embedding the cable itself, this is impractical at best. When the situation allows the connectorization of the optical cable external to the composite part, a repairable interconnect exists in some sense. The exposed cable, however, is still at risk, and should it be damaged, the entire part would also have to be replaced. To

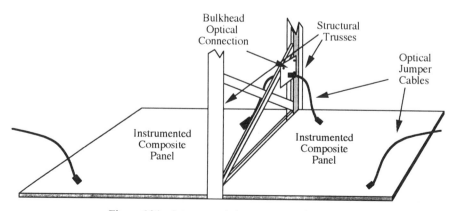

Figure 6.24. Interconnected smart composite parts.

gain the most versatility, jumper cables should be used to interconnect composite parts with embedded fibers and connectors as shown in Fig. 6.24 and to form the connector interface directly at the composite part. The discussion that follows details the design and testing of a connector capable of being embedded in composites. Although this connector can be used for either single-mode or multimode optical fiber, testing was carried out only for single-mode fiber since its use represented the more demanding requirements for use of the connection.

The optical fiber interconnection was fabricated and tested for use with graphite–epoxy composite material. It could be used with other types of composites, however, with only slight design modifications. The connector parts were adapted from existing technology and the published specifications of the parent connectors were used to evaluate the comparative performance of the embedded connection (5).

6.4.1. Design Goals

The principal design goal for the composite interconnection was to match the state of the art in terms of optical connection losses and repeatability of off-the-shelf fiber optic connectors. A secondary goal was to design an interconnection that could be utilized with only minor modification for both multimode and single-mode fibers. Initial testing was therefore carried out for connectors containing single-mode fiber since the tolerance requirements for single-mode connections are more stringent than for multimode connectors. The connector was designed for 125-μm-OD optical fiber so that various single-mode and multimode optical fibers could be used with the connection. Minimization of the number of component parts required without sacrificing performance was also an important design consideration. In addition, the design focused on reducing the profile of the external mating connector so that minimum clearance around the connection would be required. This would also minimize the chances of accidental breakage of the cable at the connection. Finally, the connection was designed to intrude as little as possible into the composite and therefore minimize the effect of the connection on the overall strength of the composite part.

6.4.2. Technology Survey

An optical connection is formed by mating two connector plugs (generally identical) within a splice bushing as shown in Fig. 6.16. The splice bushing contains a precision channel to align the opposing ferrule and therefore the optical fibers that they contain. The best connectors use ceramic sleeves with slits as the alignment device. These expand slightly when the ferrules are inserted and hold them tightly and very exactly in line with each other. The exact alignment also minimizes any tilt between fiber end faces and with proper end face finish provides a minimum of end face separation. Single-mode optical

Table 6.1   Comparison of Connectors for Development into an Embedded Connection

Connector Type	Use Multimode	Use Single Mode	Loss Spec. (dB)	Ease of Modification
SMA	Yes	No	<1.5	Single unit
Steel	Yes	No	<1.0	not easily
Ceramic				disassembled
NTT FC	Yes	Yes	<0.5	Completely
				disassembles
Biconic	Yes	Yes	<0.5	Ferrule too large
				for embedding
ST	Yes	Yes	<0.3	Single unit
				not easily
				disassembled
D-4	Yes	Yes	<1.0	Ferrule not
				amenable to
				embedding

fiber connections have such high tolerance requirements that these connections generally must be tuned to achieve maximum optical throughput. Tuning is accomplished by rotating the ferrules relative to one another to more exactly align the fiber end faces while monitoring the optical power beyond the point of connection. Once tuning is accomplished, locking rings are sometimes provided so that demating and remating the connection can be done repeatably.

For the design to gain the most from previous work, a survey of existing connector types was carried out and several candidates identified as potential baselines from which a smart structure interconnections could be developed. These connectors were evaluated based on their performance characteristics, the ease of modification for creating an embeddable connector, and the number of parts composing the connection. The goal was to make use of existing design successes to minimize development time and take advantage of the work

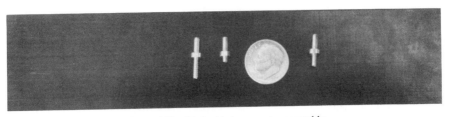

Figure 6.25.   Embedded connector assembly.

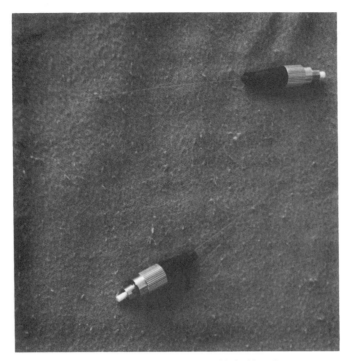

Figure 6.26. Mating connector with locking ring.

already accomplished in the field. The results are shown in Table 6.1. All of the connector types were available with highly precise ceramic ferrules that would meet the tolerance requirements for single-mode fiber. Both ST and SMA connectors were eliminated as candidates since they are only available completely assembled and sized to accept cabling. They are not easily disassembled to the component part level and the large-diameter back shells required for cable acceptance would create a large inclusion in the composite upon embedding. Biconic connectors were eliminated due to large front ferrule size and the overall size of the connection. The types of connector finally selected have the industry designation FC, and offered ease of disassembly, a locking ring to guarantee final alignment, and a superior ferrule design that would minimize any impact the part would have on structural integrity after embedding. The original FC connector ferrule back shell design provides a small-diameter sleeve to accept bare fiber (Fig. 6.25). The ferrule back shell is also small enough to embed directly into a composite, leaving only the front portion of the ferrule and the collar exposed. Although early versions of the connectors used the ferrules as received, problems occurred due to the fact that the length

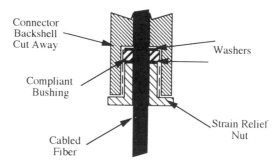

Connector
Backshell
Cut Away

Washers

Compliant
Bushing

Cabled
Fiber

Strain Relief
Nut

Figure 6.27.   Mating connector plug strain-relief system.

of the ferrule back shell was insufficient to form a strong bond with the composite resin matrix. The final form of the ferrule developed for the embedded connector featured an extended back shell that was grooved to improve the ferrule resin bond. In addition, the front of the ferrule was shortened to reduce the overall profile of the connection.

Once the ferrule was embedded successfully, a ceramic alignment sleeve was placed over the exposed ferrule and the housing was bonded in place to the

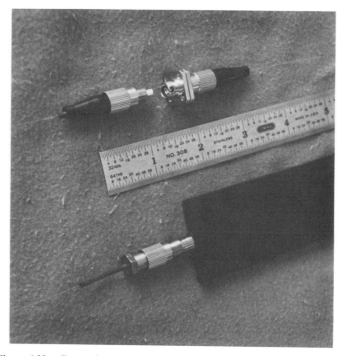

Figure 6.28.   Comparison of standard FC connection to embedded connection.

collar of the ferrule as shown in Fig. 6.25. The interconnection housing incorporated the important features of the splice bushing, used guaranteed repeatable alignment via the threaded end for securing the mating connector, and was compatible with the standard FC connector locking ring (Fig. 6.26). Early versions used the housing bore for alignment and a metal slit alignment sleeve. It was found, however, that these alignment devices wore out quickly from the abrasion resulting from multiple matings.

To meet the goal of minimizing the overall profile of the connector, the ferrule front end was redesigned (Fig. 6.25), allowing the housing and alignment sleeve both to be shortened. The mating connector was also redesigned to aid the profile minimization effort. Much of the excess length for the existing mating connector can be attributed to the strain relief boot, so that it was necessary to design a new strain relief system. The new stain concept utilized a method similar to that used on BNC-type coaxial electrical connections, where a compliant bushing is sandwiched between two metallic washers. The bushing is compressed by tightening the back shell nut, causing the bushing to expand and capture the cabling (Fig. 6.27).

In addition to the strain-relief redesign, many other parts of the mating connector were redesigned to shorten the connector's overall length profile. Key features of the final enbedded connector design are (1) decrease in the length caused by component redesign, (2) incorporation of an alternative strain relief method while still preserving the spring-loaded ferrule arrangement, and (3) the locking alignment ring. A comparison of the mated connections is provided in Fig. 6.28.

6.4.3. Interconnection Testing

Four sets of connector pairs were assembled using single-mode optical fiber optimized for use in the near infrared. The connector pairs were tested in the experimental configuration shown in Fig. 6.29. Light from a current-stabilized 780-nm diode laser was coupled into the single-mode optical fiber. The output

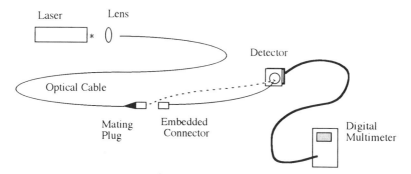

Figure 6.29. Embedded connection loss experiment.

Table 6.2   Interconnection Loss Measurement Data

Connection	Mating Plug	Embedded Receptacle	Loss (dB)
1	2	2	−0.24
2	1	1	−0.21
3	4	4	−0.30
4	3	3	−0.36
5	3	4	−0.89
6	3	2	−0.30
7	2	3	−0.43
8	2	4	−0.40
9	2	1	−5.2
10	4	1	−7.8
11	4	2	−0.56
12	4	3	−0.47
13	1	4	−2.2
14	1	2	−5.3
15	1	3	Omitted
16	3	1	Omitted

Table 6.3   Comparison of Loss Specifications for NTT Connectors

Vendor	Loss Specification (dB)
OFTI	<0.5
Seiko	<1.0
Kyocera	<0.7

from the connector ferrule was measured at a photodetector and displayed on a digital voltmeter. The connection was then assembled, tuned while monitoring the optical power, and the locking ring was then put in place. The optical fiber end facing the photodetector was then cleaved again to assure a clean endface to radiate the output optical power. No coupling gel was used between the mated ferrules, although the end faces were cleaned with alcohol prior to assembly. Intermixing the four pairs of connectors resulted in 16 unique connections that were tested for optimum coupling efficiency. These coupling results are shown in Table 6.2. Figure 6.30 is a histogram of the connection losses seen through 25 matings and demonstrates the coupling variability. From this table it can readily be seen that typical coupling losses are in the range −0.3 to −0.5 dB. These losses are comparable to those specified for NTT FC style connectors from several companies (Table 6.3).

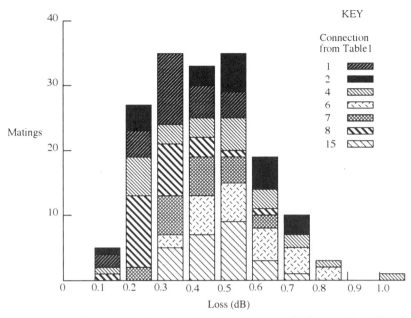

Figure 6.30. Plot of loss as a function of matings for various embedded connections, 25 matings each connection.

6.4.4. Embedded Interconnections

Connectors were embedded in graphite–epoxy composites in a number of configurations utilizing both single-mode and multimode optical fibers. Figure 6.31 shows a test panel with 50 to 125-μm multimode optical fiber embedded with the

Figure 6.31. Connectorized, fiber instrumented composite test panel.

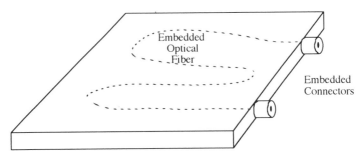

Composite Panel

Figure 6.32.   Composite panel with embedded connectors (ingress and egress on the same panel edge).

connector housings attached. Optical connection to the panel is made with laboratory NTT FC connectors. The fiber is illuminated with an argon laser. Fibers were embedded along the test panel length as shown, across the panel width, entering and exiting the same panel edge (Fig. 6.32), in tape-wound cylinders, and into panels as thin as 3.18 mm (0.125 in.) (Fig. 6.33). These panels were all made by hand-layup techniques with the exception of the tape-wound cylinder. The cylinder was done on a slow-turning lathe using 12.7-mm (0.5-in.) composite prepreg tape fed by hand.

The method for embedding these fibers and connectors required accurate knowledge of the length of fiber between the exposed portion of the ferrules

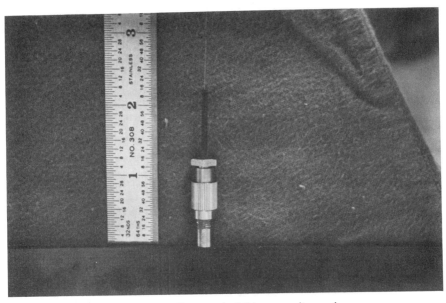

Figure 6.33.   Connectorized thin composite panel.

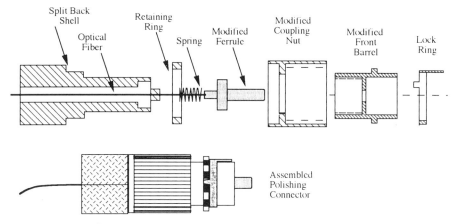

Figure 6.34. Ferrule polishing connector for modified ferrules.

(beginning at the ferrule collar). The ferrules were attached to the fiber and polished prior to construction of the test panel. For normal connector assembly only the coupling nut and the front barrel fit over the ferrule, and the remainder of the connector is made to slide into position over the cabling prior to bonding the ferrule in place. When the connector is assembled, the ferrule is captured between the front barrel and the connector back shell. The back shell parts could not fit over the ferrule at the opposite end of the fiber, so the split polishing connector shown in Fig. 6.34 was constructed. Once the fiber and bonded ferrules were assembled, polished, and inspected, the portions of the ferrules that would remain exposed after cure of the composite were wrapped in Kapton tape. The tape was used to protect the surfaces from resin leakage during the cure cycle. The fibers were placed between the middle plies of quasi-isotropic composite layups. Tiny cutouts in the composite material were made to accommodate the ferrule grooved back shells. The composite part was surrounded by silicone diking material and cutouts in the diking material were made to accommodate the exposed ferrule portions. The diking cutouts were carefully sized so that the ferrules would be held tightly during the cure and would not be allowed to shift and create fiber breakage at the rear fiber exit from the back shell. After complete assembly of the coupon and diking, a layer of Teflon sheeting was applied over the composite prepreg material and the top caul plate situated. The entire assembly was then vacuum bagged and compacted prior to placement in the autoclave for the composite cure cycle. After cure was complete and the support materials (caul plates, diking, etc.) were removed, the alignment sleeve was installed and the connector housing was bonded to the collar of the embedded ferrule.

Using methods similar to those described above, connectors were integrated successfully into test panels in several configurations, including transverse to the coupon length, entering and exiting on the same edge of the test panel, and on a tape-wound cylinder. The major limitation to incorporation of these

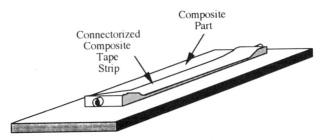

Figure 6.35. Surface-mounted connectorized composite strip.

connectors into a test edge was the panel thickness. This limitation is driven by the diameter of the ferrule retaining collar [0.450 cm (0.177 in.)]. To support the exposed portions of the ferrule adequately during cure, the panel thickness after compaction must be greater than 0.450 cm. Thinner panels could be made by machining flats onto the retaining collar. The limitation would then be the diameter of the ferrule back shell and the point where the supporting composite is significantly weakened by the embedded ferrule. From a practical point of view, making the panels much thinner than the retaining collar causes the assembled connector to be much larger than the panel thickness and reduces the utility of the configuration.

6.4.5. Surface-Mounted Connectors

One method of incorporating connectorized optical fibers onto thin composite parts is to mount the fiber and connectors exterior to the panel after fabrication and cure. Mounting the optical fiber in this manner also allows the entire connectorized fiber region to be completely contained within the boundaries of the composite part. This in turn allows adjacent parts to be fitted tightly together, and connections to these adjacent parts are made using jumper cables.

This approach was tested and resulted in the configuration shown in Fig. 6.35. A composite panel was constructed and cured and the connectorized optical fiber was added. The optical fiber was captured under just a few plies of composite material. The ends fitted with ferrules were fabricated and diked to accommodate the connectors. After cure of the composite, the support materials were removed and the embedded connector was assembled as described above.

6.5. FUTURE DEVELOPMENTS

It is critical to the creation of practical smart structures utilizing optical fibers that methods of accessing embedded fibers be created in concert with the development of fiber embedding strategies and optical sensor development. If effective means of optical fiber ingress/egress are not created, all the embedding ability and embedded sensor activity may not find application in real structural parts. The work presented forms the basis for continued development, with several important pieces of the ingress/egress puzzle addressed. First, a working embedded connector for use with either multimode or single-mode optical fiber was demonstrated. Second, the connection was integrated successfully in several different configurations. The surface mounting of the connector and optical fiber after composite cure suggests the next development step: to fabricate the connector housing from composite material (chopped fiber injection molded) and then use the composite resin to bond the connector housing to the composite part. In this manner an embedded connection is produced which is lighter, thermally stable, and compatible with the composite material.

Finally, in addition to the direct coupling to embedded fiber systems described here, other approaches to solving the ingress/egress problem are being investigated at a number of locations (various private communications). However, these investigations are currently considered to be proprietary by the parties involved, although general public disclosure is likely in the near future.

REFERENCES

1. E. Udd, ed., *Fiber Optic Sensors: An Introduction for Engineers and Scientists.* Wiley (Interscience), New York, 1991.
2. R. E. Morgan, S. L. Ehlers and K. J. Jones, Composite Embedded Fiber Optic Data Links and Related, Material/Connector Issues. *Proc. SPIE–Int. Soc. Opt. Eng.* **1588**, p. 189 (1991).
3. T. Michalski and C. Bunker, The Fracturing of Glass. *Sci. Am.* **257**, p. 122.
4. M. Schwartz, *Composite Materials Handbook.* McGraw-Hill, New York, 1984.
5. K. Nawata, Multimode and Single-Mode Fiber Connectors Technology. *IEEE J. Quantum Electron.* **QE-16**, 619 (1980).

7

Fiber Optic Sensor Overview

ERIC UDD
Blue Road Research
Troutdale, Oregon

7.1. INTRODUCTION

Fiber optic communication links have revolutionized the telecommunication industry by providing low-cost, high-fidelity, and very high transmission rate capability. In a similar manner the emerging optoelectronic industry has brought us such products as compact disk players and laser printers. As a result of these developments a second revolution is taking shape as fiber optic sensors that take advantage of components developed in association with the telecommunication and optoelectronic industry begin to enter the market. These sensors offer a series of advantages, including small size and weight; immunity to electromagnetic interference, which also reduces the cost of shielding; environmental ruggedness; high multiplexing potential; and potentially low costs due to complementary developments in the telecommunication and optoelectronic industries. Applications for these fiber optic sensors include fiber optic rotation sensors, accelerometers, vibration sensors, smoke detectors, linear and angular position sensors, and strain, temperature, and electromagnetic field sensors.

The trends for fiber optic sensor technology are illustrated by Fig. 7.1. In the early 1980s few components were available and they were expensive. This situation resulted in fiber optic sensors being used in only a few niche markets where their advantages were overwhelming. For example, these sensors were used to make temperature measurements in radio-frequency environments. By 1990 the number of components had increased dramatically, the cost of many items had dropped by an order of magnitude or more, while the quality and performance increased dramatically. Important examples include (1) the cost of laser diodes dropping from about $3000 each to $3 each while lifetimes went from a few hours to tens of thousands of hours, and (2) the cost of 1 m of single-mode fiber dropping from $10 to $0.10, with lower attenuation, greater concentricity of the core and cladding, and improved jacketing material. In the

Fiber Optic Smart Structures, Edited by Eric Udd.
ISBN 0-471-55448-0 © 1995 John Wiley & Sons, Inc.

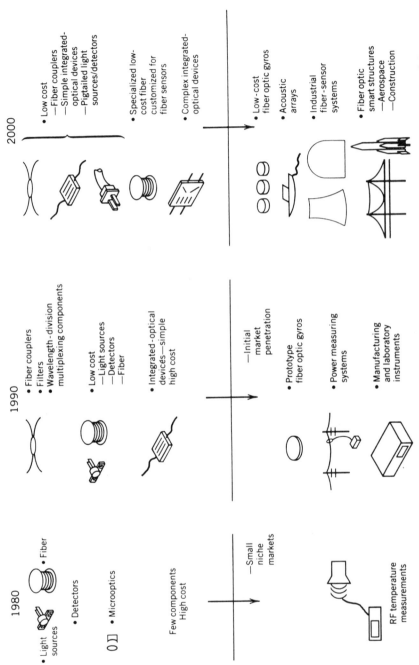

Figure 7.1. Fiber optic sensor technology trends. (From Eric Udd, ed., *Fiber Optic Sensors: An Introduction for Engineers and Scientists,* Wiley, New York, copyright 1991; used with permission.)

1980

• Light sources • Fiber
• Detectors
0 [] • Microoptics

Few components
High cost

—Small niche markets

RF temperature measurements

1990

• Fiber couplers
• Filters
• Wavelength-division multiplexing components

• Low cost
—Light sources
—Detectors
—Fiber

• Integrated-optical devices—simple high cost

—Initial market penetration

• Prototype fiber optic gyros

• Power measuring systems

• Manufacturing and laboratory instruments

2000

• Low cost
—Fiber couplers
—Simple integrated-optical devices
—Pigtailed light sources/detectors

• Specialized low-cost fiber customized for fiber sensors

• Complex integrated-optical devices

• Low-cost fiber optic gyros

• Acoustic arrays

• Industrial fiber-sensor systems

• Fiber optic smart structures
—Aerospace
—Construction

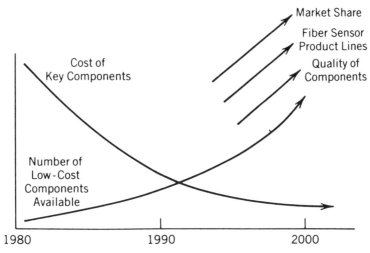

Figure 7.2. As the cost of key components drop and the number of available components increase, fiber optic sensor market opportunities rise. (From Eric Udd, ed., *Fiber Optic Sensors: An Introduction for Engineers and Scientists*, Wiley, New York, copyright 1991; used with permission.).

same time period, new components, such as fiber couplers, wavelength-division multiplexing elements, and integrated optical devices became commercially available. The net result was that many more fiber optic sensors became available, penetrating such markets as inertial rotation, power system monitoring, and manufacturing and process control. By the year 2000 it can be expected that many more components, integrated optical devices, pigtailed light sources, and fiber couplers will be available at low cost. The result will be the widespread proliferation of fiber optic sensors and their use in industrial control systems and the rapidly evolving area of fiber optic smart structures that includes health maintenance and diagnostic systems for aerospace vehicles and civil structures. Figure 7.2 graphically portrays the emergence of fiber optic sensor technology as the continually decreasing costs of an increasing number of high-quality components allows fiber optic sensor designers to produce competitive products to meet the needs of current and future markets.

This chapter provides a brief overview of fiber optic sensor technology [1–3]. Many of the fiber optic sensors that are of particular interest for fiber optic smart structures are covered in depth in subsequent chapters.

7.2. FIBER OPTIC SENSOR TECHNOLOGY

Fiber optic sensors are often categorized as being either extrinsic or intrinsic. Extrinsic or hybrid fiber optic sensors have an optical fiber carry a light beam to and from a "black box" that in response to an environment effect modulates

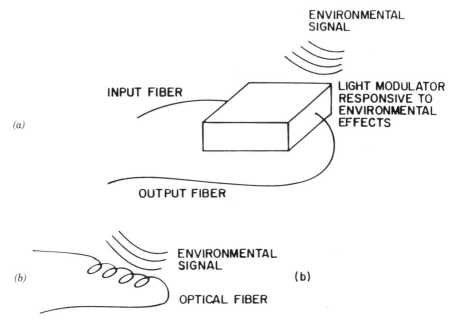

Figure 7.3. (*a*) Extrinsic or hybrid fiber sensor; (*b*) Intrinsic or all-fiber sensor. (From Eric Udd, ed., *Fiber Optic Sensors: An Introduction for Engineers and Scientists*, Wiley, New York, copyright 1991; used with permission.)

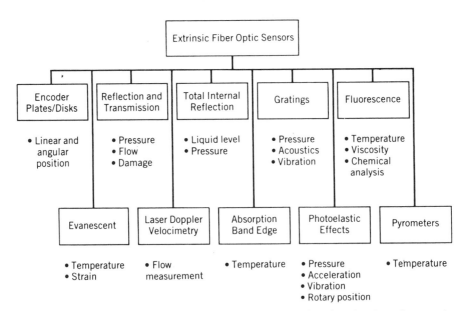

Figure 7.4. Extrinsic fiber sensor applications. (From Eric Udd, ed., *Fiber Optic Sensors: An Introduction for Engineers and Scientists*, Wiley, New York, copyright 1991; used with permission.)

the light beam. Intrinsic or all-fiber-optic sensors are sensors that measure the modulation of light by an environmental effect within the fiber. These types of sensors are shown in diagrammatic form in Fig. 7.3. A series of representative examples of extrinsic sensors and the type of environmental effects they are used to sense are listed in Fig. 7.4. These types of fiber optic sensors have found commercial use as linear and angular position sensors for such applications as fly-by-light, and in the area of temperature, pressure, liquid level, and flow measurements in process control. A corresponding chart of intrinsic fiber optic sensors is shown in Fig. 7.5. An important subclass of intrinsic fiber optic sensors are the interferometric sensors (Fig. 7.6), which often exhibit high sensitivity and are competing with conventional high-value sensors in such areas as acoustic and rotation sensing. The remainder of this chapter is devoted to a review of those classes of fiber optic sensors that are of particular interest to fiber optic smart structures. More complete discussions of fiber optic sensor technology in general can be found in Udd (1) and Dakin and Culshaw (2, 3).

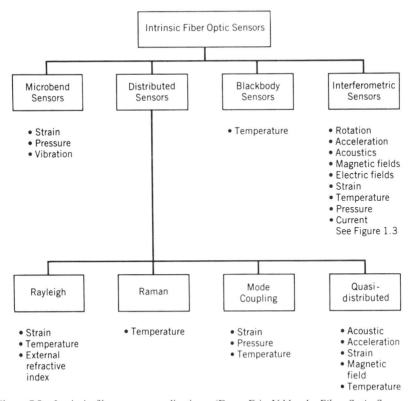

Figure 7.5. Intrinsic fiber sensor applications. (From Eric Udd, ed., *Fiber Optic Sensors: An Introduction for Engineers and Scientists*, Wiley, New York, copyright 1991; used with permission.)

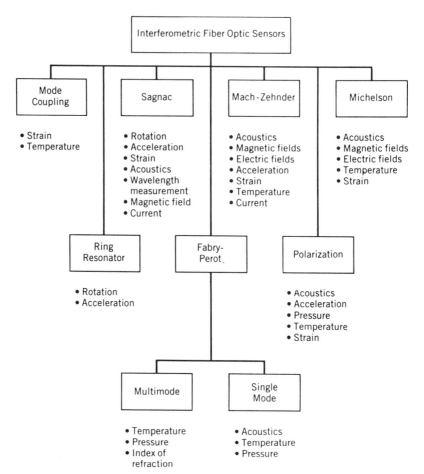

Figure 7.6. Interferometric fiber sensor applications. (From Eric Udd, ed., *Fiber Optic Sensors: An Introduction for Engineers and Scientists*, Wiley, New York, copyright 1991; used with permission.)

7.3. SENSORS FOR FIBER OPTIC SMART STRUCTURES

Fiber optic sensors have a number of advantages with respect to conventional electronic sensor technology when applied to smart structure applications. One of the most important is that these sensors are very light in weight and small enough that they can be embedded in materials in a nonobtrusive manner that does not degrade structural integrity. The space shuttle provides an example of how important this feature can be. The original space shuttle flight was made with a complete suite of electronic sensors that were used to measure many of the key performance and structural characteristics, of the first flight. The problem with the sensors and their support electronics in this case was that they weighed so much that the shuttle was not capable of flying a payload, so the sensor suites were removed so that payloads could be carried on subsequent flights. Ideally, what one wants to have is a sensing system that can support both payloads and structural monitoring on launch vehicles, and fiber optic sensor technology offers the prospect of doing just that. A second key feature of fiber sensor technology is that it is all passive. Since the fibers are dielectric glass, conductive paths are eliminated, reducing hazards due to internal electrical discharges and lightning. This is particularly important when organic composite materials are used that could literally be blown apart by a lightning strike. Strongly related to the first two features is the electromagnetic resistance of fiber sensors. This eliminates the need for the heavy electrical cable and shielding that is needed to support the very low level signal levels of electrical sensors, which can be a tremendous factor in reducing weight while mitigating electrical hazards. A third major area is the environmental rugged-ness of fiber optic sensors, which allows their performance at temperatures that would reduce many electrical sensors to a metal puddle while having intrinsic fiber strengths that compare favorably to steel. Finally, one of the most important advantages of fiber optic sensor technology is that they can be multiplexed (Chapter 15) to form many sensors along a single fiber line. Because of the intrinsic high bandwidth of optical fiber cable, this can be accomplished utilizing a single fiber without an increase in size or weight. This natural merger of fiber sensor technology with the advancing field of fiber optic communication allows the design of ever more sophisticated and capable fiber optic smart structure systems.

Some of the issues to be addressed when designing fiber optic sensors to support a fiber optic smart structure system are the parameters to be sensed (strain, temperature, viscosity, degree of cure), the gauge length over which the environmental effect must be measured, the number of fiber sensors to be multiplexed, the diameter of the fiber sensor, and the dynamic range and sensitivity of the sensor. In many cases the overall cost of the system and concerns about structural integrity are overriding issues. For the latter reasons most of the fiber sensors being investigated have low cost potential and in general are of a size that is no larger or minimally larger than the overall diameter of the optical fiber itself.

Figure 7.7 shows a microbend sensor (Chapter 12) which has been used to support several smart structure applications. In this case a light source is used to couple light into an optical fiber that becomes lossy when subject to microbends. By placing this fiber in an area that has environmentally induced microbending, the light propagating through the optical fiber is amplitude modulated. This can be done by placing the fiber in an appropriately designed transducer, or in the case of composite materials, the strength members of the composite material itself may act as microbending centers. The amplitude-modulated light then hits a detector and forms the output of the device. The microbend-sensitive sensor has the advantages of being simple and potentially very low in cost. It does, however, have a series of limitations that are

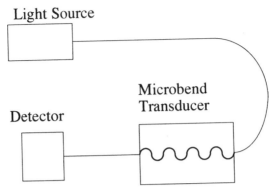

Figure 7.7. Fiber optic sensor based on microbending. (From Eric Udd, *Fiber Optic Sensor Workbook*, Blue Road Research, Troutdale, Oreg., copyright 1993 and 1994; used with permission.)

characteristic of intensity-based fiber optic sensors. These limitations are due primarily to the difficulty in distinguishing intensity variations caused by environmental effects from intensity variations due to other causes. Possible sources of variations that can result in errors include variable losses in connectors, spurious microbending loss in the cable, macrobending loss, and mechanical losses due to misalignment and creep of optical components (i.e., the light source and the fiber pigtail).

In some cases these errors may be minimized by careful design. Another approach that is commonly used is to turn to spectral techniques to reduce these error sources. For many intensity-based fiber optic sensors this involves using self-referencing techniques with dual operating wavelengths to calibrate out losses outside the sensing region. An alternative is to use fiber optic sensors that are inherently based on spectral response. Examples of these types of sensors are those based on blackbody radiation, fluorescence, and dispersive elements such as gratings and etalons.

Figure 7.8 illustrates the blackbody fiber optic sensor (Chapter 18), which is one of the simplest spectrally based sensors. In this case an optical fiber with

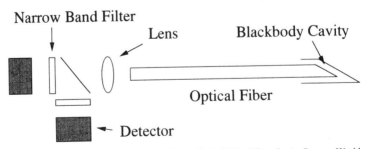

Figure 7.8. Blackbody fiber optic sensor. (From Eric Udd, *Fiber Optic Sensor Workbook*, Blue Road Research, Troutdale, Oreg., copyright 1993 and 1994; used with permission.)

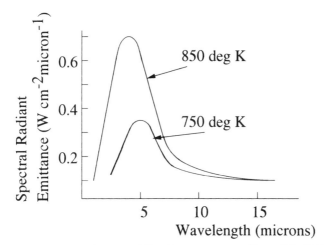

Figure 7.9. Blackbody spectral curves shift with temperature. (From Eric Udd, *Fiber Optic Sensor Workbook*, Blue Road Research, Troutdale, Oreg., copyright 1993 and 1994; used with permission.)

a blackbody cavity is placed in an area where an elevated temperature is to be measured. As the blackbody cavity heats up, infrared radiation results that is coupled through the optical fiber to a detection region that is used to measure the spectral content of the resultant light beam. In Fig. 7.8, two detectors with narrowband spectral filters have been used. By taking the ratio of the outputs of the two detectors, the blackbody spectral envelope can be inferred and the temperature determined as shown by the blackbody spectral curves of Fig. 7.9.

Using fluorescence (Chapter 13) or absorption of light that depends on the environment is another way to implement a fiber optic sensor. Figure 7.10

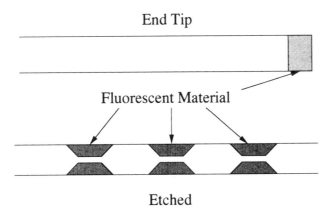

Figure 7.10. Fluorescence-based fiber sensor probes. (From Eric Udd, *Fiber Optic Sensor Workbook*, Blue Road Research, Troutdale, Oreg., copyright 1993 and 1994; used with permission.)

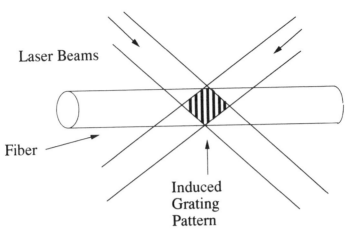

Figure 7.11. Fiber optic grating sensor. (From Eric Udd, *Fiber Optic Sensor Workbook*, Blue Road Research, copyright 1993 and 1994; used with permission.)

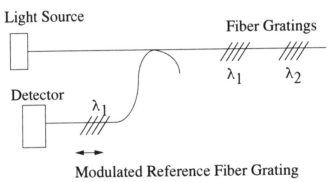

Figure 7.12. Fiber grating demodulator setup. (From Eric Udd, *Fiber Optic Sensor Workbook*, Blue Road Research, Troutdale, Oreg., copyright 1993 and 1994; used with permission.)

shows two configurations of fluorescent probes. In both cases pulses of light propagate down the optical fiber and are used to stimulate the fluorescing region. The resulting fluorescence is then measured to determine the environmental effect. In some cases this can be the time rate of decay of the fluorescence, while in others the spectral content of the fluorescing light is determined. The fluorescing material can in some cases be the resin used in organic composite materials. In many cases the stimulated material is prepared especially to enhance its fluorescing capabilities. These sensors have proven to be especially effective in measuring chemical properties of materials, including viscosity, water vapor content, and degree of cure in organic composites.

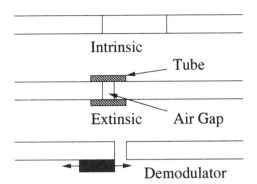

Figure 7.13. Fiber etalons. (From Eric Udd, *Fiber Optic Sensor Workbook*, Blue Road Research, Troutdale, Oreg., copyright 1993 and 1994; used with permission.)

One of the most critical parameters to be measured for structural integrity and health monitoring systems is local strain (Chapter 8). The fiber grating (Chapter 10) shown in Fig. 7.11 is a promising potentially very low cost means of making this type of measurement. The grating may be written onto the fiber by imaging two laser beams operating in the ultraviolet through the side of the optical fiber to form an interference pattern. The resultant dark and bright fringes may be used to write a grating directly onto the core region of the fiber. When the fiber is stretched or compressed under strain, the spectral content of the reflected and transmitted light varies as a function of the fiber grating period. One of the key issues is how effectively to demodulate the fiber grating remotely. A method for doing so is shown in Fig. 7.12. In this case a local grating is modulated with a piezoelectric stretching device and its filter action combined with the action of the remote grating used to generate a demodulation signal to accurately read out the fiber grating sensor position.

An alternative method is to use fiber etalons (Chapter 9), which are shown in Fig. 7.13. These sensors consist of two mirrors placed in line with the optical fiber. The mirrors may be formed by cleaving the optical fiber and placing a reflective end (usually a metallic or dielectric coating) on the fiber and then fusion splicing this end to a second cleaved fiber to form a mirror embedded in the fiber. By making two mirrors in this manner, a fiber etalon is formed. Another technique is simply to cleave the end of the optical fiber and use the air–glass interface as the reflective surface or again to coat the end. The etalon in this situation is formed by taking two such ends and placing them in a capillary tube. The former are often called intrinsic fiber etalons and the latter, extrinsic fiber etalons. By mounting one of the extrinsic etalons on a piezoelectric driver, a variable etalon may be formed that may be used to demodulate other fiber etalons or fiber gratings via its variable filter function. Figure 7.14 illustrates the optical transmission characteristics of the fiber etalon. Essentially, to be fully transmissive the wavelength of light impinging on the etalon must

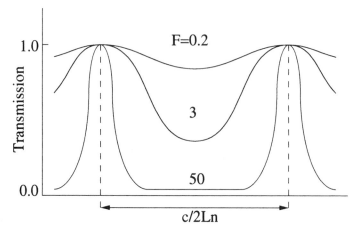

Figure 7.14. Transmission curves for fiber etalons with variable finesse *F*. (From Eric Udd, *Fiber Optic Sensor Workbook*, Blue Road Research, Troutdale, Oreg., copyright 1993 and 1994; used with permission.)

be such that an integral number of waves fit between the etalon mirrors, the resonance condition. The transmission falls off as the wavelength of the light beam moves away from this critical condition. The sharpness of the falloff is determined in the reflectivity of the mirrors, which in turn determines the cavity finesse: the higher the reflectivity, the higher the finesse.

With the exception of the microbend sensor, which may be used as a distributed sensor (discussed later below), the fiber sensors discussed above are useful primarily for measurement of short gauge length. In some cases, such as long bridge spans or measurement of total integrated strain in composite cylinders, it is desirable to have long-gauge-length sensors. Interferometric

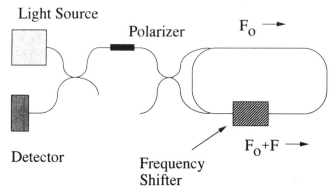

Figure 7.15. Sagnac interfereometer-based strain sensor. (From Eric Udd, *Fiber Optic Sensor Workbook*, Blue Road Research, Troutdale, Oreg., copyright 1993 and 1994; used with permission.)

sensing is one approach to the problem of long gauge lengths, and Fig. 7.15 illustrates a Sagnac strain sensor (4, 5) that could support gauge lengths up to multiple kilometers in length. In this case a light source, usually a light-emitting diode, is coupled to a fiber beamsplitter. The light beam passes through a polarizer and into a central beamsplitter, which generates counterpropagating light beams. A frequency shifter is placed in the loop so that while in the Sagnac loop the counterpropagating light beams differ in frequency by an amount F while they are at the same frequency when they recombine on the central beamsplitter. If the frequency of the shifter is adjusted so that the phase relationship of the recombined beams is held constant, a change in length dL results in a change dF, so that $dF/F = -dL/L$. For example, a frequency shifter with an offset of 100 MHz in a Sagnac strain sensor capable of resolving 1 Hz would have resolution of about 1 part in 10^8 of the overall length L.

Interferometric sensors may also be used to act as distributed sensors (Chapter 14) with extremely broad areas of coverage. The broad coverage is usually at the expense of high accuracy. Figure 7.16 shows a Sagnac distributed sensor that consists of two interleaved Sagnac interferometers operating at different wavelengths. As shown in the figure, the Sagnac loops are insensitive to time-varying environmental signals that impact the center of the loop since both counterpropagating light beams arrive simultaneously. As the signal moves along the loop toward the central beamsplitter, the signal level increases for the same input because of the time-delay increase in the arrival of the counterpropagating light beams at that location. The net result is that the response of each of the Sagnac loops varies as shown in the figure. By taking the sum of the two signals, the amplitude of the environmental effect may be measured, and by taking the ratio, the position may be determined.

More commonly distributed fiber sensors are formed using time-division multiplexing techniques based on Rayleigh, Raman, or Brillioun scattering.

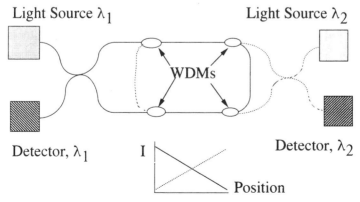

Figure 7.16. Sagnac distributed sensor with wavelength-division-multiplexed Sagnac loops. (From Eric Udd, *Fiber Optic Sensor Workbook*, Blue Road Research, Troutdale, Oreg., copyright 1993 and 1994; used with permission.)

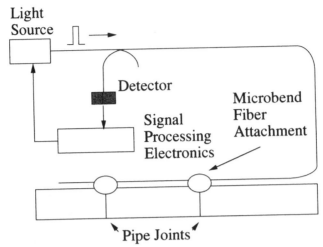

Figure 7.17. Distributed fiber sensor using microbend sensitive fiber. (From Eric Udd, *Fiber Optic Sensor Workbook*, Blue Road Research, Troutdale, Oreg., copyright 1993 and 1994; used with permission.)

Figure 7.17 illustrates a common embodiment. A light source is pulsed and propagates down an optical fiber, which in this case is optimized for microbend sensitivity. By monitoring the scattered return signal in time, the location of large losses induced by microbending can be determined and in this case be used to locate excess strain in the pipe.

7.4. SUMMARY

Fiber optic sensors have the potential to enable fiber optic smart structure systems that would be difficult or impossible to implement using conventional electronic technology. This is due to their small size and weight, electrical isolation, environmental ruggedness, and ability to be multiplexed. In this chapter we provided a brief overview of the types of fiber optic sensors used to implement smart structures. These sensors are covered in more depth in subsequent chapters.

REFERENCES

1. E. Udd, ed., *Fiber Optic Sensors: An Introduction for Engineers and Scientists.* Wiley (Interscience), New York, 1991.
2. J. Dakin and B. Culshaw, eds., *Optical Fiber Sensors: Principles and Components*, Vol. 1. Artech House, Boston, 1988.
3. B. Culshaw and J. Dakin, eds., *Optical Fiber Sensors: Systems and Applications*, Vol. 2. Artech House, Norwood, MA, 1989.

4. R. J. Michal, E. Udd, and J. P. Thériault, Derivative Fiber Optic Sensors Based on the Phase-Nulling Optical Gyro Development. *Proc. SPIE—Int. Soc. Opt. Eng.* **719**, 151 (1986).

5. E. Udd, R. G. Blom, D. M. Tralli, E. Saaski, and R. Dokka, Applications of the Sagnac Interferometer Based Strain Sensor to an Earth Movement Detection System. *Proc. SPIE—Int. Soc. Opt. Eng.* **2191** 126 (1994).

8

Fiber Optic Strain Sensing

RAYMOND M. MEASURES
Institute for Aerospace Studies
University of Toronto
Downsview, Ontario, Canada

8.1. INTRODUCTION AND NEED FOR STRAIN SENSING

The term *smart materials and structures* is finding increasing use to describe the unique marriage of material and structural engineering with fiber optic sensors and actuation control technology (17). Smart structure technology will require the development of *materials with optical nerves.* Structures constructed from such materials could continuously monitor their internal strain and thermal state. Such measurements could be used to monitor the loads imposed on the structure as well as its vibration state and deformation. In addition, strain sensing may be capable of assessing damage and warning of impending weakness in structural integrity or improving quality control of thermoset composite materials during fabrication through cure monitoring. This form of resident inspectability clearly has both safety and economic ramifications, for it could lead to greater confidence in the use of advanced composite materials and weight savings through avoidance of overdesign. This should be of particular interest to the aerospace community because of the multiplier effect of weight savings in aircraft and space structures. This technology should lead to a reduction in maintenance, repair, and downtime of future aircraft and could also find broad application in ships, submarines, and pressure vessels, where composite materials are penetrating.

In the civil engineering arena, buildings, bridges, dams, tunnels, seaports, highways, railways, pipelines, and airports represent an enormous financial investment. If resident optical fiber sensors can improve concrete evaluation, the ramifications could have enormous value to the multibillion dollar annual construction business, as aging and deterioration of many highway bridges is recognized as one of the major problems facing structural engineers in the United States, European Community, and Japan. In cold-climate countries this

Fiber Optic Smart Structures, Edited by Eric Udd.
ISBN 0-471-55448-0 © 1995 John Wiley & Sons, Inc.

deterioration has been greatly accelerated by the use of deicing salts. A large-scale research effort is currently underway to use advanced composite materials (ACMs) extensively in bridges. This research is addressing issues of corrosion, rehabilitation, and monitoring and involves new concepts and designs in bridge repair and construction. One of the most significant advances is the replacement of steel prestressed tendons with ones made of ACM. These fiber-reinforced polymers are practically immune to corrosion and also lend themselves to internal monitoring by means of embedded fiber optic sensors. A structurally integrated sensing system could monitor the state of a structure throughout its working life. It could determine the strain, deformation, load distribution, and temperature or environmental degradation experienced by the structure (64). The first Smart Bridge to be built with structurally integrated fiber optic strain and temperature sensors was opened in 1993 in Calgary (102).

In more advanced smart structures the information provided by the built-in sensor system could be used for controlling some aspect of the structure, such as its stiffness, shape, position, or orientation (36). These systems could be called *adaptive* (or *reactive*) smart structures, to distinguish them from the simpler *passive* smart structures (62). A more appropriate term for structures that only sense their state might be *sensory structures*. Smart structures might eventually be developed that will be capable of adaptive learning, and these

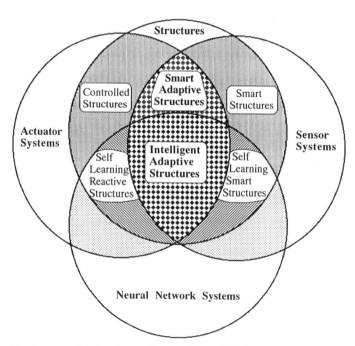

Figure 8.1. Structures possible by the confluence of four disciplines: materials and structures, sensing systems, actuator control systems, and adaptive learning neural networks. [From Measures (64).]

could be termed *intelligent structures*. Indeed, if we consider the confluence of the four fields—structures, sensor systems, actuator control systems, and neural network systems—we can appreciate that there is the potential for a broad class of structures (Fig. 8.1).

8.2. TYPES OF FIBER OPTIC SENSOR

In general, we can characterize optical fiber strain sensors as either distributed or localized. *Distributed* sensors are, in principle, very attractive, for they would permit the use of fewer sensors and represent a more effective use of optical fibers, in that *each element* of the optical fiber is used for both *measurement* and *data transmission* (18). In essence, by making a time-of-flight measurement along the optical fiber, the strain is determined along the entire length of the optical fiber. In practice, the spatial resolution attainable is quite limited. This is particularly true if the measurements have to be made in retail time. Furthermore, the need for high-resolution flexible sensing architecture and graceful degradation is not satisfied with distributed sensors. Often, a single break in the optical fiber effectively takes out of commission a large proportion of the sensing capacity. *Localized* fiber optic sensors determine the *measurand* over a *specific segment* of the optical fiber and are similar in that sense to conventional strain or temperature gauges.

Information about the state of the structure is impressed upon the transmitted (or, in many instances, the reflected) light by a number of mechanisms. These include interactions that change the intensity, phase, frequency, polarization, wavelength, or modal distribution of the radiation propagating along the optical fiber. Fiber optic sensors can be classified into two major categories: intensiometric and interferometric. The former depend on a variation of the radiant power transmitted through a multimode optical fiber. It has long been recognized that small bends in an optical fiber leads to a loss of light as the curvature permits core radiation to leak into the cladding, from which it escapes the fiber. The smaller the radius of curvature, the greater the loss. Microbend fiber optic sensors rely on this principle, using either some device or the intrinsic corrugations of the composite material into which the sensor is embedded to convert an applied force into a microbend in the optical fiber and thereby reduce its transmission of light. Unfortunately, such sensors have very limited sensitivity, measurement range, and accuracy. Furthermore, many phenomena that are unrelated to strain can also cause attenuation, and using a transduction mechanism that restricts the optical signal is not good engineering.

8.2.1. Interferometric Strain Sensors

Interferometric fiber optic sensors function by sensing the measureand-induced phase change in light propagating along a single-mode optical fiber and encompass a large class of sensitive devices. As we shall see later, one of the

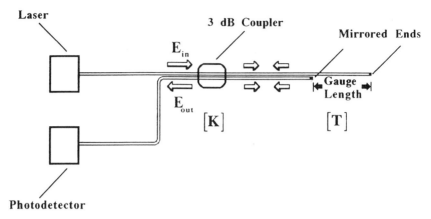

Figure 8.2.   Michelson interferometric fiber optic sensor.

most important requirements for any fiber optic sensor that is to qualify for use with smart structures is that it involve a single fiber. Nevertheless, in the past Michelson interferometric fiber optic sensor have been used in the laboratory setting to explore some of the potential applications that might be undertaken with the Fabry–Perot sensor in practical smart structures. For example, the Michelson sensor has demonstrated that damage-induced acoustic emission could be detected within various composite materials (54, 55), and that cure monitoring of thermoset composites might also be feasible. What made the Michelson fiber optic sensor attractive in the early days was its near-identical sensitivity to the Fabry–Perot sensor, combined with a much greater ease of fabrication and demodulation. However, its two-optical-fiber arrangement (Fig. 8.2) makes it quite unsuitable for smart structures, due to its rejection, greater intrusiveness, poor common mode, and required phase preservation at the structure's interface. The latter two effects are likely to represent particularly serious problems for practical structures, which are invariably subject to high levels of mechanical noise. This sensor is also vulnerable to stress-induced birefringence effects along the entire length of the embedded optical fibers (55).

There are four single-fiber sensors that are potential smart structures candidates: the Fabry–Perot, the two-mode, the polarimetric, and the intracore Bragg grating. The first three can be treated as different forms of two-path interferometers. In the Fabry–Perot fiber optic sensor, a cavity comprising two mirrors that are parallel to each other and perpendicular to the axis of the optical fiber forms the localized sensing region. A change in the optical path length between the mirrors leads to a shift in the cavity mode frequencies. In this way the reference and sensing optical fiber are one and the same, up to the first mirror, which constitutes the start of the sensing region. The Fabry–Perot sensor is normally configured to be single ended, and if built with low-reflectivity mirror will provide a sinusoidal phase-strain relationship. Fabry–

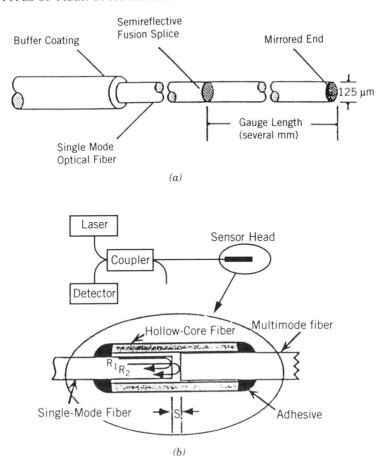

Figure 8.3. Fabry–Perot fiber optic sensor configurations: (*a*) intrinsic (35); (*b*) extrinsic (46).

Perot sensors can be configured as either *intrinsic* (Fig. 8.3(*a*) or *extrinsic* (Fig. 8.3(*b*). The former has the advantage of being the least perturbative, capable of being used in the form of a strain rosette (96), and possibly the more robust. As a consequence of the separation between gauge length and cavity length in the latter, it can be built with long gauge lengths, yet work with lasers of modest coherence length, but requires careful calibration. The extrinsic sensor has less transverse coupling and can therefore evaluate more directly the axial component of strain in a host material (90).

In general, the circular symmetry of optical fibers used to make the sensitive fiber optic interferometric sensors discussed above may not be adequate to stabilize the two orthogonal linear polarization eigenmodes. This can lead to *polarization fading* and the possibility of cross coupling between the polarization modes (85). These effects represent a source of noise in interferometric measurements based on small changes (on the order of a few fringes). Although

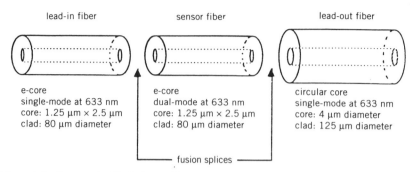

Figure 8.4. Two-mode elliptical-core fiber optic sensor configuration. [From Murphy et al. (76).] Reprinted with permission.

a number of methods have been proposed for alleviating this problem (43), one of the most effective involves the use of polarization maintaining single-mode optical fibers. The high birefringence designed into this kind of optical fiber stabilizes the polarization because the two eigenmodes are forced to have an appreciable difference in their respective propagation constants. This limits the exchange of energy between the two eigenmodes. This high birefringence is created in one of two ways. The optical fiber is fabricated with a deliberate asymmetry, such as an elliptic core, or by stress loading the core. Another method of avoiding polarization fading is to employ a high-frequency polarization scrambler. This can be a low-cost solution in the case of a network of sensors; however, this gain is at the expense of the interferometric fringe visibility, which declines to 0.5.

The best developed form of modalmetric sensor, called an *e-core two-mode sensor*, involve changes in the transverse spatial mode distribution of light within the optical fiber (10, 11, 44, 58, 76). In this sensor the lowest-order

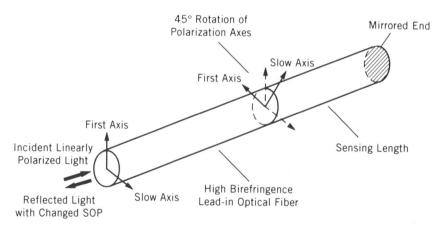

Figure 8.5. Back-reflective polarimetric fiber optic sensor using a 45° rotation of the polarization eigenaxes for localization of sensing segment.

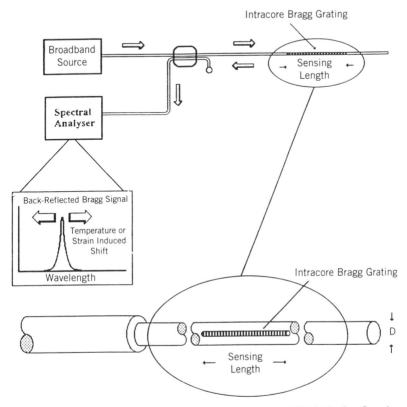

Figure 8.6. Intracore Bragg grating fiber optic sensor. Also indicated is the back-reflected narrow peak Bragg spectrum.

transverse mode $\{LP_{01}\}$ propagates along the lead-in fiber to the sensing region. Here a section of elliptic core optical fiber that permits two transverse (LP_{01} and the even LP_{11}) modes to propagate, at the wavelength for which the lead-in fiber is single mode, has been fusion spliced with its core slightly offset so that both spatial modes are excited) (Fig. 8.4). A second section of single mode fiber can be fusion spliced to the end of the sensing section or the end of the two-mode sensing section can be mirrored. Either way, the spatial variation in the transverse light distribution is converted into a nonlinear (cosine) relation between the signal and the strain in the host structure, typical of an interferometer.

In a polarimetric sensor, changes in the state of polarization of light traveling in a single-mode optical fiber are used to determine the strain imposed on the sensor. To make the sensor reliable, special polarization maintaining optical fiber is used, and it is the change in the phase of the two orthogonal polarization eigenmodes of this high-bifringence fiber that is used to evaluate the strain. Localization of the sensor is achieved by rotating the polarization eigenaxes through 45° in the sensing section relative to their orientation in the lead in/out sections of optical fiber (Fig. 8.5) and linearly polarized light is launched into one of the polarization axes of the lead-in optical fiber (12, 31, 41, 42, 97).

The intracore Bragg grating fiber optic sensor relies on the narowband reflection from a region of periodic variation in the core index of refraction of a single-mode optical fiber. In this sensor the center (Bragg) wavelength of the reflected signal is linearly dependent on the product of the scale length of the periodic variation (the period) and the mean core index of refraction. Changes in strain to which the optical fiber is subjected will consequently shift this Bragg wavelength, leading to a wavelength-encoded optical measurement

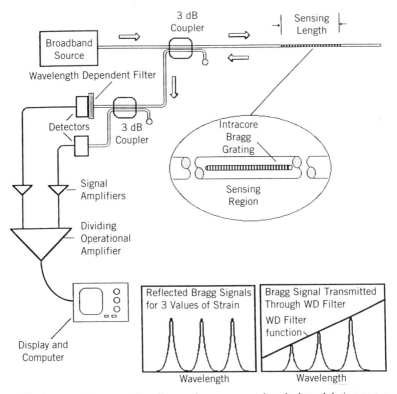

Figure 8.7. Intracore Bragg grating fiber optic sensor wavelength demodulation system. The Bragg wavelength is determined from the ratio of the signals from the two detectors indicated, one receiving its light after passage through the wavelength-dependent filter, the other receiving its light directly and serving as a reference. Small insets: bottom left indicates Bragg reflected signals for three values of strain; bottom right reveals how these three signals are affected by transmission through a wavelength-dependent linear filter.

(Fig. 8.6) (73). An inexpensive passive method of determining the Bragg wavelength based on a simple wavelength-dependent ratiometric technique has now been developed (Fig. 8.7) (66, 69) and is available from Electrophotonics Corporation.

8.3. CRITERIA AND SELECTION OF FIBER OPTIC STRAIN SENSOR

To assess the potential suitability of any fiber optic sensor for undertaking strain measurements in smart structures, the following criteria can serve as a guide (92). An ideal fiber optic strain sensor for smart structures should be:

1. Intrinsic in nature for minimum perturbation and stability.
2. Localized, so that it can operate remotely with insensitive leads.
3. Able to respond only to the strain field and discern any change in direction.
4. Well behaved, with reproducible response.
5. All-fiber for operational stabiity.
6. Able to provide a linear response.
7. A single optical fiber for minimum perturbation and common-mode rejection.
8. Single-ended for ease of installation and connection.
9. Sufficiently sensitive with adequate measurement range.
10. Insensitive to phase interruption at the structural interface.
11. Nonperturbative to the structure and robust for installation.
12. Interrupt immune and capable of absolute measurement.
13. Amenable to multiplexing to form sensing networks within structure.
14. Easily manufactured and adaptable to mass production.

Although many fiber optic sensors have been suggested and used for strain measurements, only the polarimetric (P), two-mode (TM), Fabry–Perot (FP), and Bragg grating (BG) comply with criteria 1, 2, 4, 5, 7, and 10. These are consequently selected as possible strain sensors and a further selection is made based on how they comply with the remaining criteria. This can be gauged by reference to Table 8.1.

As we shall see in the next section, the high sensitivity of both the intracore Bragg grating and the Fabry–Perot sensor make it possible to use them with short gauge lengths. However, in civil structures it is possible that large gauge lengths (on the order of 1 or more meters) might be adequate, if not desirable. Under these circumstances the lower-sensitivity modalmetric or polarimetric fiber optic sensor might be satisfactory. To some extent the modalmetric sensor is simpler to use and has slightly higher sensitivity than the polarimetric sensor.

In order that the cost of implementing smart structure technology be kept to a minimum, a single type of sensor should be developed that is capable of

Table 8.1   Comparison of Fiber Optic Sensors

	FP	BG	TM	P
Localized	Y	Y	(1)	(2)
Responds only to strain	(3)	(3)	(3)	(3)
Direction change response	(4)	Y	(4)	(4)
Linear response	(4)	Y	(4)	(4)
Single-ended	Y	Y	(5)	(5)
Adequate sensitivity and range	Y	(6)	(7)	(7)
Absolute measurement	(8)	Y	(8)	(8)
Multiplexing within structure	(9)	Y	(9)	(9)
Potential for mass production	(11)	Y	(10)	(10)

Note: Y stands for "yes."

Key:

(1) Sensing length is two-mode while lead-in/lead-out optical fibers are single mode.
(2) Polarization eigenaxes are rotated by 45° at start and sensing length ends in mirror.
(3) If temperature changes sensor must be able to compensate for thermal apparent strain.
(4) Requires *quadrature* detection and suitable signal demodulation.
(5) Requires a mirror at the end of the sensing length.
(6) Limited by demodulation system, but 1 μstrain should be achievable with laser sensor.
(7) Sensitivity is about 1% that of the FP. This would restrict spatial resolution to many centimeters.
(8) Requires special demodulation system, so absolute measurements is not available at all times.
(9) Difficult, except for large structures, where time-division multiplexing can be used.
(10) Requires at the least one fusion splice and therefore some degree of handling.
(11) Recently, Fabry–Perot sensors have been made from two Bragg gratings. Under these circumstances no fusion splice is required. Also extrinsic Fabry–Perot sensors do not require a fusion splice.

undertaking all of the requisite measurements. This suggests a sensor with as high a sensitivity as possible so it can be used in the broadest range of applications, including measurement of the internal strain and the detection of anomalies that might indicate the development of a structural weakness, monitoring of the loads applied to a structure and its subsequent deformation, evaluation of the vibration modes and resonant frequencies, cure monitoring and the measurement of residual strain in thermoset composites. The assessment of any degradation in structural integrity is certainly a most desirable capability of any resident strain sensing system. It is expected that if this is to be achieved through the detection of internal strain anomalies that arise when the structure is loaded, very careful strain measurements will have to be undertaken, and this is expected to demand a set of highly localized (almost point) sensors of high precision.

This sensor should also be capable of measuring both strain and temperature since, in fact, both will invariably be required for practical smart structures due to potential wide-temperature excursions. Based on Table 8.1 and the considerations above, the choice probably lies between the Fabry–Perot and intracore Bragg grating sensors. Both the FP and the BG sensors can certainly be used for either strain or temperature measurements, and as we shall see later, it is expected that strain and temperature will be determined simultaneously with the appropriate sensor design and demodulation. A careful comparison will have to be undertaken to determine which of these two sensors has the greatest advantage in practice.

8.4. FIBER OPTIC STRAIN SENSOR SENSITIVITY

8.4.1. Interferometric Fiber Optic Strain Sensor Sensitivity

The Michelson fiber optic sensor, illustrated in Fig. 8.2, is representative of double-pass fiber optic interferometric sensors, including the two-mode and polarimetric sensors. If the reflectivity of each mirror of a Fabry–Perot fiber optic sensor is less than 50%, this sensor can also be treated as a double-pass interferometer. In the case of a Michelson interferometric sensor the detected E-field can be related to the E-field launched into the optical fiber by a laser diode through the expression (40):

$$E_{\text{out}} = [K][T][K]E_{\text{in}} \tag{8.1}$$

where the influence of the 3-dB coupler on the E-field is described by the K-matrix:

$$[K] = \frac{1}{\sqrt{2}}\begin{bmatrix} 1 & i \\ i & 1 \end{bmatrix}$$

and the T-transduction matrix:

$$[T] = \begin{bmatrix} \exp(i\phi_s) & 0 \\ 0 & \exp(i\phi_r) \end{bmatrix}$$

describes the phase change impressed on the light propagating down each optical fiber to the mirror and back to the coupler. ϕ_s and ϕ_r represent the phase shift induced in the sensing and reference optical fibers, respectively. The resulting interference (39) leads to a cosine modulation in the intensity of light arriving at the detector:

$$I_{\text{out}} = I_{\text{in}} \frac{1 + \cos \phi}{2} \tag{8.2}$$

where $\phi \equiv \phi_s - \phi_r$.

The phase retardation between the sensing and reference fields propagating along two paths with a path imbalance, L, can be expressed in the form

$$\phi = \beta L \tag{8.3}$$

where β is the phase retardation per unit path length. This can usually be expressed in the generic form

$$\beta = kn \qquad (8.4)$$

where k is the free-space propagation constant and n is the effective index of refraction. In general, ϕ depends on the length, temperature, and operating wavelength of the sensor: that is, $\phi = \phi(L, T, \lambda)$, where T is taken as the change from some reference temperature.

The variation in ϕ due to an incremental changes in these parameters is given by the expression

$$\Delta\phi = \left(L\frac{\partial\beta}{\partial L} + \beta\right)\Delta L + \left(L\frac{\partial\beta}{\partial T} + \beta\frac{\partial L}{\partial T}\right)\Delta T + L\frac{\partial\beta}{\partial\lambda}\Delta\lambda \qquad (8.5)$$

For the simplest case of strain sensing when the temperature is kept constant (and equal to the reference value) and the wavelength is fixed (we shall see later that variations in the wavelength can lead to methods of signal recovery), Eq. (8.4), combined with a reduced form of (8.5), leads to the simple relation

$$\Delta\phi = kL(n\varepsilon + \Delta n) \qquad (8.6)$$

where the *path integrated strain*

$$\varepsilon \equiv \Delta L/L \qquad (8.7)$$

and

$$\Delta n = \frac{\partial n}{\partial L}\Delta L \qquad (8.8)$$

The sensing length, L, is assumed to be much shorter than the distance over which the strain changes appreciably. For light that is linearly polarized in the α-direction, the strain optic theory (39) allows us to write

$$\Delta n_\alpha = -\frac{n_\alpha^3}{2}P_{\alpha\gamma}\varepsilon_\gamma^f \qquad (8.9)$$

where $P_{\alpha\gamma}$ is the *strain-optic tensor* and ε_γ^f is the *strain tensor* for optical fiber, comprising three principle and three shear strain components.

For a homogeneous isotropic medium,

$$
P_{\alpha\gamma} =
\begin{bmatrix}
P_{11} & P_{12} & P_{12} & & & \\
P_{12} & P_{11} & P_{12} & & 0 & \\
P_{12} & P_{12} & P_{11} & & & \\
& & & P_{44} & 0 & 0 \\
& 0 & & 0 & P_{44} & 0 \\
& & & 0 & 0 & P_{44}
\end{bmatrix}
\tag{8.10}
$$

where $P_{44} = (P_{11} - P_{12})/2$.

We shall now derive the appropriate relations for the: *interferometric*, *polarimetric*, and *modalmetric* fiber optic sensors. For the interferometric sensor,

$$
\beta_\alpha = n_\alpha k
\tag{8.11}
$$

where n_α is the effective cavity index of refraction for the lowest-order transverse mode (LP$_{01}$), in which case

$$
\Delta\phi_\alpha = kLn_\alpha \left(\varepsilon_1^f - \frac{n_\alpha^2}{2} \sum_{\gamma=1}^{6} P_{\alpha\gamma}\varepsilon_\gamma^f \right)
\tag{8.12}
$$

where we have switched to a notation that explicitly indicates the summation of the strain optic terms. ε_1^f represents the axial strain on the fiber.

For the polarimetric sensor,

$$
\beta^p = k(n_F - n_S)
\tag{8.13}
$$

where n_F and n_S are the respective indices of refraction for the lowest-order transverse mode polarized along the fast and slow polarization eigenaxes of the optical fiber. Under these circumstances we can write the change in the phase between the two polarized states:

$$
\Delta\phi^p = kL \left[\left(n_F\varepsilon_1^f - \frac{n_F^3}{2} \sum_{\gamma=1}^{6} P_{\alpha\gamma}\varepsilon_\gamma^f \right) - \left(n_S\varepsilon_1^f - \frac{n_S^3}{2} \sum_{\gamma=1}^{6} P_{\alpha\gamma}\varepsilon_\gamma^f \right) \right]
\tag{8.14}
$$

Finally, for the elliptical-core two-mode sensor,

$$
\beta_\alpha^m = k(n_\alpha^{01} - n_\alpha^{11})
\tag{8.15}
$$

where n_α^{01} and n_α^{11} represent the respective effective indices of refraction for the LP$_{01}$ and LP$_{11}^{even}$ transverse modes that are linearly polarized in the α-direction.

This leads to an expression for the incremental change in the modal phase:

$$\Delta\phi_\alpha^m = kL\left[\left(n_\alpha^{01}\varepsilon_1^f - \frac{(n_\alpha^{01})^3}{2}\sum_{\gamma=1}^{6} P_{\alpha\gamma}^{01}\varepsilon_\gamma^f\right)\right.$$
$$\left. - \left(n_\alpha^{11}\varepsilon_1^f - \frac{(n_\alpha^{11})^3}{2}\sum_{\gamma=1}^{6} P_{\alpha\gamma}^{11}\varepsilon_\gamma^f\right)\right] \qquad (8.16)$$

where $P_{\alpha\gamma}^{01}$ and $P_{\alpha\gamma}^{11}$ represent the respective *strain optic* coefficients for the LP_{01} and LP_{11}^{even} transverse modes of the optical fiber. Egalon and Rogowski (22) have studied the influence of axial strain on the modal distribution of light propagating along a multimode optical fiber by means of a numerical model. Their analysis predicts the change in the modal pattern with increasing load and confirms that the rotation of the lobes in a two-mode fiber is due to mode coupling.

If a uniaxial longitudinal stress σ_z is applied to an isotropic, elastic optical fiber oriented in the z-direction, the resulting strain from first-order elastic theory will be

$$\varepsilon_\gamma^f = \begin{bmatrix} -v\varepsilon_1^f \\ -v\varepsilon_1^f \\ \varepsilon_1^f \end{bmatrix} \qquad (8.17)$$

where we have assumed *zero shear strain* (13). In the case of a surface adhered or embedded optical fiber sensors, Butter and Hocker made two assumptions: (1) that $\varepsilon_z = \varepsilon_1^f$, where $\varepsilon_z = \sigma_z/E$ is the *strain* in the host in the optical fiber direction, and (2) that no transverse strain from the host is coupled to the optical fiber so $-v\varepsilon_1^f$ is the only component of transverse strain. E is Young's modulus for the host and v is Poisson's ratio for fused silica. If we also assume that the optical properties are independent of the polarization state of the light, then, in general, we can write

$$\Delta\phi = SL\varepsilon_z \qquad (8.18)$$

where we have introduced S as the *phase-strain sensitivity*.

For the interferometric fiber optic sensor,

$$S^I = kn\left(1 - \frac{n^2 P_e}{2}\right) \qquad (8.19)$$

where we have introduced the *effective strain-optic coefficient*,

$$P_e = [P_{12} - v(P_{11} + P_{12})] \qquad (8.20)$$

For a polarimetric sensor the corresponding phase-strain sensitivity is given by the relation

$$S^p = k(n_F - n_S)\left[1 - \frac{N^2(n_F,\ n_S)P_e}{2}\right] \tag{8.21}$$

where we have introduced the general relation

$$N^2(n_a,\ n_b) \equiv n_a^2 + n_a n_b + n_b^2 \tag{8.22}$$

while the phase-strain sensitivity for the elliptical-core two-mode sensor can be expressed in the form

$$S_\alpha^m = k(n_\alpha^{01} - n_\alpha^{11})\left[1 - \frac{N^2(n_\alpha^{01},\ n_\alpha^{11})P_e}{2}\right] \tag{8.23}$$

assuming that the stran-optic coefficients are essentially independent of the radiation mode. This is a common assumption but one that may have to be examined more closely in future, expecially where stress-loaded polarization maintaining optical fibers are used.

Recent values of the strain-optic coefficients P_{11} and P_{12} have been reported by Bertholds and Dandliker (8) for single-mode optical fibers. They indicate that for fibers with a pure silica core and B_2O_3-doped cladding, $P_{11} = 0.113$ and $P_{12} = 0.252$, values that are about 7% lower than for bulk silica. In the case of silica optical fibers (with $n = 1.458$, $E = 70\,GPa$, $v = 0.17$), the interferometric phase-strain sensitivity is evaluated to be about $6.5°\,\mu strain^{-1}\,cm^{-1}$ (or $S = 11.3\,rad\ \mu strain^{-1}\,m^{-1}$), which is close to the value obtained in experiments. For double-pass interferometers such as the low-finesse Fabry–Perot fiber optic sensor, this figure must be multiplied by 2. A comparison of the strain sensitivities for the three kinds of fiber optic strain sensor discussed above is provided in Table 8.2. It is quite clear that the Fabry–Perot strain sensor has a much greater strain sensitivity.

For many applications it is important that the strain sensor gauge length be small compared to the scale length over which the strain changes. Some idea of the relative gauge lengths needed to achieve 1-$\mu strain$ resolution for the three different types of fiber optic sensor can be obtained from the second row of

Table 8.2   Strain Sensitivity and Gauge-Length Comparison

Strain sensitivity	Fabry–Perot	Polarimetric	Elliptical-core two-mode
Degrees $\mu strain^{-1}\,cm^{-1}$	6.5 (Valis, 1989)	0.06 [Hogg et al. (32)]	0.02 [Huang et al. (38)]
Gauge length for 1 $\mu strain$	1 cm	108 cm	325 cm

Table 8.2, where we have assumed that each sensor has a demodulation system capable of 6.5° of phase-shift resolution.

8.4.2. Bragg Grating Fiber Optic Strain Sensor Sensitivity

A Bragg grating sensor comprises a segment of optical fiber in which a periodic modulation of the core index of refraction has been formed (21, 71, 73), usually by means of exposure to an interference pattern of intense ultraviolet light (at about 245 nm). The optical back-reflected spectrum of such a Bragg grating comprises a very narrow spike (in polarization preserving optical fibers two spikes can be resolved, one for each of the two orthogonal polarization modes). This strong back-reflection occurs at the Bragg wavelength, λ_B, which can be related to the effective core index of refraction, n, and the period of the index modulation, Λ, by the relation

$$\lambda_B = 2n\Lambda \tag{8.24}$$

Since, this Bragg wavelength will shift with changes in either n or Λ, monitoring the wavelength of this narrowband spectrum will serve to determine the strain or temperature environment to which the optical fiber is subjected.

The Bragg grating possesses a number of advantages that make it very attractive for smart structures:

- **Linear Response.** The Bragg grating center wavelength is a linear function of the measurand, so there is no direction ambiguity.
- **Absolute Measurement.** The Bragg grating center wavelength determines the measurand value in terms of some reference state.
- **Spectral Encoding.** The measurand signal information is spectrally encoded.
- **Fiber Optic Strength.** The intracore fiber optic Bragg grating has inherent high strength.
- **Multiplexing Potential.** The Bragg grating sensors are readily wavelength multiplexed.

Recently, intracore Bragg gratings have been formed within otical fibers with a *single pulse* of radiation from a 248-nm KrF laser (4) and strings of these gratings have been created during the pulling of the optical fiber. This should result in a very low cost and rugged sensor.

A Taylor expansion of Eq. (8.24) permits us to determine the differential change in the Bragg wavelength resulting from an applied strain field:

$$\frac{\Delta\lambda_\alpha}{\lambda_{0,\alpha}} = \frac{1}{\Lambda_{0,\alpha}} \sum_\gamma \frac{\partial\Lambda_\alpha}{\partial\varepsilon_\gamma} \varepsilon_\gamma + \frac{1}{n_{0,\alpha}} \sum_\gamma \frac{\partial n_\alpha}{\partial\varepsilon_\gamma} \varepsilon_\gamma \tag{8.25}$$

The *strain-optic* effect allows us to write

$$\frac{1}{n_{0,\alpha}} \sum_{\gamma} \frac{\partial n_{\alpha}}{\partial \varepsilon_{\gamma}} \varepsilon_{\gamma} = -\frac{n_{0,\alpha}^2}{2} \sum_{\gamma} P_{\alpha\gamma} \varepsilon_{\gamma} \tag{8.26}$$

and if we assume that the strain-induced change in the index modulation period is independent of the state of polarization of the interrogating light and only dependent on the axial strain, we can also write

$$\frac{1}{\Lambda_{0,\alpha}} \frac{\partial \Lambda_{\alpha}}{\partial \varepsilon_{\gamma}} = 1 \tag{8.27}$$

This assumption may need further investigation but has been assumed to date (21, 71, 73). Under these circumstances Eq. (8.25) takes the form

$$\frac{\Delta \lambda_{\alpha}}{\lambda_{0,\alpha}} = \varepsilon_1 - \frac{n_{0,\alpha}^2}{2} \sum_{\gamma=1}^{6} P_{\alpha\gamma} \varepsilon_{\gamma} \tag{8.28}$$

In the case of the Butter and Hocker (13) model, we can write

$$\frac{\Delta \lambda_{\alpha}}{\lambda_{0,\alpha}} = (1 - P_{\alpha}^{\text{eff}}) \varepsilon_1 \tag{8.29}$$

which reveals that the *wavelength-strain sensitivity* of the Bragg grating sensor

$$S_B = 1 - P_{\alpha}^{\text{eff}} \tag{8.30}$$

with

$$P_{\alpha}^{\text{eff}} = \frac{n_{0,\alpha}^2}{2} [P_{12} - \nu(P_{11} + P_{12})] \tag{8.31}$$

as the *index-weighted strain-optic coefficient*.

It should be noted that we have assumed that the bulk value of the strain-optic coefficients can be used in all of the discussion above. For polarization-preserving optical fiber, this is highly questionable. In the case of germanosilicate glass, the index-weighted strain-optic coefficient, Eq. (8.31), has the value 0.22. It should also be noted that in the case of the Bragg grating sensor the strain sensitivity is independent of the length of the Bragg grating. However, the sharpness of the back-reflected narrowband spectrum is dependent on the grating length. One of the most important advantages of this sensor is the direct relation between the Bragg wavelength and the fiber strain, which makes absolute measurements of the strain possible. This makes it immune to power interruption (an intentional or unintentional break in the electric power of the system) a weakness of the other kinds of interferometer sensors (in the sense that they normally determine only the incremental strain and require a specialized demodulation system for absolute measurements).

The *wavelength strain sensitivity* of the intracore Bragg grating formed in Andrew E-type elliptal core fiber purchased from United Technology Research Centre was measured to be 0.648 pm μstrain^{-1} for the fast axis and 0.644 pm μstrain^{-1} for the slow axis when operated at 830 nm (68). A more relevant figure of merit for this kind of sensor is its *strain resolution*, and recently, this has been reduced to 8 μstrain using a wavelength-dependent ratiometric demodulation technique (70). Morey et al. (74) have determined that the *wavelength strain sensitivity* of a 1.55-μm Bragg grating is 1.15 pm μstrain^{-1}, while its *temperature sensitivity* is 1.3 pm °C^{-1}.

8.4.3. Sensor–Host Coupling and Signal Interpretation

For the more general case of a birefringent optical fiber interferometric strain sensor with "short" sensing length embedded within an infinite, isotropic, and homogeneous host, Eq (8.12) leads to phase increments, $\Delta\phi_2$ and $\Delta\phi_3$, for light linearly polarized in the 2- and 3-directions, respectively, where

$$\Delta\phi_2 = n_2 kL \left\{ \varepsilon_1^f - \frac{n_2^2}{2} \left[P_{11}\varepsilon_2^f + P_{12}(\varepsilon_1^f + \varepsilon_3^f) \right] \right\} \tag{8.32}$$

and

$$\Delta\phi_3 = n_3 kL \left\{ \varepsilon_1^f - \frac{n_3^2}{2} \left[P_{11}\varepsilon_3^f + P_{12}(\varepsilon_1^f + \varepsilon_2^f) \right] \right\} \tag{8.33}$$

while Eq. (8.28) for the Bragg grating sensor gives rise to normalized wavelength shifts for light linearly polarized in the 2- and 3-directions, respectively, of

$$\frac{\Delta\lambda_2}{\lambda_{0,2}} = \left\{ \varepsilon_1^f - \frac{n_2^2}{2} \left[P_{11}\varepsilon_2^f + P_{12}(\varepsilon_1^f + \varepsilon_3^f) \right] \right\} \tag{8.34}$$

and

$$\frac{\Delta\lambda_3}{\lambda_{0,3}} = \left\{ \varepsilon_1^f - \frac{n_3^2}{2} \left[P_{11}\varepsilon_3^f + P_{12}(\varepsilon_1^f + \varepsilon_2^f) \right] \right\} \tag{8.35}$$

assuming that the optical fiber is aligned with the 1-direction of the host principal strains. In Eqs. (32) to (35), ε_α^f represents the α-component of the optical fiber strain tensor, while it is the host strain tensor that is desired.

The essential simplifications made in the Butter and Hocker model were that the strain in the optical fiber in the 1-direction matches that of the host, while zero host strain is coupled to the optical fiber in the 2- and 3-(transverse) directions. Equations (8.32) to (8.35) clearly indicate that in the more general case the phase change for the interferometer or the wavelength shift for the Bragg grating sensor observed for each polarization eigenmode of the optical fiber depends on all three principal strain components within the optical fiber, and clearly each of these has to be related to the far-field strain in the host. Sirkis and Haslach (90) extended the Butter and Hocker (13) free-fiber uniaxial-stress phase-strain model and have shown that their results are closer to those observed in transverse loading experiments (61). Sirkis and Haslach (90) and Mathews and Sirkis (61) indicated that the Butter and Hocker (13) model leads to strain predictions with errors that increase substantially as the stiffness of the optical fiber approaches that of the host.

Recent comparison with experiments (Sirkis and Mathews, 1992) suggest that although the theoretical model of Sirkis and Haslach (90) predicts the trend in the variation of phase-strain sensitivity with fiber/host stiffness ratio, the experimental variation is stronger. However, when the optical fiber is collinear with the adjacent stiff reinforcing fibers of the host and the loading is in this ply direction, the Butter and Hocker model gives reasonable strain predictions. Interestingly, this has often been the configuration used in many experiments. For optical fiber sensors embedded within composites, the difference between the theories should not be greater than about 10% if the *stiffness* of the optical fiber is greater than 10 *times* that of the *host matrix* (i.e., $E_f/E_h > 10$). If the stiffness of the optical fiber is not that much greater than that of the host, decoupling the sensing length of the optical fiber from the transverse strains in the host allows the simpler theory of Butter and Hocker to apply. This appears to be the case for the extrinsic Fabry–Perot sensor, and to a lesser extent the polarization sensor (87). Valis and Measures (95) have suggested that both the parallel and perpendicular far-field stress components within a host can be evaluated by an embedded Fabry–Perot sensor of special design, provided that its polarization eigenaxes are suitably aligned with the principal strain axes of the host.

When considering the issue of sensor–host strain coupling, there is, however, another factor that must be taken into account. This is the possible presence of an optical fiber coating. Although it may be possible to tailor the properties of this coating to minimize the transverse coupling or greatly reduce stress concentrations in the host adjacent to the optical fiber, Sirkis and Dasgupta (88) have indicated that the sensor's phase-strain sensitivity is not likely to be influenced strongly by the coating. Clearly, much work remains to be done to fully understand the fiber–host interactions, and as will be alluded to later, the presence of a coating may be important in terms of the sensors performance life (46).

8.5. STRAIN SENSOR DEMODULATION AND CHARACTERIZATION

8.5.1. Signal Recovery for Interferometric Fiber Optic Sensors

Unlike resistive-foil strain gauges, interferometric fiber optic strain sensors require phase demodulation and suffer from interrupt ambiguity (i.e., if the phase-demodulation electronics is interrupted, the new value of the phase becomes nonunique and must be reinitialized). The generic response function, $R(\phi)$, for a two-beam interferometer can (39) be expressed in the form

$$R(\phi) = \frac{I}{I_0} = \frac{1 + V \cos \phi}{2} \qquad (8.36)$$

where I is the photodetector current in the presence of the sensor, I_0 is the photodetector current that would arise directly from the source and ϕ is the *phase retardation* introduced between the two paths of the interferometer. The *visibility factor*, V (or fringe contrast), is introduced to cover partial coherence situations.

The purpose of *signal recovery* or *demodulation* is to convert the complex optical output from a fiber optic sensor into an electrical signal proportional to the phase retardation. The problem with the fiber optic sensor response function, Eq. (8.36), is that it is sinusoidal and therefore *multivalued and nonlinear*. Furthermore, the nature of the cosine function leads to *sign ambiguity* and *signal fading*. The purpose of any signal recovery scheme is to address as many of these shortcomings as possible. One of the simplest forms of signal recovery is *fringe counting*, this is acceptable for large changes in $\Delta\phi(\gg 2\pi)$ but fails to distinguish changes in direction of the measurand. A further problem of the response function given by Eq. (8.36) is that it does not provide *interrupt immunity* (i.e., an absolute measurement). Signal recovery techniques should address all five of these issues:

1. Signal fading.
2. Interrupt immunity.
3. Sign ambiguity.
4. Nonlinearity.
5. Multivalued response.

Signal recovery techniques can be broadly classified as passive or active and they can involve heterodyne or homodyne detection (40, 53). In heterodyne detection the optical frequencies in the two interferometer arms are unequal, while in homodyne detection there is a common frequency for both arms. True heterodyne signal recovery is unlikely for sensors to be used in smart structures, as it is somewhat difficult to affect the frequency in one arm (path) of a single-fiber sensor.

8.5.2. Switched Dual-Wavelength Quadrature Technique

One of the most elegant methods of overcoming signal fading is to use a system that produces two outputs that have a phase bias of $\pi/2$. Such *quadrature* (orthogonal) outputs ensure that one output has maximum sensitivity while the other experiences minimum sensitivity. In the case of the single-fiber polarimetric (93), two-mode (57) and Fabry–Perot (32) sensors, quadrature outputs can be achieved through an active process, such as switching or shifting the laser frequency by some multiple of $\pi/2$ phase shift for the sensor. Such techniques are sometimes referred to as *quadrature-phase modulation*. If the laser wavelength is switched between two values corresponding to a $\pi/2$ phase shift, quadrature phase signals are essentially extracted from the sensor. If these are sampled by orthogonal gating pulses, summed and bandpassed filtered, the resulting sinusoidal signal can be demodulated with a phase-

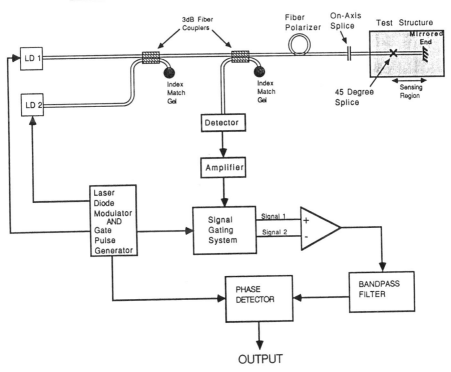

Figure 8.8. Dual-wavelength all-fiber polarimetric sensor using two lasers to provide the quadrature signals. [From Turner et al. (93).]

sensitive detector (42, 43). In the case of the Fabry–Perot sensor, the required *quadrature wavelength shift*

$$|\Delta\lambda_Q| = (2N + 1)\frac{\lambda^2}{4nL} \qquad (8.37)$$

where $N(=0, 1, 2,...)$ is termed the *quadrature order*. For gauge length $L \sim 10$ mm and wavelength 820 nm, the quadrature wavelength shift falls within the tuning range of a single longitudinal mode laser diode (i.e., $\Delta\lambda_Q = 0.01$ *nm*). By comparison, a quadrature wavelength shift of about 20 nm is required for a polarimetric sensor (birefringence beat length 1.4 mm and wavelength of 820 nm) with a gauge length of a few centimeters (93). This necessitates switching between two separate lasers and represents a problem in terms of phase control (Fig. 8.8). However, when this polarimetric sensor was operated with a gauge length of about 1 m, the quadrature wavelength shift reduced to 0.7 nm and the switching between two wavelengths could be accomplished with a single laser.

8.5.3. Pseudo-heterodyne Phase Detection

Since variation of the injection current in a semiconductor laser causes a corresponding wavelength change, the laser can directly provide optical phase modulation (40). This approach is particularly applicable to single-lead and single-ended sensors such as the Fabry–Perot interferometric fiber optic sensor. In the case of "serrodyne' or chirped wavelength modulation of the laser diode, the optical frequency is swept over a given range of frequencies. During the linear part of these sawtooth laser drive signals (of period $2\pi/\omega_m$), a constant rate of phase change is produced in the interferometer by the corresponding shift of frequency. The phase change arising within each period of the sawtooth waveform is given by the expression

$$\Theta = \frac{\phi_m \omega_m}{2\pi} t \qquad (8.38)$$

where ϕ_m represents the *depth of modulation of the phase*. If the peak amplitude of this sawtooth waveform is adjusted to be 2π (i.e., $\phi_m = 2\pi$) $\Theta = \omega_m t$ and the normalized signal will be sinusoidal with angular frequency ω_m over this linear ramp part of the sawtooth modulation. The sudden fall in the modulation frequency at the end of each ramp will give rise to a sinusoidal signal of much higher frequency (in essence a spike). If a suitable bandpass filter is used to remove this flyback spike, then to first approximation the signal is phase modulated with a carrier, ω_m, and we can write the normalized signal in

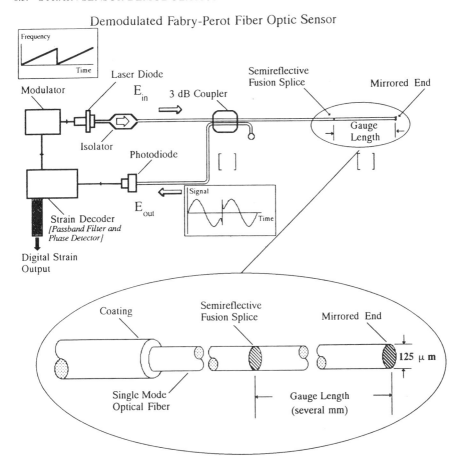

Close up of Fabry-Perot Fiber Optic Sensor

Figure 8.9. Pseudo-heterodyne Fabry–Perot interferometric fiber optic sensor demodulation system based on repeated linear frequency ramping of the laser diode and a combination of passband filtering and phase detection.

the form

$$\frac{I}{I_0} = \frac{1}{2}\cos(\omega_m t + \psi) \tag{8.39}$$

where ψ represents the sum of the measurand and nonmeasurand components of the phase shift. Clearly, this serrodyne or chirped wavelength excitation leads to a sinusoidal optical signal onto which is impressed any phase modulation arising from changes in the strain field of the sensor.

A schematic illustration of such a pseudo-heterodyne-modulated Fabry–Perot sensor system is presented in Fig. 8.9 (32). In this system the laser diode injection current was modulated by a linear sawtooth signal with an amplitude corresponding to one fringe. Figure 8.10 displays a representative laser diode driving current waveform and the resulting phase-generated optical carrier

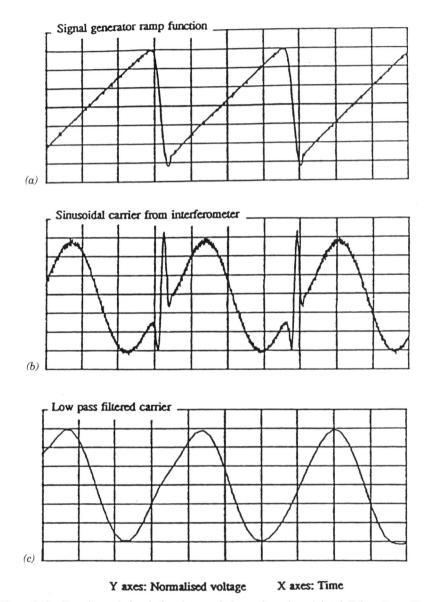

Y axes: Normalised voltage   X axes: Time

Figure 8.10. Experimental signals for the pseudo-heterodyne demodulated Fabry–Perot fiber optic sensor: (a) linear sawtooth driver signal to laser diode; (b) resulting Fabry–Perot optical signal; (c) bandpassed filtered signal. [From Hogg et al. (34).]

signal before and after filtering. After detection the carrier is fed to a digital–analog hybrid phase tracker similar to one reported by Jackson (38). The theoretical strain resolution of this system is a small fraction of a microstrain, and its time response is adequate to cover the normal loading of most practical structures. However, its limited temporal response can pose a serious problem in the event that the structure is exposed to any sudden high rate of loading, such as an impact. Under these conditions the strain sensing system can lose track of the strain.

An improved pseudo-heterodyne Fabry–Perot sensor demodulation system based on sinusoidal modulation of the wavelength of the laser diode and three-point phase sampling of the detected optical signal has been developed (60). This demodulation system provides a strain resolution of a few μstrain with the Fabry–Perot sensor with a 2 cm gauge length and it can operate to a loading rate of $10^5\,\mu$strain s^{-1}, making it suitable for a broad range of strain sensing applications. This demodulation system can also be used for demodulating an e-core two-mode fiber optic sensor.

8.5.4. Spectral Encoding/Ratiometric Demodulation

Interferometric sensors can use broadband light sources [such as light-emitting diodes, (LEDs) or superluminescent diodes (SLDs)] for spectral coding methods of modulating and demodulating sensor signals (94). This technique provides a repeatable and absolute measurement of the optical path difference and the measurand induced optical phase change. It is insensitive to optical intensity variations associated with the source, connectors, couplers or lead fibers, and fringe visibility variations due to state of polarization change in the lead fibers.

Spectral ratiometry can be used for signal recovery for sensors with an optical path length on the order of the source wavelength. In the case of a low-finesses Fabry–Perot fiber optic sensor used with a broadband LED, measurand-induced changes in the cavity will modulate its spectrum, leading to a shift in its reflection (and transmission) spectrum. To measure this shift the output beam is split in two, each detected by a dichromatic filter of different transmission bandwidth. The measurand-induced shift of the spectrum can then be deduced from the ratio of the detector signals. One of the most attractive features of this sensor demodulation is its immunity to variations in transmission losses and variations in source power.

In the case of the intracore Bragg grating a number of schemes have been devised to determine the wavelength of the narrowband back-reflected signals (74). None of them, however, are as inexpensive, simple, compact, and fast as the elegant spectral ratiometric method of wavelength demodulation demonstrated for the fiber optic intracore Bragg grating sensor by Melle et al. (68). If the narrowband back-reflected Bragg peak is modeled as a Gaussian function of spectral width, $\Delta\lambda$, and center wavelength, λ_B, and a linearized filter function of the form

$$F(\lambda) = A(\lambda - \lambda_0) \tag{8.40}$$

is assumed, the ratio of the filtered, I_F, to reference, I_R, optical signals

$$\frac{I_F}{I_R} = A\left(\lambda_B - \lambda_0 + \frac{\Delta\lambda}{\sqrt{\pi}}\right) \tag{8.41}$$

where A is the filtering slope and λ_0 is that wavelength for which $F(\lambda)$ is zero. This is clearly a linear function of the back-reflected Bragg wavelength (68). A schematic of this ratiometric wavelength-demodulated Bragg grating sensor was presented as Fig. 8.7. Fluctuations and variations of the source intensity, connector alignment, and coupling losses will not affect the output of this ratiometric detection system, as the system is self-referencing. Among the types of filters that can be used in such a system are colored-glass bandpass filters, narrowband or edge interference filters, dichromic filters, tapered optical fibers, or guided-wave Bragg gratings.

The strain sensitivity of this ratiometric wavelength demodulation system is determined primarily by the slope of the filter function, while the strain resolution is limited by the signal-to-noise ratio (69). A strain resolution of about 20 μstrain with a measurement range of about 3000 μstrain was attained in the case of such an intracore Bragg grating sensor. This resolution is limited by the small reflected optical signal and is a direct manifestation of having to use spectrally broad source (LED or SLD) to cover the wavelength range associated with the potential variation in strain, yet reflecting a very narrow band to obtain a relatively high strain resolution. The need to use this kind of low-intensity source represents the primary weakness of this approach and also restrict its multiplexing possibilities through power budget considerations.

8.5.5. Bragg Grating Laser Sensing System

A clever way around this dilemma is to use the sensing intracore Bragg grating to tune the wavelength of a laser. Intracore fiber optic Bragg gratings were first employed as an external tuning mirror for an argon ion laser by Hill et al. (30) and later for a semiconductor laser by Dunphy et al. (21). In this arrangement the lasing wavelength is controlled by the mesurand determined back-reflected Bragg grating optical signal. Monitoring the lasing wavelength with the spectral ratiometric technique allows direct determination of the measurand (Fig. 8.11). The advantage of this approach is the greatly improved signal-to-noise ratio arising from the much larger optical signals. A fiber laser can be used as an alternative to a semiconductor laser, and tuning of this kind of laser has already been demonstrated (5). We have continuously tuned an

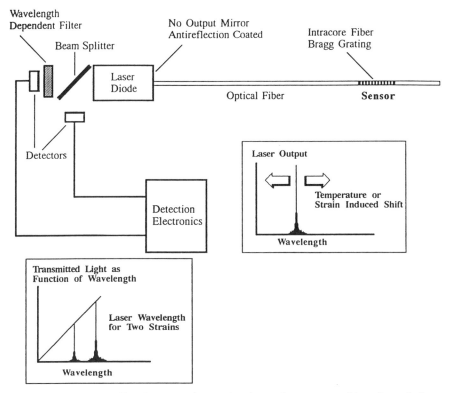

Figure 8.11. Intracore fiber Bragg grating semiconductor laser sensor with ratiometric laser wavelength demodulation. [From Measures et al. (67).] Reprinted with permission.

erbium-doped optical fiber laser over an interval of about 5 nm by varying the temperature of its intracore Bragg grating. We have also achieved a strain resolution of about 5 μstrain and a frequency response of 13 kHz, with this type of fiber laser strain sensor (70). More recently, we have shown that three Bragg grating fiber laser sensors can be multiplexed along a single optical fiber (Fig. 8.12) (2).

The real advantage of this intracore Bragg grating laser sensor approach, however, is the potential for the demodulation system to be fabricated on an optoelectronic chip and be very inexpensive if used in sufficient numbers (66). It would also be compact, robust, and able to be integrated with the structure that is serving as host to the sensing system (Fig. 8.13). Under these circumstances this optoelectronic chip would form the structural interface and could permit communication via a single output channel (which could be electrical or optical) due to the electronic multiplexing possible within this chip. This could solve many of the critical issues facing the practical implementation of smart structure technology, discussed in the next section. In certain situations a noncontact mode is feasible for this optoelectronic smart structure interface chip, and this lends itself to the concept of an optical synapse (67).

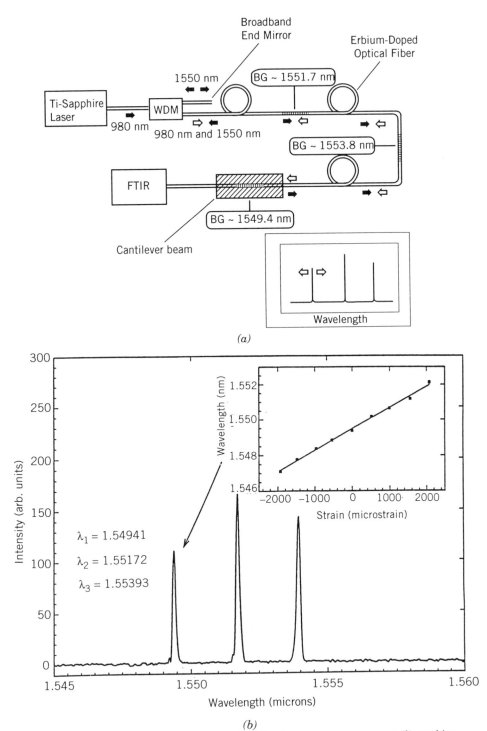

Figure 8.12. (a) Experimental three Bragg grating fiber laser sensor arrangement; (b) resulting spectrum with three laser wavelengths, inset showing linear tuning of one of the fiber lasers when its Bragg grating is strained on a cantilever beam. [From Alavie et al. (2).] Reprinted with permission.

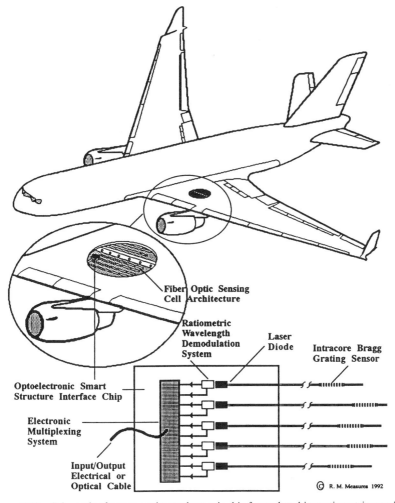

Figure 8.13. Schematic of a proposed optoelectronic chip for undertaking ratiometric wavelength demodulation of many Bragg grating laser sensors.

8.6. CRITICAL ISSUES FOR SMART STRUCTURE STRAIN SENSING

Implementation of smart structure technology will require that a number of critical issues be addressed. These are indicated in Fig. 8.14 and can be divided into micromechanic issues and system architecture issues (66).

8.6.1. Influence of Embedded Optical Fibers on the Material Properties

If optical fibers are to be embedded within advanced composite materials or civil engineering concrete structures, they must not compromise the tensile or compressive strength, increase the damage vulnerability, or reduce the fatigue

Fiber Optic Smart Structure Critical Developments

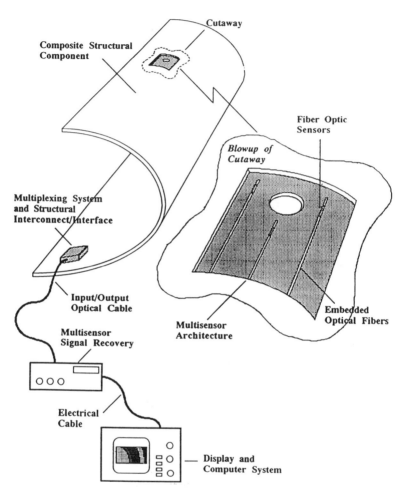

Figure 8.14.   Critical development areas for fiber optic smart structures.

life of these materials. Preliminary evidence suggests that the presence of embedded optical fibers has a minimal degradation of composite material properties (9, 18, 56, 83) provided that the diameter of the optical fiber is less than about 125 μm. Nevertheless, micrographic studies reveal that if the optical fibers are embedded at an angle to the adjacent ply directions, resin cavities (termed "resin eyes" because of their shape) are created. The formation of resin eyes leads to high stress concentrations at the host–optical fiber interface which may over a period of time, and under occasional high loading conditions, lead to debonding of the optical fiber from the host. Clearly, more definitive research will be needed before optical fibers can be embedded with confidence within structures intended to have a 20 (plus)-year working life.

Recently, we have developed a new sensing modality based on Bragg intra-grating sensing that permits any degradation in the Bragg grating sensor/host bonding to be detected (103).

8.6.2. Sensor–Host Interface and Performance Life

The interfacial shear strength between the optical fiber and the host resin matrix determines the degree of adhesion of the optical fiber within the composite material and thereby the reliability and performance life of embedded fiber optic sensors. One method of trying to improve our understanding of the micromechanical behavior of optical fibers embedded within composites is to use instrumented macromodels. Lesko et al. (52) have compared extrinsic Fabry–Perot fiber optic sensors with conventional resistive foil strain gauges in such studies. Their initial results demonstrated that the embedded fiber optic strain sensor provides results that were not too dissimilar from those obtained with the embedded conventional strain gauges, which in turn appeared to confirm the theoretical predicted matrix stress concentration created in the vicinity of a fiber fracture (Fig. 8.15).

It is possible that the high stress concentration observed around embedded optical fibers could cause debonding between the optical fiber and the host matrix. This represents a potentially serious concern in terms of sensor

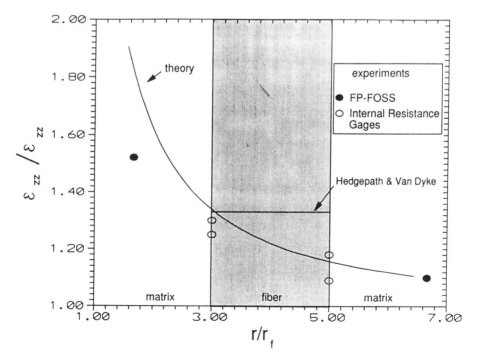

Figure 8.15. Strain distribution measured with embedded fiber optic sensors and conventional strain gauges for a macromodel of fiber-reinforced composite material. [From Lesko et al. (52).] Reprinted with permission.

performance. Careful consideration will have to be given to the diameter of optical fibers and their type of coating if they are to be embedded within composite structures and function correctly with no performance degradation for the useful life of the structure. The presence of coating could have a profound influence on this sensor–host interface and could permit the sensor's sensitivity and performance life to be optimized by reducing the high stress concentration in the vicinity of the optical fiber. Sirkis and Dasgupta (88, 89) have considered the effects of coating size and stiffness, while DiFrancia et al. (21) employed a pullout test {used for some time to study the fiber–matrix interphase for composite materials (80, 81) to compare the properties of different polyimide coated optical fibers embedded within neat resins. Pak (79) has modeled a coated optical fiber embedded within a host composite as a set of concentric circular inclusions. The influence of the coating modulus and its thickness on the stress and strain concentration factors have been calculated for the case of a far-field longitudinal shear stress applied parallel to both composite reinforcing fibers and optical fiber. It is shown that maximum strain is induced when the coating shear modulus is the geometric mean of the moduli of the optical fiber and the matrix. It is also predicted that when the coating is softer (stiffer) than the matrix, a thinner (thicker) coating will induce more shear transfer. An extensive study of the micromechanical properties of embedded optical fibers in graphite–epoxy has been undertaken by Roberts and Davidson (82). Although they reveal cases of debonding between the optical fiber and the resin in the case of an acrylate coating, their general conclusion is that an 80-μm optical fiber with a 6.5-μm polyimide coating is quite suitable for incorporation into composite materials.

When optical fibers are embedded collinear with the ply direction, no resin eye forms and minimal stress concentration is expected. However, for this configuration a resin cavity is formed at the tip of the optical fiber, and this could lead to initiation of sensor debonding from the host material. This suggests that it may be prudent to locate the sensing region some distance from the end of an embedded optical fiber intended for extensive use. We are currently investigating the seriousness of this problem and have shown that a shear lag model can be used to determine the interfacial shear stresses on the fiber–host boundary. We have also shown that it is possible to design coatings for the optical fiber sensors that would alleviate the high stress concentration and greatly reduce the prospect of debonding (LeBlanc and Measures, 47). We also plan to undertake cyclic loading of embedded fiber optic strain sensors to evaluate the improvement in the performance life of these sensors made possible with specially designed coatings.

8.6.3. Sensing System Damage Vulnerability and Degradation

A sensing system within a practical smart structure will have to be fairly robust and degrade gracefully when the structure suffers modest damage. Special coatings and a judicious choice of location and orientation may help to reduce premature fracture of the embedded optical fiber and use of a cellular sensing architecture could minimize the degradation of sensing capability associated with the loss of any particular set of optical fibers.

8.6.4. Sensing System Architecture

The type of measurement to be undertaken will dictate whether the fiber optic sensors should be localized or distributed, while the nature of the structure will determine if they are multilayered or limited to form a single layer. Optical fiber orientation and placement, especially in an advanced composite material layup, spatial resolution, and constraints imposed by the optical fiber finite bend radius are all important factors to be considered. It will also be important to identify any special structural features or regions of high stress concentration. The power budget and signal-to-noise factors will certainly play a key role in defining the sensing architecture (62). Sensing system damage vulnerability and ease of fabrication represent other considerations to be taken into account.

8.6.5. Multiplexing Strategies

Multiplexing is the merging of data from several channels into one channel, while demultiplexing is the inverse. The primary parameters used in optical multiplexing schemes are, wavelength, time, frequency, phase, and space; consequently, there are five multiplexing techniques. In the case of large civil structures, time-division multiplexing can be practical and is also the least expensive. Our wavelength-demodulation approach can also be applied in this case to determine the strain from a number of intracore Bragg gratings spaced along a suitably long length of optical fiber (66). However, there are other reasons for not placing all the sensors on a single optical fiber and winding it throughout the structure. It may make good sense to have a few sensors on each optical fiber and use a cellular sensing architecture for flexibility in terms of geometrical considerations and graceful degradation. In terms of modest scale structures, such as the section of an aircraft wing, it may not be possible to accomplish the multiplexing as part of the structure, in which case an array of optical fibers would form the structural interface. The Bragg laser sensing system does, however, offer the prospect of undertaking sensor demodulation and multiplexing electronically right at the structural interface. This is discussed further in the next section. For large structures, such as pipe lines and bridges, serially multiplexed sensors might be better.

8.6.6. Structural Interconnect/Interface

The nature of the structural interconnect problem hinges on the form of the input–output. Current thinking is predicated on extracting the optical signals from the array of single-mode optical fiber sensors via some form of structural interface. In general, this interface must have minimal structural perturbation and be easy to build and introduce during the fabrication of the structure. If multiplexing is not used, each fiber optic sensor has its own input–output and a ribbon or bundle of single-mode optical fibers would have to egress from the structure. This represents an extremely difficult task and is one of the major challenges facing implementation of the smart structures concept to practical structural components, like the wings of an aircraft or the blades of a helicopter. The structural interconnect problem would be greatly simplified if each major sensing cell were connected through a single electrical (or multi-mode optical) cable to the computer facility for interpreting the data. The development of a multisensor signal processing optoelectronic chip would simplify the interconnect issue and provide a number of advantages (66). One of the most important is the great reduction of unit cost when so much of the system is reduced to the form of an optoelectronic chip. This includes multiplexing, multisensor signal processing, and conversion to either an elec-trical or optical output on a single cable. Another important aspect of the structural interconnect is its need to be extremely user friendly. This is particularly true if smart structure technology is to be implemented broadly, such as on aircraft, space platforms, helicopters, and ships, where someone may have to assemble and connect the system in a harsh environment or an awkward location.

8.7. THERMALLY INDUCED APPARENT STRAIN

The change in the phase of light propagating along a structurally integrated optical fiber subject to a change in temperature, but no applied force, is another issue that has to be addressed when considering the use of optical fiber sensors for strain measurements in practical smart structures. This is especially so in the aerospace field, where structures are likely to be subjected to considerable temperature excursions. Strain measurements made with optical fiber sensors embedded within such structures would thus be complicated by the appearance of apparent strain due to the appreciable thermal sensitivity of the index of refraction and thermal expansion of the optical fiber and the host (32, 63).

8.7.1. Thermal Sensitivity and Apparent Strain

In the case of interferometric sensors the incremental change in the phase associated with an incremental change in both the local stress, $\Delta\sigma$, and the temperature, ΔT, is given by

$$\Delta\phi = \left[\frac{\partial\phi}{\partial\sigma}\right]_T \Delta\sigma + \left[\frac{\partial\phi}{\partial T}\right]_\sigma \Delta T \qquad (8.42)$$

This can be expanded in the form

$$\Delta\phi = k\left(n\left[\frac{\partial L}{\partial \sigma}\right]_T + L\left[\frac{\partial n}{\partial \sigma}\right]_T\right)\Delta\sigma + k\left(L\left[\frac{\partial n}{\partial T}\right]_\sigma + n\left[\frac{\partial L}{\partial T}\right]_\sigma\right)\Delta T \quad (8.43)$$

which can also be written

$$\Delta\phi = nkL\left[\left(\left[\frac{\partial\varepsilon}{\partial\sigma}\right]_T + \frac{1}{n}\left[\frac{\partial n}{\partial\varepsilon}\right]_T\left[\frac{\partial\varepsilon}{\partial\sigma}\right]_T\right)\Delta\sigma + \left(\frac{1}{n}\left[\frac{\partial n}{\partial T}\right]_\sigma + \frac{1}{L}\left[\frac{\partial L}{\partial T}\right]_\sigma\right)\Delta T\right] \quad (8.44)$$

which can be expressed in terms of Young's modulus, E, and the coefficient of thermal expansion for the optical fiber, α_F:

$$\Delta\phi = nkL\left[\left(1 + \frac{1}{n}\left[\frac{\partial n}{\partial\varepsilon}\right]_T\right)\frac{\Delta\sigma}{E} + \left(\frac{1}{n}\left[\frac{\partial n}{\partial T}\right]_\sigma + \alpha_F\right)\Delta T\right] \quad (8.45)$$

where the second term in (8.45) corresponds to the *strain-optic effect*, and the third term accounts for the *thermooptic effect*. Note that the coefficient of thermal expansion in (8.45) corresponds to that of the free fiber.

We introduce the *bonded fiber phase-strain coefficient*,

$$g = nkL\left(1 + \frac{1}{n}\frac{\partial n}{\partial\varepsilon}\right)_T \quad (8.46)$$

and the *free fiber thermal phase coefficient*,

$$f = nkL\left(\frac{1}{n}\frac{\partial n}{\partial T} + \alpha_F\right)_\sigma \quad (8.47)$$

In the case of an optical fiber bonded to a composite host material, under conditions of no applied load other than the incremental stress arising from the difference in the coefficient of thermal expansion between the host material and the optical fiber, we can write

$$\Delta\sigma = E(\alpha_H - \alpha_F)\Delta T \quad (8.48)$$

where α_H is the *coefficient of thermal expansion* for the host material. Equation (8.45) would thus indicate that the incremental change in the phase associated with an increase of temperature, ΔT, can be expressed in the form

$$\Delta\phi = g(\alpha_H - \alpha_F)\Delta T + f\,\Delta T \quad (8.49)$$

Table 8.3   Strain and Temperature Coefficients for Two Types of Optical Fiber

	York HB 600 [Hogg et al. (33)]	830-nm PANDA [Hogg et al. (35)]
g (deg $\cdot$ μstrain^{-1} cm^{-1})	13.3	8.2
f (deg $\cdot$ K^{-1} cm^{-1})	55.0	83.1
κ (μstrain K^{-1})	27.4	33.4

If we rewrite Eq. (8.49) in terms of an *apparent strain*, ε_{app}, we have

$$\Delta\phi = g\varepsilon_{app} \qquad (8.50)$$

and we see that the apparent strain is given by

$$\varepsilon_{app} = \left(\frac{f}{g} + \alpha_H - \alpha_F\right)\Delta T \qquad (8.51)$$

The *apparent strain sensitivity* can thus be defined by the relation

$$\kappa = \frac{f}{g} + \alpha_H - \alpha_F \qquad (8.52)$$

A similar analysis undertaken for the intracore Bragg grating sensor leads to comparable results.

For a double-pass Fabry–Perot fiber optic sensor with $\alpha_F \sim 0.5$ (μstrain K^{-1}), representative values of g, f, and κ are provided for two types of optical fiber in Table 8.3, assuming that the sensor is bonded to an aluminum plate [$\alpha_H = 23.8$ (μstrain K^{-1})].

8.7.2.   Temperature-Compensated Strain Sensing and Cross Sensitivity

One method of dealing with the appearance of thermal apparent strain is to measure the temperature at the same time as the strain is monitored. This approach is often employed with conventional strain gauges and is fairly easy to implement for fiber optic strain sensors by introducing a compensating sensor that is exposed to the same temperature, but not the strain (Fig. 8.16a).

For an optical fiber bonded to a host material subject to an incremental change in temperature, ΔT, with no applied load, we saw from Eq. (8.48) that the incremental change in the stress,

$$\Delta\sigma = E(\alpha_H - \alpha_F)\Delta T$$

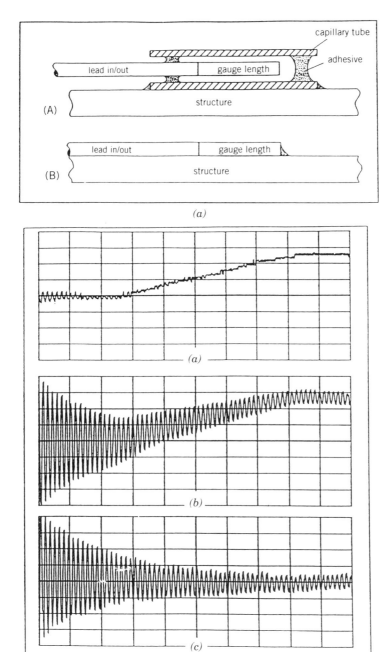

Figure 8.16. Temperature–compensated strain sensing with two surface-adhered Fabry–Perot fiber optic sensors. (*a*) (A) temperature sensor that is supported only at one end so as to be decoupled from any external strain, (B) sensor for measuring both strain and temperature of host. (*b*) Set of three Fabry–Perot fiber optic sensor temporal outputs from a vibrating cantilever beam subject to a sudden rise of temperature: (A) temperature sensor output; (B) strain and temperature sensor output; (C) strain signal with temperature influence removed (see the text).

where α_H is the *coefficient of thermal expansion* for host material, and α_F is the *coefficient of thermal expansion* for the optical fiber. The changes in phase experienced by two sensors, one bonded to the host, the other free (i.e., uncoupled to host strain) subject to an incremental change in temperature, ΔT, are

$$\Delta\phi_{\text{bonded}} = [g(\alpha_H - \alpha_F) + f]\,\Delta T\,g\,\Delta\varepsilon_{\text{stress}} \tag{8.53}$$

and

$$\Delta\phi_{\text{free}} = f\,\Delta T \tag{8.54}$$

where $\Delta\varepsilon_{\text{stress}}$ represent the stress-induced strain to be evaluated.

If we introduce the *temperature compensation factor,*

$$C_T \equiv [g(\alpha_H - \alpha_F) + f]/f \tag{8.55}$$

the stress-induced strain

$$\Delta\varepsilon_{\text{stress}} = [\phi_{\text{bonded}} - C_T\phi_{\text{free}}]/g \tag{8.56}$$

Hogg et al. (34) have determined experimentally that for an 830-nm Panda optical fiber adhered to aluminum,

$$C_T = 3.32 \quad \text{and} \quad g = 8.2\,(\text{deg}\cdot\mu\text{strain}^{-1}\,\text{cm}^{-1})$$

They have also demonstrated that this means of temperature compensation permits load-induced strain sensing to be undertaken even in the presence of a rapidly varying temperature (Fig. 8.16b), at least in the case of a sur-face-adhered sensor. Note that use of a similar sensor for both the temperature and strain measurement is in keeping with the strategy of using one kind of sensor for all measurements.

In certain situations it may be possible to deconvolute the strain and temperature by making two measurements with the same sensor. For example, Blake et al. (10, 11) and Huang et al. (37) suggested that strain and temperature could be evaluated simultaneously by monitoring the LP_{01} and LP_{11} spatial mode interference on the two eigenpolarization axes of an elliptic-core two-mode sensor. The two polarization modes in a highly birefringent Fabry–Perot fiber optic sensor were used by Farahi et al. (23) to evaluate the strain and the temperature experienced by the sensor when heated. Simultaneous measurements of strain and temperature were also undertaken by Vengsarkar et al. (98) using two-wavelength excitation of an elliptical-core two-mode sensor. The sensor performed as a polarimeter when operated at a wavelength greater than the cutoff wavelength and as a two-mode sensor when operated below the cutoff. A resolution of $10\,\mu\text{strain}$ and $5°\text{C}$ was attained when a 28-cm gauge length was employed. A similar approach should be possible with a Bragg grating sensor.

In the case of a polarimetric sensor, Dakin and Wade (19) and Bock and Wolinski (12), have shown that the use of a 90° rotation of the polarization axes can lead to temperature compensation that diminishes the temperature sensitivity of a free fiber optic sensor. In reality, however, it is necessary to bond or embed the sensor, and the difference in the thermal expansion coefficients between the optical fiber and the host material is still likely to give rise to an appreciable apparent strain (see the above-calculated relative contributions to the apparent strain sensitivity for an optical fiber bonded to aluminum.

In the case of optical fiber sensors embedded within composite material structures, the thermal apparent strain can depend quite strongly on the orientation of the optical fiber relative to the reinforcing fibers due to the highly anisotropic properties of these materials. These factors might have contributed to the errors observed in trying to make simultaneous strain and temperature measurements within a heated composite specimen with embedded polarimetric and two-mode fiber optic sensors (72). The relative importance of transverse coupling to these considerations also has to be resolved. Conventional resistive foil strain gauges expected to work over appreciable temperature ranges have to be tailored to the material to which they are adhered. A similar procedure may be required for fiber optic strain sensors, and this would necessitate the development of special thermally compensated optical fibers, probably based on suitable doping.

Finally, the issue of strain and temperature *cross-sensitivity* terms may have to be taken into account. However, Farahi et al. (23) and Vengsarkar et al. (98) have shown that for free optical fibers these terms are negligible unless large strain or temperature excursions are to be considered.

8.8. STRAIN MEASUREMENT APPLICATIONS

8.8.1. Sensing Strategy for Strain Field Measurements

If a scalar field, such as pressure or temperature, is to be mapped, a grid of localized line-integrated fiber optic sensors can be used in conjunction with an inverse radon transform. Indeed, direct measurements of angular orientation of a structure can be made directly with such line-integrated sensors (Fig. 8.17) (93). Such measurements have specific value in certain forms of structural control situations. However, a grid of line-integrated optical fiber sensors, in general, cannot be used to map a vector or tensor field, such as strain. Exceptions can arise where the field is constrained in such a way that it is not truly two-dimensional in nature. Where the field is two-dimensional, three independent measurements are required at each point to specify the field uniquely. This can be accomplished, in certain situations, by using the optical analog of the conventional resistive foil strain rosette, comprising three small identical high-sensitivity fiber optic sensors set at different orientations

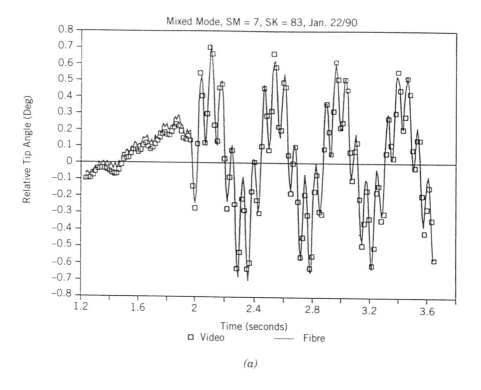

(a)

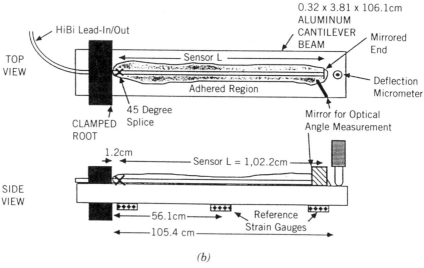

(b)

Figure 8.17. (a) Side and top view of a polarimetric fiber optic sensor bonded to surface of an aluminum cantilever beam; (b) temporal variation of the beam tip angle as measured by the polarimetric fiber optic sensor (line) and an impinging system that measured the deflection of a laser beam reflected from a small mirror mounted on the end of the beam (squares).

UNIVERSITY OF TORONTO
INSTITUTE FOR AEROSPACE STUDIES

Figure 8.18. Comparison of a conventional foil electrical strain rosette with that of a fiber optic strain rosette.

(Fig. 8.18) (96). Under these conditions high spatial resolution is important, as each sensor must have a gauge length that is small compared to the scale length for the field gradient. This tends to rule out two-mode and polarization sensors, except for very large structures (Table 8.2).

The optimum sensing strategy is likely to be highly dependent on the specific situation and the types of measurement to be undertaken. For the relatively small complex structures found in aircraft, helicopters, spacecraft, pressure vessels, and so on, a highly flexible "sensing cell architecture" would be appropriate (Fig. 8.19). This is particularly so for multilaminate composite structures, which have regions of different ply layups and thickness. An example is presented in Fig. 8.20. This may also be true for large structures such as a bridge if load distribution or strain anomalies are to be recorded, structural vibrations monitored, the traffic flow tracked, and corrosive degradation assessed. On the other hand, if wind loading on a tall building is to be monitored, the best sensing strategy might be to use a distributed (or quasi-distributed) fiber optic sensor system that is bonded to the wall of the building. This kind of sensor would also be well suited for other large structures, such as bridges, large flexible space structures, the hulls of ships, and pipelines.

The structural interface and interconnect aspects of any particular situation are likely to be key factors that must be taken into account when considering the sensing system architecture. In the case of structural components that have

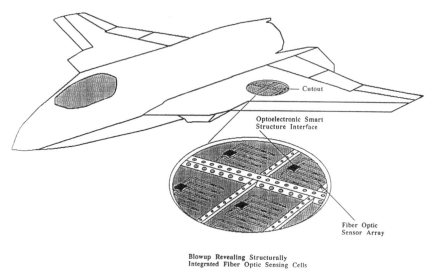

Figure 8.19. Proposed fiber optic "sensing cell" architecture's potential flexibility in terms of structural integration.

to be inspected, tested, and serviced on a regular basis (e.g., aircraft) an extremely robust and user-friendly interface/interconnect system will be required. The multisensing system based on the intracore Bragg grating laser sensor described earlier (Fig. 8.11) has the potential to be integrated onto an optoelectronic chip (Fig. 8.13) to form a part of such a structural interface (66) and greatly simplify the interconnect problem. This kind of structural interface would also lend itself to the kind of flexible sensing cell architecture discussed above.

Since it will be necessary to make the unit cost of this sensing technology as low as possible for its implementation to be economically attractive, one type of sensor should be capable of undertaking as many different measurements as possible, for then it could be used in a diverse range of applications. Some of the most important measurements include mapping strain distribution, monitoring load history, evaluating mechanical vibration frequencies, identifying spatial vibration modes, measuring pressure and temperature, tracking acceleration, determining elongation or deformation, cure monitoring for advanced composite materials or concrete, detecting corrosion degradation, locating impact sites and or pressure leaks, and assessing damage such as cracks in metals or concrete or delaminations in composite materials.

The Fabry–Perot fiber optic sensor is a very sensitive strain gauge for bonding to, or embedding within, a structure. However, the Bragg grating fiber optic sensors may have sufficient sensitivity to determine strain distribution, acceleration, vibration frequency, spatial mode, pressure, temperature, and elongation measurements. For many practical situations, the temperature has

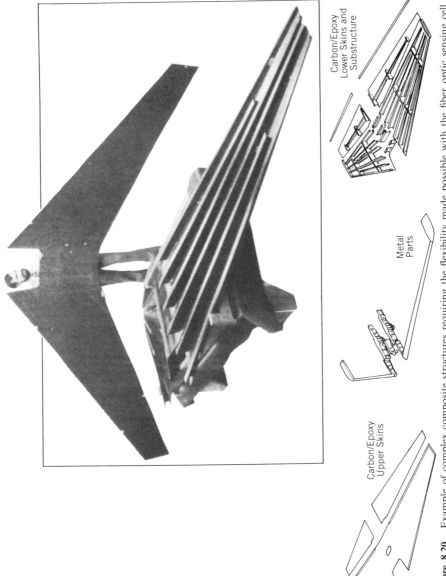

Carbon/Epoxy
Lower Skins and
Substructure

Metal
Parts

Carbon/Epoxy
Upper Skins

Figure 8.20. Example of complex composite structures requiring the flexibility made possible with the fiber optic sensing cell architecture.

213

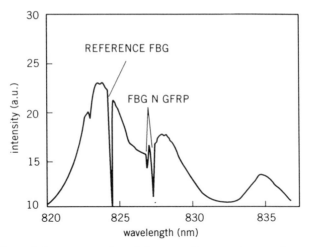

Figure 8.21. Transmission spectrum through intracore Bragg grating revealing notches arising from the reference and sensing gratings. Also seen is a small additional notch (to the left of the sensing notch) which arises from strain-induced birefringence. [From Simonsen et al. (86).] Reprinted with permission.

to be measured to account for thermally induced apparent strain. This can be quite considerable (of the order of $1000\,\mu$strain or more) for structures that may be subject to 80°C temperature excursions. Both the Fabry–Perot and Bragg grating fiber optic sensor can measure the temperature directly by preventing them from experiencing strain coupling from the structure (21, 34). It is also possible to measure both strain and temperature simultaneously using one or two sensors, even when they are strain coupled to the host material (see Section 8.7.1).

8.8.2. Basic Strain Measurements

The strain field within a structure represents a primary parameter that can be related to any load applied to the structure and is clearly relevant when considering the response of the structure in terms of its vibration and deformation. We shall see that such measurements are crucial for active control of adaptive structures and that precise measurements of strain may also permit other parameters, such as the degree of cure of a thermoset composite or the extent of delamination damage within a multilaminate composite material structure, to be evaluated.

Simonsen et al. (86) have investigated the strain sensitivity, linearity, and reproducibility of fiber optic intracore Bragg grating sensors BGS embedded within glass-fiber-reinforced polyester (GFRP). The shift in the Bragg wavelength notch of the sensor was determined by comparison to the wavelength notch of a reference BGS in a free segment of the optical fiber. Figure 8.21 displays a spectra recorded on a $\frac{1}{2}$-m monochromator, showing the

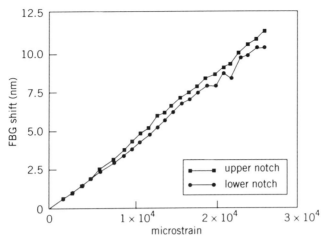

Figure 8.22. Variation of the shift in the wavelength of the intracore Bragg grating with strain revealing the creation of second notch at high strain values. [From Simonsen et al. (86).]

two notches corresponding to a load applied to the specimen that registers 5000 μstrain on the conventional strain gauge. The appearance of an additional small notch on the shoulder of the embedded sensing notch is associated with strain-induced birefringence in the low-birefringent optical fiber. The development and separation of this second notch is even more apparent in Fig. 8.22, in which the BGS was tested to destruction. At 26,000 μstrain on the embedded BGS, the optical fiber fractured. The measured wavelength sensitivity was 0.65 nm per 1000 μstrain, which is in agreement with the calculated value and the experiments of Melle et al. (68).

The more relevant parameter is the strain resolution, which for the Simonsen system was about 100 μstrain, compared to the 8 μstrain achieved recently using a passive, ratiometric, wavelength-demodulation technique on a reflective Bragg grating sensor (71). Furthermore, in the case of the ratiometric wavelength demodulation system it was the very low optical signal that limited the strain resolution to 8 μstrain. It is expected that this limit could easily be exceeded by at least an order of magnitude when the ratiometric wavelength demodulation system is used with an intracore Bragg grating laser sensor (66) (Fig. 8.11).

Vengsarkar et al. (99), have demonstrated that it is possible to create fiber optic sensors in which the strain sensitivity is a function of position along the optical fiber. One of the ways they achieved this weighted strain sensitivity is through modulating the germanium-doped core index of refraction of an elliptic–core optical fiber by exposure to intense optical radiation from an argon-ion laser. During exposure the structure with the optical fiber attached is forced into the specific vibration-mode shape to which the sensor is to selectively respond. The modal interference of this radiation creates an index

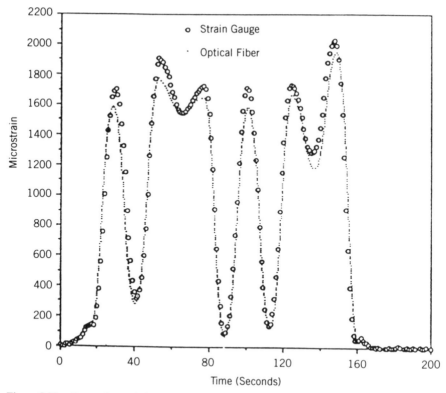

Figure 8.23. Comparison of the strain measured on the wing of an F-15 aircraft by a conventional foil strain gauge (open circle) and extrinsic Fabry–Perot fiber optic sensor (dot). [From Murphy et al. **(77)**.] Reprinted with permission.

grating along the length of the optical fiber with a beat length that is directly related to the strain imposed on the optical fiber. Although these weighted strain sensors would have merit for vibration suppression or control of specific modes in simple structures, they are too specialized for general strain sensing and cannot be used for measurements of the local strain field.

The first fiber optic measurements of strain from the wing of an aircraft have been reported by Murphy et al. (79). They affixed an extrinsic Fabry–Perot sensor to the underside of the wing of an F-15 aircraft mounted on a test facility at the Wright Patterson Airforce, Base, Ohio. The wing was loaded from zero to 100% load (representative of a fully loaded F-15 at an altitude of about 6000 m performing a 7g manoeuver) and also cyclically loaded. A comparison of the quadrature phase-shifted fiber optic sensor with that of a conventional foil strain gauge during part of the cyclic loading is presented as Fig. 8.23. The comparison is quite reasonable, and although this represents a significant first, the complexity of the dual fiber extrinsic Fabry–Perot sensor head, required for quadrature detection (Fig. 8.24), leaves much to be desired in terms of a practical, robust, and unobtrusive sensor for embedding within

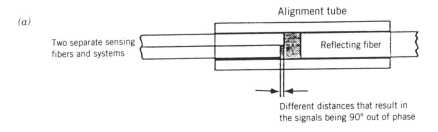

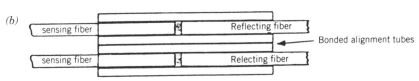

Figure 8.24. Two quadrative phase-shifted sensor configurations for the extrinsic Fabry–Perot fiber optic sensor; each involves essentially two sensors, with a 90° phase shift between them. [From Murphy et al. (77).] Reprinted with permission.

composite material structures. For this reason the "intrinsic" fiber optic Fabry–Perot interferometer pioneered by Lee and Taylor (48) constitutes an alternative sensor. We have been able to fabricate such intrinsic Fabry–Perot sensors using a reflective fusion splice based on a metal evaporation technique (96). The advantage of this approach is that it leads to the formation of internal mirrors of high reflectivity, which is important from a power budget standpoint. Recently, Lee et al. (50, 51) have demonstrated that internal mirrors of high reflectivity could be fabricated using multilayers of TiO_2 and SiO_2 and

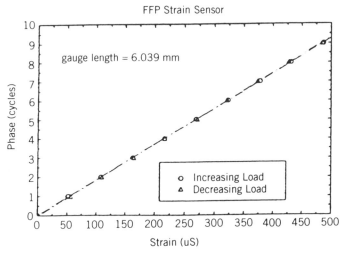

Figure 8.25. Load–unload phase-strain response of a graphite–epoxy embedded Fabry–Perot fiber optic strain sensor. [From Valis et al. (96).] Reprinted with permission.

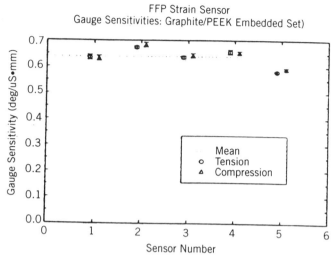

Figure 8.26. Fabry–Perot fiber optic sensor phase-strain sensitivity for a number of sensors of different gauge length embedded within graphite–epoxy. [From Valis et al. (96).] **Reprinted with permission.**

that these Fabry–Perot sensors can be embedded in thermoplastics and even cast aluminum.

We have measured the strain sensitivity of embedded intrinsic Fabry–Perot strain gauges and shown that their strain response is linear when embedded within several different composite materials—Kevlar–epoxy, graphite–epoxy,

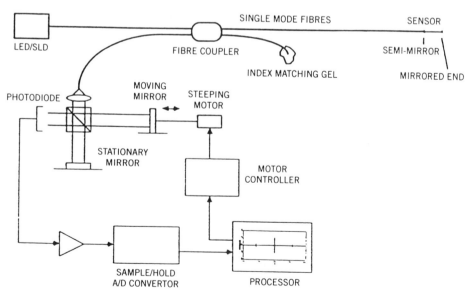

Figure 8.27. Active coherent detection scheme for use with a Fabry–Perot fiber optic sensor. The interrogating bulk Michelson interferometer is driven in a feedback mode.

Table 8.4 Fabry-Perot Fiber Optic Strain Gauge Characterization

Typical gauge length (mm)	3-20
Sensor diameter (μm)	125
Operating wavelength (nm)	632.8
Strain sensitivity (at 633 nm) (deg·μstrain^{-1}cm^{-1})	13.3
Temperature sensitivity (deg·K^{-1}cm^{-1})	55.0
Tensile strength (μstrain)	2,000

and graphite–PEEK—and that it is free of hysteresis during load and unload cycles (Fig. 8.25). We have also shown that the strain sensitivity of the Fabry–Perot sensor is not a function of its gauge length (Fig. 8.26) (96). We have embedded uniaxial FP sensors in graphite–PEEK and Kevlar–epoxy to assess sensor survival and response. Table 8.4 summarizes the Fabry–Perot fiber optic strain sensor characterization.

The performance of these devices indicates that they can match, and probably surpass, the performance of conventional resistive-foil electrical strain gauges eventually. However, unlike the conventional resistive-foil devices, they require phase demodulation and suffer from interrupt ambiguity. This weakness can be addressed by using white-light (low-coherence) interferometry to perform absolute sensing. In this approach two interferometers are used in sequence. The sensing interferometer is the Fabry–Perot fiber optic sensor, while the interrogating can be either another fiber Fabry–Perot (49) or a bulk Michelson (101). This method of demodulation can also be combined with active feedback to match the path imbalance of the two interferometers to provide high-frequency quadrature detection for the fiber Fabry–Perot sensor (53) (Fig. 8.27) that will permit them to be used in optoacoustic work. Unfortunately, none of these low-coherence techniques are likely to be convenient when used with sensors embedded within practical structures. A demodulation system has been developed recently for the Fabry–Perot sensor that may be suitable from a smart structures perspective (60).

Unfortunately, the semireflective metal fusion splice represents a source of structural weakness for the Fabry–Perot sensors. It is susceptible to failure in bending and tension at loads significantly less than those of bare fibers. The reason for this weakness is that at present the entire end face of each optical fiber is coated with a metallic film prior to fusion. This approach is unwarranted, as only the core needs to be reflective and has the effect of preventing complete fusion of the two fibers. Ideally, the cladding regions of the end faces should be glass-to-glass fused. To improve the strength of our FP sensors we have developed a method of creating localized semireflective mirrors (96), an example of which is displayed in Fig. 8.28. As can be seen, this was achieved in the case of a bowtie polarization maintaining optical fiber, needed to avoid signal fading through strain-induced birefringence.

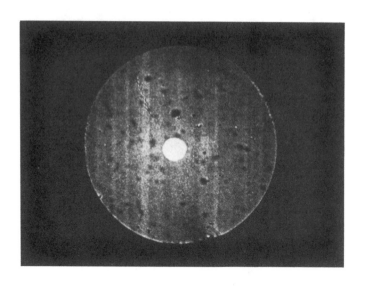

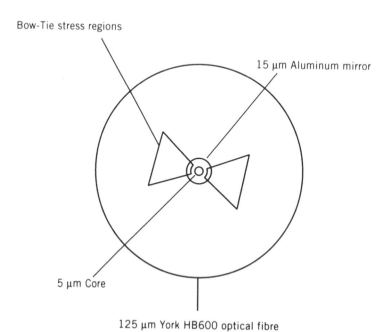

Bow-Tie stress regions

15 μm Aluminum mirror

5 μm Core

125 μm York HB600 optical fibre

Figure 8.28. (*a*) Micrograph of a 15-μm-diameter localized-mirror evaporated onto a 125-μm-diameter York HB 600 fiber; (*b*) schematic to illustrate mirror placement with respect to the core and the bowties of the polarization preserving optical fiber. [From Valis et al. (**96**).]Reprinted with permission.

Where it is necessary to measure the in-plane strain tensor at a point within a structure, the fiber optic analog of a resistive foil strain rosette is required. We were the first to recognize the importance of such a fiber optic strain rosette and to demonstrate its viability based on three Fabry–Perot fiber optic sensors (Fig. 8.29) (96). Test optical strain rosettes have constructed and bonded to the

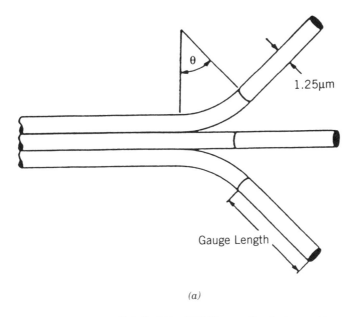

(a)

Detail of the FFP fiber-optic strain rosette

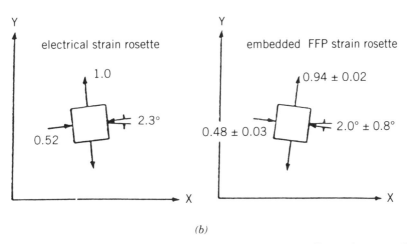

(b)

Figure 8.29. (*a*) Fiber optic strain rosette based on three Fabry–Perot fiber optic sensors; (*b*) comparison of the laminar strain tensor measured by the fiber optic strain rosette embedded within a composite cantilever beam with the resualts obtained from a surface-adhered conventional foil strain rosette. [From Valis et al. (96).] Reprinted with permission.

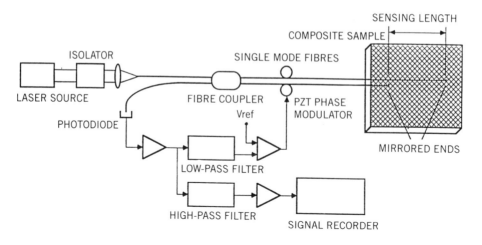

Figure 8.30. Schematic diagram of the piezoelectric phase demodulated Michelson fiber optic sensor employed for acoustic emission detection within composite materials. [From Liu et al. (55).] Reprinted with permission.

surface of an aluminum cantilever beam and embedded within a carbon–PEEK cantilever beam. The laminar strain tensor measured with the embedded fiber optic strain rosette was found to be in good agreement with the results obtained from a surface-adhered electrical strain rosette (Fig. 8.29). A picture of an illuminated optical strain rosette and a conventional resistive strain rosette is presented as Fig. 8.18. Recently, we have extended this work to the more general case of an arbitrary three-dimensional strain field (53).

Measurements Relevant to Aerospace Applications. Although, in the next section we discuss strain measurements under the aerospace heading it should be borne in mind that many of the measurements are generic in nature and would find broad application to other mobile platforms or stationary structures. A selection of some of the more important measurements for aircraft, helicopters, space platforms, and so on, includes:

- Vibration frequencies of flexible space structures, flight control surfaces, propellers
- Spatial vibration modes of flexible space structures, flight control surfaces, propellers
- Thermal deformation of space structures caused by shadows and one-sided solar heating
- Excessive loading, pressures leaks, impacts
- Long-term health monitoring, including mechanical and thermal load history

- Internal strain distribution and pressure loads (especially for human-occupied structures)
- Onset of internal crack formation (especially for human-occupied structures)
- Impact detection and localization
- Damage assessment (delamination in ACM)
- Corrosion degradation of aircraft and helicopters
- Oxygen erosion of composite space structures
- Real-time aerodynamic loads for aircraft and helicopters
- Shape evaluation for actuation control and monitoring of vibration damping
- Flap and engine control position monitoring
- Flutter and noise measurement for active suppression
- Temperature distribution and anomalous hot-spot detection

For convenience we discuss the application below under three headings.

8.8.3. Damage Assessment

Even though the Michelson sensor is not suitable for practical smart structures, it has served as a useful demonstrator of what should be possible with Fabry–Perot sensors, as their strain sensitivity is comparable. For example, the Michelson fiber optic sensor has verified that acoustic energy released as a result of matrix cracking, fiber breakage, or interlamina debonding (delamination) within various composite materials (Kevlar–epoxy and graphite–epoxy) could be it detected by an embedded interferometric fiber optic sensor. A schematic of the Michelson fiber optic sensor system used for this purpose is displayed as Fig. 8.30. The sensor can be seen to comprise a pair of unbuffered optical fibers with mirrored ends embedded within the composite material. In this work, quadrature detection was ensured by a low-frequency piezoelectric (PZT) phase modulation feedback system that also eliminated drifts and slowly varying strains (54). In these experiments the composite panels were subjected to out-of-plane loading and the high-frequency signals from the fiber optic sensor recorded. We have been able to demonstrate that embedded Michelson fiber optic strain gauges can detect acoustic emission associated with the generation of *threshold* delaminations. A representative pair of acoustic emission signals and associated frequency spectra is presented as Fig. 8.31a. The corresponding enhanced backlighted images Glossop et al. (29) revealing the delaminations association with the detected acoustic signals are displayed below the frequency spectra. If the ratio of acoustic energy in the frequency range 100 to 300 kHz and 300 to 600 kHz is plotted as a function of delamination size (Fig. 8.31b), a trend is suggested which might lead to a method of assessing the extent of delamination detected by means of acoustic

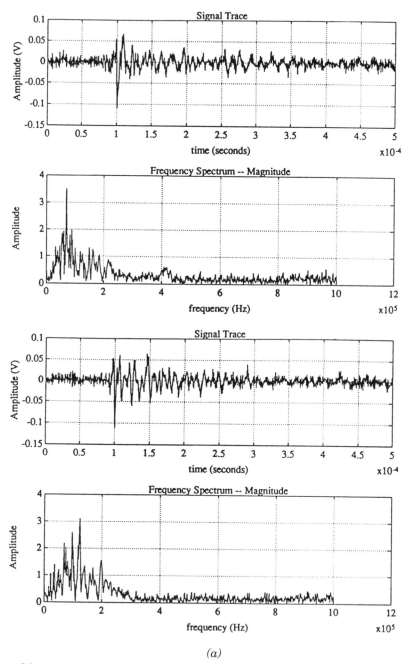

Figure 8.31. Acoustic emisssion signals (upper traces) and corresponding frequency spectra (lower traces) for the two delaminations revealed in the lower pictures through image-enhanced backlighting of the eight-ply Kevlar–epoxy coupon subjected to out-of-plane loading. The lower graph displays the ratio of acoustic energy in the frequency range 100 to 300 kHz and 300 to 600 kHz, plotted as a function of delamination size.

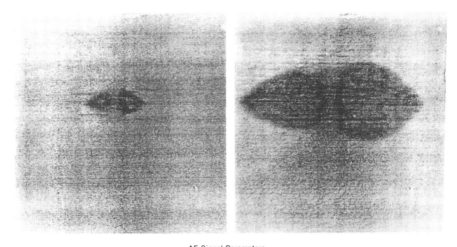

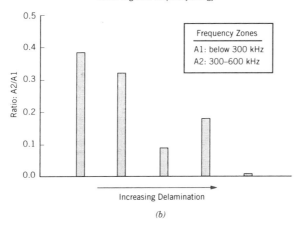

(b)

Figure 8.31. (*Continued*).

emission (55). However, the complexity of composite materials may frustrate this approach, and in the end the use of *neural networks* may be required to help locate and assess the extent of damage from acoustic emission signals. Furthermore, acoustic wave-generated birefringence induced in the embedded optical fibers can prevent the amplitude of the acoustic emission signals from conveying any useful information unless polarization maintaining optical fibers are employed (55). Recently, Lee et al. (50, 51) have demonstrated that intrinsic Fabry–Perot fiber optic sensors can indeed detect acoustic waves with frequencies of 0.1 to 8 MHz within cast aluminum and graphite–epoxy composite structural elements.

Another possible method of assessing damage (specifically, delamination) within composite material structures is based on vibration analysis (3, 59). It is

THROUGH-WIDTH DELAMINATION

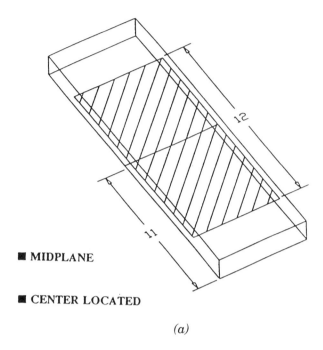

■ MIDPLANE

■ CENTER LOCATED

(a)

Effect of the Size of Delamination on
the Natural Frequencies of Cantilever
GRE Beam (Experimental Results)

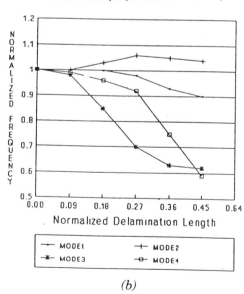

(b)

Figure 8.32. (*a*) Schematic of composite cantilever beam with full-width delamination; (*b*) normalized change in the measured frequency of a composite cantilever beam's first four vibration modes as a function of the normalized (full-width) delamination length.

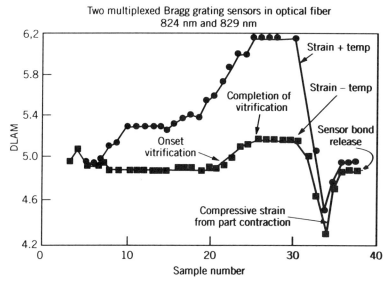

Figure 8.33. Variation in wavelength of two multiplexed Bragg grating sensors, one embedded within the curing composite specimen, the other positioned outside. Both are on the same optical fiber. [From Dunphy et al. (21).] Reprinted with permission.

anticipated that an embedded optical fiber strain sensing system might detect shifts in the frequency and/or changes in the damping constants of one or more mechanical vibration modes of a structure. This concept is illustrated in Fig. 8.32. However, initial work (28) suggests that vibration analysis cannot serve as the sole indicator of delamination damage within composite structures, although it may well complement other approaches. Optoacoustic probing and strain anomaly measurements represent two such techniques to be explored.

8.8.4. Composite Material Cure Monitoring

Cure monitoring of thermoset advanced composites with the same embedded sensing system designed for in-service structural monitoring would improve quality control of many composite material structures and thereby represent another very important application for the resident optical strain sensors. Dunphy et al. (21) indicated that an intracore Bragg grating sensor embedded within a six-layer graphite–epoxy specimen detected the onset of vitrification (Fig. 8.33) and survived the high temperature and pressure conditions within the autoclave. Their measurements also appeared to pick up the thermally induced residual strain as the specimen cooled. In these experiments a second, reference, intracore Bragg grating sensor along the same optical fiber was strain decoupled from the specimen and served to compensate for the temperature changes. However, any approach that uses the magnitude of the strain field coupled to the optical fiber to evaluate the cure state may be limited in its accuracy and relevance, as it samples the resin properties only in the immediate vicinity of the optical fiber.

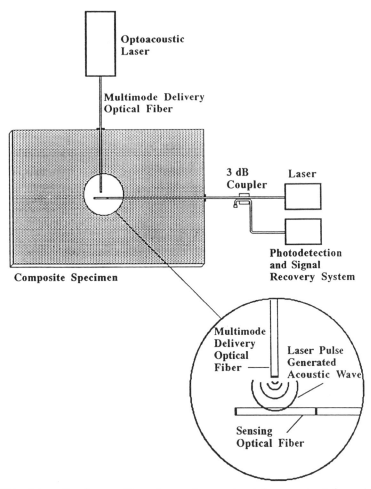

Figure 8.34. Schematic of a possible optoacoustic experimental arrangement for resin cure monitoring involving an embedded Fabry–Perot fiber optic sensor and a multimode laser pulse delivery optical fiber.

Sun and Winfree (91) suggested that measurement of spatially averaged bulk moduli should serve as good indicators of the cure state. We have undertaken some initial research to determine if cure monitoring of thermoset advanced composites might be feasible through the use of optoacoustic probing. In this approach the high sensitivity of embedded interferometric fiber optic sensors is used to measure the velocity of acoustic waves averaged over the distance between the launch and sensing sites. This should provide a precise measurement of the average cure state over a specific region of the structure. If it is desirable to keep to a purely optical technique, intense laser pulses can be employed to create the probing acoustic waves. A schematic of such a possible laser-generated optoacoustic sensing arrangement involving a Fabry–

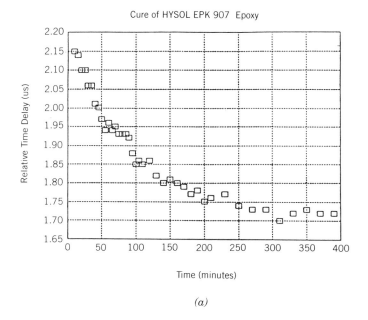

(a)

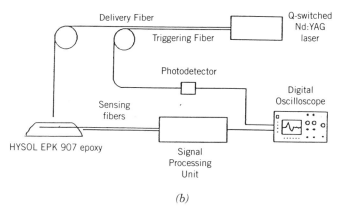

(b)

Figure 8.35. (*a*) Preliminary results of the variation in the arrival time for optically generated acoustic pulses as a function of cure time for Hysol resin: (*b*) experimental setup is indicated in the inset.

Perot fiber optic sensor and an embedded laser pulse delivery optical fiber is presented in Fig. 8.34. Preliminary results of the variation in the arrival time for acoustic pulses as a function of cure time for Hysol resin is presented in Fig. 8.35*a* (65). The experimental setup is indicated in Fig. 8.35*b*, where it is seen that the acoustic waves were generated on the surface of the specimen by laser pulses delivered from a Nd–YAG laser through an optical fiber, and

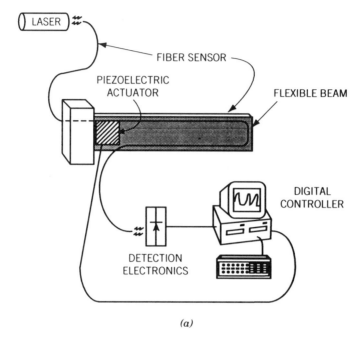

(a)

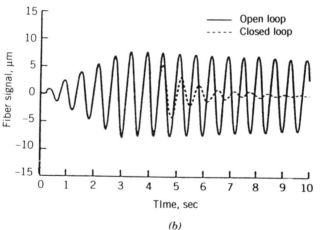

(b)

Figure 8.36. Active structural control experiment: (a) schematic of system involving an ellipti-cal-core two-mode fiber optic sensor and a piezoelectric patch actuator at the root of the cantilever beam; (b) sensor signal when experiment is run in closed- and open-loop modes [From Cox (15).] Reprinted with permission.

a Michelson fiber optic sensor was used. These and later results (81) are quite encouraging, and it is expected that autoclave tests should be shortly conducted.

It is also possible that this optoacoustic approach might be applicable to cure monitoring of concrete with an embedded fiber optic sensing system (26).

However, it is more probable in this situation that the high-frequency acoustic waves used to interrogate the material will be generated mechanically, possibly by means of a surface-mounted piezoelectric transducer.

8.8.5. Active Structural Control

In smart "adaptive" structures the information provided by the built-in sensor system is used for controlling some aspect of the structure, such as its stiffness, shape, position, or orientation. Cox (15) used an elliptical-core two-mode sensor with lead-in polarization-preserving fiber to monitor active vibration damping of a steel cantilever beam. The sensor was bonded around the periphery of the beam, with the localized region extending from behind the clamp to 4 cm from the free end. Actuation was performed with two piezoceramic patches of G-1195 mounted on opposite sides of the beam near the clamped end. The arrangement is shown in Fig. 8.36a, and an example of the controlled damping achieved with this system is indicated in Fig. 8.36b.

Active acoustic control using an elliptical-core two-mode fiber optic sensor was demonstrated by Fogg et al. (24). A simply supported thin plate was excited at both resonance and nonresonant frequencies by means of a shaker mounted to the center of the plate. Two piezoceramic plates were mounted on opposite sides of the plate and wired out of phase to produce uniform bending about the neutral axis. Attenuation of 25 and 10 dB in sound radiation was achieved for the resonance and nonresonance excitation cases, respectively, using the fiber optic sensor to provide the error signal for the control actuator (Fig. 8.37).

Adaptive (variable-geometry) truss structures possess multiple degrees of freedom, which makes them ultradexterous and capable of adopting complex configurations impossible for a conventional multijoint multilink anthropo-morphic manipulator of comparable size and mass. They are superior in terms of deployment and contraction, minimal storage space, strength to mass and stiffness to mass, telescopic motions, redundancy of actuator function, and resistance to longitudinal and lateral buckling. Smartruss represents a generic smart adaptive structure element (Fig. 8.38) that could find application in space structures as diverse as the space-based radar satellite (SBRS) and the space station remote manipulator system (SSRMS) and serve as the forerunner of flexible robotic manipulators.

Accurate and fast strain measurements will be required by any resident sensing system to be used for smart adaptive structures. We have developed a dual Fabry–Perot fiber optic strain sensor system to be used with the actuation control system of Smartruss (60). Each of the Fabry–Perot strain gauges was formed in SM800 optical fiber and had a *phase-strain sensitivity* $S = 1.02° \mu\text{strain}^{-1} \text{mm}^{-1}$ (33) at 819 nm. In this dual Fabry–Perot sensor system one modulated laser was used to excite both Fabry–Perot sensors. The back-reflected signal from each sensor was directed to its own photodiode and strain decoder electronics (Fig. 8.39). This system has a linear response

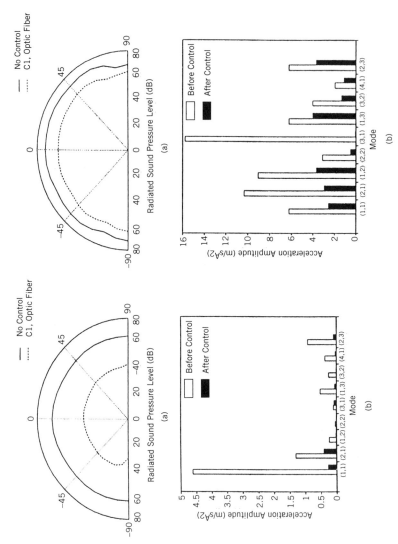

Figure 8.37. Left side indicates (*a*) acoustic directivity pattern of the plate–baffle system excited on-resonance for the (1, 1) mode at 88 Hz; (*b*) corresponding modal response of the plate with and without control. Right side indicates (*a*) acoustic directivity pattern of the plate–baffle system excited off-resonance at 320 Hz; (*b*) corresponding modal response of the plate with and without control. [From Fogg et al. (**24**).] **Reprinted with permission.**

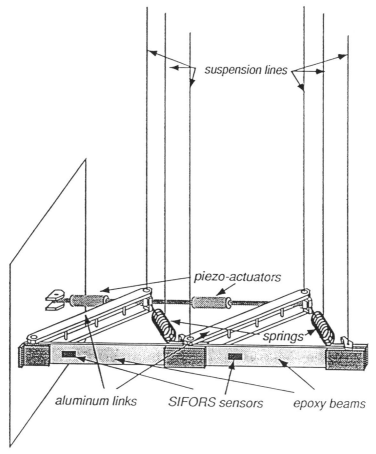

Figure 8.38. Schematic of Smartruss revealing-fiber optic sensor and piezoelectric actuator control configuration. Note in final configuration that both Fabry–Perot fiber optic sensors were placed on the left segment of the beam. (**Courtesy of Dynacon Enterprises Limited © Canadian Space Agency, 1992**)

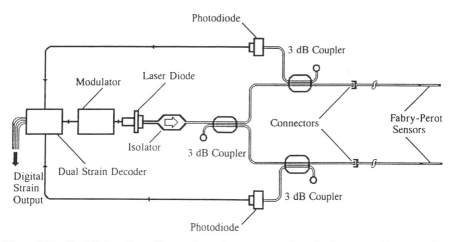

Figure 8.39. Dual Fabry–Perot fiber optic sensing system used on the Smartruss. (Courtesy of **Dynacon Enterprises Limited © Canadian Space Agency, 1992**)

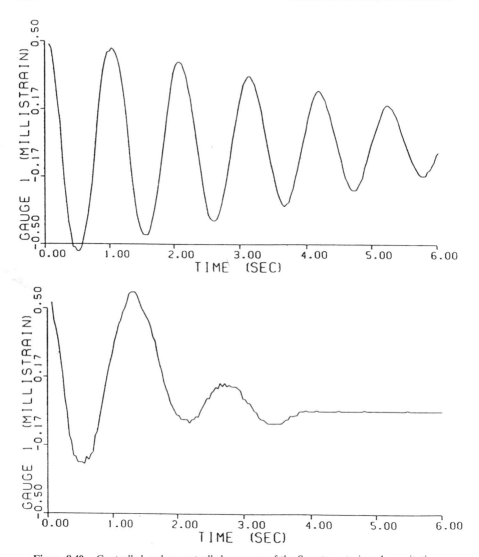

Figure 8.40. Controlled and uncontrolled response of the Smartruss to impulse excitation.

and provides a strain resolution of better than $10\,\mu$strain for a 9 mm gauge length. Furthermore, this system can respond to strain changes of $10^5\,\mu$strain s^{-1}, which is important when strain measurements are to be attained against a mechanically noisy background (vibrations and impacts). An example of the controlled and uncontrolled response to step excitation is presented as Fig. 8.40

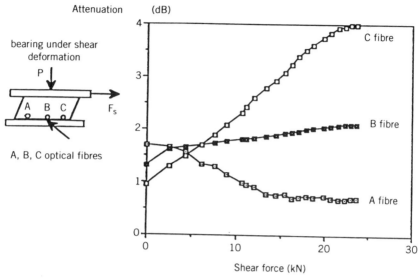

Figure 8.41. Attenuation observed in three fiber optic sensors embedded within a shear-loaded bridge bearing. Small inset indicates location of sensors and loads applied to the bearing. [From Caussignac et al. (14).] Reprinted with permission.

8.8.6. Integrated Sensing for Bridges and Other Civil Structures

A selection of some of the more important measurements that should be undertaken with the implementation of smart structure sensing to large civil structures:

- Vibration frequencies of support columns, floors, windows, bridge decks, and cables
- Spatial vibration modes of walls, floors, bridge decks, and cables
- Thermal strain and deformations caused by sunlight to one side of the structure
- Construction loads (excessive loading, pressures, impacts)
- Wind monitoring and wind pressure on bridges, buildings, and so on
- Long-term health monitoring, including load history and excessive loads
- Shear forces on bridge bearings
- Internal strain distribution and hydropressure for dams
- Onset of internal crack formation in concrete structures
- Impact detection and localization
- Damage assessment (delamination in ACM)

- Debounding of reinforcing bars and prestressing tendons in concrete structures
- Corrosion degradation
- Chemical sensor for acid rain, smog, bird droppings, deicing salt solution
- Real-time truck/automobiles weight and load distribution
- Traffic flow patterns (number of vehicles, weight, and velocity)
- Ground creep or seismic movement
- Temperature distribution and anomalous hot-spot detection

Some of the characteristic problems facing the designer of a structurally integrated sensing system for monitoring concrete civil structures are:

- Small strains, so need for high strain sensitivity
- Rough handling and a hard casting process
- Harsh environment (high levels of moisture, large temperature excursions)
- Large dimensions and exposure to electrical interference

Most bridges comprise three basic structural components: one or more deck spans, piers (or supports), and bearings. The bearings are constructed of alternating layers of elastomer sheet and steel plates and are subject to vertical compression, horizontal shear, and rotation. Special hard cladding, multimode

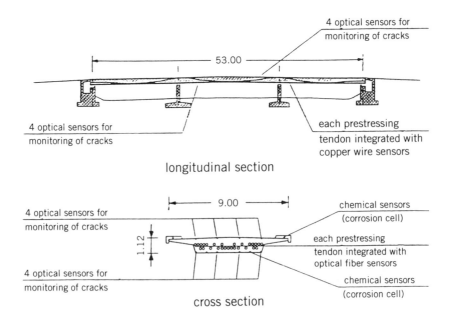

Figure 8.42. Arrangement of optical fiber sensors and chemical sensors in the triple-span Schiessbergstrasse road bridge. [From Wolf and Miessler (100).] **Reprinted with permission.**

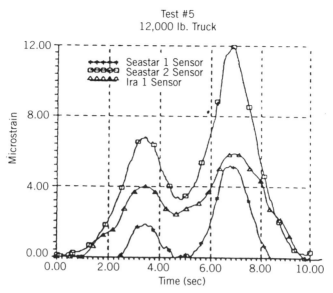

Figure 8.43. Microstrain signals versus time recorded by a pair of extrinsic Fabry–Perot fiber optic sensors adhered to a concrete bridge deck subjected to a 12,000-lb truck with an asymmetric load. This strain variation is compared with a conventional strain gauge. [From Kruschwitz et al. (45).] Reprinted with permission.

microbend optical fiber sensors were integrated into the elastomer sheets and the sensors were used both in transmission and back-reflecting modes (14). Although the attenuation of light in these optical fibers varied with both normal and shear forces applied to the bearing, its nonlinear and inconsistent dependence limits its usefulness (Fig. 8.41). Nevertheless, this initial work clearly indicates the value of pursuing this research with reliable, calibrated optical strain gauges having a linear response.

The Schiessbergstrasse triple-span road bridge (53 m long, 10 m across) in Leverkusen is one of the first of a new generation of civil structures which will use ACM tendons and built-in fiber optic sensors (Fig. 8.42). A combination of microbend sensors and elongation sensors are designed to evaluate the load on the tendons and monitor strain within the concrete (100). The bridge is designed with limited prestressing using 27 glass fibers prestressing tendons, three of which are instrumented with fiber optic microbend strain sensors to monitor excessive loading and deformation. These sensors check the integrity of the tendons or locate the damage. The tensile zone above the piers and the spans are monitored permanently with integrated optical fiber sensors with an accuracy of ±0.15 mm. Information extracted by an on-site computer is transmitted to client along a phone line. This constant surveillance permits early detection of problems and reduced cost of maintenance and repair.

Detailed knowledge of the reinforcement strains in reinforced concrete structures is a prerequisite to a thorough understanding of how such structures actually behave. Strain distributions are profoundly affected by the formation of cracks and can change during the life of a structure due to effects such as creep, shrinkage, and load history. To avoid affecting the steel–concrete interface, thereby weakening the structure, a set of about 100 electrical strain gauges were mounted within the rebars by milling two solid bars and bonding them together with the gauges and wiring included (84). These sensors were able to pick up cracks in the concrete before they reached the surface. It is expected that strain gauge instrumented concrete structures could warn if any component was approaching its overload stress and in general monitor the state of internal strain. This might provide insight into weakness developing due to corrosion. Although this research has been undertaken to date using conventional foil strain gauges, it demonstrates the kind of information that could be attained more easily using fiber optic strain gauges.

Excessive displacements or deformations in large civil structures can lead to critical states of stability that can endanger public safety. The use of electrical-based measurement systems in hydraulic plants is undesirable, as they are susceptible to corrosion of the metallic guides (due to the high atmospheric humidity), and electromagnetic interference (from lightning strikes, overvoltage). These problems are magnified by the vast distances involved (km) in major civil structures such as dams. Two types of optical fiber system for monitoring span joints and structural stability in dams is currently under consideration (35).

Extinsic Fabry–Perot optical fiber sensors have been developed and used with short gauge lengths to measure strain and temperature within concrete specimens (45). Although these sensors can be spatially multiplexed along a linear sensor data bus, they are likely to be limited to short gauge lengths and represent somewhat of a perturbation if embedded within composite materials. Nevertheless, strain direction ambiguity can be overcome in these sensors using two slightly different wavelengths or two adjacent sensors with different cavity lengths. Physically decoupling one end of the sensor head from the structure also permits the temperature to be determined. Two such sensors were embedded in a 15 cm × 20 cm × 122 cm reinforced-cement concrete specimen during fabrication. Cylindrical metal washers ensured effective longitudinal strain coupling for sensors embedded directly in concrete, while the other sensors were attached to metal reinforcing rods prior to embedding. A 5% difference in the strain was observed between the rod-mounted sensors and those directly embedded, suggesting some slip between the sensor, the rod, and the matrix, or nonparallel alignment of the sensor axes. Four of these sensors were attached, using two-part epoxy adhesive, to the undersurface of a simple span bridge made of reinforced concrete with a wooden bottom. A maximum reading of 12 μstrain was recorded when a 12,000-lb truck was driven slowly over the bridge and an asymmetry between the front and rear wheel loading was evident (Fig. 8.43) (45).

The University of Vermont is in the process of constructing a five-story, 65,000-ft^2 concrete (Stafford) Building, equipped with fiber optic and conventional sensors embedded in the concrete superstructure (25). Specifically, these sensors were embedded into the concrete itself—in the floor, the load-bearing support columns, the walls, and the ceilings. This network of sensors will provide information about the building's structural integrity, its response to both microvibrations (traffic) and macrovibrations (earthquakes), the strain levels of the reinforcement bars within the concrete, local wind pressures being applied to the outer surface of the building, and any thermal deformations of the building caused by sun-induced heating of only a part of the building. Both conventional sensors (e.g., strain gauges and accelerometers) and fiber optic and optical sensors are being used. Recently, we embedded a set of Bragg grating sensors into the precast concrete deck support girders of a unique new highway bridge in Calgary. These sensors have monitored the strain relief experienced by the carbon fiber and steel prestressing tendons over an extended period of time. The sensors have also been shown to respond to the static and dynamic loading of the bridge by a truck (102).

8.9. CONCLUDING REMARKS AND FUTURE PROSPECTS

Structurally integrated fiber optic sensors (SIFORS) will form the optical nerves of future smart structures, and strain sensing will constitute one of the most important measurement to be undertaken. Strain measurements will provide valuable information on the use and loading of such structures and will also indicate the response of the structure of these loads in terms of shape change and vibration. Strain information would obviously be vital to the structurally integrated actuation control system of any smart adaptive structure. Strain measurements may also be used to detect and assess any damage sustained by the structure and even more crucially ascertain if the structure has suffered an acute or chronic loss of structural integrity. Continuous monitoring of the applied loads could also be used to determine the actual fatigue life of the structure at any point in its operation. In the case of composite materials, improved quality control during fabrication may be possible from such a built-in sensing system, and this could lead to a more consistent and reliable product. Overall, the introduction of an embedded fiber optic strain sensing system to composite structures may enhance confidence in their use and lead to an expansion in their range of application, especially as primary structures in the aerospace field.

Two types of short gauge fiber optic sensor appear to qualify for the broad range of potential smart structures. Intracore fiber optic Bragg gratings have been formed during the fiber pulling process by exposure to intense pulses of laser (or other) radiation. This makes possible automated fabrication (4) and ensures no loss in the strength of the optical fiber. Furthermore, if they are used to control the wavelength of (fiber or semiconductor) lasers and used with the new ratiometric wavelength demodulation system (66), they can form robust,

te measurement, low-cost multisensor systems with the potential
d on an optoelectronic chip that can form part of the structural
ɪ essence solve the critical interconnect problem. There are two
ᵧᵤₑₛᵢₒₙs that remain to be proven: (1) Can the intracore Bragg grating sensor
perform all of the tasks required? and (2) Can Bragg gratings of "sufficient
quality" be fabricated in an automated process? The Fabry–Perot sensor can
undertake most transient measurements likely to be required of an optical
strain gauge. For the Fabry–Perot sensor to be seriously considered for smart
structures, however, a low-cost, fast method of fabrication will have to be
developed and one that does not compromise the strength and fatigue life of
the optical fiber. It will also need to be demodulated in a manner that allows
the absolute kind of measurement intrinsic to the Bragg grating.

Although the sensing architecture is strongly dependent on the specific
application and the structure to be instrumented, it is clear that a sensing cell
architecture offers great flexibility and is particularly attractive for complex,
multilaminate structures. In the case of very large structures, long gauge,
distributed or quasi-distributed fiber optic sensors maybe desirable. For
embedded fiber optic sensors, interpretation of the optical signals in terms of
the host strain field, thermal apparent strain, performance lifetime, structural
interface, and interconnection represents serious issues that remain to be
addressed.

An enormous financial investment is associated with the established civil
infrastructure: bridges, highways, railways, dams, pipelines, port facilities,
airports and buildings. Aging of these structures represents a financial problem
of immense proportions. Instrumentation of these structures with short and
long gauge fiber optic sensors that could accurately assess their useful remaining life and schedule repair or replacement when required could lead to
substantial cost savings. Furthermore, we are witnessing a major change in the
design and fabrication of large civil structures like bridges, based on the
replacement of steel with composite materials. It just happens that these new
materials are quite conductive to the incorporation of resident fiber optic
sensing systems which will allow the future custodians of these engineering
investments to monitor the structure in a comprehensive way (102). This
built-in inspectability will enhance confidence in using these new materials so
that the two advances (use of composites and sensing) will reinforce each other
and greatly accelerate the implementation of smart structure technology.

In the twenty-first century, smart structure technology could lead to
structures that are self-monitoring and even self-scheduling of their maintenance and repair. This revolution in technology could usher in a new age of
engineering. This era would see the marriage of fiber optic technology and
artificial intelligence with material science and structural engineering. Major
structures constructed in this period would be constructed with built-in optical
neurosystems and active actuation control that would make them more like
living entities than the inanimate edifices we are familiar with today. To carry
this biological paradigm one step further, it may even be possible to contem-

plate future structures that have the capacity for limited self-repair. Indeed, if we think of the new frontiers for engineering as being in space or underwater, self-diagnosis, self-control, and self-healing may not be so much esoteric as vital.

ACKNOWLEDGMENTS

The work reported from the UTIAS Fiber Optic Smart Structures Laboratory was supported by the Ontario Laser and Lightwave Research Centre, the Natural Science and Engineering Research Council of Canada, the Institute for Space and Terrestrial Science, and the Ontario Centre for Materials Research. The author would like to acknowledge the excellent work of T. Alavie, A. Davis, S. Ferguson, D. Hogg, D. Janzen, M. Le Blanc, K. Liu, B. Mason, S. Melle, M. Ohn, R. Turner, and T. Valis, who contributed to this research.

REFERENCES

1. A. Ahmad, A. Crowson, C. A. Rogers, and M. Aizawa, *US–Japan Workshop on Smart/Intelligent Materials and Systems.* Technomic Publishing, Lancaster, PA, 1990.

2. A. T. Alavie, S. E. Karr, A. Othonos, and R. M. Measures, Fiber Laser Sensor Array. *Proc. SPIE–Int. Soc. Opt. Eng.* **1918** (1993).

3. J. M. Alper, Two-Mode, Elliptical-Core, Weighted Fiber Sensors for Vibration Analysis. *Proc. Conf. Opt. Fiber Sens.-Based Smart Mater. Struct.*, Blacksburg, VA, *1991*, pp. 102–106 (1991).

4. C. G. Askins, T. E. Tsai, G. M. Williams, M. A. Putman, M. Bashkansky, and E. J. Friebele, Fiber Bragg Reflectors Prepared by a Single Excimer Pulse. *Opt. Lett.* 17, 833–835 (1992).

5. G. A. Ball and W. W. Morey, Continuously Tunable Single-Mode Erbium Fiber Laser. *Opt. Lett.* **17**, 420–422 (1992).

6. G. A. Ball, W. W. Morey, and J. P. Waters, Nd^{3+} Fibre Laser Utilizing Intra-Core Bragg Reflectors. *Electron. Lett.* **26**, 1829–1830 (1990).

7. C. H. Beuchamp, R. H. Nadolink, S. C. Dickinson, and L. M. Dean, Shape Memory Alloy Adjustable Camber (SMAAC) Control Surfaces. *Proc. SPIE—Int. Soc. Opt. Eng.* **1777**, 189–192 (1992).

8. A. Bertholds and R. Dandiliker, Determination of the Individual Strain-Optic Coefficients in Single-Mode Optical Fibers. *IEEE J. Lightwave Technol.* **LT-6**, 17–20 (1988).

9. B. Blagojevic, W. Tsaw, K. McEwen, and R. M. Measures, The Influence of Embedded Optical Fibers on the Interlamina Fracture Toughness of Composite Materials. *Rev. Prog. Quant. Nondestr. Eval.*, Brunswich, ME, 1989 (1989).

10. J. N. Blake, S. Y. Huang, and B. Y. Kim, Elliptical Core Two-Mode Fiber Strain Gauge. *Proc. SPIE—Int. Soc. Opt. Eng.* **838**, 332 (1987).

11. J. N. Blake, S. Y. Huang, B. Y. Kim, and H. J. Shaw, Strain Effects on Highly Elliptical Core Two-Mode Fibers. *Opt. Lett.* **12**, 732 (1987).

12. W. J. Bock and T. R. Wolinski, Temperature-Compensated Fiber-Optic Strain Sensor Based on Polarization-Rotated Reflection. *Proc. SPIE—Int. Soc. Opt. Eng.* **1370**, 189–196 (1991).

13. C. D. Butter and G. P. Hocker, Fiber Optics Strain Gauge. *Appl. Opt.* **17**, 2867–2869 (1978).

14. J. M. Caussignac, A. Chabert, G. Morel, P. Rogez, and J. Seantier, Bearing of a Bridge Fitted with Load Measuring Devices Based on Optical Fiber Technology. *Proc. SPIE—Int. Soc. Opt. Eng.* **1777**, 207–210 (1992).

15. D. E. Cox, Active Vibration Control Using a Modal-Domain Fiber Optic Sensor. *Opt. Eng.* **31**, 8–12 (1992).

16. B. Culshaw, A. McDonach and P. Gardiner, eds. *European Conference on Smart Structures and Materials*, Proc. SPIE, Vol. 1777. SPIE, Bellingham, WA, 1992.

17. R. Czarnek, Y. F. Guo, K. D. Bennett, and R. O. Claus, Interferemetric Measurements of Strain Concentrations Induced by an Optical Fiber Embedded in a Fiber Reinforced Composite. *Proc. SPIE—Int. Soc. Opt. Eng.* 43–54 (1989).

18. J. P. Dakin, (1989). Distributed Optical Fiber Sensor Systems. In *Optical Fiber Sensors: Systems and Applications* (B. Culshaw and J. Dakin, eds.), Vol. 2, pp. 575–598. Artech House, Boston, 1989.

19. J. P. Dakin and C. A. Wade, Compensated Polarimetric Sensor Using Polarization-Maintaining Optical Fibre in a Differential Configuration. *Electron. Lett.* **20**, 51–53 (1984).

20. C. DiFrancia, R. O. Claus, J. W. Hellgeth, and T. C. Ward, (1991). Discussion of Pullout Tests of Polyimide-Coated Optical Fibers Embedded in Neat Resin. *Proc. Cont. Opt. Fiber Sens.-Based Smart Mater. Struct.*, Blacksburg, VA, *1991*, pp. 70–82.

21. J. R. Dunphy, G. Meltz, F. P. Lamm, and W. W. Morey, Fiber-Optic Strain Sensor Multi-Function, Distributed Optical Fiber Sensor for Composite Cure and Response Monitoring. *Proc. SPIE—Int. Soc. Opt. Eng.* **1370**, 116–118 (1991).

22. C. O. Egalon and R. S. Rogowski, Model of Axially Strained Weakly Guiding Optical Fiber Modal Pattern. *Proc. SPIE—Int. Soc. Opt. Eng.* **1588**, 241–254 (1992).

23. F. Farahi, D. J. Webb, J. D. C. Jones, and D. A. Jackson, Simultaneous Measurement of Temperature and Strain: Cross-Sensitivity Considerations. *IEEE J. Lightwave Technol.* **LT-8**, 138–142 (1990).

24. B. R. Fogg, W. V. Miller, III, A. M. Vengsarkar, R. O. Claus, R. L. Clarke, and C. R. Fuller, Optical Fiber Sensors for Active Structural Acoustic Control. *Opt. Eng.* **31**, 28–33 (1992).

25. P. L. Fuhr, D. R. Huston, P. J. Kajenski and T. P. Ambrose, Performance and Health Monitoring of the Stafford Medical Building Using Embedded Sensors. Private communication (1991).

26. P. L. Fuhr, D. R. Huston, P. J. Kajenski, T. P. Ambrose, and W. B. Spillman, Installation and Preliminary Results from Fiber Optic Sensors Embedded in a Concrete Building. *Proc. SPIE—Int. Soc. Opt. Eng.* **1777**, 409–412 (1992).

27. N. Furstenau, W. Schmidt, and H. Goetting, Simultaneous Interferometric and Polarimetric Strain Measurements on Composites Employing a Fiber Optic Strain Gauge. *Appl. Opt.* **37** (1992).

28. P. Gao, A. T. Alavie, B. Mason, and R. M. Measures, Effect of Delamination on the Natural Frequencies of Composite Beam as Measured by Fiber Optic Fabry–Perot Sensors. *Proc. SPIE—Int. Soc. Opt. Eng.* **1798** (1993).

29. N. D. Glossop, J. D. Lymer, R. M. Measures, and R. C. Tennyson, Image Enhanced Backlighting: A New Method of Non-destructive Evaluation in Translucent Composite Materials. *J. Nondestr. Eval.* **8**, 181–193 (1989).

30. K. O. Hill, Y. Fujii, D. C. Johnson, and B. S. Kawasaki, Photosensitivity in Optical Fiber Waveguides: Application to Reflection Filter Fabrication. *Appl. Phys. Lett.* **32**, 647–649 (1978).

31. D. Hogg, R. D. Turner, and R. M. Measures, Polarimetric Fibre Optic Structural Strain Sensor Characterization. *Proc. SPIE—Int. Soc. Opt. Eng.* **1170**, 542–550 (1990).

32. D. Hogg, D. Janzen, T. Valis, and R. M. Measures, Development of a Fiber Fabry–Perot Strain Gauge. *Proc. SPIE—Int. Soc. Opt. Eng.* **1588**, 300–307 (1992).

33. D. Hogg, D. Janzen, B. T. Mason T. Valis, and R. M. Measures, Development of a Fiber Fabry–Perot (FFP) Strain Gauge with High Reflectivity Mirrors. *Proc. Act. Mater. Adapt. Struct. Conf.*, Alexandria VA, 4–8 Nov. *1991*, p. 000 (1990).

34. D. Hogg, T. Valis, B. Mason, and R. M. Measures, Temperature Compensated Strain Measurements Using Fiber Fabry–Perot Strain Gauges. *Opt. Fiber Sens.-Based Smart Mater. Struct.*, Blacksburg, VA, *1992* (1992).

35. A. Holst, W. Habel, and R. Lessing, Fiber Optic Intensity Modulated Sensors for Continuous Observations of Concrete and Rock-Fill Dams. *Proc. SPIE—Int. Soc. Opt. Eng.* **1777**, 223–226 (1992).

36. G. W. Housner, S. F. Masri, and T. T. Soong, An Overview of Active Structural Control. *Proc. SPIE—Int. Soc. Eng.* **1777**, 201–206 (1992).

37. S. Y. Huang, J. N. Blake, and B. Y. Kim, Perturbation Effects on Mode Propagation in Highly Elliptical Core Two-Mode Fibers. *IEEE J. Lightwave Technol.* **LT-8**, 23–33 (1990).

38. D. A. Jackson, A Prototype Digital Phase Tracker for Fiber Interferometer. *J. Phys. E.* **14**, 1274–1278 (1981).

39. D. A. Jackson and J. D. C. Jones, Fibre Optic Sensors. *Opt. Acta* **33**, 1469–1503 (1986).

40. D. A. Jackson, A. D. Kersey, M. Cork, and J. D. C. Jones, Pseudo-heterodyne Detection Scheme for Optical Interferometers. *Electron. Lett.* **18**, 1081–1083 (1982).

41. A. D. Kersey, A. C. Lewin, and D. A. Jackson, Pseudo-hetrodyne Detection Scheme for the Gyroscope. *Electron. Lett.* **20**, 368–370 (1984).

42. A. D. Kersey, M. Corke, and D. A. Jackson, Linearised Remote Sensing Using a Monomode Fibre Polarimetric Sensor. *Proc. SPIE—Int. Soc. Opt. Eng.* **514**, 247–250 (1984).

43. A. D. Kersey, A. Dandridge, and A. B. Tveten, Elimination of Polarization Induced Signal Fading in Interferometric Fiber Sensors Using Input Polarization Control. *Opt. Fibre Sens. '88*, WCC2-1-4 (1988).

44. B. Y. Kim, J. N. Blake, S. Y. Huang, and H. J. Shaw, Use of Highly Elliptical Core Fibers for Two-Mode Fiber Devices. *Opt. Lett.* **12**, 729–731 (1987).

45. B. Kruschwitz, R. O. Claus, A. Murphy, R. G. May, and M. F. Gunther, Optical Fiber Sensors for the Quantitative Measurement of Strain in Concrete Structures. *Proc. SPIE—Int. Soc. Opt. Eng.* **1777**, *Conf. Smart Struct. Mater.*, Glasgow, 241–244 (1992).

46. M. LeBlanc and R. M. Measures, Impact Damage Assessment in Composite Materials with Embedded Fiber Optic Sensors. *J. Compos. Eng.* **2**, 573–596 (1992).

47. M. LeBlanc and R. M. Measures, Micromechanical Considerations for Embedded Single-Ended Sensors. *Proc. SPIE—Int. Soc. Opt. Eng.* **1918** (1993).

48. C. E. Lee and H. F. Taylor, Interferometer Fiber Optic Sensors Using Internal Mirrors. *Electron. Lett.* **13**, 193–194 (1988).

49. C. E. Lee and H. F. Taylor, Interferometric Fiber Optic Temperature Sensor Using a Low Coherence Light Source. *Proc. SPIE—Int. Soc. Opt. Eng.* **1370**, 356–364 (1991).

50. C. E. Lee, J. J. Alcoz, Y. Yeh, W. N. Gibler, R. A. Atkins, and H. F. Taylor, Optical Fiber Fabry–Perot Sensors for Smart Structures. *Smart Mater. Struct.* **1**, 123–127 (1992).

51. C. E. Lee, W. N. Gibler, R. A. Atkins, and H. F. Taylor, In-Line Fiber Fabry–Perot Interferometer with High Reflectance Mirrors. *IEEE J. Lightwave Technol.* (1992).

52. J. J. Lesko, G. P. Carman, B. R. Fogg, W. V. Miller, III, A. M. Vengsarkar, K. L. Reifsnider, and R. O. Claus, Embedded Fabry–Perot Fiber Optic Strain Sensors in the Macromodel Composites. *Opt. Eng.* **31**, 13–20 (1992).

53. K. Liu and R. M. Measures, Signal Processing Techniques for Localized Interferometric Fiber-Optic Strain Sensors. *J. Intell. Mater. Syst. Struct.* (1992).

54. K. Liu, S. M. Ferguson, and R. M. Measures, Damage Detection in Composites with Embedded Fiber Optic Interferometric Sensors. *Proc. SPIE—Int. Soc. Opt. Eng.* **1170**, 205–210 (1990).

55. K. Liu, S. Ferguson, K. McEwen, E. Tapanes, and R. M. Measures, Acoustic Emission Detection for Composite Using Embedded Ordinary Single-Mode Fiber Optic Interferometric Sensors. *Proc. SPIE—Int. Soc. Opt. Eng.* **1370**, 316–323 (1991).

56. D. Loken, Effect of Fiber Straightness on Fatigue of Aligned Continuous CFRP Composites. M.A.Sc. Thesis, Chemical Engineering Dept., University of Toronto (1990).

57. Z. J. Lu and F. A. Blaha, A Fiber Optic Strain and Impact Sensor System for Composite Materials. *Proc. SPIE—Int. Soc. Opt. Eng.* **1170**, 239–242 (1990).

58. Z. J. Lu and F. A. Blaha, Application Issues of Fiber Optics in Aircraft Structures. *Proc. SPIE—Int. Soc. Opt. Eng.* **1588**, 276–281 (1992).

59. S. L. Marple, Jr., Digital Spectral Analysis with Applications. Prentice-Hall, Englewood Cliffs, NJ, 1987.

60. B. Mason, D. Hogg, and R. M. Measures, Fiber Optic Strain Sensing for Smart Adaptive Structures. *Proc. SPIE—Int. Soc. Opt. Eng.* **1777** (1992).

61. C. T. Mathews and J. S. Sirkis, The Interaction of Interferometric Optical Fiber Sensors Embedded in a Monolithic Structure. *Proc. SPIE—Int. Soc. Opt. Eng.* **1370**, 142–153 (1991).

62. R. M. Measures, Smart Structures with Nerves of Glass. *Prog. Aerosp. Sci.*, **26**, 289–351 (1989).

63. R. M. Measures, Smart Composite Structures with Embedded Sensors. *Compos. Eng.* **2**, 597–618 (1992).

64. R. M. Measures, (1992b). Smart Structures: A Revolution in Civil Engineering, Keynote Address. *ACMBA-1 Conf.*, Sherbrook, Quebec, Canada, *1992* (1990).

65. R. M. Measures, K. Liu, A. Davis, and M. Ohn, Composite Cure Monitoring with Embedded Optical Fiber Sensors. *SPIE OE/Aerosp. Sens. Conf. 1489, Struct. Sens. Control*, Orlando, FL, *1991* (1991).

66. R. M. Measures, S. M. Melle, and K. Liu, Wavelength Demodulated Bragg Grating Fiber Optic Sensing Systems for Addressing Smart Structure Critical Issues. *J. Smart Mater. Struct.* **1**, 36–44 (1992).

67. R. M. Measures, T. Alavie, S. Karr, and T. Coroy, Smart Structure Interface Issues and Their Resolution: Bragg Grating Laser Sensor and the Optical Synapse. *Proc. SPIE—Int. Soc. Opt. Eng.* **1918** (1993).

68. S. M. Melle, K. Liu, and R. M. Measures, Strain Sensing Using a Fiber Optic Bragg Grating. *Proc. SPIE—Int. Soc. Opt. Eng.* **1588** (1991).

69. S. M. Melle, K. Liu, and R. M. Measures, A Passive Wavelength Demodulation System for Guided-Wave Bragg Grating Sensors. *IEEE Photon. Technol. Lett.* **4**, 516–518 (1992).

70. S. Melle, T. Alavie, S. Karr, T. Coroy, K. Liu, and R. M. Measures, A Bragg Grating-Tuned Fiber Laser Strain Sensor System. *IEEE Photon. Technol. Lett.* **5**, 263–266 (1993).

71. G. Meltz, W. W. Morey, and W. H. Glam, Formation of Bragg Grating in Optical Fibers by a Transverse Holographic Method. *Opt. Lett.* **14**, 823–825 (1989).

72. W. C. Michie, B. Culshaw, S. J. Roberts, and R. Davison, Fibre Optic Technique for the Simultaneous Measurement of Strain and Temperature Variations in Composite Materials. *Proc. SPIE—Int. Soc. Opt. Eng.* **1588**, Sept. (1991).

73. W. W. Morey, G. Meltz, and W. H. Glenn, Fiber Optic Bragg Grating Sensors. *Proc. SPIE—Int. Soc. Opt. Eng.* **1169**, 98–107 (1989).

74. W. W. Morey, J. R. Dunphy, and G. Meltz, Multiplexed Fiber Bragg Grating Sensors. *Proc. SPIE—Int. Soc. Opt. Eng.* **1586**, 216–224 (1991).

75. A. A. Mufti, M.-A. Erki, L. G. Jaeger, eds., Advanced Composite Materials with Application to Bridges. *Anchorages for Prestressing Tendons and Cables*, pp. 238–239. Canadian Society for Civil Engineering, 1991.

76. K. A. Murphy, M. Miller, A. M. Vengsarkar, R. O. Claus, and N. E. Lewis, Embedded Modal Domain Sensors Using Elliptical Core Optical Fibers. *Proc. SPIE—Int. Soc. Opt. Eng.* **1170**, 566–573 (1990).

77. K. A. Murphy, M. F. Gunther, A. M. Vengsarkar, and R. O. Claus, Fabry–Perot Fiber Optic Sensors in Full-Scale Fatigue Testing on an F-15 Aircraft. *Proc. SPIE—Int. Soc. Opt. Eng.* **1588**, 134–144 (1992).

78. M. Ohn, A. Davis, K. Liu, and R. M. Measures, Embedded Fiber Optic Detection of Ultrasound and Its Application to Cure Monitoring. *Opt. Fiber Sens.-Based Smart Mater. Struct. Conf.*, Blacksburg, VA, *1992* (1992).

79. Y. E. Pak, Longitudinal Shear Transfer in Fiber Optic Sensors. *Smart Mater. Struct.* **1**, 57–62 (1992).

80. L. S. Penn and S. M. Lee, Interpretation of Experimental Results in the Single Pull-out Filament Test. *J. Compos. Technol. Res.* **11**, 23–30 (1989).

81. M. R. Piggott, The Effect of the Interface/Interphase on Fiber Composite Properties. *Polym. Compos.* **8**, 291–297 (1987).

82. S. S. J. Roberts and R. Davidson, Mechanical Properties of Composite Materials Containing Embedded Optic Sensors. *Proc. SPIE—Int. Soc. Opt. Eng.* **1588**, 326–341 (1992).

83. S. S. J. Roberts and R. Davidson, Short Term Fatigue Behaviour of Composite Materials Containing Embedded Fiber Optic Sensors and Actuators. *Proc. Eur. Conf. Smart Struct. Mater., 1st,* Glasgow, *1992,* pp. 255–262 (1992).

84. R. H. Scott and P. A. T. Gill, Possibilities for the Use of Strain Gauged Reinforcement in Smart Structures. *Proc. SPIE—Int. Soc. Opt. Eng.* **1777**, 211–214 (1992).

85. S. K. Sheem, T. G. Giallorenzi, and K. Koo, Optical Techniques to Solve the Signal Fading Problem in Fiber Interferometers. *Appl. Opt.* **21**(4), 689 (1982).

86. H. D. Simonsen, R. Paetsch, and J. R. Dunphy, Fiber Bragg Grating Sensor Demonstration in Glass-Fiber Reinforced Polyester Composite. *Proc. SPIE—Int. Soc. Opt. Eng.* **1777**, 73–76 (1992).

87. J. S. Sirkis, A Unified Approach to Phase-Strain-Temperature Models for Smart Structure Interferometric Optical Fiber Sensors. *Opt. Eng.* (1991).

88. J. S. Sirkis and A. Dasgupta, Optimal Coatings for Intelligent Structure Fiber Optic Sensors. *Proc. SPIE—Int. Soc. Opt. Eng.* **1370**, 129–140 (1991).

89. J. S. Sirkis and A. Dasgupta, What Do Embedded Optical Fibers Really Measure? *Proc. SPIE—Int. Soc. Opt. Eng.* **1777**, 69–72 (1992).

90. J. S. Sirkis and H. W. Haslach, Jr., Full Phase-Strain Relation for Structurally Embedded Interferometric Optical Fiber Sensors. *Proc. SPIE—Int. Soc. Opt. Eng.* **1370**, 248–259 (1991).

91. K. J. Sun and W. P. Winfree, Propagation of Acoustic Waves in a Copper Wire Embedded in a Curing Epoxy. *IEEE Ultrasonics Symposium*, 439–442, (1987).

92. R. D. Turner, T. Valis, W. D. Hogg, and R. M. Measures, Fiber Optic Strain Sensors for 'Smart' Structures. *J. Intell. Mater. Syst. Struct.* **1**, 26–49 (1990).

93. R. D. Turner, D. G. Laurin, and R. M. Measures, Localized Dual-Wavelength Fiber-Optic Polarimeter for the Measurement of Structural Strain and Orientation. *Appl. Opt.* **31**, 2994–3003 (1992).

94. R. Urlich, Theory of Spectral Encoding for Optic Sensors. *NATO ASI Ser., Ser. E* **132** (1987).

95. T. Valis and R. M. Measures, Far-Field In-Plane Stress Measurement with an Embedded Eigenaxis-Flipped Fiber Fabry–Perot Strain Gauge. In preparation.

96. T. Valis, D. Hogg, and R. M. Measures, Composite Material Embedded Fiber Optic Fabry–Perot Strain Rosette. *Proc. SPIE—Int. Soc. Opt. Eng.* **1370**, 154–161 (1991).

97. M. P. Varnham, A. J. Barlow, D. N. Payne, K. Okamoto, Polarimetric Strain Gauges Using High Birefringes Fibers. *Electron. Lett.* **19**, 699–700 (1983).

98. A. M. Vengsarkar, W. C. Michie, L. Jankovic, B. Culshaw, and R. O. Clause, Fiber Optic Sensor for Simultaneous Measurement of Strain and Temperature. *Proc. SPIE—Int. Soc. Opt. Eng.* **1367**, 249–260 (1990).

99. A. M. Vengsarkar, K. A. Murphy, B. R. Fogg, W. V. Miller, J. Greene, and R. O. Clause, Two-Mode, Elliptical-Core Weighted Fiber Sensors for Vibration Analysis. *Proc. Conf. Opt. Fiber Sens-Based Smart Mater. Struct.* Blacksburg, VA, *1991*, pp. 40–45 (1991).

100. R. Wolf and H. J. Miesseler, Monitoring of Prestressed Concrete Structures with Optical Fiber Sensors. *Proc. SPIE—Int. Soc. Opt. Eng.* **1777**, 23–29 (1992).

101. G. Zuliani, D. Hogg, K. Liu, and R. M. Measures, Demodulation of a Fiber Fabry–Perot Strain Sensor Using White Light Interferometry. *Proc. SPIE—Int. Soc. Opt. Eng.* **1588**, 308–313 (1992).

102. R. M. Measures, A. T. Alavie, R. Maaskant, M. Ohn, S. Karr, and S. Huang, Structurally Integrated Bragg Grating Laser Sensing System for a Carbon Fiber Prestressed Concrete Highway Bridge. Accepted for Journal for Smart Materials & Structures, 1995.

103. S. Huang, M. Ohn, M. LeBlanc, R. Lee, R. M. Measures, Fiber Optic Intra-Grating Distributed Strain Sensor. SPIE Vol. 2294, Distributed and Multiplexed Fiber Optic Sensors, 27–28 July 1994, San Diego, 1994.

9

Sensors for Smart Structures Based on the Fabry–Perot Interferometer

CHUNG E. LEE* AND HENRY F. TAYLOR
Texas A&M University
College Station, Texas

9.1. INTRODUCTION

The Fabry–Perot interferometer (FPI), sometimes called the Fabry–Perot etalon, consists of two mirrors of reflectance R_1 and R_2 separated by a distance L, as shown in Fig. 9.1. Since its invention in the late nineteenth century (1), the bulk-optics version of the FPI has been widely used for high-resolution spectroscopy. In the early 1980s, the first results on fiber optic versions of the FPI were reported. In the late 1980s, fiber Fabry–Perot interferometers (FFPIs) began to be used in the sensing of temperature, strain, and ultrasonic pressure in composite materials.

The FFPI would appear to be an ideal transducer for many smart structure sensing applications, particularly those in which the sensor must be embedded in a composite or metal. As with all fiber interferometers, the FFPI is extremely sensitive to environmental perturbations. The sensing region, which consists of the portion of the fiber core (or in some cases an air gap) between the two mirrors, can be very compact—equivalent to a "point" transducer in many applications. Unlike other fiber interferometers (Mach–Zehnder, Michelson, Sagnac) used for sensing, the FFPI contains no fiber couplers—components that can greatly complicate the embedding process and the interpretation of data produced by an embedded sensor. Finally, FFPI sensors are amenable to both time-division and coherence multiplexing.

In later sections of this chapter we review the theory of the FPI and previous work on FFPIs, describe demonstrated application of the FFPI to smart structures, discuss multiplexing techniques that might be employed, and speculate upon future directions for research and development in this area.

Present address: FFPI Industries Inc., 909 Industrial Boulevard, Bryan, Texas.

Fiber Optic Smart Structures, Edited by Eric Udd.
ISBN 0-471-55448-0 © 1995 John Wiley & Sons, Inc.

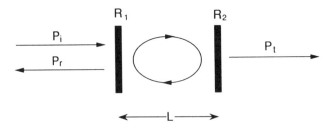

R_1 R_2

P_i

P_r

P_t

L

Figure 9.1. Configuration for Fabry–Perot Interferometer.

9.2. THEORY OF THE FABRY–PEROT INTERFEROMETER

In this section, expressions for the transmittance and reflectance of the FPI that can be used to characterize the performance of FFPI sensors are introduced.

The individual mirrors in the FPI can be characterized by transmittance T_i and excess power losses A_i, $i = 1, 2$, such that $A_i + R_i + T_i = 1$. The excess loss corresponds to the portion of the incident power that is absorbed by the mirror or scattered out of the beam. The Fabry–Perot reflectance R_{FP} and transmittance T_{FP} are found to be (2)

$$R_{FP} = \frac{R_1 + R_2(1 - A_1)^2 + 2\sqrt{RZ_1 R_2}(1 - A_1)\cos\phi}{1 + R_1 R_2 + 2\sqrt{R_1 R_2}\cos\phi} \tag{9.1}$$

$$T_{FP} = \frac{T_1 T_2}{1 + R_1 R_2 + 2\sqrt{R_1 R_2}\cos\phi} \tag{9.2}$$

where R_{FP} represents the ratio of the power reflected by the FPI to the incident power, T_{FP} is the ratio of the transmitted power to the incident power, and ϕ, the round-trip phase shift in the interferometer, is given by

$$\phi = \frac{4\pi n L}{\lambda} \tag{9.3}$$

with n the reflective index of the region between the mirrors and λ the free-space optical wavelength.

It is evident from Eq. (9.2) that T_{FP} is a maximum for $\cos\phi = -1$, or $\phi = (2m + 1)\pi$, with m an integer. If we define $\Delta = \phi - (2m + 1)\pi$, then near a maximum in T_{FP}, $\cos\phi \cong -(1 - \Delta^2/2)$, with $\Delta \ll 1$. In the case that the mirrors are lossless ($A_1 = A_2 = 0$) and the mirror reflectances are equal and approach unity, Eq. (9.2) simplifies to

$$T_{FP} = \frac{T^2}{(1 - R)^2 + R\Delta^2} \tag{9.4}$$

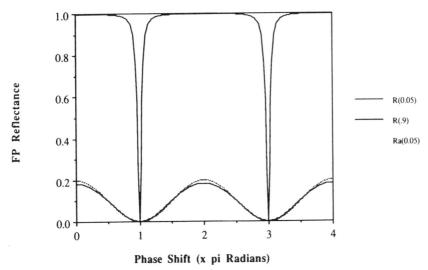

Figure 9.2. Dependence of reflectance R_{FP} of a lossless FPI on round-trip phase shift for two values ($R = 0.9$ and 0.05) of the individual mirror reflectance. The dotted line represents the approximation of Eq. (9.6).

where $R = R_1 = R_2$ and $T = 1 - R$. The maximum transmittance occurs when $\Delta = 0$. The finesse F, a frequently used figure of merit for the FPI, is defined as the ratio of the phase change between adjacent transmittance peaks to the phase change between half-maximum points on either side of a peak. From Eq. (9.2) it follows that T_{FP} is a periodic function of ϕ with period 2π, so that a phase change of 2π radians is required to tune from one peak to the next. But it follows from Eq. (9.4) that T_{FP} is half its maximum value for $\Delta = \pm(1 - R)/\sqrt{R}$. This implies that the finesse can be written

$$F = \frac{\pi\sqrt{R}}{1 - R} \tag{9.5}$$

Thus, in an interferometer with lossless mirrors, $F = 29.8$ for $R = 0.9$ and $F = 312.6$ for $R = 0.99$.

Another limiting case, with low-reflectance mirrors, is of particular interest in the case of FFPI sensors. Assuming once again that the mirrors are lossless and have equal reflectances, with $R = R_1 = R_2$, it follows from Eqs. (9.1) and (9.2) that if $R \ll 1$, then

$$R_{FP} \cong 2R(1 + \cos\phi) \tag{9.6}$$

$$T_{FP} \cong 1 - 2R(1 + \cos\phi) \tag{9.7}$$

It should be noted that the concept of finesse is not intended to apply to FPIs with $R \ll 1$. In fact, it follows from our definition of finesse and Eq. (9.2) that $F = 1$ for $R = 0.172$ and F is undefined for $R < 0.172$.

The reflectance for a lossless FPI given by Eq. (9.1) is plotted as a function of round-trip phase shift in Fig. 9.2 for $R = 0.9$ and $R = 0.05$. the approximate expression from Eq. (9.6) is also plotted for comparison.

9.3. FIBER FABRY–PEROT INTERFEROMETERS

The first reports on FFPIs began to appear in the early 1980s. In these early experiments the interferometer cavity was a single-mode fiber with dielectric mirrors (3,4) or cleaved fiber ends (5) to serve as mirrors. Reflectance or transmittance was monitored as light from a He–Ne laser was focused into the fiber interferometer. The expected high sensitivity of these interferometers to strain and temperature was confirmed.

The use of internal fiber mirrors in FFPIs was first reported in 1987. Internal mirrors are reflectors formed as an integral part of a continuous length of fiber. They have been produced by "bad" fusion splices between uncoated fibers (6) and by fusion splicing of an uncoated fiber to a fiber with a thin dielectric (7) or metallic (8) coating on the end. An FFPI with internal mirrors is illustrated schematically in Fig. 9.3.

Internal mirrors formed from dielectric coatings have shown the best mechanical properties, lowest excess optical loss, and widest range of reflectance values. The most commonly used mirror material is TiO_2, which has a refractive index of 2.4 (versus 1.46 for fused silica). The reflection results from the refractive index discontinuities at the two film–fiber interfaces. The TiO_2 films have been produced by sputtering in a radio-frequency (RF) planar magnetron system or by electron beam evaporation. Typical film thicknesses are in the neighborhood of 1000Å. The fusion splicer is operated at lower arc current and duration than for a normal splice, and several splicing pulses are used to form a mirror. The mirror reflectance generally decreases monotonically as a function of the number of splicing pulses, making it possible to select a desired reflectance over the range from much less than 1% to about 10% during the splicing process. Fabrication of an internal mirror can typically be accomplished in less than 1 min. To achieve internal mirror reflectances greater

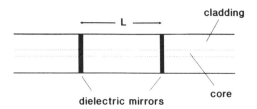

Figure 9.3. Fiber Fabry–Perot interferometer with two internal mirrors.

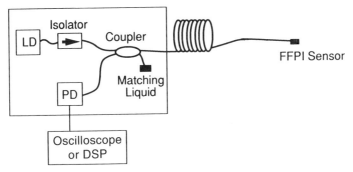

Figure 9.4. Arrangement for monitoring the reflected power from a FFPI.

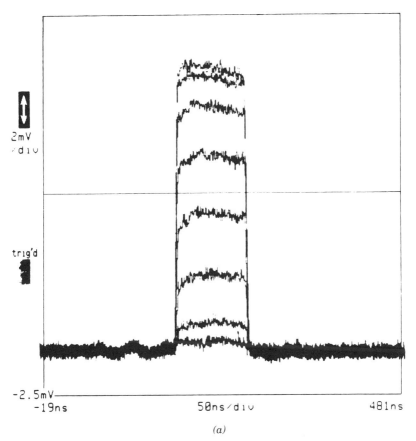

(a)

Figure 9.5. Oscilloscope tracings showing reflected power from FFPI at different temperatures for a pulsed laser input. In (a), the temperature was changed from 77°C to 56°C in 3°C increments. The interferometer was short (1.5 mm), the pulse width was short (100 ns), and the modulating pulse amplitude was small, so that the effect of laser chirping was small. In (b), the temperature was changed from 29°C to 23°C in 1.5°C increments. A longer (1-cm) interferometer, longer pulse duration (0.8 μs), and larger modulating pulse amplitude causes the interferometer output to sweep through fringes in response to chirping of the laser.

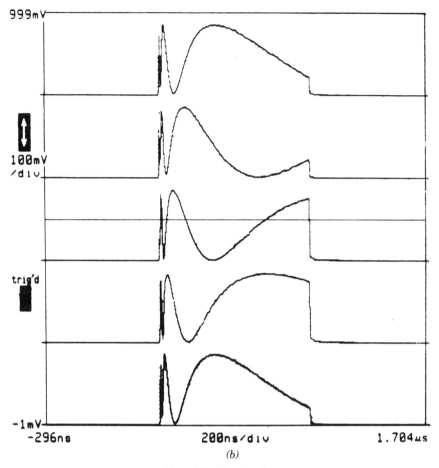

Figure 9.5. (*Continued*).

than 10%, multilayer TiO_2–SiO_2 films produced by magnetron sputtering have been used. The best result to date with multilayer mirrors is a reflectance of 86% in the FFPI with a finesse of 21 at a wavelength of 1.3 μm (2).

An experimental arrangement to monitor the reflectance of an FFPI sensor is shown in Fig. 9.4. Light pulses from a 1.3-μm fiber-pigtailed laser diode pass through an optical isolator to a fiber coupler and are reflected by the FFPI. After passing through the coupler again, the reflected pulses are converted by a photodiode to an electrical signal that is displayed on an oscilloscope. Some typical waveforms obtained in this manner are shown in Fig. 9.5. If the laser pulse is short, the driving current amplitude is small, and the FFPI cavity is short, the reflected amplitude is almost constant for the duration of the pulse,

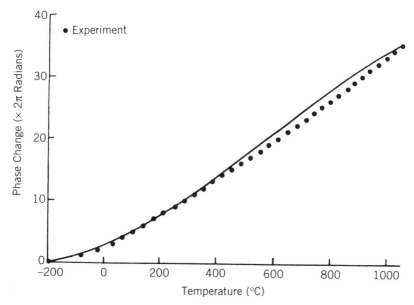

Figure 9.6. Phase shift in FFPI determined by fringe counting as a function of the temperature as measured with a thermocouple. Each dot represents a 2π radian phase shift. The solid line is a theoretical plot determined from data on the temperature dependence of refractive index and thermal expansion coefficient for fused silica.

as in Fig. 9.5*a*. Increasing the pulse width, driving current amplitude, and interferometer length causes the reflected signal from the FFPI to vary with time due to chirping of the laser. As the laser heats during the pulse, its frequency and consequently the reflected power from the interferometer changes with time, as in Fig. 9.5*b*. Thus the temporal variation in the reflected pulse amplitude represents a sweeping through fringes of the interferometer.

An FFPI with internal mirror reflectances of about 2% and excess losses of 0.1 to 0.2 dB per mirror was used to sense temperature over the range -200 to $1050°C$ (9). No hysteresis or change in the mirror reflectances were observed over the temperature range of the experiment, although it was noted that the fiber became brittle at the higher temperatures. As illustrated by the data of Fig. 9.6, the sensitivity is lowest at cryogenic temperatures and the response is fairly linear at room temperature and above.

The FFPI sensors described above make use of a fiber propagation path between the mirrors, and thus sense parameters that change the refractive index or length of the fiber. These are termed *intrinsic* Fabry–Perot sensors. Other sensors where the propagation path between the mirrors is external to the fiber are termed *extrinsic* sensors. One example of an extrinsic FFPI sensor uses a cavity formed by the fiber end and a membrane located beyond the fiber

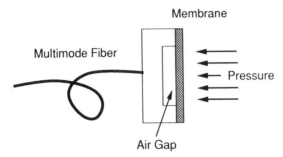

Figure 9.7. Extrinsic FFPI sensor with a cavity formed by the air gap between the fiber end and a membrane.

end (10, 11) (Fig. 9.7). The reflectance from the interferometer changes in response to a motion of the membrane that affects the length of the external cavity. Typical separation between the membrane and the fiber end is on the order of $1\,\mu$m. Such short cavity lengths make it possible to operate these sensors with multimode-fiber and low-coherence light-emitting diode (LED) light sources. Sensors of this type have been used to measure temperature, pressure, and refractive index of a liquid, but their use in smart materials applications has not yet been reported in the literature.

Finally, FFPIs have found application in optical communications as wavelength filters and as frequency discriminators in carrier-frequency-modulated systems (12). In this case, high-quality dielectric mirrors deposited on the fiber ends or on external mirrors have made it possible to achieve finesse values as high as 1000 (13). The cavity length, and hence the wavelengths of peak transmittance can be tuned continuously using piezoelectric elements to stretch the fibers. The effect of diffractive beam spreading on the performance of FFPIs with an air gap between the end of the fiber and a mirror has recently been investigated (14).

9.4. EXTERNALLY MOUNTED FFPI SENSORS

Surface-mounted transducers are generally used for monitoring parameters such as strain, temperature, and acoustic pressure in structural materials. Although embedding is the ultimate goal in some smart materials development efforts, externally mounted FFPI sensors can also be very useful in this field. Unlike the embedded FFPI, the surface-mounted sensor can readily be repositioned or replaced. Furthermore, surface-mounted sensors can be used with materials that must be processed at such high temperatures that embedding is impossible.

The conventional way to monitor strain in a structural part is to bond electrical strain gauges to its surface. The high sensitivity of the FFPI to

FIBEROPTIC PRESSURE SENSOR

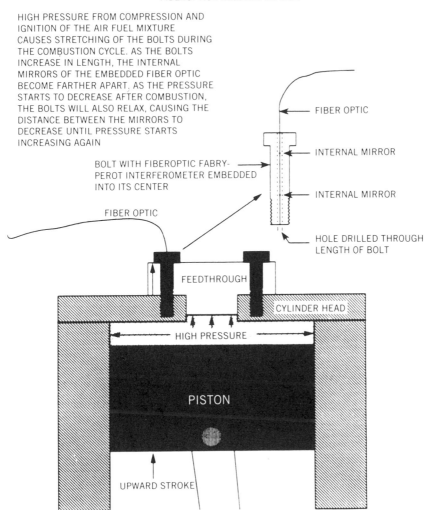

HIGH PRESSURE FROM COMPRESSION AND
IGNITION OF THE AIR FUEL MIXTURE
CAUSES STRETCHING OF THE BOLTS DURING
THE COMBUSTION CYCLE. AS THE BOLTS
INCREASE IN LENGTH, THE INTERNAL
MIRRORS OF THE EMBEDDED FIBER OPTIC
BECOME FARTHER APART. AS THE PRESSURE
STARTS TO DECREASE AFTER COMBUSTION,
THE BOLTS WILL ALSO RELAX, CAUSING THE
DISTANCE BETWEEN THE MIRRORS TO
DECREASE UNTIL PRESSURE STARTS
INCREASING AGAIN

FIBER OPTIC

INTERNAL MIRROR

BOLT WITH FIBEROPTIC FABRY-
PEROT INTERFEROMETER EMBEDDED
INTO ITS CENTER

INTERNAL MIRROR

FIBER OPTIC

HOLE DRILLED THROUGH
LENGTH OF BOLT

FEEDTHROUGH

CYLINDER HEAD

HIGH PRESSURE

PISTON

UPWARD STROKE

Figure 9.8. Arrangement for using a FFPI for measuring pressure in an internal combustion engine.

longitudinal strain can be utilized in a similar manner. In the first experiment of this type, a FFPI with a cavity formed with a silver internal mirror and a silvered fiber end was bonded to the surface of a cantilevered aluminum beam (8). A conventional resistive foil strain gauge was also bonded to the beam as a reference. Strain was introduced by loading the beam in flexure. The optical phase change in the FFPI, determined by monitoring the reflected power and counting fringes, was found to be a linear function of the strain reading from the resistive device over the range 0 to 1000 $\mu\varepsilon$.

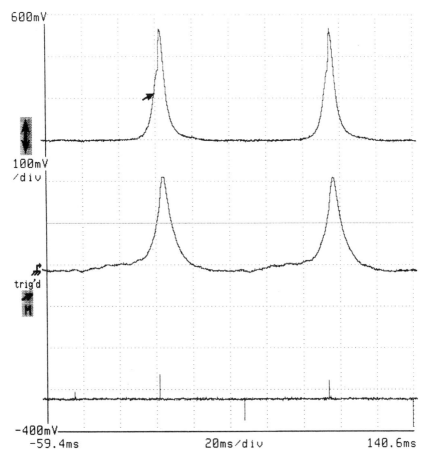

Figure 9.9. Combustion chamber pressure in a diesel test engine as measured with a conventional piezoelectric pressure sensor (upper trace) and a FFPI (middle trace). The lower trace indicates the "top dead center" piston position.

The high strain sensitivity of FFPI sensors has also been utilized in the measurement of gas pressure in internal combustion engines (15). Many engines are constructed such that an element such as a fuel injector valve which is exposed to the combusion chamber pressure is bolted to the cylinder head, as in Fig. 9.8. The variation in longitudinal strain on the bolts during an engine cycle is approximately proportional to the combusion chamber pressure. An FFPI epoxied into a hole drilled in one of these bolts can thus be used for engine pressure measurements. Data obtained with a diesel test engine comparing the response of a FFPI with that of a conventional pressure transducer are shown in Fig. 9.9. Unlike the piezoelectric sensors presently used to monitor engine pressure, the FFPI does not need cooling and should operate for long periods of time without recalibration.

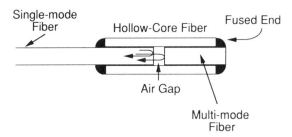

Figure 9.10. Extrinsic FFPI with cavity formed by the air gap between the ends of a single mode and a multimode fiber.

Surface-mounted extrinsic FFPIs with the configuration of Fig. 9.10 have also been used for monitoring temperature, strain, and acoustic pressure (16). In this case, an air gap separating the cleaved ends of a single-mode fiber and a multimode fiber forms the FFPI cavity, typically a few microns to a few hundred microns in length. The fiber alignment is maintained by a silica tube in which the fibers are inserted. Each of the fibers is free to move longitudinally in the region near the cavity, but is constrained at some point along its length by bonding either to the silica tube or to the sample being monitored. The distance between bonding points is the length of the region over which a perturbation affects the sensor output, known as the *gauge length*. The FFPI is monitored in reflection, with the input and reflected light carried by the single-mode fiber.

In one experiment, two of the extrinsic FFPIs were bonded on the surface of a ceramic material in close proximity to one another (16). The cavity lengths in the two sensors were slightly different, so that the round-trip phase shifts were different by about $\pi/2$ radians. By monitoring the two sensor outputs simultaneously, it is possible to determine the direction of change of the phase shift and avoid sensitivity nulls. When the ceramic was temperature-cycled from 25 to 600°C, the sensor data provided information on temperature change and on the expansion of a crack in the material.

In another experiment, an extrinsic FFPI bonded to the surface of an aluminum block was used to detect surface acoustic waves launched by a piezoelectric transducer at a frequency of 1 MHz (17). A dual-wavelength scheme with lasers at 0.78 and 0.83 μm was used to obtain in-phase and quadrature signals from the sensor. The FFPI also detected the acoustic noise burst produced by breaking a pencil lead on the surface of the aluminum sample.

9.5. EMBEDDED FFPI SENSORS

One of the most desirable attributes of a sensor technology for smart structures is the ability to embed the sensor within the structural material. Embedding makes it possible to measure parameters at locations not accessible to ordinary sensors which must be attached to the surface. The embedded sensor can in some cases continue to function properly under conditions (e.g., elevated temperatures) which epoxies for bonding a sensor to the structural material will not survive. The embedded sensor is protected from damage and isolated from extraneous environmental effects by the structure itself.

For a FFPI sensor to be embedded successfully in a composite or metal part, it must withstand the mechanical and thermal stresses experienced when the part is formed. Curing of composites generally requires a combination of elevated temperature and applied pressure. During the casting of metals of structural interest, the FFPI will experience high temperatures combined with severe compressional stresses as the part cools to room temperature.

The FFPI with dielectric internal mirrors formed by fusion splicing is a candidate for embedding in both composites and metals because the mechanical properties of the mirrors themselves, as well as those of the silica fibers, are excellent. Tensile tests on several fiber containing internal mirrors indicated an average tensile strength of 40 kpsi, about half that of ordinary splices made with the same equipment (9). The mirrors readily survive the stresses of ordinary handling in the laboratory.

A graphite–epoxy composite was the material used in the first experiments on embedding of FFPIs (18). The composite was formed from eight coupons of graphite–epoxy panel, 15 cm square by 1.1 mm thick. The panel had a sequence of 0/90/0/90/FFPI/90/0/90/0, where 0 or 90 indicates parallel or perpendicular orientation, respectively, of the graphite fibers in each coupon relative to the top layer. The FFPI with dielectric mirrors was embedded in the middle of the sample and was oriented in the 90 direction. The sample was cured in vacuum for 2 h at 180°C under 5.3 atm pressure.

The temperature sensitivity of the embedded FFPI was tested by monitoring fringes at a wavelength of 1.3 μm as the panel was heated from room temperature to 200°C. The value of the quantity Φ_T, defined as

$$\Phi_T = \frac{d\phi}{\phi dT} \tag{9.8}$$

was measured to be $8.0 \times 10^{-6} °C^{-1}$, slightly less than the value $8.3 \times 10^{-6} °C^{-1}$ for the same FFPI in air prior to embedding. From these data, a thermal expansion coefficient of $2.1 \times 10^{-7} °C^{-1}$ was estimated for the composite.

The ability to embed FFPIs in structural materials suggests application as a pressure transducer in ultrasonic nondestructive testing (NDT). In conventional NDT studies, piezoelectric transducers (PZTs) are positioned on the surface of the sample to launch and detect ultrasonic waves. The ability to locate the receiving transducer deep within the material make it possible to obtain new information on the properties of a bulk sample. For example, it may be possible to better detect and locate delamination sites in composites or cracks in metals. Experiments with a PZT transducer positioned on the surface

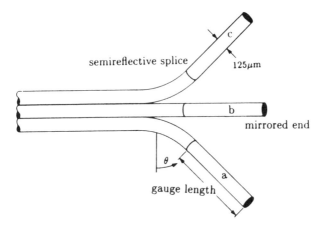

Figure 9.11. Three-axis strain rosette with FFPI sensors.

of the composite sample described above yielded an acoustic response from the FFPI over the frequency range 100 kHz to 5 MHz (19). The compressional acoustic wave generated by the transducer interacts with light in the fiber to produce a phase shift via the strain-optic effect. Only the acoustic wave in the region between the mirrors contributes to the sensor output.

Another application for the fiber sensor is in the strain monitoring. In one such experiment uniaxial FFPIs, each with a single aluminum internal mirror, were embedded in graphite–PEEK and Kevlar–epoxy coupons, and a three-axis strain rosette was embedded in Kevlar–epoxy (20). The strain rosette

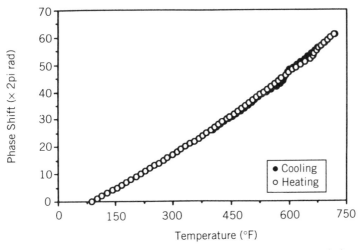

Figure 9.12. Dependence of interferometer phase shift ϕ on temperature during the curing process for a graphite–PEEK composite.

configuration is illustrated in Fig. 9.11. A deposited aluminum film on the fiber
end formed the second mirror for each of the interferometers. The sensors were
tested by monitoring the reflectance as a flexural force was applied to the
coupon. The phase change in a number of sensors tested was found to be a
linear function of strain over the range 0 to 500 $\mu\varepsilon$, and no significant hysteresis
was observed.

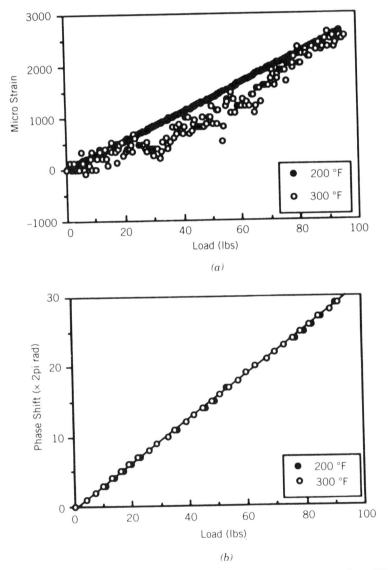

Figure 9.13. Response of (*a*) electrical strain gauge and (*b*) FFPI sensor to strain at 200°F and
300°F.

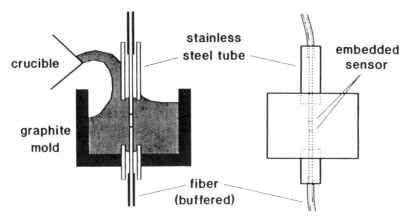

Figure 9.14. Process for embedding FFPI in a cast aluminum part.

In another experiment, an FFPI with dielectric mirrors and a thermocouple were embedded between the second and third layers of a 16-layer graphite–PEEK panel (21). The panel was then cured in a hot press using standard temperature and pressure profiles. Figure 9.12 shows the dependence of round-trip optical phase shift in the interferometer on temperature (determined using the embedded thermocouple) during the curing process under a constant pressure of 16.9 atm. An abrupt change in the slope of the curve is observed at around 700°F for heating and 600°F for cooling. This phenomenon is thought to be associated with the strain that results from decrystallization (for heating) or crystallization (for cooling) of the polymer in the coupon. This kind of information is important in optimizing the curing process, since the crystallization and decrystallization temperatures can be adjusted by changing the

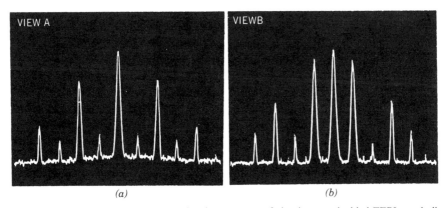

Figure 9.15. Spectrum analyzer trace showing response of aluminum-embedded FFPI to a bulk ultrasonic wave at a frequency of 1.8 MHz. The laser frequency is thermally tuned to maximize the first harmonic in (b) and the second harmonic in (a). (Horizontal scale: 2 MHz/div, vertical scale: 10 dB/div.)

temperature and pressure profiles. For strain measurements, two electric strain gauges (ESGs) were bonded to opposite sides of the completed coupon, above and below the embedded FFPI. A strain was induced by applying a load to the center of the coupon, which was supported at opposite ends. The strain sensitivity of the FFPI, $\Delta\phi/\Delta L$, was determined to be 9.1×10^6 rad/m by comparing the optical phase shift was the ESG readings. This is about 18% less than the value of $\Delta\phi/\Delta L$ measured for a similar fiber in air (22).

Strain measurements were also made at elevated temperatures. Both the fiber optic and ESG sensors showed good linearity at 200°F. However, as indicated by the data of Fig. 9.13, the ESG response was unstable at 300°F. By contrast, the data of Fig. 9.13 show the same linear load profiles at 200 and 300°F for the FFPI sensor.

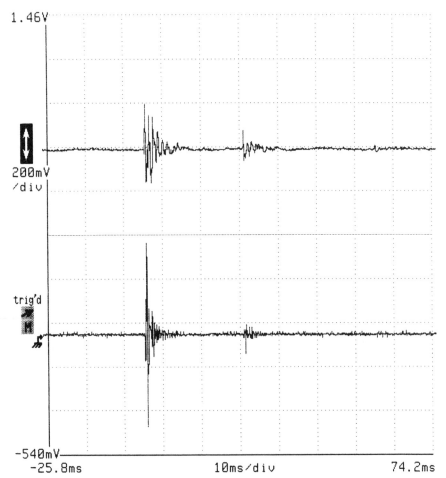

Figure 9.16. Response of aluminum-enbedded FFPI sensor (upper trace) to broadband burst of acoustic energy generated by breaking a pencil on the surface of the sample. For comparison, the response of a PZT transducer bonded to the surface of the sample is also shown (lower trace).

Recently, the sensing of temperature and of ultransonic pressure with FFPIs embedded in aluminum was demonstrated (21, 23). The aluminum parts were cast in graphite molds in air, as illustrated in Fig. 9.14. Breakage of the fibers at the air–metal interface during the casting process was avoided by passing the fiber through stainless-steel stress-relief tubes, which extended a short distance into the finished part. Thermal expansion of the aluminum caused the optical phase in the embedded FFPI to be 2.9 times more sensitive to temperature than for the same interferometer in air. The same FFPI was also used to detect ultrasonic waves launched by a surface-mounted PZT transducer over the frequency range 0.1 to 8 MHz. The spectral response of the FFPI for an ultrasonic frequency of 1.8 MHz is shown in Fig. 9.15. The response of the fiber sensor to a noise burst generated by breaking a pencil on the surface of the cast aluminum plate is shown in Fig. 9.16.

9.6. MULTIPLEXING AND SIGNAL PROCESSING TECHNIQUES

Although little has been published to data on multiplexing techniques for smart structures, multiplexing will be necessary in many cases to reduce cost and the number of fiber leads and connectors. Here we define a multiplexed system as one in which two or more sensors utilize the same light source, photodetector, and signal processor.

The reflectivity monitored FFPI is well suited for time-division multiplexing using a pulsed light source (7). In such a system, fiber delay lines of different length are provided between transmitter and receiver such that for each laser pulse, the photodetector receives reflected pulses from each of the sensors in different time slots. The FFPI signals can be processed by digital means under microprocessor control. In such a system the reflected waveforms are sampled for analog-to-digital conversion at fixed time delays relative to the start of the pulse, and the samples are averaged digitally. In-phase and quadrature signals from the interferometers needed for high sensitivity can be obtained by adjusting the dc bias current to the laser between pulses. A reference FFPI isolated from the smart structure environment and time-multiplexed with the FFPI sensors can be used to correct for fluctuations and drift in the laser wavelength.

The sensors can be deployed in either a linear or a star arrangement, as in Fig. 9.17, or a combination of the two can be used. These are equivalent to linear and star buses commonly used for communications. With a star configuration, the sensors can be positioned as desired to monitor a smart structure, using different lengths of fiber delay line between the star coupler and the sensors. With the linear arrangement, however, the sensors must be far enough apart for the reflected pulses to be separated in time at the photodetector. For example, if the laser pulse is 10 ns wide, the FFPIs must be separated by at least 1 m of fiber.

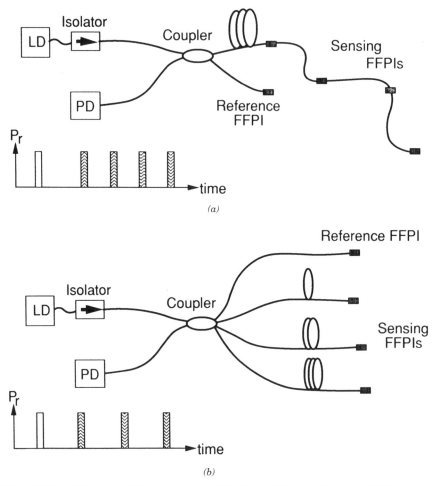

Figure 9.17. (a) Linear and (b) star arrangements for time-division multiplexing of FFPI sensors.

Other multiplexing techniques allow for the sensor to be located close together on a linear bus. One such scheme uses sensor FFPIs of different lengths and a semiconductor laser light source driven with sawtooth current waveform to produce a linear chirp in the laser frequency (24). Demodulation uses bandpass filters to separate the signals from the multiplexed sensors.

Another form of multiplexing in which the sensors can be arbitrarily close together on a linear bus is coherence multiplexing (25). In this case a broadband light source such as a multimode laser or (preferably) a light-emitting diode (LED) or superluminescent diode (SLD) is used instead of a narrowband laser. Coherence multiplexing also requires the use of a reference FFPI matched in length to the sensor FFPI to within the coherence length of

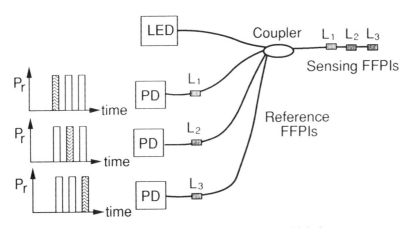

Figure 9.18. Arrangement for coherence multiplexing.

the light source ($\cong 10\,\mu$m for a typical LED or SLD). Light from the broadband source must be transmitted or reflected by both the sensor and reference FFPI before reaching the photodetector. If these conditions are satisfied, the sensor response will be similar to that obtained with a narrow-band laser source and no reference FFPI (26). An experimental arrangement for coherence multiplexing is shown in Fig. 9.18.

Coherence-multiplexed sensors can be arbitrarily close together in a linear arrangement provided that their cavity-length differences are substantially greater than the coherence length of the light source (e.g., the sensors could have lengths of 1.0 and 1.1 mm with an LED light source). The reference FFPI will then demodulate only the signal associated with the corresponding sensor of the same length.

9.7. CONCLUSIONS

The fiber Fabry–Perot interferometer (FFPI) is a strong candidate for smart structure applications because it is extremely sensitive, provides point-sensing capability, is simple to produce, has excellent mechanical properties, and is amenable to time-division and coherence multiplexing. Externally mounted FFPIs have been used to sense strain, temperature, ultrasonic pressure, and gas pressure in an internal combustion engine. FFPIs have been embedded in composites and in aluminum, where they have sensed temperature, strain, and ultrasonic pressure. Both time and coherence-division multiplexing of the FFPIs can be employed in smart structure applications. Further advances in manufacturing methods and signal processing to reduce the cost of individual sensing elements will be needed before widespread practical application of FFPIs in smart materials can be realized.

REFERENCES

1. C. Fabry and A. Perot, *Ann. Chim. Phys.* **16**, 115 (1899).

2. C. E. Lee, W. N. Gibler, R. A. Atkins, and H. F. Taylor, In-Line Fiber Fabry–Perot Interferometer with High-Reflectance Internal Mirrors. *IEEE J. Lightwave Technol.* **LT-10**, 1376 (1992).

3. S. J. Petuchowski, T. G. Giallorenzi, and S. K. Sheem, A Sensitive Fiber-Optic Fabry–Perot Interferometer. *IEEE J. Quantum Electron.* **QE-17**, 2168 (1981).

4. T. Yoshino, K. Kurosawa, and T. Ose, Fiber-Optic Fabry–Perot Interferometer and Its Sensor Applications. *IEEE J. Quantum Electron.* **QE-18**, 1624 (1982).

5. A. D. Kersey, D. A. Jackson, and M. Corke, A Simple Fibre Fabry–Perot Sensor. *Opt. Commun.* **45**, 71 (1983).

6. P. A. Leilabady and M. Corke, All-Fiber-Optic Remote Sensing of Temperature Employing Interferometric Techniques. *Opt. Lett.* **12**, 773 (1987).

7. C. E. Lee and H. F. Taylor, Interferometric Optical Fibre Sensors using Internal Mirrors. *Electron. Lett.* **24**, 193 (1988).

8. T. Valis, D. Hogg, and R. M. Measures, Fiber Optic Fabry–Perot Strain Gauge. *IEEE Photon. Technol. Lett.* **2**, 227 (1990).

9. C. E. Lee, R. A. Atkins, and H. F. Taylor, Performance of a Fiber-Optic Temperature Sensor from −200 to 1050°C. *Opt. Lett.* **13**, 1038 (1988).

10. R. A. Wolthuis, G. L. Mitchell, E. Saaski, J. C. Hartl, and M. A. Afromowitz, Development of Medical Pressure and Temperature Sensors Employing Optical Spectrum Modulation. *IEEE Trans. Biomed. Eng.* **BME-38**, 974 (1991).

11. G. L. Mitchell, Intensity-Based and Fabry–Perot Interferometer Sensors. In '*Fiber Optic Sensors: An Introduction for Engineers and Scientists* (E. Udd, ed.), p. 139. Wiley (Interscience), New York, 1991.

12. J. Stone and D. Marcuse, Ultrahigh Finesse Fiber Fabry–Perot Interferometer. *IEEE J. Lightwave Technol.* **LT-4**, 382 (1986).

13. J. Stone and L. W. Stulz, High-Performance Fibre Fabry–Perot Filters. *Electron. Lett.* **27**, 2239 (1991).

14. K. Ogusa, Analysis of End Separation in Single Mode Fibers and a Fiber Fabry–Perot Resonator. *IEEE Photon. Technol. Lett.* **4**, 602 (1992).

15. M. Oakland, C. E. Lee, W. N. Gibler, R. A. Atkins, M. Spears, V. Swenson, H. F. Taylor, J. McCoy, and G. Beshouri, Fiber Optic Pressure Sensor for Internal Combustion Engines. *Appl. Opt.* **33**, 1315 (1994).

16. K. A. Murphy, C. E. Kobb, A. J. Plante, S. Desu, and R. O. Claus, High Temperature Sensing Applications of Silica and Sapphire Optical Fibers. *Proc. SPIE—Int. Soc. Opt. Eng.* **1370**, 169 (1991).

17. T. A. Tran, W. V. Miller, III, K. A. Murphy, A. M. Vengsarker, and R. O. Claus, *Proc. SPIE—Int. Soc. Opt. Eng.* **1584**, 178 (1992).

18. C. E. Lee, H. F. Taylor, A. M. Markus, and E. Udd, Optical-Fiber Fabry–Perot Embedded Sensor. *Opt. Lett.* **14**, 1225 (1989).

19. J. J. Alcoz, C. E. Lee, and H. F. Taylor, Embedded Fiber-Optic Fabry–Perot Ultrasound Sensor. *IEEE Trans. Ultrason., Ferroelectr., Freq. Control* **UFFC-37**, 302 (1990).

20. T. Valis, D. Hogg, and R. M. Measures, Composite Material Embedded Fiber-Optic Fabry–Perot Strain Gauge. *Proc. SPIE—Int. Soc. Opt. Eng.* **1370**, 154 (1991).

21. C. E. Lee, W. N. Gibler, R. A. Atkins, J. J. Alcoz, H. F. Taylor, and K. S. Kim, Fiber Optic Fabry–Perot Sensors Embedded in Metal and in a Composite. *Opt. Fiber Sens. Conf.*, 8th IEEE Monterey, CA, *1992*.

22. A. Bertholds and R. Dandliker, Deformation of Single-Mode Optical Fibers Under Static Longitudinal Stress. *IEEE J. Lightwave Technol.* **LT-5**, 895 (1987).

23. C. E. Lee, W. N. Gibler, R. A. Atkins, J. J. Alcoz, and H. F. Taylor, Metal-Embedded Fiber Optic Fabry–Perot Sensors. *Opt. Lett.* **16**, 1990 (1991).

24. F. Farahi, T. P. Newson, P. A. Leilabady, J. D. C. Jones, and D. A. Jackson, A Multiplexed Remote Fibre Optic Fabry–Perot Sensing System. *Int. J. Optoelectron.* **3**, 79 (1988).

25. C. M. Davis, C. J. Zarobila, and J. D. Rand, Fiber-Optic Temperature Sensor for Microwave Environments. *Proc. SPIE—Int. Soc. Opt. Eng.* **904**, 114 (1988).

26. C. E. Lee and H. F. Taylor, Fiber-Optic Fabry–Perot Temperature Sensor Using a Low-Coherence Light Source. *IEEE J. Lightwave Technol.* **9**, 129 (1991).

10

Optical Fiber Bragg Grating Sensors: A Candidate for Smart Structure Applications

JIM R. DUNPHY, GERALD MELTZ, AND WILLIAM W. MOREY
United Technologies Research Center
East Hartford, Connecticut

10.1. INTRODUCTION

The purpose of this chapter is to introduce a new optical fiber sensor called a fiber Bragg grating sensor (FBGS). The content of this chapter is specifically tutorial in nature and is not intended as an exhaustive technology review nor as a presentation of the most current development activities. With the tutorial goal in mind, we provide information concerning the basic concept of the FBGS, how it is fabricated, its physical and optical properties, its advantages, how one might utilize it in a smart structure sensing system, and some basic demonstrations of its operation.

United Technologies Research Center (UTRC) has developed a method to write permanent optical Bragg grating filters into the core of conventional optical fibers. The outstanding feature of the UTRC devices is that the index perturbations are exposed into the core of the fiber from the side, extending over a limited length of fiber. This approach provides a finite length of modulated disturbance in the index of refraction of the core to create the Bragg grating (Fig. 10.1). Input and output leads become inherent features with no added labor involved in splicing or coupling. The optical signal processing potential of such a fiber-compatible filter can now be interfaced easily with an appropriate optical system with minimal power penalty.

The fiber grating filters introduced above are of interest to sensor systems because

- Optical fiber gratings respond to strain and can be applied as all-dielectric replacements for conventional strain gauges.

Fiber Optic Smart Structures, Edited by Eric Udd.
ISBN 0-471-55448-0 © 1995 John Wiley & Sons, Inc.

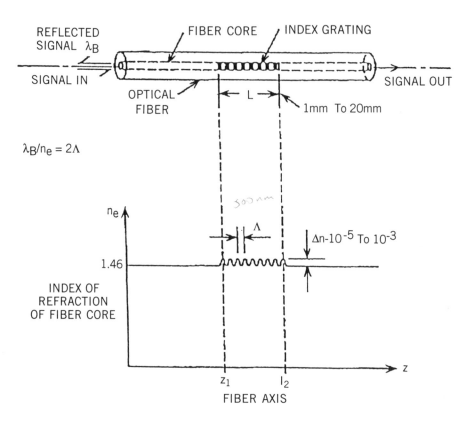

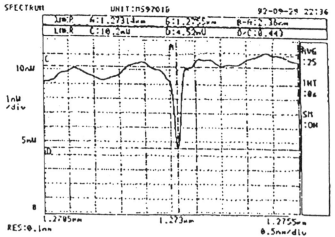

Figure 10.1. Fiber optic sensors for composites: candidate optical fiber sensor.

- A large number of FBGSs can be placed at predetermined locations on a single fiber string to create an array of quasi-distributed, quasi-point sensors.
- The FBGSs on the fiber string can be interrogated in transmission (a two-connector system) or in reflectance (a one-connector system).
- WDM and TDM multiplexing instrumentation can be configured to address a number of fiber grating sensors on a fiber string simultaneously.
- Optical fiber grating filters can be used as components in instrumentation for demultiplexing and signal processing of signals from remotely located fiber grating sensors.
- The FBGS fabrication method can be extended to provide relatively low-cost, highly available devices.
- FBGS devices have material and geometry characteristics that make them compatible with many diagnostics applications, such as embedded in composite smart structures.

With this brief introduction to the FBGS, it is appropriate to establish for this chapter a working definition of smart structures so that the potential scope of utility for the FBGS can be assessed. For the present, we suggest that *smart structures* is a design philosophy—in fact, a design philosophy that emphasizes integration and performance. It is through the integration of specialized sensors, materials, artificial intelligence and control systems, and actuators that the enhanced performance will be achieved. The improvements will be realized in more efficient manufacturing, enhanced operational performance, increased durability, and reduced lifetime costs associated with built-in diagnostic functions that support maintenance for cause rather than scheduled maintenance. With this perspective it is desirable to have a single sensor component in the smart structure as a functional, contributing element during composite structure fabrication, quality control evaluation, design testing, qualification testing, and in-service diagnostics. The FBGS device described in this chapter represents such a candidate.

10.2. PHYSICAL PROPERTIES

The physical properties of a candidate sensor are important in smart structure applications. The device must provide significant function without significant compromise. Hence any sensor embedded in a composite should live as long as the structure and not induce premature failure. The FBGS technology leads directly to satisfaction of these requirements. To demonstrate this, the basic physical properties are described and rule-of-thumb smart structure constraints are addressed.

Optical fibers are fabricated from fused silica, one of the most ideal, high-temperature elastic materials known. Freshly drawn silica filaments exhibit ultimate tensile strengths of the order of 1 Mpsi for short test sections, approaching the inherent strength of the basic constituent. Manufacturing compromises imposed by mass production constraints and nonideal coatings generally lead to a generic fiber with a functional tensile strength of about 800 kpsi. When this parameter is considered along with the Young's modulus (10 Mpsi), one finds that a high-quality optical fiber can survive strain approaching 8%. These are important reference figures because they imply that the basic materials of the FBGS are stronger than the matrix material used in most composite structures and that the basic FBGS material can operate to a higher strain-to-failure than many of the reinforcing filaments utilized in high-performance composite structures.

The next consideration is the problem of integration. FBGS devices can be attached to structures with appropriate adhesives. Recent experience indicates that such installations are longer lived than conventional strain gauge installations. For instance, adhesively bonded optical fiber can be used for delamination detection on a full-scale, graphite-reinforced helicopter tail. During laboratory simulation tests, the optical fiber installation survived more than 1 million test cycles without failure, whereas conventional strain gauging required new installations every few thousands cycles. Keep in mind that this durable performance strongly depends on proper selection of adhesives and optical fiber coatings.

Coatings for optical fibers is a very important issue to address. Generally speaking, one can select from low-temperature/low-modulus polymers, high-temperature/high-modulus polymers, ceramic-like, and metal coatings. Our experience indicates that low-temperature/low-modulus polymers such as acrylates are completely inappropriate—especially considering that they degrade, both chemically and mechanically, during the curing conditions for most epoxies used for aerospace structures. In addition, acrylates are usually applied in thicknesses of about 100 μm. This produces a total outside diameter of about 250 μm for the coated fiber, which is unacceptably large for most embedded smart structure applications. Finally, when one realizes that thick acrylate coatings yield and then fall in shear during strain measurement, one must conclude that there is a better alternative.

A better choice for smart structure applications is the category of high-temperature/high-modulus polymer coatings such as polyimides. These coatings typically increase the fiber diameter by only 10 μm while providing good adhesion to epoxies and demonstrating no significant chemical or mechanical degradation during typical epoxy curing programs. In addition, our experience indicates that strain is adequately transferred through the thin, hard polyimide coating to provide good diagnostic measurements. For smart structure applications, the fiber grating sensors are protected by a polyimide coating.

To create a FBGS, the polyimide coating is removed over the fiber length required for exposure. Using the exposing system highlighted in Fig. 10.2, a

SET-UP FOR MAKING BRAGG GRATINGS IN OPTICAL FIBERS

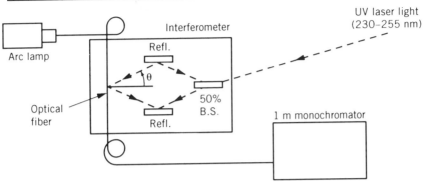

Figure 10.2. Setup for making Bragg gratings in optical fibers.

series of index perturbations is written into the core with a total pattern length that is programmable between 1 mm and several centimeters. The period of the index perturbations is adjusted to produce a strong Bragg reflection at the optical wavelength of interest, and the strength of the high-spatial-frequency perturbations on the order of 1 part in 1000 or less. After exposure, the stripped portion of fiber is recoated with polyimide to restore its durability.

As can be expected, the process of polyimide stripping, grating exposure, and polyimide recoating compromises some of the fiber strength. An ultimate strength of the order of 200 kpsi (a strain capability of 2%) is practical after all processing, and we anticipate that processing refinements will yield even greater strengths in the near future.

One final physical feature must be addressed, and it concerns the practical issue of connectorizing splicing of the sensor fiber. The FBGS is based on a concept that utilizes single-mode fiber. In the 1300-nm band this means that the waveguides core of the fiber is about 9 μm in diameter. For proper system operation and efficiency, high-quality single-mode connectors and splicing methods are required.

10.3. OPTICAL PROPERTIES

The purpose of this section is to introduce the optical properties of the FBGS. Understanding these basic aspects of the sensor will help when setting up instrumentation to conduct measurements.

In general nature, the FBGS is simply a tuned optical filter. The UTRC fabrication procedure enables us to write the grating with resonant wavelengths between 400 and 2000 nm. The resonant filter has a typical bandwidth on the order of 0.1 nm in the 1300-nm band and transmission loss

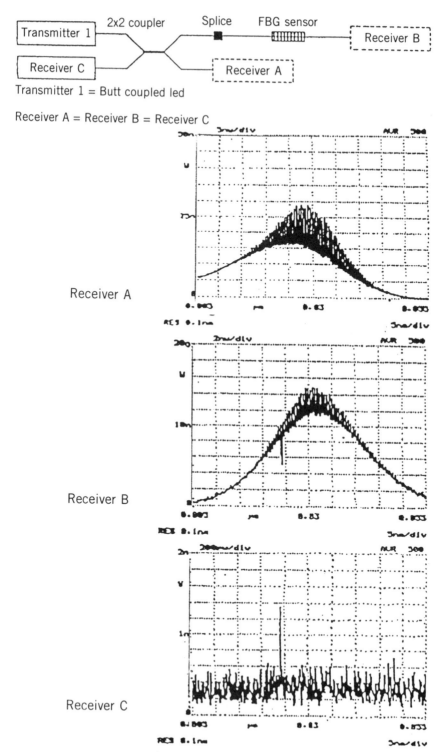

Figure 10.3. FBG instrumentation.

of up to 100%. Since the optical filter is indeed a Bragg grating, this transmission loss is due to a 100% reflection of the optical signals in the bandwidth of the filter back along the axis of the fiber to the transmitter. Example spectra are provided in Fig. 10.3. Again, it is important to note that the reflectivity (or transmission loss) can be programmed to as little as a few percent by reducing the accumulated exposure of the fiber. The excess loss due to the filter in the fiber has been measured and shown to be less than 0.001%, implying excellent efficiency for use in complex sensing configurations.

The width of the filter function is, to a large degree, dependent on the length of the grating through an inverse relationship. Hence simple methods associated with changing the length of exposed fiber can be used to modify the effective Q value of the filter. Using such methods, gratings as broad as 2 nm have been fabricated. The inverse dependency of the filter Q value on length is not a perfect relationship, due to interaction between the fundamental filter perturbations and a background change in average index of refraction of the exposed region of the fiber core.

Further, the shape of the filter function is dependent on a large number of index perturbation parameters and is simple only for weak gratings. In the case of a weak exposure over a short length of fiber, the intensity distribution of the exposing irradiation pattern will have a strong impact on the filter function shape. For instance, a uniform illumination pattern will yield a filter function with significant sidebands associated with the sinc function, while a Gaussian-like illumination function can be associated with minimum sidebands. Asymmetric features can develop in the shape of the filter function for strong gratings over short lengths of fiber or for weaker gratings over long lengths. These issues will not be addressed in this chapter since they are not an issue for the development of basic instrumentation methods. For some advanced instrumentation methods, the exact filter function shape can be very important.

When the grating region of the optical fiber is disturbed in a fashion that modifies the distribution of index perturbations, the filter function changes. For small, uniformly distributed, longitudinally aligned strain changes, the peak in the filter function will shift to longer wavelengths or shorter wavelengths corresponding to either tension or compression in the spacing of elemental perturbations. The response of a typical filter in the 1300-nm range used as a strain gauge is about 1 nm per 1000 μstrains. Note that the devices also respond to temperature changes with a responsivity of about 0.01 nm per °C. These responsivity parameters scale with the peak wavelength of the filter function. Care must be exercised in establishing the geometry of the measurement layout because weak, nonuniform strain distributions will be sensed as the integrated strain level over the length of the grating. On the other hand, strong, nonuniform strain distributions can significantly disturb the basic uniformity of the index perturbations in the fiber and cause distortions in the filter function. Such distortions can affect the operation of some instrumentation methods. Hence, as a precautionary procedure for strain measurements, one should use only straight layouts of the optical fiber gauge section. Outside the gauge region, the fiber can bend with a radius of curvature approaching a few millimeters to accommodate complex input–output lead routing requirements.

One simplifying aspect of the grating sensor is the fact that it tends to reject the effect of strain fields not aligned with the longitudinal axis. We estimate its response to unidirectional, transverse components to be attentuated by a factor of 500 or more compared with the longitudinal response.

Advanced sensor and instrumentation methods are presently under development to address the mixing of strain and temperature responses. For instance, some fiber designs minimize one or the other parameter, and other fiber designs provide differing sensitivities at different wavelengths. Combined with advanced instrumentation, one parameter will be rejected or the two parameters will be resolved separately.

During most measurements, the temporal response limitation is dependent on the transmitting and receiving instrumentation. For instance, the grating sensors slowest strain response is limited by the acoustic transit time across the gauge length, implying a rise time of less than $1\,\mu s$. Or for a uniform transverse field, one might expect a response as fast as 10 n. So for all but ultrasonic applications, the bandwidth of the detector system and associated signal-to-noise limitations will determine the measurement performance for many systems.

10.4. SENSING CONCEPTS

The simplest arrangement of apparatus is very effective for static strain measurement with moderate resolution and is configured around a transmitter consisting of a broadband light source and a narrowband-tunable receiver, a conventional laboratory spectrometer, as shown in Fig. 10.4. In this arrangement, arc lamps, light-emitting diodes (LEDs) and superluminescent diodes (SLDs) have been used successfully to produce adequate strain measurement for many initial smart structures testing requirements. A typical spectrometer provides a resolution of about 0.1 nm, corresponding to about $100\,\mu$strains for a FBGS in the 1300-nm band. Depending on the type of signal processing used to analyze the resulting data, the reflection spectrum of Fig. 10.4 may be easier to process. In addition, when one tries to interpret strain signals resulting in less than 0.1-nm perturbations of the spectral peak, the noise of the light-source spectrum becomes important. One should select LEDs and SLDs with a minimum of spectral ripples from residual longitudinal-mode effects.

Future smart structure applications anticipate the need for multiple sensors. Hence an expansion of the basic measurement method is required. The concepts utilizing wavelength-division multiplexing (WDM) are easily accommodated by the strain response of the FBGS and the ability to fabricate the sensors with a programmable center wavelength. As one extends the concept to an array of FBGS devices, some limitations are encountered. For the broadband transmitter case, one limitation is associated with the spectral width of the transmitter light source. Suppose that each sensor in the array is to experience $\pm 5000 \, \mu$strains (this strain range is representative of that required for an aggressively loaded aerospace structure, including both static and dynamic disturbances). The sensor filter wavelength spacing must be 10 μm. If the spectral width of the source, say an LED, is 100 μm, only 10 sensors can be unambiguously measured. To extend the array beyond this basic set of 10 requires either a combination of other light sources (such as a combination of 1300- and 1500-nm LEDs) incorporated through fiber couplers or sensor string switching to sequentially address different strings of fiber each with 10 sensors.

Further, temporal response issues arise with the configuration described above. Most laboratory spectrometers provide a single, recordable spectrum every minute or so, yielding an effective bandwidth of less than 0.01 Hz. This is a serious limitation when attempting to track dynamic strain measurements.

One simple method of alleviating the dynamic constraint of the spectrometer is to change the measurement configuration by replacing the narrow-band-scanned receiver spectrometer by a fixed array of wavelength-selective transmission filters. Each filter must be tuned such that the sensor signal is decoded properly from a wavelength change to a signal magnitude change (Fig. 10.4). This configuration can provide strain measurements with a temporal bandwidths to several kilohertz. On the other hand, widening the temporal bandwidth will increase the system noise, leading to a reduction in the minimum resolved strain.

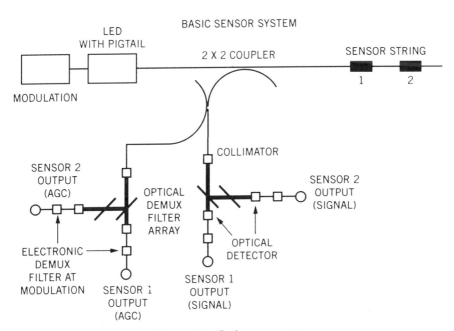

Figure 10.4. Basic sensor system.

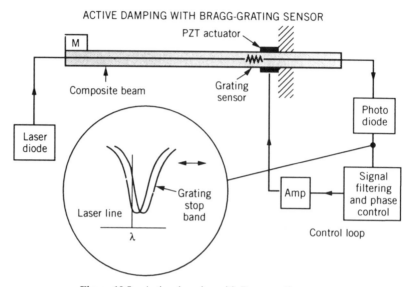

Figure 10.5. Active damping with Bragg grating sensor.

An alternative instrumentation method can implement a rapid temporal response measurement without compromising the resolution of low-level strains. It utilizes a narrowband-tunable transmitter with a spectral width less than the grating sensor. In this configuration the receiving detector is very simple and has a broad temporal response. Operation is based on the perspective that the remote sensor is a dynamic filter used to program the loss of the narrowband transmitter signal through the sensing fiber (Fig. 10.5). Vibrations in the structure change the peak of the sensor filter function and modulate transmission losses, yielding dynamic strain resolution to about 1 μstrain. Demonstration of an active damping system with this configuration is addressed in the next section. In the case of the narrowband transmitter arrangement, the system limitation is usually associated with the tuning range of the transmitter. Special LD transmitters are limited to tuning ranges of less than 50 nm, indicating that five or fewer sensors can be addressed before special switching or coupling arrangements are necessitated.

In this introduction we have discussed three simple configurations: broadband transmitter with a narrowband-tunable receiver; broadband transmitter with a narrowband fixed-filter receiver; and narrowband-tunable transmitter with a broadband receiver. All utilize the strain-to-wavelength encoding feature of the fiber grating sensor. The systems acommodate 10 or fewer sensors before special switching or coupling methods are required for aggressively loaded structures. On the other hand, for systems that do not require rapid response, dynamic strain sensing, time-domain multiplexing (TDM) methods can be used to increase the number of addressable sensors. In this alternative case, the dynamic response is compromised by the necessity for multiple pulse sampling techniques required by optical time-delay reflectometry (OTDR) techniques. However, as many as 100 low-reflectivity FBGSs can be connected serially to provide a large measurement array. A new issue that arises in this case is that the pulsed light source must have a small, tunable range, as it is used in the OTDR arrangement to adjust to the disturbed wavelengths of the various strained FBGSs.

In this section we have introduced several straightforward instrumentation methods for conducting strain measurements with the FBGS. Basic concepts have been highlighted to emphasize the principles of operation. More advanced methods are under development and can be explored by consulting the current literature.

10.5. EARLY DEMONSTRATIONS OF SMART STRUCTURE POTENTIAL

In this section we provide several examples of the FBGS operating in diagnostic functions associated with smart structures. The demonstrations focus on laboratory test samples, not full-scale structures.

A FBGS was installed between the plies of a six-layer graphite–epoxy prepreg layup in a hot press. One grating sensor was positioned within the uncured composite, while another remained outside as a reference. The response of the gratings was monitored throughout the 350°F/3500 lb curing program. When the temperature response of the internal sensor was processed

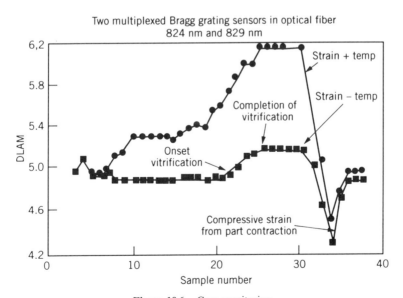

Figure 10.6. Cure monitoring.

out of the data as identified in Fig. 10.6, we observed significant information concerning the internal state of the composite. For instance, a very clear indication of the onset of vitrification was seen. In addition, a buildup of thermally induced residual strain was detected as the part was cooled. Tracking such internal parameters can lead to optimized curing programs and greater manufacturing productivity.

Once structure fabrication is completed, the embedded FGBS can be used for part qualification testing. Stiffness parameters can be assessed by conducting static load tests. As an example, the sample described above was mounted in a cantilevered beam geometry. The FBGS response to beam displacement is shown in Fig. 10.7 for both up and down directions. Tensile and compressive strains are resolved. Tests like this have been conducted on more flexible samples, demonstrating operation of the FBGS to about 2% strain.

To measure the dynamic strain associated with potential in-service application, the instrumentation was modified to include a LD as introduced in Section 10.4. The output of the LD was tuned to be coincident with the edge of the grating filter response. Then as the cantilevered beam was bent up or down, the shifting stopband of the grating filter function moved closer to or farther from the laser wavelength, modulating the signal transmitted. This very sensitive dynamic strain signal was fed into a control system for proper filtering, phasing, and gain adjustments before driving a piezoelectric actuator pair. The actuator pair was mounted at the root so as to induce a bending moment to counteract displacement of the beam from its resting position. With the feedback signal from the embedded sensor driving the control system, excellent active damping of the 60-Hz fundamental mode of vibration was

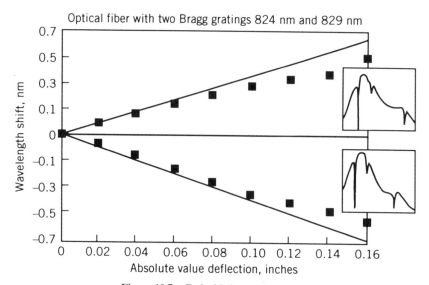

Figure 10.7. Embedded sensor response.

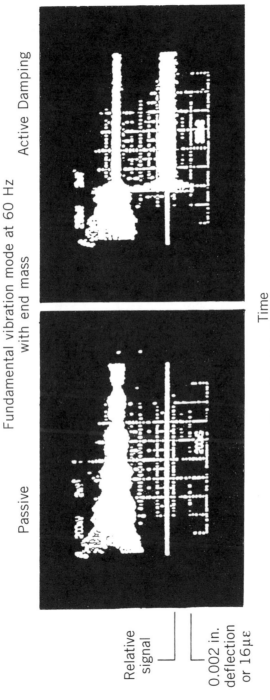

Figure 10.8. Embedded sensor response.

283

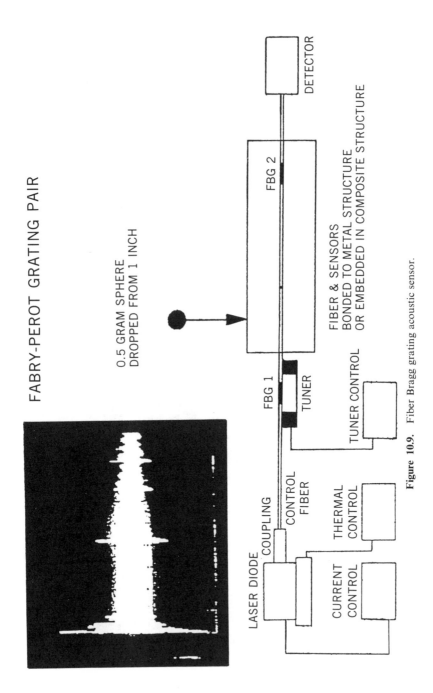

Figure 10.9. Fiber Bragg grating acoustic sensor.

achieved, as demonstrated in Fig. 10.8. This arrangement was effective for strains as low as 1 μstrain.

An additional change in the instrumentation provided even greater dynamic sensitivity. Figure 10.9 shows a wavelength-agile laser diode transmitter. The output wavelength is controlled by an external cavity consisting of a FBGS attached to a piezoelectric tuning device. The transmitter was adjusted so that its output wavelength coincided with that of the remote FBGS attached to a composite panel. When tuned in this manner, the two grating devices create a Fabry–Perot pair with greatly enhanced strain sensitivity. The unique feature of this setup is that the transmitter was sympathetically tuned to the grating pair to yield a very strong signal for recording low-level disturbances of the panel. The example shows the acoustic disturbances resulting from dropping a 0.5-g sphere from a height of 1 in.—a 120-μJ impact.

10.6. CONCLUSIONS

This chapter has provided a tutorial introduction to the optical fiber Bragg grating sensor as a candidate for smart structure applications. This all-dielectric replacement for the conventional strain gauge is compatible with most attaching and embedding techniques. Furthermore, it can be utilized to monitor composite fabrication in addition to providing valuable diagnostics during parts qualification testing and in-service monitoring. The potential significance of this device is underscored further when these capabilites are added to its versatile combination with a wide variety of instrumentation techniques. Development is continuing to increase its availability, lower the unit cost, further enhance its durability, and advance associated instrumentation concepts.

11

Elliptical-Core Two-Mode Optical Fiber Sensors

KENT A. MURPHY, ASHISH M. VENGSARKAR, AND RICHARD O. CLAUS
Fiber & Electro-Optics Research Center,
Virginia Polytechnic Institute and State University,
Blacksburg, Virginia

11.1. INTRODUCTION

Some of the most sensitive optic sensing systems use single-mode fibers in phase-modulated schemes. While ultrahigh sensitivities can be obtained with such single-mode fiber sensors, there are practical drawbacks associated with implementation techniques. In the Michelson and the Mach–Zehnder configurations, the reference and sensing arms are physically at different locations, and hence a minuscule change in temperature in the reference arm may affect the phase difference detected at the output of the sensor. This may lead to a misinterpretation of the observed change in phase due to thermal drifts being attributed to the external perturbation being measured. For example, a temperature gradient of 10^{-2}°C between the two arms of a 1-m-length homodyne Michelson arrangement can lead to a $\pi/2$ phase change in the output.

One way to circumvent the problem of requiring a reference arm is to include the sensing and reference arms within the same fiber. A two-mode fiber sensor can be considered as an equivalent system. In such a sensing technique, the phase difference between the first two modes is modulated by the parameter being measured, and the intensity monitored at the output is a measure of the perturbation. This approach eliminates the need for a reference arm since both modes propagate within the same fiber. The trade-off is that a slightly lower sensitivity (30 dB less than the theoretically achievable single-mode fiber sensor sensitivity) is obtained. The major advantage of such a sensor is that it can be ruggedized and maximum performance levels can be achieved without invoking the need for stabilization schemes that are an essential ingredient of the single-mode Mach–Zehnder and Michelson arrangements.

Fiber Optic Smart Structures, Edited by Eric Udd.
ISBN 0-471-55448-0 © 1995 John Wiley & Sons, Inc.

Few-mode, circular-core fibers were first used as acoustic sensors by Layton and Bucaro (1). Two- and three-lobe far-field intensity distributions obtained at the output by exciting appropriate modes within the fibers were later used for measurement of pressure, strain, and vibrations (2, 3). The standard method of analysis involves a measurement of the exchange of optical power between these lobes to determine the nature of the external perturbation. The major obstacles in practical implementation of sensing systems that use circular-core fibers have been the strict input conditions required to launch specific modes and the instability of the cross-sectional intensity distribution of the second-order modes (4).

11.2. THEORY

In the weakly guiding approximation, single-mode circular core fibers can support the two orthogonal polarizations of the LP_{01} mode as well as the four degenerate, higher-order, LP_{11} modes when operated just below the single-mode cutoff wavelength. Two-lobe patterns can be obtained in the far field at the output of the fiber, and the oscillation of the power distribution between the lobes can be used to sense strain or vibration, as described below. However, environmental conditions can also introduce differential phase shifts between the almost degenerate four eigenmodes, leading to an instability of the second-order mode pattern. This limits the practical implementation of such sensors.

The use of optical fibers with highly elliptical cores has been shown to remedy this situation (4). Since the circular symmetry of the fiber has been eliminated, only two second-order modes, the LP_{11}^{even} modes, are guided by the elliptical-core fibers just below the single-mode cutoff wavelength. The intensity

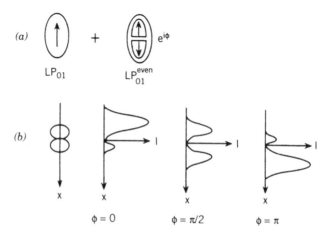

Figure 11.1. Evolution of the far-field two-lobe pattern in dual-mode elliptical-core fibers.

distribution of the second-order modes is stable, and practical operation of a sensor system is possible since there is a considerable range of the optical spectrum over which the LP_{11}^{odd} modes are unguided.

The operation of the sensor can be described with reference to Fig. 11.1. When the LP_{01} and the LP_{11}^{even} modes are excited equally in an elliptical-core fiber, the output radiation pattern will be a superposition of the contributions from the two modes and will be a function of the phase difference between them. The evolution of the two-lobed output pattern for different values of the phase difference, ϕ, is shown in Fig. 11.1b, where the ovals represent the bright intensity regions of the far-field pattern. For a change in ϕ of 2π there will be one complete oscillation of the intensity pattern. Measurement of the deformation ΔL required in a specific type of fiber for a 2π change in ϕ is the first step toward calibrating a strain/vibration sensor based on this principle.

The two-mode elliptical-core fiber used in a differential interferometric sensing scheme can be operated in two regimes:

1. The linear range where the induced perturbation is small enough to keep the phase shift between the LP_{01} and the LP_{11} modes to less than π. The resulting output is therefore a varying intensity pattern which depends on the parameter being measured. No fringes are observed.

2. The nonlinear range in which large perturbations force the output into the nonlinear range, thereby inducing fringes. Most fiber optic sensors, being highly sensitive, operate in this mode and fringe counting techniques need to be used to process the output signal.

A common problem associated with the practical implementation of such sensors is that linearly polarized light launched at the input of the elliptical-core fiber may change its orientation with respect to the major/minor axis of the core, depending on the magnitude and type of strain induced on the clamps of the sensing fiber. The input polarization orientation may also change, due to a rotational misalignment between the lead-in fiber and the sensing fiber. An effect of such misalignments is that the output intensity pattern does not remain truly sinusoidal but undergoes an amplitude modulation over large dynamic range measurements. Also, the oscillation period varies with the launch angle. In the calibration of any sensor, it is essential that the oscillation period (also termed the beat length, which we define later) be known accurately for the measurement of small strain fields.

We first consider some of the basic operating principles of two-mode fibers. The V number, or normalized frequency, of a circular-core optical fiber is defined by the relation

$$V = \frac{2\pi a}{\lambda}\sqrt{n_1^2 - n_2^2} \qquad (11.1)$$

where a is the core radius, n_1 the refractive index of the core, n_2 the refractive index of the cladding, and λ the wavelength of operation (5). The fiber operates in its single-mode propagation regime when the V number is less than 2.405. For $2.405 < V < 3.832$, two linearly polarized modes can propagate through the fiber.

Consider a fiber operating with a single mode at its nominal wavelength λ_0. If this fiber is operated at a wavelength λ_1, where $\lambda_1 < \lambda_0$, the first two linearly polarized modes, denoted by LP_{01} and LP_{11}, respectively, can propagate through the fiber. The electromagnetic field descriptions for the two linearly polarized modes can be combined to derive an expression for the total intensity at the output end of the sensing fiber (6). Two-lobe patterns can be obtained in the far field at the output of the fiber and the oscillation of the power distribution between the lobes can be used to sense changes in strain or vibration, as described below.

We first describe the operation of a two-mode fiber sensor in terms of a circular-core fiber. For a weakly guiding circular-core fiber, the modal field amplitude of the linearly polarized LP_{lm} mode can be expressed in the cylindrical coordinate system as

$$\psi(r, t) = J_l(\kappa_{lm}\rho)\cos(l\phi)\exp(j\beta_{lm}z)\exp(-j\omega t) \qquad (11.2)$$

where $\kappa_{lm}^2 = \omega^2 n_1^2/c^2 - \beta_{lm}^2$, ω is the angular frequency of operation and β_{lm} describes the longitudinal propagation constants of the mode. For an elliptical-core fiber, the transverse field expression, given by $J_l(\kappa_{lm}\rho)\cos(l\phi)$ for circular-core fibers, is expressed in terms of the Mathieu functions in the elliptical coordinate system. In the model developed here we try to maintain the analytical tractability of the ϕ variation in the circular-core case; at the same time we consider only the LP_{01} and the LP_{11} modes, described by the equations

$$\psi_{01}^{xy} \approx \psi_{01t}A_{01}\exp(j\beta_{01}^{xy}z)(\cos\theta, \sin\theta)(\mathbf{a}_x\mathbf{a}_y) \qquad (11.3a)$$

and

$$\psi_{11}^{xy} \approx \psi_{11t}A_{11}\exp(j\beta_{11}^{xy}z)\cos\phi(\cos\theta, \sin\theta)(\mathbf{a}_x\mathbf{a}_y) \qquad (11.3b)$$

respectively, where $\psi_{01t,11t}$ denote the transverse variation of the scalar fields in the core, $\mathbf{a}_x$ and $\mathbf{a}_y$ are unit vectors in the x and y directions, respectively, and θ is the angle the linearly polarized light makes with the major axis of an elliptical-core fiber. In Eq. (11.3), ρ and ϕ correspond to the radial and angular coordinates in a cylindrical coordinate system, respectively. The $\cos\phi$ variation maintained in our expressions will help us visualize the two-lobe pattern typically obtained in the far field of the output. The transverse fields, $\psi_{01t,11t}$, are normally comprised of a product of a radial component and an angular component in the elliptical coordinate system, as we describe later. In Eq. (11.3) we have assumed that the angular component described by the

Mathieu function $N(\eta)$ can be approximated by the $\cos \phi$ variation in the weakly guiding case. This approximation is valid to a certain extent and can be proved by plotting the field variations of the LP_{01} and the LP_{11} modes in the core for an elliptical-core fiber. The elliptical-core fiber assumption implies that the LP_{11} mode is necessarily even when the fiber is operated at a wavelength just below the single-mode cutoff wavelength of the fiber; we thus remove the degeneracy between the odd and even LP_{11} modes in a circular core fiber. Also the x- and y-polarized modes are nondegenerate, as seen from Eq. (11.3). Note that the time dependence ($e^{-j\omega t}$) has been ignored in the expressions and that A_{01} and A_{11} are constants that depend on the relative strengths of the input fields launched. The transverse term ψ_t is useful in determining, theoretically, the propagation constants of the two modes. For an intuitive evaluation of the far-field pattern due to axial strains along the z-direction, this term does not play any role. The differences in the propagation constants, $\Delta\beta_x = \beta_{01}^x - \beta_{11}^x$ and $\Delta\beta_y = \beta_{01}^y - \beta_{11}^y$, can be evaluated experimentally.

When both the LP modes are launched at a polarization angle θ, the x- and y-polarized fields can be expressed as

$$\psi^x \approx [A_{01} \exp(j\beta_{01}^x z) + A_{11} \cos \phi \exp(j\beta_{11}^x z)] \cos \theta \, \mathbf{a}_x \qquad (11.4a)$$

and

$$\psi^y \approx [A_{01} \exp(j\beta_{01}^y z) + A_{11} \cos \phi \exp(j\beta_{11}^y z)] \cos \theta \, \mathbf{a}_y \qquad (11.4b)$$

respectively. When there is no polarizer at the output, the total intensity at the output is given by a sum of the individual components:

$$I = |\psi^x|^2 + |\psi^y|^2$$

$$= A_{01}^2 + A_{11}^2 \cos^2 \phi + 2A_{01}A_{11} \cos \phi [\cos^2 \theta \cos \Delta\beta_x z + \sin^2 \theta \cos \Delta\beta_y z]$$
$$(11.5)$$

where we have dropped the transverse field terms and $\Delta\beta_x$ and $\Delta\beta_y$ are as defined earlier.

The first term gives rise to a static two-lobe pattern (with maxima at $\phi = 0$ and $\phi = \pi$) that does not change with external perturbations. Since the second term contains a $\cos \phi$ factor, when the term ($\cos^2\theta \cos \Delta\beta_x z + \sin^2\theta \cos \Delta\beta_y z$) increases, the intensity of one of the lobes (with maxima at $\phi = 0$) in the two-lobe pattern increases, whereas that of the other lobe decreases. Thus an external perturbation that changes $\Delta\beta_x$, $\Delta\beta_y$ or z leads to an intensity oscillation between the lobes. This effect has been observed by several researchers and has been used for the fabrication of two-mode fiber sensors.

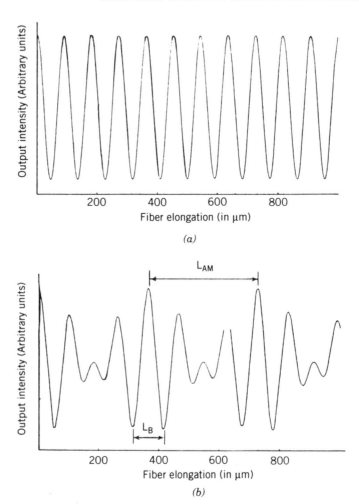

Figure 11.2. Theoretical prediction of output intensity pattern for linearly polarized light launched into the elliptical-core fiber: (a) $\theta = 90°$; (b) $\theta = 45°$.

From Eq. (11.5) we see that the dependence of the output intensity on axial strains is dictated by the relation

$$I(z) \propto \cos^2\theta \cos \Delta\beta_x z + \sin^2\theta \cos \Delta\beta_y z \qquad (11.6)$$

which shows that when $\theta = 0°$ or $90°$, we have a pure sinusoidal variation with the beat length L_B (defined as the elongation required for the sinusoidal intensity variation to undergo a full 2π phase shift) given by $2\pi/\Delta\beta_x$ or $2\pi/\Delta\beta_y$, respectively. The output intensity variation for $\theta = 90°$ is shown in Fig. 11.2a.

When $\theta > 45°$, the beat length will be dictated by the dominant term $\cos \Delta\beta_y z$, and hence L_B will be closer to $2\pi/\Delta\beta_y$. Similarly, for $\theta < 45°$, the beat length will be closer to $2\pi/\Delta\beta_x$. Since typical values of $\Delta\beta_x$ and $\Delta\beta_y$ are on the

order of $100\,\mu m$ for elliptical-core fiber manufactured by Andrew Corporation
and differ from each other by approximately tens of micrometers, one will also
observe a significant amplitude modulation and a variation in the beat length
for launch angles in the range $30° < \theta < 45°$. An example of amplitude
modulation is shown in Fig. 11.2b for a launch angle of 45°. Near $\theta = 0°$ and
$\theta = 90°$, the amplitude modulation will be less visible. The beat lengths L_B and
L_{AM} associated with a typical output pattern are thus functions of the
differences in the beat lengths, $\Delta\beta_x$ and $\Delta\beta_y$, as well as the angle of launch.

For experimental verification, LP light ($\lambda = 633\,nm$) launched at 0°, 30°,
45°, 60°, and 90°, into an elliptical-core fiber (dual mode at 633 nm, core
dimensions $1.5\,\mu m \times 2.5\,\mu m$, manufactured by Andrew Corporation) has

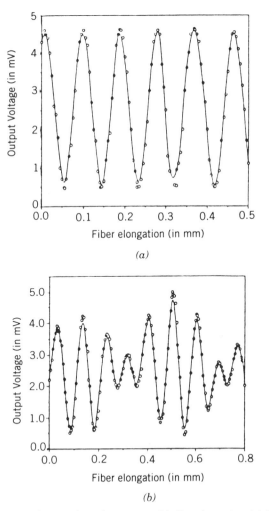

Figure 11.3. Variation of output intensity pattern with fiber elongation: (a) $\theta = 90°$; (b) $\theta = 45°$.

shown good correlation with the simple model described by Eq. (11.6). For a specific example, LP_{01} and LP_{11}^{even} modes excited equally within the fiber have given experimental values for $L_B = 121.1\,\mu m$ (for $\theta = 0°$) and $91\,\mu m$ (for $\theta = 90°$). Experimental plots for launch angles of $90°$ and $45°$ are shown in Fig. 11.3. A comparison of experimentally observed values for different angles of launch and those predicted by Eq. (11.6) for $\theta = 30°, 45°$, and $60°$ are tabulated in Table 11.1. The terms L_{AM} and L_B have been defined in Fig. 11.2b. It must be emphasized that the values for $\theta = 0°$ and $\theta = 90°$ used in Eq. (11.6) were those obtained experimentally. For a more accurate analysis, one would have to solve the eigenvalue equation for the elliptical-core waveguide and obtain the β_{01} and β_{11} values for the x- and y-polarized modes for the given ellipticity; also, the variation in the propagation constants due to the anisotropy introduced within the fiber by the strain may have to be considered.

From Eqs. (11.4) and (11.5), we can also see that by placing a polarizer oriented at $0°$ at the output end, the intensity is given by

$$I = |\psi^x|^2 = \cos^2\theta(A_{01}^2 + A_{11}^2\cos^2\phi + 2A_{01}A_{11}\cos\phi\cos\Delta\beta_x z) \qquad (11.7)$$

which shows that the amplitude modulation is eliminated and the beat length is given by $2\pi/\Delta\beta_x$. The effect of the misoriented launch is to reduce the intensity at the output by a factor $\cos^2\theta$. Similarly, a polarizer oriented at $90°$ can lead to a pure sinusoidal variation with a beat length $2\pi/\Delta\beta_y$. Only in cases where a polarizer cannot be used at the output end (e.g., when an offset splice is used to fuse a lead-out fiber) will one observe a modulation of the amplitude and a variation in the beat length.

A basic requirement for the performance analysis of any of these devices or sensors is a theoretical model for the elliptical-core fiber. The propagation characteristics of the elliptical-core fiber must be determined numerically since closed-form solutions represented by Mathieu functions are difficult to handle. Previous work in this area has included an analysis based on solving the determinant of a truncated infinite-dimensional matrix or on the expansion of the solutions in terms of a series of Bessel functions (7, 8). In a recent paper, Kumar and Varshney have developed a perturbation model based on the analyses of rectangular waveguides (9).

Table 11.1 Comparison of Theoretical Model and Experimental Results

	L_{AM}			L_B		
θ	Theory (μm)	Experiment (μm)	Error (%)	Theory (μm)	Experiment (μm)	Error (%)
$30°$	364	399.5	9.75	111	107.5	3.15
$45°$	364	402.5	10.57	105	99.6	5.14
$60°$	364	399.0	9.62	95	93.0	2.10

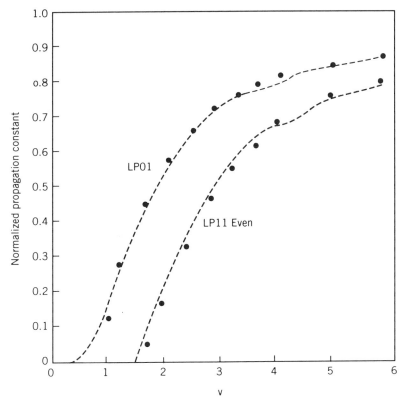

Figure 11.4. Characteristic propagation curves for the first two modes: filled circles, our results: dashed curves, from Ref. 8. $\Delta n = 0.21$.

Shaw et al. (10) have presented a direct solution of the scalar equation in terms of the elliptical coordinate system without taking recourse to infinite-dimensional matrices or infinite series. The scalar wave equation is written in terms of the elliptical coordinate system and decomposed into two ordinary differential equations using the method of separation of variables. The continuity of the ratio of the scalar field to its normal derivative is used as the boundary condition and values of the propagation constant β are calculated for a user-specified range of λ. In Fig. 11.4 we plot the normalized propagation constants b^*, defined by $b^* = [(\beta/k)^2 - n_2^2]/(n_1^2 - n_2^2)$, versus the normalized frequency, $V = (2\pi\beta/\lambda)\Delta n$, where $\Delta n = (n_1^2 - n_2^2)^{1/2}$ and b is the semiminor axis of the fiber core. The sensitivity of the fiber sensor is directly related to the difference in the propagation constants between the LP_{01} and the LP_{11}^{even} modes. Figure 11.5 shows the variation of the effective index difference $\Delta\beta[\Delta\beta = (\beta_{01} - \beta_{11})/k]$ between the modes as a function of the normalized frequency V for three values of the ellipticity.

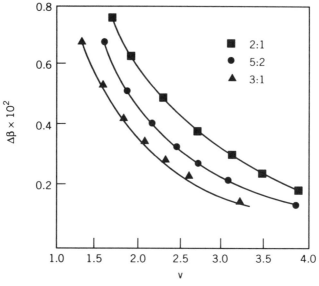

Figure 11.5. Variation of the effective index difference ($\Delta\beta$) as a function of the normalized frequency for three values of the fiber ellipticity. $\Delta n = 0.21$.

11.3. RUGGEDIZED SENSOR DESIGNS

Several detection techniques for few-mode fiber sensors have been suggested. An elementary technique involves the placement of a spatial demodulator in the form of a pinhole, which samples only part of the far-field pattern (2). An optical signal processing technique that uses a CCD array for the analysis of the speckle pattern at the output of a highly multimode fiber sensor has recently been demonstrated (11). This technique can be adapted for a two-mode fiber sensor. We have also used two large-core fibers to pick up parts of the two-lobe, far-field pattern and have obtained satisfactory results. Most such methods used in the past require the output fiber to display the far-field pattern on a monitoring system. This is not optimal for remote sensing systems in which the sensing region may be restricted to hard-to-reach environments. The use of misaligned splices seems to eliminate the need to obtain far-field patterns. In such a scheme, a misaligned splice taps light from each mode equally into a single-mode fiber, and relative phase shifts between the LP_{01} and the LP_{11}^{even} modes manifest themselves as intensity variations at the output of the single-mode, lead-out fiber (12).

The increasing focus on attached or embedded fiber sensors in smart structure applications demands that ruggedized designs of these sensors be available. In the application of two-mode fiber sensors in harsh environments, it is necessary that the fiber that acts as a conduit, or lead-in, for the transfer of optical power to the sensing element is not affected by the parameter being measured. Similarly, at the output end of the sensing fiber, it is desirable to have the detection electronics/optics immediately following the sensing portion, or to carry the information to a remote location via an insensitive, lead-out fiber.

The ruggedized sensor consists of an elliptical-core single-mode fiber as the lead-in fiber and a circular-core single-mode fiber as the lead-out fiber. Linearly polarized light is launched parallel to the major axis of the ellipse of the single-mode lead-in fiber. The polarization-preserving properties of the elliptical-core fiber make this fiber section relatively insensitive to external perturbations. At the first fusion splice, the ellipses of the single- and dual-mode fibers may or may not be aligned. This depends entirely on the sensitivity desired from the sensor. It has been shown that the elongation required for a complete oscillation of the two-lobe pattern is smaller for linearly polarized light launched parallel to the minor axis of the ellipse. Hence for maximum sensitivity of the sensor, the major axes of the ellipses of the two fibers being spliced should be perpendicular to each other. However, a trade-off is the loss of optical power, which could lead to a smaller signal-to-noise-ratio and, possibly, a restriction on the maximum length of the lead fibers being used in remote sensing applications. At the second fusion splice, the axes of the sensing fiber and the circular-core single-mode fiber are offset from each other, so the lead-out fiber picks up equal amounts of contribution from the LP_{01} and the LP_{11} modes from only one of the lobes. This results in a varying intensity pattern at the output of the single-mode fiber. The fused lead-out fiber thus acts as a ruggedized spatial demodulator.

In the practical implementation of any interferometric scheme, it is essential to maintain the quadrature point (Q-point) of operation for maximum sensitivity. That is, the phase bias that determines the initial point of operation should lie within the linear range of the sinusoidal output curve. Since signal fading is a common problem in an interferometer, it is advantageous to have two signals out of phase by 90°, so that when one of the signals has a low sensitivity to the phase shift (and hence to the parameter being measured), the other signal has a high sensitivity. Several techniques to acquire such phase quadrature signals have been developed for single-mode interferometric systems. Specifically, the use of incorporating frequency or phase modulators to generate heterodyne signals (13), and the development of passive waveguide devices to obtain phase quadrature signals (14–16) have been reported. A scheme using the effect of Gouy phase shifts has recently been suggested for two-mode elliptical-core fiber sensors (17).

In the detection scheme for a strain sensor, it is desirable to monitor both the amplitude and the direction of a dynamically varying strain. For example, in the vibration of cantilever beams, it is essential to measure the amplitude of the vibration as well to indicate whether the strain on parts of the beam is tensive or compressive. This is practically realizable if two output signals, out of phase by 90°, are obtained simultaneously in the multiple-fringe mode of operation and their lead–lag properties are monitored.

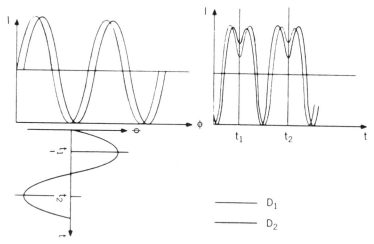

Figure 11.6. Transfer function curve for dual-mode fiber sensor and the evolution of the lead–lag properties of the sensor signals.

Figure 11.6 shows the sinusoidal variation of the intensity at the output of one of the lobes of the far-field pattern with respect to the phase difference ϕ between the LP_{01} and the LP_{11} modes. This is the transfer function curve. If the phase difference ϕ varies sinusoidally with time and the peak-to-peak variation is large enough to push the sensor out of its linear range, we observe fringes at the output of the detector as shown in the figure. The basic principle of operation of the detection scheme for a dynamic strain measurement system can now be described by considering two sinusoidal transfer functions out of phase by 90°. Assuming that the transfer function for the first detector (D_1) leads that of the second (D_2), we find that the output waveform for D_1 leads that of D_2 until time t_1. At time t_1 the phase ϕ changes direction (because of a change in direction of the strain) and the output waveforms switch their lead–lag properties. We see that the output from D_2 now leads that of D_1 until time t_2, when the strain changes direction again. Keeping track of the lead–lag phenomenon between the two detectors gives us unambiguous information about the direction of the strain.

If only one detector had been used, the switch in direction would not be noticeable if the strain changed direction at a peak of the transfer function curve. Since we now have two signals out of phase by 90°, if the direction change occurs at one peak (of either D_1 or D_2), the other transfer function curve will provide information about the direction change.

A 45° splitter–recombiner (S-R) was constructed to obtain two outputs that would be out of phase by 90° to make the desired dynamic strain measurement scheme feasible. The construction procedure was as follows:

1. A stripped elliptical-core fiber was placed between four glass slides (each of dimensions 10 mm × 30 mm × 1 mm) as shown in Fig. 11.7a. The fiber

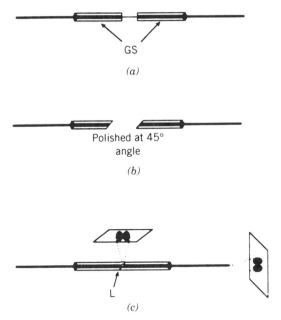

Figure 11.7. Fabrication procedure for the 45° splitter recombiner. L, liquid with refractive index = 1.4; GS, glass slides.

was bonded to the glass slides with an epoxy. A small section of the fiber, denoted by length l, was not bonded, to allow cleaving.

2. The open section of the fiber was cleaved to create two parts, each comprised of two glass slides with an elliptical-core fiber sandwiched between them.

3. Both the aforementioned parts were polished at a 45° angle to give a matched pair, shown in Fig. 11.7b).

4. The two parts were realigned with the help of a micropositioner, and a liquid with a refractive index of 1.4 was introduced between them to produce a partial reflection.

5. When positioned actively, a part of both the modes was reflected to project the two-lobe pattern on a screen placed directly above the splitter–recombiner. The other portion of the two modes was propagated through the fiber in the second part of the device and a similar two-lobe pattern was obtained at the output of this fiber. This phenomenon is shown schematically in Fig. 11.7c.

The two far-field patterns so obtained are not necessarily out of phase by 90°. However, this can be arranged by attaching the elliptical-core fiber of the second part of the device to a static strain controller, which, in effect, introduces the desired phase shift between the LP_{01} and LP_{11} modes. A description of this

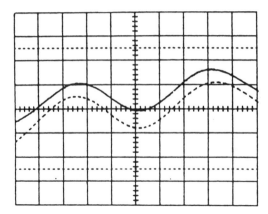

Figure 11.8. Signals from the detectors are nearly in phase.

method for obtaining a 90° phase shift between the far-field patterns is given in the following section.

A He–Ne laser ($\lambda = 633$ nm) launched linearly polarized light into a two-mode elliptical-core fiber that was wrapped around a piezoelectric cylinder excited at its resonant frequency of 24 kHz. The dynamic strain induced by the piezoelectric cylinder was monitored and the change in direction of the strain was determined by the lead–lag properties of the two signals, as described earlier. A static strain controller placed before the 45° S-R set the initial operating point. Another static strain controller was placed after the 45° S-R to adjust the phase difference between the two detectors, D_1 and D_2. Figure 11.8 shows the outputs of the two detectors before any adjustment of the second static strain controller has been made. In this case the two signals were very nearly in phase with each other. Minor changes in the second controller led to a shift in the phase between the two outputs. Figure 11.9 shows the tweaked outputs for an almost quadrature-phase-shifted system. The phase-adjusted

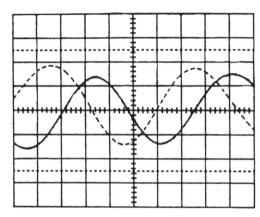

Figure 11.9. Tweaked output leads to quadrature-phase signals.

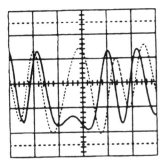

Figure 11.10. Leading and lagging of the tweaked sensor operated in the nonlinear range.

system was then used to determine the strain induced in the fiber due to the piezoelectric cylinder. Figure 11.10 shows that portion of the signal where a change in the direction of the strain takes place. We see that, as predicted, the signals switch their leading–lagging roles when the direction of the strain changes. Counting the number of fringes on each side of these waveforms gives the magnitude of the strain.

A compact detection scheme can now be envisioned, consisting of a detection box which includes two detectors, one placed very close to the top surface of the 45° S-R and the other positioned at the output end of the fiber in the second half of the device. Both these detectors would be preceded by spatial demodulators to access information from one of the lobes at the far-field pattern. The extra length of the fiber at the output would have to be adjusted to get the desired 90° phase shift between the signals from D_1 and D_2. This can be permanently achieved simply by grinding and polishing the fiber between the glass slides to the desired length. The outputs from the detection box are taken into a processing unit that contains a comparator and a fringe counter. The comparator performs two functions: (1) keeps track of the leading–lagging nature of the signals and thus indicates the direction of the strain, and (2) informs the fringe counter to display a count of the fringes whenever a change in direction takes place. The number of fringes are converted into the induced strain by a process of precalibration. A splitter–recombiner (S-R) is used to obtain two outputs that are out of phase by 90°, to make the desired dynamic strain measurement scheme feasible.

11.4. SPATIALLY WEIGHTED FIBER SENSORS

The motivation for fabricating a sensor with a variable sensitivity along its length can be understood when the problem is attached from a control systems viewpoint. Current research in vibration sensing and control has shown that variable-sensitivity spatially distributed transducers may be more suitable than point sensors for optimal control architectures (18, 19). Discrete point sensors have been shown to suffer from actuator/observer spillover, leading to instabilities in the closed-loop system (20). The development of piezoelectric spatially weighted sensors has led to extensive research on the advantages of modal sensors and actuators. Modal sensors, which sense the modal coordinate of a particular vibration mode of a structure, can be operated within a control system without extensive on-line real-time computation requirements. In their first description of modal sensors, Lee and Moon made sensing elements out of polyvinylidene fluoride (PVDF) films shaped in the form of specific modes of a structure (20). The fundamental difference in operation principles between distributed model sensors and conventional point sensors, as viewed by control systems analysts, is that flexible structures and distributed sensors are both infinite-dimensional systems and hence compatible, whereas conventional point sensors are finite-dimensional systems and hence require extensive signal-processing techniques.

We now consider the development of distributed modal sensors using optical fiber techniques. The variable sensitivity of the fiber sensors has been achieved by utilizing the feature that the differential propagation constant in a two-mode fiber is directly dependent on the normalized frequency (or V-number). Tapering the fiber changes the V-number and hence can change the sensitivity of the sensor along its length. These sensors are fiber optic analogs of shaped piezoelectric modal sensors that have emerged recently in the area of structural control.

11.4.1. Beam Mechanics

The Euler equation for transverse vibration of beams is given by

$$\frac{\partial^2}{\partial x^2}\left[EI(x)\frac{\partial^2}{\partial x^2}y(x,t)\right] + \rho_l(x)\frac{\partial^2}{\partial t^2}y(x,t) = Q(t) \qquad (11.8)$$

where E is Young's modulus, $I(x)$ the moment of inertia, ρ_l the linear density, $Q(t)$ the forcing function, and $y(x,t)$ the transverse displacement of the beam. Assuming the beams to approximately isotropic, Eq. (11.8) can be simplified to

$$EI\frac{\partial^4}{\partial x^4}y(x,t) + \rho_l\frac{\partial^2}{\partial t^2}y(x,t) = Q(t) \qquad (11.9)$$

Using the method of separation of variables, we attempt a solution of the form

$$y(x,t) = \sum_{n=1}^{\infty}\psi_n(x)\eta_n(t) \qquad (11.10)$$

where $\psi_n(x)$ represents the mode shapes of the beam and $\eta_n(t)$ represents the modal amplitudes. This approach necessitates an infinite sum of natural modes to describe the behavior of the beam completely. For practical implementation, this model can later be truncated to the first few modes of interest. Substituting Eq. (11.10) into Eq. (11.9) and setting the forcing function, $Q(t)$, to zero, we can solve independently for the functions $\psi_n(x)$ and $\eta_n(t)$ and arrive at a solution for the different modes of vibration of the beam. The equation for $\psi_n(x)$, given by

$$\frac{\partial^4}{\partial x^4} \psi_n(x) - \frac{\rho_l a_n^2}{EI} \psi_n(x) = 0 \tag{11.11}$$

has a closed-form solution determined by the boundary conditions of the beam. The key to the weighted sensing approach is that the mode shapes are orthogonal, that is,

$$\int_0^L \psi_m(x) \psi_n(x) \, dx = \delta_{mn} \tag{11.12}$$

where δ_{mn} is the Kronecker delta and L is the length of the beam. This property will be used in the design of fiber-based vibration-modal sensors, as shown below.

The output signal from a two-mode fiber sensor is sinusoidal and can be expressed as

$$I(t) = I_0 + I_{ac} \cos \phi(t) \tag{11.13}$$

where ϕ is the phase difference between the LP_{01} and the LP_{11}^{even} modes and can be written as

$$\phi(t) = \int_a^b \Delta\beta(x) \varepsilon(x, t) \, dx \tag{11.14}$$

where ε is the strain experienced by the fiber, $\Delta\beta$ is the difference in the propagation constants of the LP_{01} and the LP_{11}^{even} modes, x denotes the longitudinal direction along the fiber axis, and a and b denote the two endpoints of the two-mode sensing region of the fiber. In Eq. (11.14), the explicit dependence of $\Delta\beta$ on x implies that the strain sensitivity can be a function of the length along the beam.

To evaluate the vibration modes of the beam, we express strain as

$$\varepsilon(x, t) = \frac{\partial^2 y(x, t)}{\partial x^2} \tag{11.15}$$

where $y(x, t)$ denotes the deflection of the beam away from its equilibrium point. For a conventional two-mode fiber [with $\Delta\beta(x) = \Delta\beta = $ constant] we can obtain the expression for the sensor output as a function of the slopes of

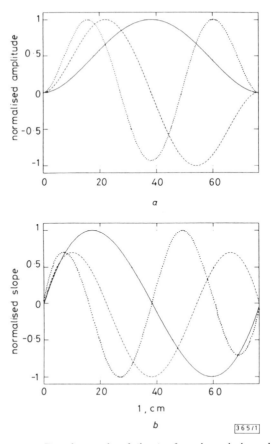

First three modes of vibration for a clamped–clamped beam

a Deflections of beam
b Slopes of deflections
——— first mode
– – – second mode
······· third mode

Figure 11.11. First three modes of vibration for a clamped beam: (*a*) deflection of beam; (*b*) slopes of deflections. Solid curve, first mode; dashed curve, second mode; dotted curve, third mode.

the deflections at the endpoints by substituting Eq. (11.15) into Eq. (11.14):

$$I(t) = I_0 + I_{ac} \cos \left\{ \Delta\beta \left[\left(\frac{\partial y}{\partial x} \right)_{x=b} - \left(\frac{\partial y}{\partial x} \right)_{x=a} \right] \right\} \tag{11.16}$$

The plots of the first three natural modes of vibration for a clamped–clamped beam are shown in Fig. 11.11. We note that at a normalized length of 0.5, the slopes of the first and third modes are zero. Hence a sensor with endpoints at $x = 0$ and $x = 0.5$ should not be sensitive to these modes. Similarly, a sensor placed between $x = 0.75$ and $x = 1$ will not pick up the second mode and will be sensitive to the first and third modes of vibration. An exponential verification of this effect has been presented by Vengsarkar et al. (18). Two sensors were placed along the length of a cantilever beam at the locations described above. Output signals from piezoelectric patches attached to the beam were compared to the signals obtained from the fiber sensors. The first three modes of vibration of the beam (at 14, 38 and 76 Hz, respectively) were excited and the outputs from both the sensors were monitored. In Fig. 11.12a, we show the fast Fourier transform (FFT) from the output of the sensor placed in configuration 1, between $x = 0$ and $x = 38$ cm. The plot shows that the sensor enhances

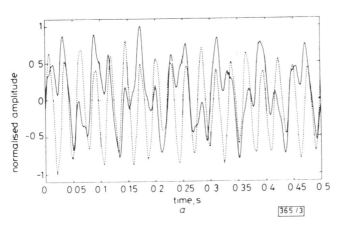

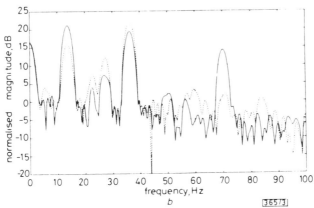

Output waveforms from configuration 1

- - - - fibre sensor
———— piezoelectric sensor
First and third modes of vibration are suppressed
a Time-domain trace
b Fast Fourier transform of time-domain waveform

Figure 11.12

the second mode of vibration in comparison with the piezoelectric sensor. On the other hand, the first and third modes of vibration are suppressed by 5 and 10 dB, respectively. In Fig. 11.12b we plot the FFT from the sensor in configuration 2, placed between $x = 54$ cm and $x = 76$ cm. The FFT shows that the second mode of vibration has been selectively filtered out by the sensor and suppressed by 12 dB. The detection of the first and the third modes of vibration, on the other hand, has been left unaltered.

While proper placement of fiber sensors can thus lead to an elimination of some of the modes for a clamped–clamped beam, filtering out information from individual modes (or achieving the same effect for a clamped–free beam where the slope of the deflection is not zero at the free end) is still not possible.

It is possible to weight the information actually present in the strucure by using a priori knowledge of the mode shapes of the structure. Instead of relying on the sensor placement to enable structural mode analysis, it would be advantageous if the sensor itself could provide the weighting capability. To analyze such a sensor fabricated from tapered two-mode fibers, the differential propagation constant, $\Delta\beta$, should now be considered to be an explicit function of x. Substituting Eq. (11.15) into (11.14) and integrating by parts leads to the equation

$$\phi(t) = \eta_n(t) \left[Q(a, b) + \int_a^b \Delta\beta''(x)\,\psi_n(x)\,dx \right] \qquad (11.17a)$$

where

$$Q(a, b) = [\Delta\beta(x)\,\psi_n'(x)]_a^b - [\Delta\beta'(x)\,y_n(x)]_a^b \qquad (11.17b)$$

and the primes indicate spatial derivatives with respect to x. Comparing Eqs. (11.17a) and (11.12) leads one to pick a possible weighting function given by $\Delta\beta''(x) = \psi_m(x)$. Except for the contributions of $Q(a, b)$, $\phi(t)$ would filter out all but the mth mode for a fiber sensor spanning the entire length of the beam. Hence fairly mode-specific information can be acquired without resorting to conventional analog or digital postacquisition processing. The function $Q(a, b)$ is essentially a constant once the fiber sensor has been attached to (or embedded in) the structure of interest. For a clamped–clamped beam, if the sensor gauge length spans the entire beam, $Q(a, b)$ is identically equal to zero, irrespective of the weighting function.

A comparison of Figs. 11.13 and 11.14 shows that linear or exponential tapers can produce $\Delta\beta''(x)$ profiles that resemble some of the vibration mode shapes of one-dimensional beams. Hence, by using the orthogonality property of the modes and Eq. (11.14), we should be able to tailor fiber profiles that will lead to vibration selective weighted sensors. Direct superposition of the mode shapes and the $\Delta\beta''(x)$ function for different taper profiles shows good correla-

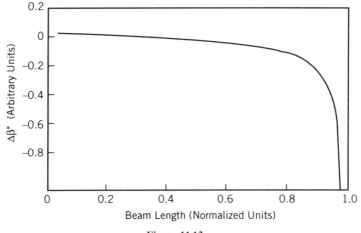

Figure 11.13

tion. In Fig. 11.15a we have considered a taper with the minimum radius of the fiber, $a_{min} = 1\,\mu$m and the maximum radius, $a_{max} = 1.15\,\mu$m, with a linear taper over the length of the beam. Similarly, the exponential taper can be tailored such that the weighting function matches the first mode of the clamped–clamped beam. This possibility is depicted in Fig. 11.15b, for a radius profile described by the equation

$$a = 3.02 \times 10^{-7}\exp(-5x) + 8.98 \times 10^{-7} \tag{11.18}$$

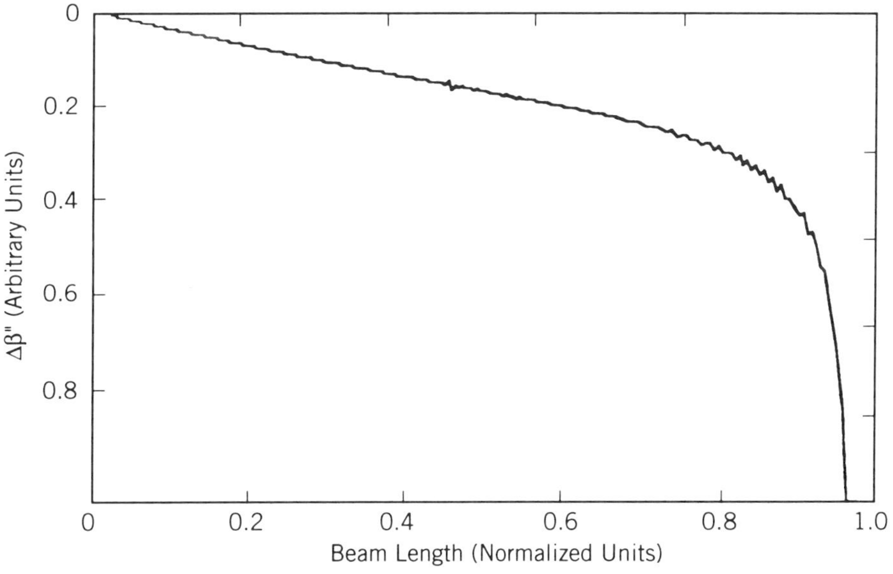

Figure 11.14

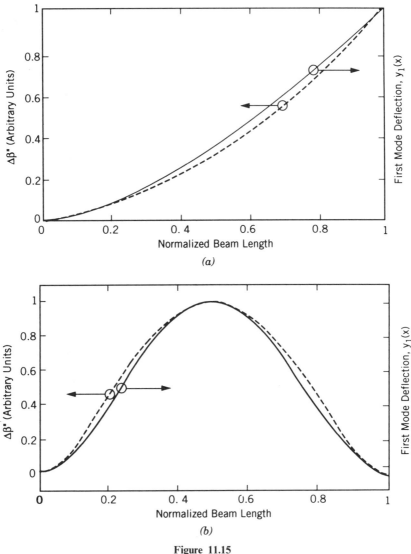

Figure 11.15

over half the beam normalized length. These two examples demonstrate the possibility of tailoring the weighting profiles as desired for any mode of the one-dimensional beam.

A conventional elliptical-core sensor was adhered to a clamped–free beam ($a = 0$, $b = 0.61$). Output signals from piezoelectric patches attached to the beam were compared to the signals obtained from the fiber sensor. The first three modes of vibration of the beam were excited and the outputs from the fiber optic sensor and the piezoelectric patch were monitored a fast Fourier transform (FFT) of an oscilloscope waveform can be seen in Fig. 11.16a. In

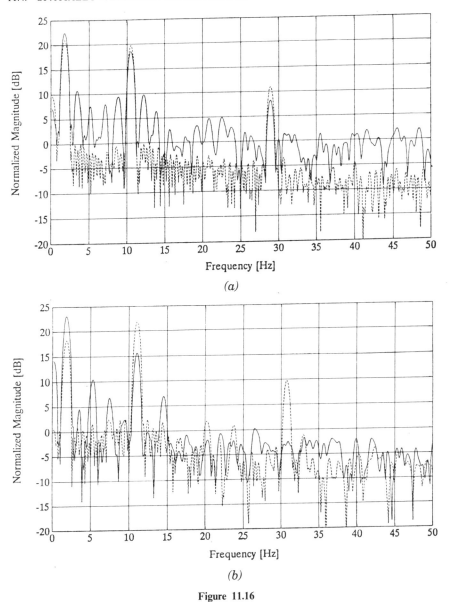

(a)

(b)

Figure 11.16

Fig. 11.16*b* we show a fast Fourier transform (FFT) of an oscilloscope waveform which shows that the sensor enhances the first mode of vibration in comparison with the piezoelectric sensor, and picks up other modes of vibration fairly well, with reduced sensitivities on the order of 2 and 4 dB for the second and third modes, respectively. A comparison between the theoretically expected values for the beam frequencies and those detected by the fiber sensor is given in Table 1.

11.5. GRATING-BASED TWO-MODE FIBER SENSORS

Hill gratings were first formed in 1978 by sustained exposure of germanosilicate fibers to a standing interference pattern resulting from two counter-propagating beams within the core (21). Meltz and co-workers subsequently showed that Bragg gratings could be formed in germanium-doped fibers by exposing the core through the side of the cladding by using a transverse holographic method (22). Recently, permanent index gratings in two-mode elliptical-core optical fibers have been demonstrated, and several applications such as their use in intermodal switching and chirped filters for dispersion compensation have been proposed (23, 24). We should like to evaluate the effect of permanent-index gratings on the differential-phase modulation between the LP_{01} and LP_{11}^{even} modes in two-mode fibers and investigate if these grating-based sensors have special properties that would make them useful in weighted-sensing applications.

We first describe the process of writing the grating within the fiber. When a high-intensity argon-ion laser beam ($\lambda = 514.5$ nm) is launched into a germanosilicate elliptical-core two-mode fiber such that both modes are excited equally, a permanent refractive-index change is created due to the photo-sensitivity of the germanium-doped core with the same profile and periodicity as the two-lobe pattern. Such a two-mode fiber, with a permanent grating induced within its core, can no longer be described in terms of the original waveguide equations.

In the theory presented below we take recourse to the weakly guiding approximations as applied to a circular waveguide and combine the results

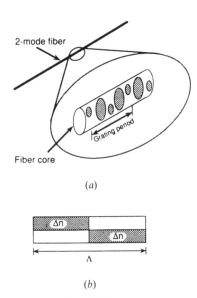

(a)

(b)

Figure 11.17

with conventional coupled-mode theory; this leads to a simplification of the resulting expressions, maintains analytical tractability, yet provides a good physical understanding of the underlying processes. Figure 11.16 shows the simplified model of the periodic-index grating (23).

Assuming the simplified model shown in Fig. 11.17, the slowly varying amplitudes of the LP_{01} and LP_{11} modes can be written as

$$A_{01}(z) = \exp\left(i\frac{\delta z}{2}\right)\left(\cos \eta z - i\frac{\delta}{2\eta}\sin \eta z\right) \tag{11.19a}$$

and

$$A_{11}(z) = -\exp\left(-i\frac{\delta z}{2}\right)\left(\frac{\gamma}{\eta}\sin \eta z\right) \tag{11.19b}$$

respectively, where

$$\eta^2 = |\gamma|^2 + \left(\frac{\delta}{2}\right)^2 \tag{11.20a}$$

$$\gamma = \left(\omega\varepsilon_0 n_{av}\frac{\Delta n}{\pi}\right)\int_0^a d\rho \int_0^\pi d\varphi\, \psi_{01}\psi_{11}^* \tag{11.20b}$$

and

$$\delta = \Delta\beta - \frac{2\pi}{\Lambda} + (C_{01} - C_{11}) \tag{11.20c}$$

In Eqs. (11.19) and (11.20), $\Delta\beta$ is the differential propagation constant defined earlier, ω the angular frequency associated with the optical beam, ε_0 the free-space permittivity, a the fiber core radius, n_{av} the average refractive index of the fiber core, and Δn the induced refractive index change in the fiber. The grating spacing is denoted by Λ, and C_{01} and C_{11} are self-coupling coefficients given by

$$C_{lm} = \left(\omega\varepsilon_0 n_{av}\frac{\Delta n}{\pi}\right)\int_0^a d\rho \int_0^\pi d\varphi\, |\psi_{lm}|^2 \tag{11.21}$$

From Eqs. (11.19) we can show that the modes will exchange power with a spatial frequency of 2η. If we define the beat length, L_B, in a strain sensor as the elongation required to induce a complete oscillation in one of the lobes at the output, then $2L_B = 2\pi/\eta$.

Note that in a two-mode fiber without the induced refractive-index changes, $\gamma = 0$, $C_{lm} = 0$, and the beat length is dependent (inversely proportional) only to $\Delta\beta$. Equations (11.20) imply that in a two-mode fiber grating, the beat length, which determines the sensor sensitivity, is a function of the induced grating spatial frequency as well as the coupling coefficients. When a fiber with

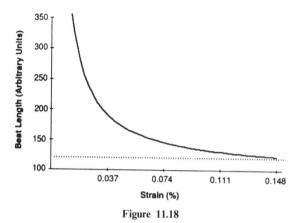

Figure 11.18

a permanent-index grating written in its core is strained, the grating period, Λ, will increase with increasing strain. The theoretical variation of the sensor beat length as a function of the induced strain is shown in Fig. 11.18, where we have assumed that the strain is equally distributed along the length of the fiber and that the induced refractive-index change is on the order of 10^{-5}. The plot indicates that as the fiber is strained, the successive beat length should become smaller as the grating period moves away from its "resonance" condition with $\Delta\beta$. The horizontal dotted line on the plot represents the beat length of the fiber before the grating was written. Exact values for the vertical axis in Fig. 11.18 can be calculated by knowing the original beat length, the intensity of incident optical power, and the glass constants that govern the two-photon absorption process. The behavior of the beat length with varying strain is important to this analysis; for very small amounts of strain ($\approx 0.1\%$) applied to the fiber, the effect of the grating becomes negligible and the beat length of the two-mode fiber approaches its original value. In Fig. 11.18 we assumed that the fiber was unstrained when the grating was written; this implied that the grating spacing would match the differential phase shift between the two modes in the unstrained state. If the fiber were held in a strained condition while being exposed to the high-intensity argon-ion beam, one would expect that upon the release of tension, the grating formed would relax and possess a period somewhat less than the original two-lobe profile. During low-power probing of the grating, a differential-phase matching condition may then be obtained when the fiber is subjected to the same amount of strain as during exposure.

In Fig. 11.19a we show the variation of the beat length for a fiber that was kept unstrained during the writing process. The gradual decline of the beat length with increasing strain is clearly visible. It is interesting to note that low-power, y-polarized probing of the fiber along the minor axis resulted in no noticeable change in beat length. This confirms earlier hypotheses that the induced refractive-index changes are polarization dependent. For fibers that were held at constant strain levels of 0.7%. 1.4%, and 1.8% while being exposed

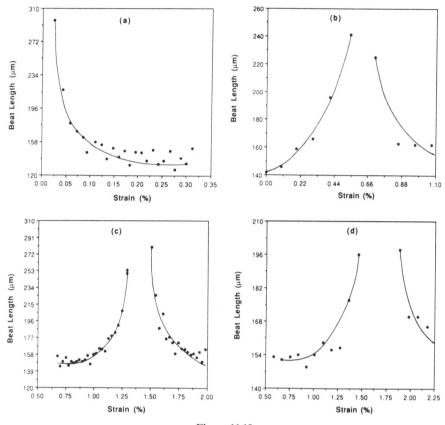

Figure 11.19

to the high-power beam, we have observed resonant (differential phase matching between the grating spacing and $\Delta\beta$) conditions at the corresponding strain
levels, as shown in Fig. 11.19*b–d*. These results show that strained fibers
exposed to high-power writing beams can be used as sensing elements with
varying beat lengths and hence, varying sensitivities. An extension of this
concept to weighted sensors makes this effect more significant.

As described in Section 11.4, Eq. (11.18), one should pick a possible
weighting function given by $\Delta\beta''_{eq}(x) = \psi_n(x)$. Here $\Delta\beta''_{eq}(x)$ is the equivalent $\Delta\beta$,
which is not calculated, and $\psi_n(x)$ is the nth mode shape of the cantilever beam.
Except for the contributions of $Q(L)$ from Eq. (11.18), $f(t)$ would then filter
out all but the nth mode for a fiber sensor spanning the entire length of the
beam. Hence fairly mode-specific information can be acquired without resorting to conventional analog or digital postacquisition processing. Again, the
function $Q(L)$ is essentially a constant once the fiber sensor has been attached
to (or embedded in) the structure of interest.

The parameter $\Delta\beta''_{eq}$ can be made to vary as a function of x by including an
x-dependent strain in a grating-based two-mode elliptical-core fiber. The

method demonstrated here consists of attaching a fiber to a one-dimensional beam and exposing the germanium-doped core to a high-intensity argon-ion laser beam. During exposure, the beam is placed under static strain in a shape that is determined by the beam dynamics and the desired variation of the equivalent differential propagation constant. Results from above have shown that the beat length varies almost inversely with strain ($L_B \propto 1/\varepsilon$); this variation implies that $\Delta\beta_{eq}$ would be directly proportional to the induced strain ($\Delta\beta_{eq} \propto \varepsilon_{induced} = \psi''_{induced}$) as a result of the inverse dependence of the beat length and $\Delta\beta_{eq}$. The problem is thus reduced to finding the appropriate strain field to be applied to the beam while the grating is being written in a fiber. Since $\Delta\beta$ is directly proportional to the strain induced in the fiber, its second derivative, $\Delta\beta''_{eq}$, is directly proportional to the fourth derivative of the induced mode shape, $\psi''''_{induced}$. From Eq. (11.18) and the foregoing argument, to sense a specific vibration-mode of order n selectively, the shape of the fiber attached to the beam, $\psi_{induced}$, should be such that its fourth derivative, $\psi''''_{induced}$, is proportional to ψ_n. Hence the actual shape of the beam is given by (24)

$$\psi_{induced} \propto \int\int\int\int \psi_n(x)\,dx \qquad (11.22)$$

For a cantilever beam, the mode shapes ψ_n's are expressed in terms of a linear combination of sine, cosine, hyperbolic-sine, and hyperbolic-cosine functions. From the mathematical properties of these functions, one can see that a quadruple integral restores the original functions, except for a scaling factor. As a consequence, the beam needs to be positioned in exactly the same shape as that of the specific vibration mode one wishes to enhance selectively.

Prior to writing a grating in the fiber, low-power ($1\,mW, l = 514.5\,nm$) probing of a two-mode elliptical-core fiber sensor provided the vibration-mode information of the cantilever beam under investigation. Fast Fourier trans-

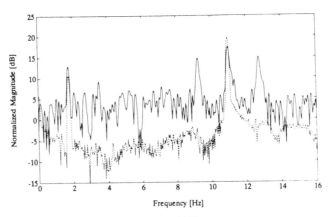

Figure 11.20

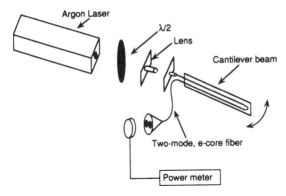

Figure 11.21

forms (FFTs) of the fiber sensor signal and an adjacent piezoelectric sensor signal
attached to the beam are shown in Fig. 11.20. A high-intensity argon-ion laser beam
(40 mW, l = 514.5 nm) was launched into the germanosilicate elliptical-core
two-mode fiber attached to a cantilever beam using the experimental setup shown
in Fig. 11.21. Note that since the sensing fiber is looped around the beam length, the
phase parameter given by Eq. (11.18) will be doubled. The launch conditions were
set such that both modes were excited equally in order to create a permanent
refractive-index change with the same profile and periodicity as the two-lobe
pattern. The beam was held in its second vibration-mode shape during the
high-power exposure of the fiber core. Postexposure analysis of the fiber sensor with
a low-power beam at 514.5 nm resulted in the output signal shown in Fig. 11.22. A
comparison of Figs. 11.20 and 11.22 indicates that the fiber sensor has suppressed
the first mode of vibration by 15 dB. Note that the comparison is made between the
relative magnitudes of the FFTs of the fiber output and the piezoelectric sensor
signal in Fig. 11.2 and then compared to the difference in the relative amplitides of

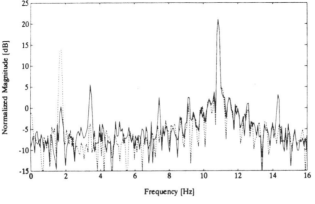

Figure 11.22

the corresponding signals in Fig. 11.20 to eliminate any discrepancies that may arise due to excitation conditions.

In conclusion, the behavior of a two-mode elliptical-core fiber gratings has been analyzed experimentally, and a two-mode elliptical-core fiber with a chirped grating, with the sensitivity varying as a function of length, has been demonstrated as a selective vibration-mode sensor.

REFERENCES

1. M. R. Layton and J. A. Bucaro, Optical Fiber Acoustic Sensor Utilizing Mode–Mode Interference. *Appl. Opt.* **18**, 666 (1979).

2. B. D. Duncan, Modal Interference Techniques for Strain Detection in Few-Mode Optical Fibers. M. Sci. Thesis, Department of Electrical Engineering, Virginia Polytechnic Institute and State University (1988).

3. A. Safaai-Jazi and R. O. Claus, Synthesis of Interference Patterns in Few-Mode Optical Fibers. *Proc. SPIE—Int. Soc. Opt. Eng.* **986**, 180 (1989).

4. B. Y. Kim, J. N. Blake, S. Y. Huang, and H. J. Shaw, Use of Highly Elliptical Core Fibers for Two-Mode Fiber Devices. *Appl. Opt.* **12**, 729 (1987).

5. J. Dakin and B. Culshaw, eds., *Optical Fiber Sensors: Principles and Components*, Vol. 1. Artech House, Boston, 1988.

6. A. Kumar and R. K. Varshney, Propagation Characteristics of Dual-Mode Elliptical-Core Optical Fibers. *Opt. Lett.* **14**, 817 (1989).

7. C. Yeh, *J. Appl. Phys.* **33**, 3235 (1962).

8. P. D. Gianino, B. Bendow, and P. Wintersteiner, Propagation Characteristics of Irregularly Shaped Step-Index Fiberguides. *Advan. Ceram.* **2**, *Physics of Fiber Optics*, B. Bendow and S. Mitra, Eds., American Ceramic Society, Ohio, p. 330 (1981).

9. A. Kumar and R. K. Varshney, Propagation Characteristics of Dual-Mode Elliptical-Core Optical Fibers. *Opt. Lett.* **14**, 817 (1989).

10. J. K. Shaw, A. M. Vengsarkar, and R. O. Claus, Direct Numerical Analysis of Dual-Mode Elliptical-Core Optical Fibers. *Opt. Lett.* **16**, 135 (1991).

11. W. B. Spillman, Jr., B. R. Kline, L. B. Maurice, and P. L. Fuhr, Statistical-Mode Sensor for Fiber Optic Vibration Sensing Uses. *Appl. Opt.* **28**, 3166 (1989).

12. K. A. Murphy, M. S. Miller, A. M. Vengsarkar, and R. O. Claus, Elliptical-Core Two-Mode Optical-Fiber Sensor Implementation Methods. *J. Lightwave Technol.* **30**, 1688 (1990).

13. J. H. Cole, B. A. Danver, and J. A. Bucaro, Synthetic Heterodyne Interferometric Demodulation. *IEEE J. Quantum Electron.* **QE-18**, 694 (1982).

14. B. Y. Kim and H. J. Shaw, Phase-Reading All-Fiber-Optic Gyroscope. *Opt. Lett.* **9**, 378 (1984).

15. T. Niemeier and R. Ulrich, Quadrature Outputs from Fiber Interferometer with 454 Coupler. *Opt. Lett.* **11**, 677 (1986).

16. D. W. Stowe and T. Y. Hsu, Demodulation of Interferometric Sensors Using a Fiber-Optic Passive Quadrature Demodulator. *IEEE J. Lightwave Technol.* **LT-1**, 519 (1983).

17. S. Y. Huang, H. G. Park, and B. Y. Kim, Novel Quadrature Phase Detector Using Two-Mode Waveguides. *Opt. Lett.* **14**, 1380 (1990).

18. A. M. Vengsarkar, B. R. Fogg, W. V. Miller, K. A. Murphy, and R. O. Claus, Elliptical-Core, Two-Mode Optical Fibre Sensors as Vibration Mode Filters. *Electron. Lett.* **27**, 464–466 (1991).

19. T. Bailey and J. E. Hubbard, Distributed Piezoelectric-Polymer Active Vibration Control of a Cantilever Beam. *J. Guidance, Control Dyn.* **8**, 605–611 (1985).

20. C.-K. Lee and F. C. Moon, Modal Sensors/Actuators. *J. Appl. Mech.* **57**, 434–441 (1990).

21. K. O. Hill, Y. Fujii, D. C. Johnson, and B. S. Kawasaki, Photosensitivity in Optical Fiber Waveguides: Application to Reflection Filter Fabrication. *Appl. Phys. Lett.* **32**, 647 (1978).

22. G. Meltz, W. W. Morey, and W. H. Glenn, Formation of Bragg Gratings in Optical Fibers by a Transverse Holographic Method. *Opt. Lett.* **14**, 823 (1989).

23. H. G. Park, S. Y. Huang, and B. Y. Kim, All-Optical Intermodal Switch Using Periodic Coupling in a Two-Mode Waveguide. *Opt. Lett.* **14**, 877 (1989).

24. F. Oulette, All-Fiber filter for Efficient Dispersion Compensation. *Opt. Lett.* **16**, 303 (1991).

12

Microbend Fiber Optic Sensors

TIMOTHY CLARK
Amphenol Fiber Optic Products
Lisle, Illinois

AND

HERB SMITH
McDonnell Douglas Aerospace
St. Louis Missouri

12.1. INTRODUCTION

Intensity losses resulting from optical fibers undergoing bending have long been of concern to the telecommunications industry. The pioneering and most extensive work on the effects of mode coupling caused by bending was done by such investigators as D. Marcuse and D. Gloge. As the field of optical fiber sensing developed, microbending losses were transformed from a nuisance into a useful transducing effect. A number of innovative techniques have been employed in the effort to harness microbending to measure many different physical phenomena.

The simplicity of the microbending process makes it highly attractive for a wide variety of applications. Intruder detection utilizing fibers placed within floor mats is commercially available, as are pressure and displacement sensors. These sensors typically incorporate a mechanism that imposes a microbending with an appropriate spatial period upon the fiber, thereby optimizing the response.

Smart structure applications of optical fiber sensors usually focus on situations in which the fibers are embedded within the composite material at the time of its manufacture. The fibers can, in some cases, be used to monitor the processing variables during the manufacturing stage as well as the health of the part after an aircraft or other vehicle enters service.

Fiber Optic Smart Structures, Edited by Eric Udd.
ISBN 0-471-55448-0 © 1995 John Wiley & Sons, Inc.

Optical fibers have been embedded successfully into several types of composite laboratory specimens and used as sensors, both during the curing process and in tests that simulate the service environment of typical aircraft components. It was found that microbend sensitive fibers show attenuation of optical intensity in response to the strain field in the material. We have investigated the use of this phenomenon as a sensor for strain due to quasi-static loads from aircraft maneuvers, strain due to buffet loads in the frequency range 15 to 45 Hz, and impact from low-velocity and ballistic projectiles.

We begin with an overview of some of the theory behind microbend losses, followed by a summary of experimental results obtained since 1980, when the effect was first put to use as a fiber sensing technique. Variations on the technique are discussed, as well as problems encountered utilizing this method. Smart structure applications will also be reviewed, including experimental results obtained in our investigation of microbend sensors embedded in composite materials. One clarification in terminology can be made; we use the term *extrinsic microbend sensor* to indicate a situation in which microbending is induced by an external set of grips, jaws, or other mechanical means, which possesses a known, regular spatial period. *Intrinsic microbend sensor* then refers to a fiber embedded within, or bonded to, a component, and which relies only upon the irregularities of its embedded or bonded state to induce the microbending. In both cases the actual microbend losses occur within the fiber, the only real difference between the two having to do with the regularity and size of the microbend stimuli.

12.2. OVERVIEW OF MICROBEND LOSS THEORY

Transmitted light intensity is lost in response to microbending. This loss occurs most frequently when the highest-order guided mode in the fiber core is coupled to the first cladding (radiation) mode, which then is rapidly attenuated (1–9). The coupling can be induced by environmental effects, such as temperature, pressure, impact, or acoustic waves, which produce structural strain. Mechanical coupling of structural changes to an embedded fiber modifies the fiber geometry and allows light to be coupled between modes, thus modulating the light intensity transmitted by the fiber. These curvatures are very small, and very abrupt, in this way differing from the more familiar phenomenon, macrobending (Fig. 12.1).

Effectively employing this phenomenon requires some understanding of mode propagation as well as the ways in which these modes are affected by the microbending process. A great body of literature concerning microbending has been generated; due to space limitations, we merely present some of the applicable results. The interested reader is urged to consult the references listed for a more complete understanding of the topic.

Macrobending   **Microbending**

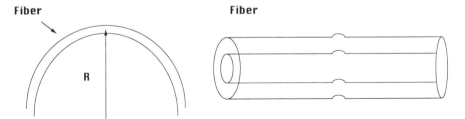

Figure 12.1. Macrobending versus microbending.

The index profile of an optical fiber can be described in the following (3) form:

$$n^2(r) = n_0^2\left[1 - 2\Delta\left(\frac{r}{a}\right)^\alpha\right] \tag{12.1}$$

with

$$\Delta = \frac{n_0^2 - n_{cl}^2}{2n_0^2} = \frac{(NA)^2}{2n_0^2} \tag{12.2}$$

where Δ is the relative index difference between core and cladding, a the core radius, n_0 the peak core refractive index, n_{cl} the cladding refractive index, NA the fiber numerical aperture, and $\alpha = 2$ for graded index (parabolic) fibers, $\alpha = \infty$ for step-index fibers.

It has been shown (3,4,8,9) that for modes of optical fibers to be coupled together by external perturbations, the disturbance must have spatial wavelength Λ, related to the difference in propagation constants, β, of the modes by

$$\beta' - \beta = \frac{2\pi}{\Lambda} \tag{12.3}$$

The spacing in propagation constants between groups possessing principal mode numbers m and $m + 1$ is (1,3,10)

$$\Delta\beta_m = \frac{2}{a}\left(\frac{\alpha\Delta}{\alpha + 2}\right)^{1/2}\left(\frac{m}{M}\right)^{(\alpha - 2)/(\alpha + 2)} \tag{12.4}$$

with M the number of modal groups guided by the fiber, and M^2 the total number of modes guided by the fiber,

$$M^2 = \left(\frac{\alpha}{\alpha + 2}\right)(n_0 ka)^{(2)}\Delta \tag{12.5}$$

where $k = 2\pi/\lambda$ is the free-space propagation constant for the wavelength of light used, λ. There are two significant cases (11):

(a) Step-index Fiber: $\alpha = \infty$

$$\beta_{m+1} - \beta_m = \frac{2\sqrt{\Delta}}{a}\left(\frac{m}{M}\right) \tag{12.6}$$

In step-index fibers, the spacings of adjacent modes vary depending on the mode number. The perturbing influence will couple only those pairs of modes with the correct spacing, not all the modes at once. A perturbing influence at a large spatial wavelength will couple adjacent lower-order modes; a larger mode number results in a smaller spatial wavelength required to couple the modes. Since microbend loss occurs at the transition from highest-order core mode ($m = M$) to a radiation mode, the wavelength at which microbending loss occurs, from (12.3) and (12.6), is

$$\Lambda = \frac{\pi a}{\sqrt{\Delta}} = \frac{\sqrt{2}\,\pi a n_0}{NA} \tag{12.7}$$

(b) Graded Index Fiber: $\alpha = 2$

$$\beta_{m+1} - \beta_m = \frac{\sqrt{2\Delta}}{a} \tag{12.8}$$

Thus the same spacing exists between all adjacent modes (10, 11), and (12.3) and (12.8) yield (4):

$$\Lambda = \frac{\sqrt{2}\,\pi a}{\sqrt{\Delta}} = \frac{2\pi a n_0}{NA} \tag{12.9}$$

In graded-index fibers, the same perturbation wavelength will cause each adjacent mode to couple, from the lowest guided mode, in turn, to the radiative mode. Thus there is no mode number dependency, and lower-order modes will be coupled into higher-order modes (10, 11) while higher-order modes are being lost to radiation.

There are other factors to take into account in addition to the periods of the disturbing forces. Sensor considerations can be divided into two general areas, the fiber microbending sensitivity, represented by $\Delta T/\Delta X$, and the mechanical design of the device, associated with $\Delta X/\Delta F$ (12, 13).

$$\frac{\Delta T}{\Delta F} = \frac{\Delta T}{\Delta X}\frac{\Delta X}{\Delta F} \tag{12.10}$$

where T is the fiber transmission, F the applied force, and X the deformation of fiber normal to its axis.

Sensitivity to microbending, or large $\Delta T/\Delta X$, is increased by small values of numerical aperture, a large number of deformations, optimized perturbation period, and a large core diameter relative to cladding diameter. Mechanical design considerations may militate against some of those selections, since to maximize $\Delta X/\Delta F$, a small-diameter fiber, few bends, and a large perturbation wavelength are desired (13).

One example of the many influences on the total sensitivity of the microbend sensor is given by the length dependence of microbend sensitivity. It has been found (14) that the microbend sensitivity of the sensor

$$\frac{\Delta T}{\Delta X} \propto l^q \tag{12.11}$$

where l is the length of the microbent fiber, and $0 < q \leqslant 1$. A fiber with a perfectly absorbent jacket would have $q = 1$ and a sensitivity that depended linearly upon the sensing length. Silicone possesses a value of $q \approx 0.2$, and for black paint, $q \approx 1$.

Perturbations to the fiber having either a random (or a very large number) or periodic spatial frequency can cause light to be coupled out of the core and into the cladding. Periodic disturbances of the correct frequencies are much more efficient at mode coupling (i.e., when the bend periodicity matches the optical mode separation Λ) (1–4). This matching condition can be used as a selection criterion for fibers for specific applications (determining numerical aperture, core diameter, step or graded index). Environmental effects on the structure can then be sensed by monitoring the optical transmission of the embedded fiber.

Smaller losses can also appear due to random microbending, in which no single spatial frequency dominates. In some cases, fibers in smart structures may be embedded within composites such as carbon epoxy. These materials have carbon reinforcing fibers in an epoxy matrix where the periodicity is greatly diminished. The bending environment in which the waveguides lie may then be more like a random-frequency situation, similar to that to which cabled fibers are exposed. On the other hand, a spatial periodicity may be present in the event that a prepreg (a weave comprising orderly bundles of structural fibers) is used, or it may be induced artificially in a number of ways. Thus elements of intrinsic or extrinsic microbend sensing may be employed in smart structure applications.

Microbending losses first became of concern to workers in the telecommunication industry, when early graded-index fibers displayed higher losses than had been measured in the lab. Two sources of this loss were the bending induced during reeling or respooling the fiber, when jacketed fiber was wound over itself, and the bending experienced by the fiber within a cable (15, 16). This loss was often exacerbated by low temperatures, when the organic and metallic materials around the fiber (including its jacket) shrank, imposing further bends upon the fiber (17, 18). Single-mode fibers also were subject to higher losses than anticipated (19–25). Various doping schemes, combinations of soft inner and hard outer jacket, and modified cabling and respooling techniques were applied as solutions to the problem of microbending. Models and standardized tests were also developed to predict the likelihood of their occurrence. Microbend losses could be predicted on the basis of spot size measurements (26), or measured by official fiber optic test procedures (FOTPs; see Section (12.3.1). Most of the telecommunications problems were caused by random microbending, in which the fiber was subjected to axial directional perturbations occurring over a range of spatial frequencies. This random perturbation usually has little effect upon short lengths of fiber, as only a few of the frequencies involved are near the critical frequencies for causing cross-coupling, but can cause substantial losses over longer spans.

12.3. OVERVIEW OF MICROBENDING EXPERIMENTS

These problems in the telecommunication industry were in turn applied as solutions by researchers in the nascent fiber sensor community. Perhaps the first recorded instance of employing the microbend effect in an intensity-modulated sensor was by Fields, who demonstrated its use in pressure measurements (27). Sound, vibration, acceleration, impact, strain, temperature, structural distortion, and even water leakage into cables have been measured by means of microbending-related phenomena (28–34). De Jongh et al. (31) constructed a sensor in which a material that swells upon exposure to water then presses the fiber in the cable against a periodic V-groove structure, thereby inducing microbend losses. Attenuation increases of from 0.01 to 0.05 dB/cm were then easily located by means of optical time-domain reflectometry. With their horizontal accelerometer for seismic measurements, Freal et al. (32) have demonstrated a minimum detectable acceleration of $3\,\mu g$, a dynamic range of 90 dB, and a frequency range of 1 to 100 Hz. The same group had previously measured displacements down to $0.04\,\text{Å}$ at a frequency of 800 Hz (33, 34).

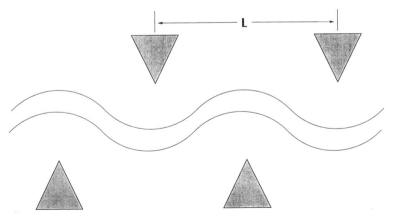

Figure 12.2. Microbending structure.

With microbend sensors, the allowable loss per length of fiber is usually much larger than in the case of cabled fibers, both for ease in measurements and because the lengths used are so much shorter. Unfortunately for sensor researchers, fiber manufacturers have, of necessity, concentrated their efforts on reducing the susceptibility of fibers to microbending losses. Thus in most cases, intrinsic microbend sensors have shown small response to strain, hence limited application in sensing (some examples of this type of microbend sensing are illustrated later). The more sensitive type of microbending, the extrinsic, with an imposed spatial perturbation matching the correct frequency linking the guiding and radiative modes, is usually desired. This is accomplished by the addition of structures external to the fiber, which then provide the necessary periodic deformations. Figure 12.2 displays a typical construction (35) that is used to perturb the sensor fiber.

12.3.1. Standardized Test Procedures and Experimental Techniques

Various fiber optic test procedures (FOTPs) for fiber bending have been proposed, including microbend (36) (FOTP-68, Optical Fiber Microbend Test Procedure, EIA-455-68) and macrobend (37) (FOTP-62, Optical Fiber Macrobend Attenuation, EIA-455-62). FOTP-68 employed two flat plates, one or both having a sheet of sandpaper attached. The fiber was laid between the plates, with weights then being placed on the top plate to form the *sandpaper* or *lateral load test* (Fig. 12.3). Another test used, known as the *basket weave* (38), involves fiber wound on a drum over itself in numerous layers, with tension being applied to the fiber. In a serpentine bending fixture, or pin array (39), which tests for both microbending and macrobending susceptibility, several small pins were placed 5 mm or so apart in a row, with the fiber snaked between them (Fig. 12.4). It should be noted that there have been concerns expressed regarding the reproducibility of some of these measurements, at least

Figure 12.3. Random-frequency sandpaper or lateral load test.

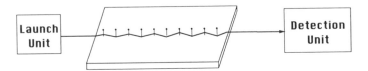

Fiber Bent by .43 mm D Pins Spaced 5 mm Apart

Figure 12.4. Pin array test sketch.

Figure 12.5. Tunable frequency test setup.

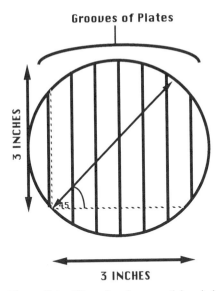

Figure 12.6. Effect of angle on spatial period.

in the case of single-mode fiber. For that reason, one group used ASTM standard sieves in place of the sandpaper and reported more repeatable results (40). [The microbend test procedure, EIA-455-68, was withdrawn in August 1991 and no replacement has yet been adopted (41).]

For the matched-frequency case, the tunable period turntable has proven to be of great value. To determine the appropriate spacings for a microbend sensor, a test can be made by scanning across candidate frequencies, in a manner similar to that described in Fields (10). The apparatus, a set of interlocking (upper and lower) grooved plates (Fig. 12.5) is machined with a specific spatial period between the teeth. The fiber to be tested is then placed between the plates, with additional weights being placed on the upper one. (An alternative version of this technique monitors the displacement of the upper plate.) The effective period of perturbation imposed on the fiber sample can be

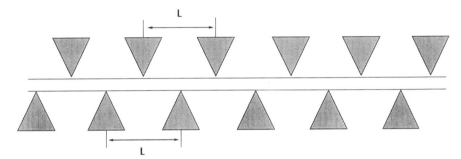

Figure 12.7. Side view of tunable spatial period apparatus.

varied by rotation of the plates (Figs. 12.6 and 12.7) while keeping the length of perturbed fiber constant. In this manner, a fiber can be rapidly tested, to determine its vulnerability to certain spatial periods. The relation between the base spatial period and the angle of rotation is

$$\Lambda = \frac{\Lambda_0}{\cos \theta} \tag{12.12}$$

where Λ_0 is the period machined into the plate and θ is the angle of the fiber with respect to the normal to the grooves.

Results for both types of microbending are given in Section 12.3.5.

12.3.2. Brightfield/Darkfield

While the majority of experiments performed have utilized the decrease in intensity of core transmitted light (Fig. 12.8), a few have measured the resultant intensity increase of light in the cladding. The two techniques have been labeled *brightfield* and *darkfield*, respectively. The latter approach, first used by Lagakos et al. (42), attaches either a detector or another fiber to the cladding at the point where microbending is occurring, or shortly afterward. The cladding light is then measured, with the increase in intensity being proportional to the loss in the microbent fiber. The darkfield technique, measuring the increase in light in the cladding (which was mode stripped just before the sensing section), can provide increased sensitivity for small displacements, at the price of higher complexity in the optical setup (35, 43). The darkfield technique must either have the detector (which detects the cladding light) very close to the sensing region or utilize another fiber to carry a portion of the cladding light from the sensing region to the detector (Fig. 12.9). Figure 12.10 (35) demonstrates the application of darkfield detection to examine more than one sensor. The use of an integrating cube made up of photosensitive materials to measure radiative or scattering losses was first put forward by Tynes (44). In practice, the fiber would be mode stripped immediately before each sensing region, and be stripped of its jacket within the cube, which would be filled with index-matching liquid. As many as 25 microbend sensors have been employed in a dark field system (35).

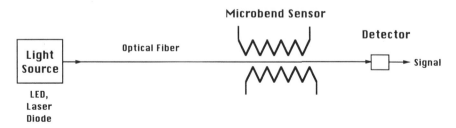

Figure 12.8. Standard microbend light field transmission setup.

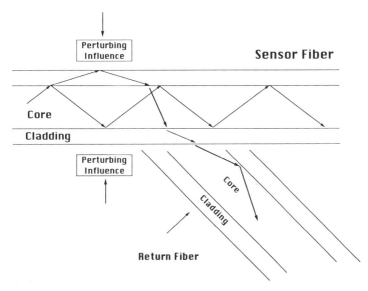

Figure 12.9. Darkfield; use of fiber to carry cladding light to the detector.

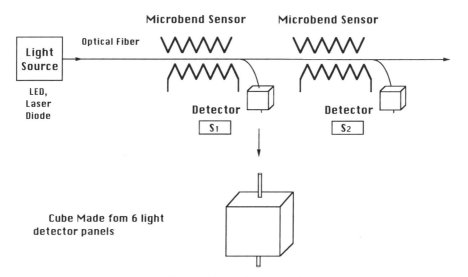

Figure 12.10. Application of darkfield detection to examine more than one sensor.

12.3.3. Transmission/Reflection

When utilizing microbend sensitive fiber, one standard technique is to inject light into the fiber core and monitor the loss in the forward direction, the decrease in transmitted light intensity being proportional to the measurand (Fig. 12.8). This can be done either with a continuous-wave (CW) source or by pulsing the light source, which increases the momentary power and provides for the use of sophisticated timing techniques that can also yield distance information (see below).

While the majority of the experiments in microbending have been performed in the forward direction (i.e., monitoring the change in either transmitted or coupled light, as illustrated in Sections 12.3.1 and 12.3.2), there have been several efforts to utilize the changes in reflected intensity due to microbending. Most of these employ an optical time-domain reflectometer (OTDR) or similar techniques. Pulsed light (usually a laser source) is transmitted down the fiber core, and the presence of any microbending loss attenuates the amount of light in the fiber, and thus the return signal. The signal may be examined in two ways: Rayleigh scattered light, arising from microscopic variations in refractive index, is sent in all directions, including backward; the intensity is inversely proportional to the fourth power of the wavelength and dependent on the intensity of the light pulse at that point in the fiber. Fresnel reflections, which occur at significant index changes such as at cleaves and connections, result in light being launched backward down the fiber. These reflections are much greater in intensity than the Rayleigh scatter. Use of OTDRs is covered in greater detail elsewhere in this book.

A high-resolution OTDR is an example of an instrument that can be used either by itself or with other sensors. By examining intensities from Fresnel back-reflections and the change in time delay between successive reflections, strain can be measured in optical fibers. Fresnel reflections are sometimes put into the fiber in order to have large-intensity back-reflections, monitoring the change in relative timing of back-reflected pulses to determine longitudinal stress. Microbending techniques, however, examine the intensity of either Fresnel or Rayleigh reflected light, to yield information regarding the radial stresses on the fiber. The OTDR method thus allows the examination of radial strains along the length of the fiber, in this manner creating a quasi-distributed sensor. The intensity of light propagating in the rearward direction will be attenuated twice by microbending losses which do not themselves induce reflections. Faltin (45) has demonstrated the use of a high-reflectivity fiber face located at the far end of a path in order to observe both sides of a fiber component, or link, at once, from a single end of the link. Employing this

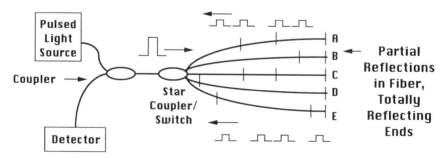

Figure 12.11. TDM/OTDR technique with overlap.

technique ensures that regions of fiber possessing suitable microbend losses can be monitored, even though they might not generate index differences potent enough to be observed as Fresnel reflections. Figure 12.11 depicts such a measurement setup.

Even a single microbend can provide useful information to an OTDR; Rourke measured the loss in a step-index (84/126 μm) fiber from a single microbend of 10 μm displacement and found it to be 0.026 dB (46), while Mickelson et al. (47) used backscatter to determine microbending loss in three serial sensors and showed the loss of one sensor to be nearly independent of that of the other sensors (given correct modal power distribution). Griffiths has demonstrated measuring strain and other structural parameters by means of the distributed microbending associated with a continuous length of fiber (48).

12.3.4. Multiplexing

A perpetual desire to save space, fibers, and/or money has led to the pursuit of sensor multiplexing, in which more than one signal is transmitted down a fiber. The utilization of fiber optic sensors in certain areas depends on the capability to multiplex several sensors onto a small number of fiber leads (49). Embedded sensors in smart structures is obviously an application in which the number of ingress/egress leads is to be minimized.

At least three multiplexing techniques appear to be applicable to microbending. The simplest, spatial multiplexing, involves utilization of a fiber switch, which may be an optomechanical or integrated optics device. Switching from one fiber path to another allows the investigator to interrogate one sensor, or group of sensors, at a time, and to economize on numbers of detectors and sources (Fig. 12.12).

Another method, time-division multiplexing (TDM), is closely related to the application of OTDRs. By pulsing the light source and employing accurate timing and nonequal selected lengths of fiber between sensors, the various sensors' responses can be discriminated by the length of time required for the light to be transmitted from or reflected by the sensors (50–52). The output of this system is a train of pulse groups, which must then be time gated to extract the information from each sensor channel. The delay between groups is the same as the input pulse delay, and the delay within a group corresponds to the

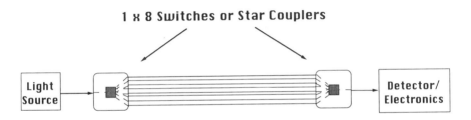

1 × 8 Switches or Star Couplers

Light Source

Detector/ Electronics

Microbend Sensitive Fibers

Figure 12.12. Switching SDM.

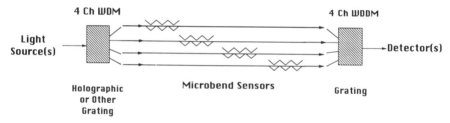

Figure 12.13. WDM technique with gratings.

time delay between sensors. From this there arises the requirement that the optical pulse duration must be less than the time of flight between any two sensors along the fiber. In addition, the time between pulses must exceed the maximum differential delay time in the fiber, the difference between the longest and shortest paths. Assuming a pulse duration Δt of 50 ns, to prevent output pulse overlap, the sensors must be separated by a length L, where

$$L \geqslant \frac{c}{n} \Delta t \approx 10 \, \text{m} \tag{12.13}$$

Figure 12.11 illustrates this technique. Note that the speed of source, detector, and electronics is also a limiting factor on how closely the sensors can be spaced. Use of this approach yields the potential of great savings in fibers, as well as information regarding position along the fiber. Unfortunately, there is a concomitant power loss; the optical intensity available at the nth sensor has

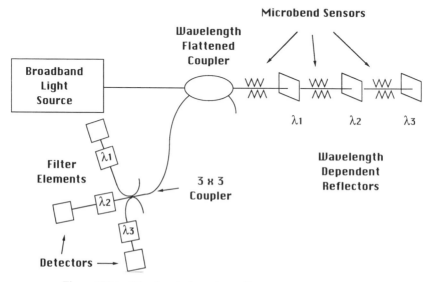

Figure 12.14. Wavelength-dependent reflectors; WDM microbending.

been attenuated by the microbend losses of the first $n - 1$ sensors (the problem is twice as large in the case of the single-ended sensor string with a highly reflective end face).

A final method, wavelength-division multiplexing (WDM), employs a coupler or other element with large wavelength dependence to inject light of two or more wavelengths into the sensing string. Application of the WDM technique to both SM and MM fibers is possible, as demonstrated by Fig. 12.13, in which gratings, which may be bulk optic or holographic in nature, are employed to split and recombine the various wavelengths, in a manner similar to that of the spatial multiplexing case. Microbending in single-mode fibers shows strong wavelength dependence, which has been used to reference the sensor behavior (see Section 12.3.6). Graded-index multimode fibers possess virtually no wavelength dependence in microbending, unlike single mode, but the WDM technique can be put to use to distinguish sensors in either case. Figure 12.14 illustrates another application of the WDM technique, in which wavelength-dependent reflectors are inserted into the fiber after microbend sensitive sections, so that the signals from the various sections can be distinguished (53).

12.3.5. Random and Multiple Frequency/Matched Spatial Frequency

The response from random frequency, or intrinsic, microbending has been examined both in laboratory test fixtures (sandpaper, basketweave tests), and in situ, with fibers embedded in composites possessing no strong spatially periodic structures. We have embedded fibers in carbon–epoxy and

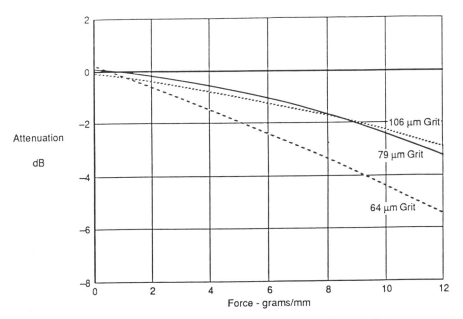

Figure 12.15. Fibers' random responses as a function of force applied.

thermoplastic materials, and have surveyed fibers' random responses. Figure 12.15 displays this behavior as a function of force applied. Note that, depending on the size of the grit in the sandpaper, the jacket diameter, and the softness of the jacket material, results may be changed substantially for the same fiber exposed to different grit sizes.

One problem with the lateral load test is that it is basically destructive, as the grit tends to saw through the jackets of most fibers; experiments with high levels of force cannot always be completed, due to fiber breakage.

The most sensitive microbending sensors can be constructed when the difference in guiding constants between guiding and radiative modes is matched by the spatial periodicity of the perturbation imposed by the sensing device. Sensors have used mechanical jaws, wire wrapped in helical fashion around the fiber, serpentine arrangements of pins, machined cylindrical shells, and other devices to impose the proper periodic disturbance on the fiber. The tunable frequency test (Fig. 12.5) is useful in checking predictions of spatial wavelength vulnerability. Some results obtained with the tunable grating are presented in Fig. 12.16.

In addition to the fundamental of the spatial frequency matching the separation between the modes, it has been demonstrated (54, 55) that high-

Attenuation Versus Wavelength for Corning 1521 Single Mode Fiber With Polymide Buffer
Run Number 3

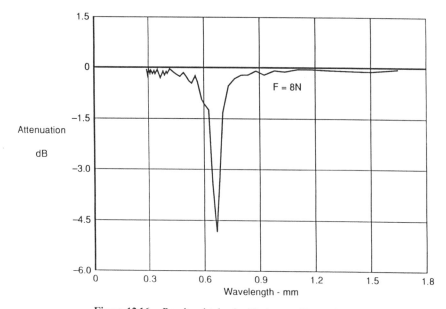

Figure 12.16. Results obtained with the tunable grating.

er-order harmonics exist. The third harmonic, in particular, is useful in mode coupling in parabolic-index multimode fibers. Higher-order harmonics also have proven valuable in the case of single-mode fibers, where the fundamental spatial wavelength may be less than 1 mm. Figure 12.40 also displays this higher harmonic behavior.

12.3.6. Limitations

The major problem with the use of microbending as a sensing means is that it is an intensity- or amplitude-modulated sensor. Although this eliminates the complications associated with phase that workers in interferometric measurements endure, it introduces problems involved with variations in received optical power. An interferometric sensor does not give a false reading merely because the output power has temporarily dipped 5%, but an amplitude-modulated sensor will register this as a 5% change unless a referencing system is in place to normalize the measurement. These power differences may arise from source fluctuations due to temperature or drive current changes, or device aging, from connector instabilities, or from effects on the fiber leads on either side of the sensing region desired.

Source problems can be dealt with by the addition of a reference coupler and detector to monitor the input power (Fig. 12.17). Variations in connector loss during operation, or between connections, are more difficult to ameliorate. The system can perhaps be calibrated between operations, in the event that connectors are disengaged, and parallel reference lines can be set up with the aid of fiber taps or couplers; this will come at the expense of system dynamic range, or number of sensors per system. Wlodarczyk et al. (56) have investigated the application of wavelength referencing using two or more wavelengths through the same sensor. This has generally been thought applicable only to single-mode fibers, where one wavelength may be unaffected by the microbending, and serve as a throughput reference, but (57) wavelength

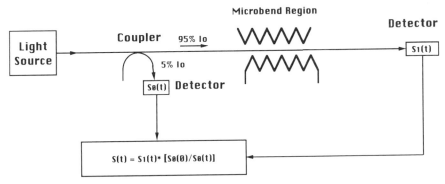

Figure 12.17. Reference coupler and detector, to monitor the input power.

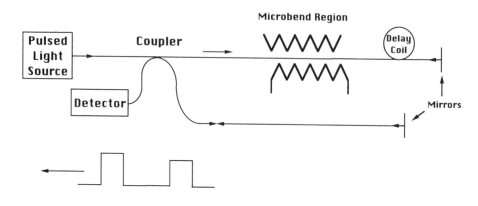

Pulses Arriving at Detector for Ratio Measurements

Figure 12.18. TDM normalization.

dependence has been demonstrated in step-index multimode fibers. The two wavelengths can be obtained by filtering the output of a single LED, and can be as little as 20 nm apart, while yielding 10 dB or more in loss difference (56). The unaffected wavelength can then be used to normalize the source and link-dependent variations. Figure 12.18 demonstrates the use of TDM normalization of sensor response (58), in which the ratios of pulses reaching the detector are analyzed.

Spillman and Lord (59) have described a self-referencing and multiplexing method for intensity sensors based on recirculating loops. The intensity sensor is incorporated into the loop, which is situated between parallel transmitting and receiving optical lines. Trains are formed by the pulses being tapped off successively by the loops and through the sensors; the ratio of the nth pulse height to that of pulse $n - 1$ (for the same loop) will be independent of variations of source and connectors.

The best light source for a microbending fiber sensor is a light-emitting diode (LED) (60). This is so despite the fact that a laser diode launches more light into a fiber. The chief benefit of the LED is that the leads going to and from the sensor are made less sensitive to bending losses or other signals imposed on the light being transmitted by the fiber. Speckle resulting from mode–mode interference is reduced drastically, due to the much shorter coherence length of the LED. In addition, the intrinsic noise is lower for LEDs (60) than for lasers.

LEDs are readily available that are already pigtailed to $100/140 \mu$m fiber, and which inject over 300μW of light into the core (center wavelength about 835 nm). LEDs pigtailed to multimode or single-mode fiber are also available in the 1300-nm range, albeit with less power. Superluminescent diodes (SLDs) are now available with over 300 uW of light injected into single-mode fibers at 1300 nm.

Sensors ingress and egress leads should be constructed from fiber relatively unaffected by moderate bending effects, with higher numerical apertures (NAs) (60, 61). Single-mode fibers with larger (0.13 to 0.16) NAs, designed for microbend insensitivity, are available from more than one manufacturer. Multimode fibers are also available with NAs of 0.29 or higher.

Problems have been caused by light being guided by cladding modes. The guiding properties of these modes are extremely sensitive to wavelength and bend radius (62, 63) and can be coupled back into the core modes. Lipscomb et al. (63) obtained good sensor sensitivity and much lower noise by using black absorbing paint on the outside of fiber bends to eliminate these "whisper gallery" modes.

12.4. OVERVIEW OF SMART STRUCTURE MICROBENDING EXPERIMENTS

The number of smart structure experiments is much smaller than the total number of microbending sensing trials but is growing rapidly. An early experiment by Asawa and co-workers attached a microbending fixture to a large structure to detect bending and vibrations (30). Multiple sensors were evaluated by means of OTDR techniques, and bend radii larger than 5 km were measured. Bruinsma et al. (64) also dealt with the attachment of fibers to a structure, the fibers constrained by a serpentine arrangement of pins so that any axial stress would produce a lateral load, thus causing microbending. In this way, microbending was used to monitor axial strains, attaining a resolution of $100\,\mu$strain. Schoenwald et al. (65) also used OTDR techniques to investigate microbending sensors attached to a structure. The microbending structure was about 1 in. in length and $\frac{1}{8}$ to $\frac{1}{4}$ in. in thickness.

12.4.1. Standard Fibers

Urruti and his co-workers examined the effects of different coatings upon the microbend sensitivity of a standard fiber (66) (Corning 1521) as measured by the lateral load test. The coatings comprised an acrylate, a silicone, and two varieties of polyimide, which proved to be the best for retaining microbend vulnerability. Udd et al. (67) investigated the performance of commercially available Corning fibers embedded in composites, under strain. These fibers were embedded in carbon–epoxy coupons. Figure 12.19 shows some of the results obtained.

The aforementioned experiments made use of standard telecommunication fibers, both single and multimode. Both localized and integrated microbend sensors can be fabricated from commercially available fibers by splicing microbend sensitive fiber to microbend insensitive leads. An example of such an implementation, using an OTDR, is given in Fig. 12.20.

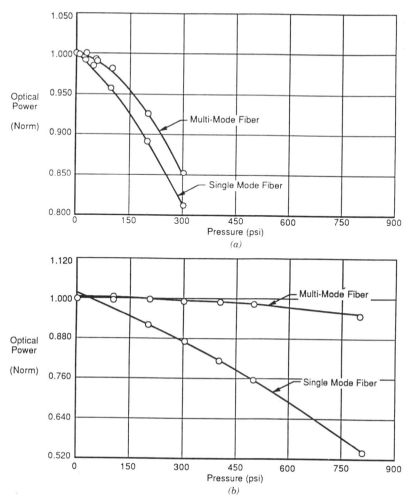

Figure 12.19. Fibers embedded in carbon epoxy coupons (a) loss before curing; (b) loss after curing.

12.4.2. Integrated Strain Sensing

Microbend sensitive fiber can be used to monitor the strain level in composite parts when embedded, or in metallic parts when bonded to the surface. At McDonnell Douglas, tests were conducted on composite and aluminum coupons with embedded and bonded microbend sensitive fibers, respectively. The fibers were oriented in several ways (Fig. 12.21).

Data were taken during fully reversed longitudinal loading to 16,000 psi. The result for an embedded optical fiber oriented parallel with the load is shown in Fig. 12.22. It should be noted that the peaks correspond to the compression cycles and the troughs to the tension cycles. Several interesting

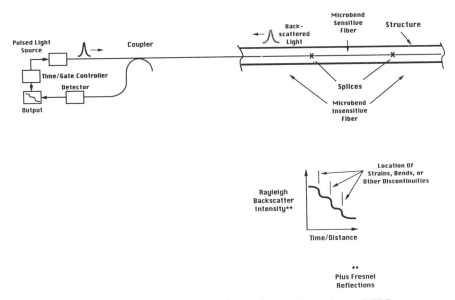

Figure 12.20. Localized microbending implementation, using an OTDR.

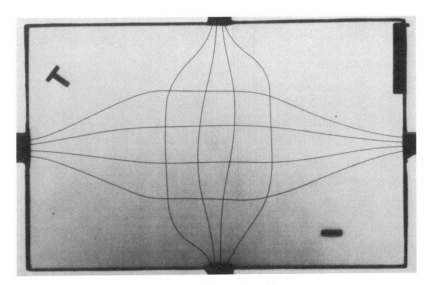

Figure 12.21. X-ray of fibers oriented in test coupon.

features can be seen in this figure. Under compressive load, the composite specimen buckles slightly at maximum loading. This can be seen clearly in the peaks, where the signal drops off slightly as a result of the buckling. The signal also has a fairly low overall peak-to-peak value. Therefore, the smaller vibrations make it difficult to locate the peaks and valleys precisely. Signal filtering removes these higher-frequency vibrations (Fig. 12.23). The static

SPECIMEN 1 –12000 TO 12000 LBS

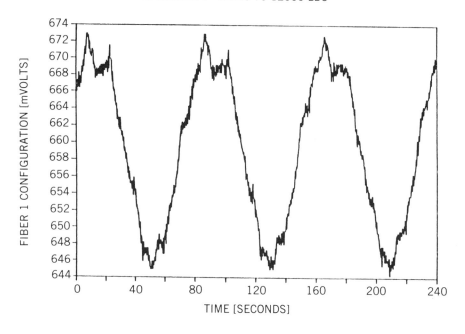

Composite Specimen–Embedded Fiber

Figure 12.22. Result for an embedded optical fiber oriented parallel with the load.

± 10,000 psi

Band Pass Filter 0.5 - 50.0 Hz

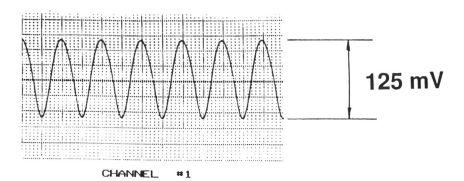

Figure 12.23. Signal filtering removes higher-frequency noise.

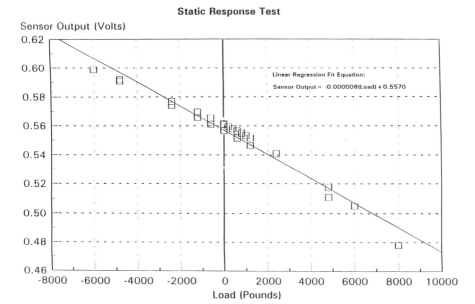

Figure 12.24. Static response test for integrated strain specimen.

response test for one of these specimens is displayed in Fig. 12.24. Specialty fiber, with enhanced sensitivity to microbending, would generate a much larger response, improving the signal-to-noise ratio. Identification of peaks and valleys is important to some structural health monitoring efforts, such as fatigue tracking. Damage detection methods, on the other hand, may utilize the frequency content; therefore, the output signal from the sensor should be split so that the quasi-static and low-frequency loads can be separated from the high-frequency effects.

Microbend sensitive fiber bonded to aluminum test coupons produced a signal very similar to that from embedded sensors (Fig. 12.25). The buckling behavior is not exhibited by the aluminum coupon, but in all respects the bonded sensor performs as well as the embedded fiber. Bonded sensors provide the capability for monitoring metallic structures as well as for use in repair situations when panels with embedded sensors suffer ballistic damage. The testing apparatus for both types of panels is shown in Fig. 12.26, with an aluminum coupon in place. Figure 12.27 shows a close-up of a composite panel, along with special fiber protection developed to ensure that the fiber is not subject to stress concentrations during ingress and egress from the panel.

12.4.3. Integrated Impact Sensing

In many composite structures, the integrity of the part can be severely compromised as the result of impact damage, which causes delaminations to occur between the plies of the laminate. In the case of this low-velocity impact,

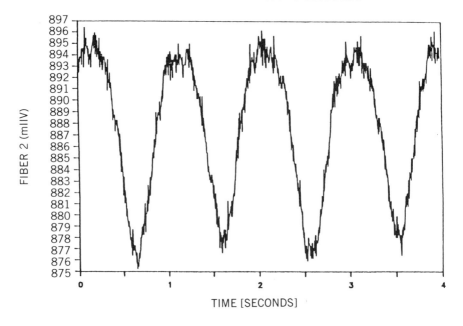

Aluminum Specimen–Bonded Fiber

Figure 12.25.   Fiber bonded to aluminum coupon: response similar to embedded sensors.

there will often be little or no apparent physical evidence of the impact on the surface. These impacts can cause significant delaminations, which can grow to castastrophic failures under continued loading of the laminate.

At McDonnell Douglas, optical fibers were embedded into composite panels for impact testing. Figure 12.28 displays the x-ray of a typical panel used in these tests. The panel is 7.0 by 11.0 in. and is comprised of 40 plies of carbon epoxy material. The horizontal fibers were placed between different layers than the vertical fibers, to prevent high contact stresses between the fibers and subsequent potential breakage. Both layers of fibers were near the center of the material layup. As the figure indicates, a matrix of fibers was placed in the panel on 1.0-in. centers. One of the original goals of this effort was to locate the damaged area and determine its approximate size from the fibers that might be broken during impact. Although excellent transient data were obtained during the impact events, the fibers remained stubbornly intact through the maximum of 20.0 ft-lb of impact energy; thus no damage location or size information was obtained in this manner. This was despite the fairly significant delamination that occurred during the 20-ft-lb impact.

The impact test fixture is shown in Fig. 12.29. It is a free-fall impactor that can be raised to a maximum height of 8 ft. An automatic pneumatically

Figure 12.26.   The testing apparatus for both types of panels.

actuated catch activates when the impactor rebounds from the panel and prevents additional impacts from occurring. A heavy steel cover plate is bolted over the composite specimen, leaving open a 5.0- by 5.0-in. test region.

Data from the sensors (68) are shown in Fig. 12.30 and 12.31, from a 1.0-ft-lb and a 20.0-ft-lb impact, respectively. In the first figure, the classic exponentially decaying sinusoidal behavior so typical of damped vibration situations is evident. In the second figure, the clean behavior is lost, and the signal is clearly much more energetic.

The fast fourier transforms (FFTs) of these signals reveal their frequency content. In Fig. 12.32, a definite concentration of energy is shown at approximately 3300 Hz, which is the primary bending frequency of the 5.0- by 5.0-in. test section of the panel. This clean response is lost in Fig. 12.33 in the FFT of the 20-ft-lb impact.

A neural network was utilized to correlate transient signals to impact energy. After being trained on a number of signals, the neural network was very successful in identifying impacts at a variety of energy levels (Fig. 12.34). Some confusion exists at the lower energy levels, where the signals are very similar. At these low levels, no damage is being done, so the confusion presents no major concern. Utilizing the neural network with the FFT of the impact signal might provide better discrimination at the lower energy levels.

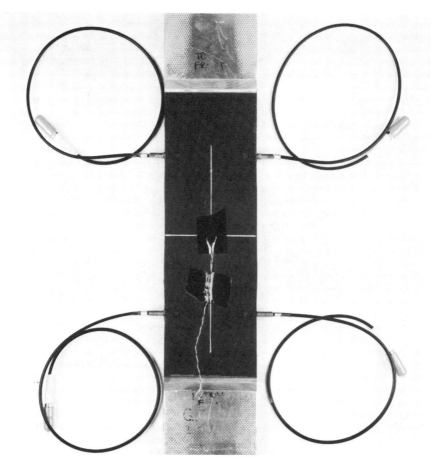

Figure 12.27. Composite panel showing ingress/egress fiber protection.

12.4.4. Modified Fibers

Several fiber manufacturers sell microbend insensitive fiber, useful for sensor ingress and egress (both multimode and single mode). Taking a different approach, Morse outlined a scheme to fabricate microbend sensitive fiber by introducing hard spherical particles with diameters $\approx 10\,\mu$m into an otherwise soft jacket (69). Radial stresses would impart strong localized bending to the fiber (Fig. 12.35).

Efforts have been made to attach or embed fibers with periodic structures within a composite, thus subjecting the fiber to a far more potent mode coupling effect, preferably without the need for external structures. Harmer wrapped wire in a helical fashion around the fiber, the wire periodically perturbing the fiber radially (70) (Fig. 12.36). The best choices of jacket

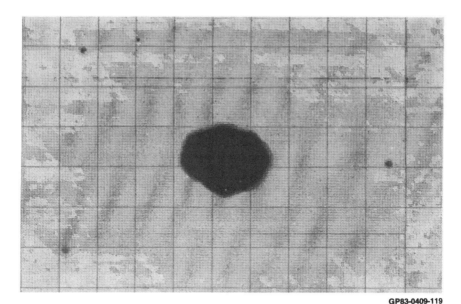

GP83-0409-119
TZ-321

a) Without Optical Fibers

b) With Optical Fibers

Figure 12.28.   X-ray of typical panels used in impact tests, (*a*) without optical fibers, (*b*) with optical fibers.

Figure 12.29. Impact test fixture with coupon in place.

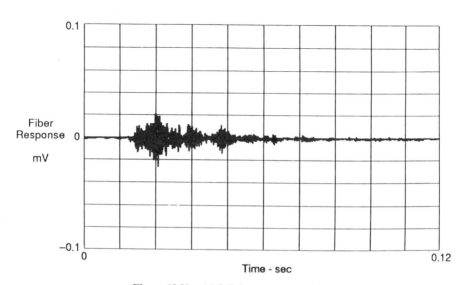

Figure 12.30. 1.0-ft-lb impact test results.

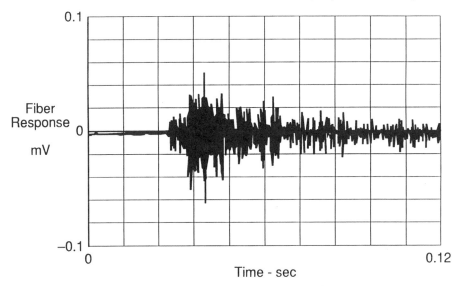

Figure 12.31.   20.0-ft-lb impact test results.

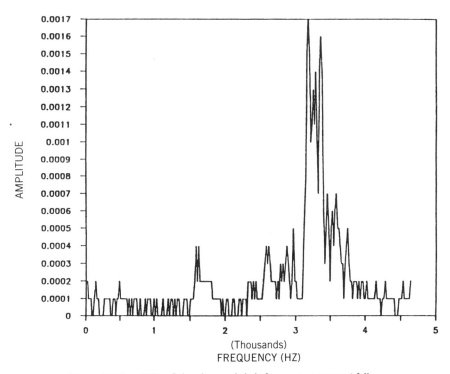

Figure 12.32.   FFTs of signals reveal their frequency content; 1 ft lb.

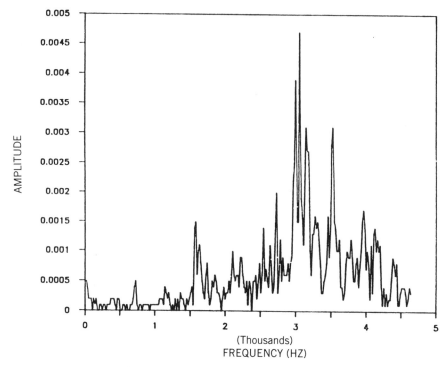

Figure 12.33.   FFT of the 20.0-ft-lb impact.

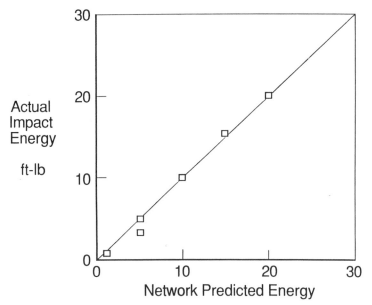

Figure 12.34.   Neural network successful in correlating signals/energy.

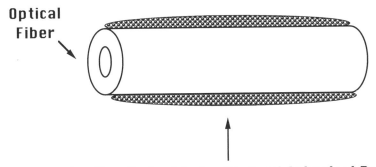

Soft Coating Embedded with Hard Spherical Particles

Figure 12.35. Radial stresses from embedded particles; localized bending to the fiber.

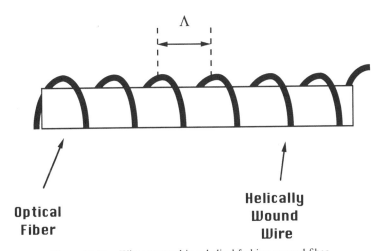

Figure 12.36. Wire wrapped in a helical fashion around fiber.

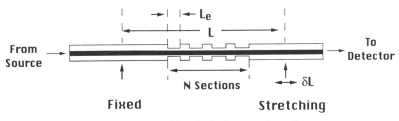

Figure 12.37. Etching the jacket away in sections.

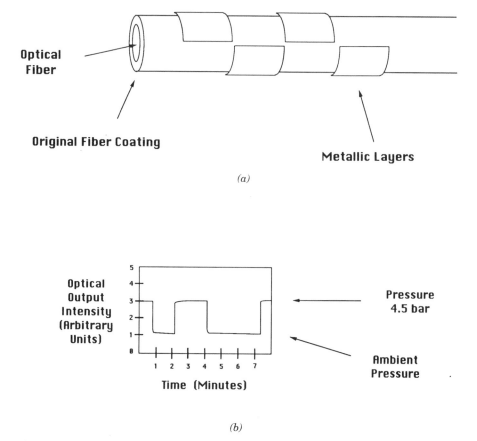

Figure 12.38. (*a*) Adding a metal periodic structure to the fiber's jacket; (*b*) Some of the results, for pressure cycles from (*a*).

materials gave static sensitivities reduced by a factor of 7 relative to responses obtained from a matched period case. A sensitivity of 0.1 dB/g (force applied)/cm length was reported, with a minimum detectable displacement of $5 \times 10^{-5} \mu$m. This has resulted in a commercial product.

Vaziri and Chen incorporated a periodic structure into the fiber's embedded environment by removing the jacket, then etching the cladding away in sections. The spatial variation in glass diameter then acts to concentrate strain and increase fiber loss when tension is applied (71, 72) (Fig. 12.37). Strains as large as 0.4% were measured with this technique. Falco and Parriaux added a metal periodic structure to the fiber's jacket (73), thereby accomplishing the same end. Their arrangement is illustrated in Fig. 12.38*a*, and some of the results, for a pressure cycle, in Fig. 12.38*b*.

Weiss fabricated strain gauges from plastic step-index fibers by building permanent microbends into them (Fig. 12.39). This was accomplished by first

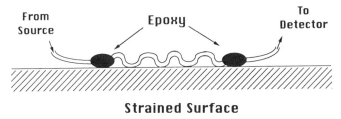

Figure 12.39. Plastic fibers: building permanent microbends into them.

heating them to the softening point and then cooling the fibers while under compression. These sensors, which can be bonded to surfaces, exhibit an increase in optical transmission when put under tension and are estimated to have a minimum detectable strain level of about 2.2×10^{-9} (74).

2.4.5. Coating Concerns

Depending on the application, the role of fiber coatings may be that of protective layers, cushions distributing lateral forces, hermetic seals, or force transducers (75). In the case of microbending sensors, the second use is excluded; communications-grade fibers possess a soft inner coating and hard outer jacket, intended to *minimize* the effects of microbending (76).

Coatings safeguard fibers both from incurring surface scratches and the growth of preexisting flaws. To prevent surface microcrack growth, the combination of moisture (OH^-) and strain must be kept from occurring at the location of any microcracks. While a bare fiber surface may be protected if embedded, its jacket in the regions external to the composite must certainly survive for the fiber to do as well. For communication fibers, a UV-cured acrylate jacket (or jackets) typically is deposited on-line, as the fiber is being drawn from the preform. Nylon may be employed as a final layer, and the total diameter of the jacketed fiber may be anywhere from 200 to 900 μm, the latter size far too large for embedding purposes. In addition, the materials fail at about 130°C or below, woefully short of processing temperatures required for composites.

Polyimide coating, also usually applied in-line (a layer $\approx 15\,\mu$m thick), protects the glass up to about 350 to 400°C. Polyimides have the additional virtue of possessing moduli ranging from that of the hard outer acrylate jackets, and up; hence they do not shield the fiber from the imposed microbending. Aluminum coating is a higher-temperature material but is rated only to about 425°C, although aluminum melts at close to 660°C. Berthold (77), working on microbend pressure sensors for power plants, reported pressure measurements at 425°C to better than 1% accuracy. However, when aluminum-coated fiber has been embedded in carbon–epoxy composites, the combination has been observed to cause a galvanic reaction, making it unsuitable for many embedded applications. Gold is the highest-temperature jacketing commercially available, and is deposited on-line in a 12- to 20-μm-thick layer. Although pure gold melts at about 1050°C, this fiber coating is rated to 750°C.

The buffer, jacket, and cable materials of fibers are also at risk from low temperatures, although some jacketed fibers are claimed to perform down to −310°F, and some have been tested to −300°F. The microbending loss of fibers increases at low temperatures as the buffer and jacket shrink. Polyimide coatings are less subject to this temperature-related loss, as they are already below the glass transition temperature in normal environments. Researchers from VPI have recently demonstrated repeated thermal cycling of polyimide-jacketed fibers while testing a Fabry–Perot sensor (78), down to temperatures of ≈ 4 K.

Another concern with coatings is the effect they have on microbending loss measurements. The original theory assumed that the fiber claddings were finite, and ignored complications with the transmission and leakage of light due to the jacket index. Lipscomb et al. (63) have investigated the problems of cladding–core mode interference for single and multimode fibers, and demonstrated the improvement in stability that occurs when a heavily absorbent

Attenuation Versus Wavelength for Corning 1521 Single Mode Fiber No Buffer

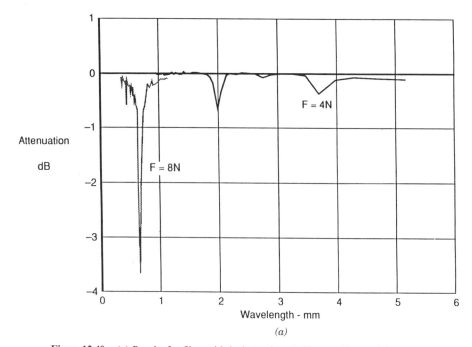

(a)

Figure 12.40. *(a)* Results for fiber with jacket stripped; *(b)* same fiber, polyimide.

Attenuation Versus Wavelength for Corning 1521 Single Mode Fiber With Thermocoat Buffer

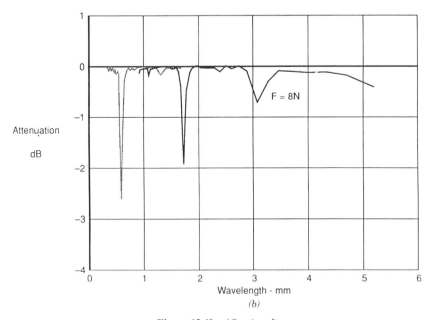

Figure 12.40. (*Continued*).

jacket is used, thereby eliminating the oscillatory behavior of the transmitted light. Lagakos et al. (60) have shown the effect the absorbance of the jacket has on the efficiency of microbending over various lengths and the problems resulting from incorrect choice of jacket index.

Romero-Borja et al. have examined the effects that coating exerts on microbending response and found that hard coatings such as nylon give good sensitivity (79). We have also tested fibers and their jackets; a few of our results are illustrated by Fig. 12.40. Figure 12.40a displays the results for a fiber with the jacket stripped, tested in the tunable-frequency fixture. Figure 12.40b has the same fiber, but with a polyimide jacket. The hard coatings or bare fibers usually suffer the most loss, as expected.

12.4.6. Embedded Sensor Issues

There are several questions that arise as a result of embedding optical fibers into composite materials, some of which can only be answered empirically. These include the possibility of delamination of composites due to embedding of fibers (and any dependence upon fiber–fiber orientation), the resultant bending (or lack of) due to relative stiffness of composite to optical fibers, and the effects of the layering of the composites on the bending performance of the sensor.

Delamination problems directly related to the optical-composite fiber orientation have been observed. Fibers laid at 90° to each other caused difficulties such as fiber buckling and nonuniformity of carbon fiber density around the optical fiber, while those parallel did not. Those tests were, however, conducted with epoxy–carbon panels and do not necessarily apply to all composites. Note that these difficulties exist for any optical fiber embedded in a composite laminate; they are not unique to microbend sensitive fibers. A more complete discussion of these topics is offered in Chapter 22.

Compromises have also to be made with respect to preparation of microbending samples. Thus while the best performance would occur with optical fibers woven between the structural fibers to imitate the bending of an extrinsic sensor (in cases with a fiber prepreg), in most instances applications are restricted to the easier technique of insertion between layers, with the understanding that the ultimate sensitivity will therefore be lower. Bocquet (80) has reported weaving fibers into a composite in such a manner.

There are practical limitations to how well a waveguide's properties can be matched with any periodic spatial disturbance that it may be expected to experience. Although random frequency microbending does not have the ultimate sensitivity of well-matched tuned fiber microbending, it enjoys a certain immunity to mishap, as the fiber is vulnerable to at least some of the frequencies contained in the random conglomerate, whether of sandpaper grains or of composite fibers.

Another issue which must be taken into account is that the microbending losses are linear over a certain region; losses at the low end of a strain regime may not fall into this linear region unless steps are taken to prebias the sensor in the composite or to illuminate the higher-order modes more fully. The latter approach has been discussed by Chen, who suggested depositing thin-film gratings (81) on the ends of graded-index fibers.

12.5.   OTHER APPLICATIONS

It is worth noting that microbend-caused modal coupling has also been employed in several other areas. The process has served as a method of injecting and detecting light locally (82), as an aid to optical fiber splicing (83, 84), to emulate long-distance effects by mixing the modes in optical fibers under test, and as a mode coupler, to couple optical power from one mode into another, in two-mode fibers (85, 86). Microbending has also been demonstrated to have applications in interferometric sensors; it has been utilized for phase shifting in single-mode fibers, achieving phase shifts of up to 560° per centimeter of transducer length when used with a grating of 508-μm period. A fiber stretched by being wrapped around a PZT cylinder would have required 50 times the length for the same phase shift (87).

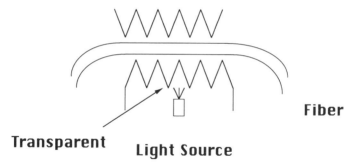

Figure 12.41.   Local injection module sketch.

Hybrid networks incorporating both intensity sensors (including microbending) and communication systems have been described. These systems employed WDM techniques, with the communication signals at 820 nm and the microbending signals carried at 1300 nm. The goal was to demonstrate that a hybrid network would allow multiple nodes of distributed memory computers within smart structures to process and transmit data from sensors as well as to control information (88).

Development of microbending for other sensor applications continues. Fibertek Inc. (Herndon, Virginia) performed research for NUSC on a microbend fiber optic pressure sensor (89) to determine induced pressure on spinning gears, up to 3000 rpm.

There are several microbend-type commercial devices available but only a few embedded microbend sensors. Some devices utilizing microbending are the previously mentioned local injection and detection modules (82–84) (as illustrated by Fig. 12.41). These devices provide efficient light injection into single-mode fiber cores, and keep stresses from bending lower than 50 kpsi, compared to the 125 to 625 kpsi caused by macrobend-based light injectors (83). Discrete microbend strain sensors are offered by Focal Marine and OPTECH, which also has a microbend accelerometer on the market.

Hergalite (Allentown, Pennsylvania) represents one of the few embedded commercial offerings. An optical fiber is helically wound with plastic line (90,91), and then embedded in a mat (Fig. 12.36). Light transmission through the fiber is monitored and a signal produced when the loss goes over a preset value. The market being addressed here is primarily safety; mats can be set around dangerous equipment or physical sites, and the equipment shut off, or an alarm set off, when the weight of a person on the mat drops the transmitted light intensity. Obviously, such a system can be used to signal other intrusions as long as the mat is triggered successfully.

Smart structure projects involving embedded or attached microbending sensors are continuing, especially in the civil structures area. Microbending is being used to monitor the health and diagnose potential problems of a bridge (92). Such a simple, easy-to-implement technique must surely have many more applications in its future.

REFERENCES

1. C. N. Kurtz and W. Streifer, Guided Waves in Inhomogeneous Focussing Media. Part II. *IEEE Trans. Microwave Theory Tech.* **MTT-17**, 250–253 (1969).
2. D. Marcuse, Derivation of Coupled Power Equations. *Bell Syst. Tech. J.* **51**(1) 229–237 (1972).
3. D. Gloge and E. A. J. Marcatili, Multimode Theory of Graded-Core Fibers. *Bell Syst. Tech. J.* **52**, 1563–1578 (1973).
4. D. Marcuse, Losses and Impulse Responses of a Parabolic Index Fiber with Random Bends. *Bell Syst. Tech. J.* **52**(8), 1423–1432 (1973).
5. D. Marcuse, Coupling Coefficients for Imperfect Asymmetric Slab Waveguides. *Bell Syst. Tech. J.* **52**(1), 63–82 (1973).
6. W. B. Gardner, Microbending Loss in Optical Fibers. *Bell Syst. Tech. J.* **54**(2), 457–465 (1975).
7. D. Marcuse, Microbending Losses of Single-Mode, Step-Index and Multimode, Parabolic-Index Fibers. *Bell Syst. Tech. J.* **55**(7), 937–955 (1976).
8. R. Olshansky, Propagation in Glass Optical Waveguides. *Rev. Mod. Phys.* **51**, 341–368 (1979).
9. L. Jeunhomme and J. P. Pocholle, Mode Coupling in a Multimode Optical Fiber with Microbends. *Appl. Opt.* **14**(10), 2400–2405 (1975).
10. J. N. Fields, Attenuation of a Parabolic-Index Fiber with Periodic Bends. *Appl. Phys. Lett.* **36**(10), 799–801 (1988).
11. D. Marcuse, Mode Conversion Caused by Surface Imperfections of a Dielectric Slab Waveguide. *Bell Syst. Tech. J.* **48**(10), 3187–3215 (1969).
12. C. M. Davis, Fiber Optic Sensors: An Overview. *Opt. Eng.* **24**(2), 347–351 (1985).
13. J. A. Bucaro, N. Lagakos, J. Cole, and T. Giallorenzi, Fiber Optic Acoustic Transduction. In *Physical Acoustics* (R. N. Thurston and W. P. Mason, eds.), Vol. 16, Chapter 7. Academic Press, New York, 1982.
14. N. Lagakos and J. A. Bucaro, Optimizing Fiber Optic Microbend Sensor. *Proc. SPIE—Int. Soc. Opt. Eng.* **718**, 12–20 (1986).
15. Y. Murakami, K. Ishihara, Y. Negishi, and N. Kojima, Microbending Losses of P_2O_5-Doped Graded-Index Multimode Fiber. *Electron. Lett.* **18**(18), 774–775 (1982).
16. P. L. François, J. F. Bayon, F. Alard, and D. Groto, Characterization Procedure of Fiber Packagings Relative to Microbends. *Electron. Lett.* **21**(11), 471–472 (1985).
17. N. Yoshizawa, M. Ohnishi, O. Kawata, K. Ishinara, and Y. Negishi, Low Temperature Characteristics of UV-Curable Resin Coated Optical Fiber. *IEEE J. Lightwave Technol.* **LT-3**(4), 779–784 (1985).
18. H. Grebel and G. J. Herskowitz, Multimode Cabled Optical Fibers at Low Temperatures: An Investigation. *Appl. Opt.* **25**(23), 4426–4433.
19. H. Grebel, G. J. Herskowitz, and M. Mezhoudi, Characterization of Single-Mode Coated Fibers at Low Temperatures Using Periodic Distortion of the Fiber Axis. *IEEE J. Lightwave Technol.* **LT-5**(8), 1111–1116 (1987).
20. G. Grasso and F. Meli, Microbending Losses of Cabled Single Mode Fibers. 526–532 *14th European Conference on Optical Communication Proceedings,* 526–532, Brighton, England, (1988).

21. D. Marcuse, Microdeformation Losses of Single-Mode Fibers. *Appl. Opt.* **23**(7), 1082–1091 (1984).

22. A. Bjarklev, Microdeformation Losses of Single-Mode Fibers with Step-Index Profiles. *IEEE J. Lightwave Technol.* **LT-4**(3), 341–346 (1986).

23. S. Das, C. G. Englefield, and P. A. Goud, Power Loss, Modal Noise and Distortion due to Microbending of Optical Fibers. *Appl. Opt.* **24**(15), 2323–2334 (1985).

24. R. Yang, C. Yu, and G. L. Yip, Approximate Formulas for the Microbending Loss in Single-Mode Fibers. *Opt. Lett.* **12**(6), 428–430 (1987).

25. H. Grebel and M. Mezhoudi, Effect of Periodic Strain on Leaky Modes in Single-Mode Fiber. *Appl. Opt.* **27**(23), 4819–4821 (1988).

26. K. Petermann, Fundamental Mode Microbending Loss in Graded-Index and W Fibers. *Opt. Quantum Electron.* **9**, 167–175 (1977).

27. J. N. Fields et al., Fiber Optic Pressure Sensor. *J. Acoust. Soc. Am.* **67**(3), 816–818 (1980).

28. J. N. Fields and J. H. Cole, Fiber Microbend Acoustic Sensor. *Appl. Opt.* **19**(19), 3265–3267 (1980).

29. N. Lagakos, L. S. Schuetz, J. H. Cole and J. A. Bucaro, Optimized Microbend Fiber-Optic Displacement Sensor. *Optical Fiber Communication Conference Technical Digest*, 56–58, New Orleans (1984).

30. C. K. Asawa et al., High-Sensitivity Fiber-Optic Strain Sensors for Measuring Structural Distortion. *Electron. Lett.* **18**(9), 362–364 (1982).

31. A. G. W. M. De Jongh et al., Simple Fiber-Optic Sensor for Detecting Water Penetration into Optical Fiber Cables. *Electron. Lett.* **19**(23), 980–982 (1983).

32. J. Freal et al., A Microbend Horizontal Accelerometer for Borehole Deployment. *IEEE J. Lightwave Technol.* **LT-5**, 993–996 (1987).

33. C. J. Zarobila and C. Davis, A Microbend-Modulated Fiber-Optic Accelerometer. *Sens. Expo Proc.* Chicago, IL (1986).

34. C. Davies et al., Fiber-Optic Sensors for Geophysical Applications. *Proc. SPIE—Int. Soc. Opt. Eng.* **985**, 26–32 (1988).

35. C. M. Davis et al., Dark-Field Microbend-Modulated Fiber Optic Sensor System. *OFS 1 Proc.*, London pp. 127–131 (1983).

36. EIA Std. FOTP-68, *Optical Fiber Microbend Test Procedure*, EIA-455-68. Electronic Industries Association, Washington, DC, 1989.

37. EIA Std. FOTP-62, *Optical Fiber Microbend Attenuation*, EIA-455-62. Electronic Industries Association, Washington, DC, 1988.

38. A. Tomita, P. F. Glodis, D. Kalish, and P. Kaiser, Characterization of the Bend Sensitivity of Single Mode Fibers Using the Basket-Weave Test. *Tech. Dig. Symp. Opt. Fiber Meas.*, Boulder, pp. 89–92 (1982).

39. J. A. Dixon, M. S. Giroux, A. R. Isser, and R. V. Vandewoestine, Bending and Microbending Performance of Single-Mode Optical Fibers. *Opt. Fiber Commun. Conf. Proc.,* p. 40, Reno (1987).

40. A. O. Garg and C. K. Eoll, New Measurement Technique for Measurement of Microbend Losses in Single Mode Fibers. *Tech. Dig. Symp. Opt. Fiber Meas.,* pp. 125–128 (1986).

41. EIR, JEDEC, & TIA Standards and Engineering Publications, *1993 Catalog*, Global Engineering Documents.

42. N. Lagakos et al., Multimode Optical Fiber Displacement Sensor. *Appl. Opt.* **20**(2), 167–168 (1981).

43. M. Krigh et al., Fiber-Optic Dark-Field Microbend Sensor. *Proc. SPIE—Int. Soc. Opt. Eng.* **586**, 387–394 (1985).

44. A. R. Tynes, Integrating Cube Scattering Detector. *Appl. Opt.* **9**, 2706–2710 (1970).

45. L. Faltin, Two-Sided OTDR Measurements from Same Fiber End. *J. Opt. Commun.* **9**(1), 24–26 (1988).

46. M. D. Rourke, Measurement of the Insertion Loss of a Single Microbend. *Opt. Lett.* **6**(9), 440–442 (1981).

47. A. A. Mickelson et al., Backscatter Readout from Serial Microbending Sensors. *IEEE J. Lightwave Technol.* **LT-2**(5), 700–709 (1984).

48. R. W. Griffiths and R. C. Lamson, Adaptation of an Electro-Optic Monitoring System to Aerospace Structures. *Proc IEEE—Int. Soc. Opt. Eng.* **838**, 173–181 (1987).

49. T. E. Clark et al., Overview of Multiplexing Techniques for All-Fiber Interferometric Sensor Arrays. *Proc. SPIE—Int. Soc. Opt. Eng.* **718**, 80–91 (1986).

50. A. R. Nelson et al., Passive Multiplexing System of Digital Fiber-Optic Sensors. *Appl. Opt.* **20**(6), 915 (1981).

51. A. R. Nelson and D. H. McMahon, Passive Multiplexing System for Fiber-Optic Sensors. *Appl. Opt.* **19**(17), 2917–2920 (1980).

52. M. E. Curran and T. E. Clark, Fiber-Optic Applications for Launch Vehicles. *Proc. SPIE—Int. Soc. Opt. Eng.* **989** (1988).

53. E. Udd and J. P. Thériault, Fiber Optic Microbending Sensors Sensitive over Different Bands of Wavelengths of Light. U.S. Pat. 5,118,931 (1992).

54. M. B. J. Diemeer and E. S. Trommel, Fiber-Optic Microbend Sensors: Sensitivity as a Function of Distortion Wavelength. *Opt. Lett.* **9**(6), 260–262 (1984).

55. M. T. Wlodarczyk and S. R. Seshradi, Analysis of Microbending Losses on a Single-Mode, Step-Index Fiber at Higher-Order Harmonics of Deformation Spectrum. *Proc. SPIE—Int. Soc. Opt. Eng.* **566**, 387–394 (1985).

56. M. T. Wlodarczyk et al., Wavelength Referencing in Single-Mode Microbend Sensors. *Opt. Lett.* **12**(9), 741–743 (1987).

57. M. T. Wlodarczyk, Microbending Losses of Metal Coated Single-Mode, Multimode, and Cladding-free Fibers. *Proc. SPIE—Int. Soc. Opt. Eng.* **985**, 320–326 (1988).

58. T. A. Lindsay et al., Standard Fiber Optic Sensor Interface for Aerospace Applications: Time Domain Intensity Normalization. *Proc. SPIE—Int. Soc. Opt. Eng.* **989** (1988).

59. W. B. Spillman and J. R. Lord, Self-Referencing Multiplexing Techniques for Intensity Modulating Fiber Optic Sensors. *Proc. SPIE—Int. Soc. Opt. Eng.* **718**, 182–191 (1986).

60. N. Lagakos et al., Microbend Fiber-Optic Sensor. *Appl. Opt.* **26**(11), 2171–2180 (1987).

61. J. P. Dakin, Lead-Insensitive Optical Fiber Sensors. *Opt. Fiber Sensors Conf. Proc.,* pp. 323–327 (1988).

62. R. Morgan et al., Wavelength Dependence of Bending Loss in Monomode Optical Fibers: Effect of the Fiber Buffer Coating. *Opt. Lett.* **15**(17), 947–949 (1990).

63. G. F. Lipscomb et al., Stabilization of Single and Multimode Fiber-Optical Microbend Sensors. *Opt. Fiber Sens. Conf. #1 Proc.*, pp. 117–121 (1983).

64. A. J. A. Bruinsma, P. van Zuylen, C. W. Lamberts and A. J. T. de Krijger, Fiber-Optic Strain Measurement for Structural Integrity Monitoring. *Opt. Fiber Sens. Conf. #2 Proc.*, pp. 399–402 (1984).

65. J. S. Schoenwald et al., Evaluation of an OTDR Microbend Distributed Sensor. *Proc. SPIE—Int. Soc. Opt. Eng.* **986**, 150–157 (1988).

66. E. H. Urruti et al., Optical Fibers for Structural Sensing Applications. *Proc. SPIE—Int. Soc. Opt. Eng.* **986**, 158–163 (1989).

67. E. Udd, J. P. Thériault et al., Microbending Fiber Optic Sensors for Smart Structures. *Proc. SPIE—Int. Soc. Opt. Eng.* **1170**, 478–482 (1990).

68. H. Smith et al., Smart Structures Concept Study. *Proc. SPIE—Int. Soc. Opt. Eng.* **1170**, 224–229 (1990).

69. T. F. Morse, Overview of Fiber Optic Research at Brown University. *Proc. SPIE—Int. Soc. Opt. Eng.* **1169**, 30–41 (1990).

70. A. L. Harmer, Distributed Microbending Sensor. *Opt. Fiber Sens. Proc.*, pp. 126–128 (1985).

71. M. Vaziri and C-L. Chen, Strain Sensing with Etched Fiber Intensity Sensors. *Opt. Fiber Sens. Proc.*, pp. 281–284 (1991).

72. M. Vaziri and C.-L. Chen, Etched Fibers as Strain Gauges. *IEEE J. Lightwave Technol.* **LT-10**(6), 836–841 (1992).

73. L. Falco and O. Parriaux, Structural Metal Coatings for Distributed Fiber Sensors. *Opt. Fiber Sens. Conf. Proc.*, pp. 254–257 (1992).

74. J. D. Weiss, Fiber Optic Strain Gauge. *IEEE J. Lightwave Technol.* **LT-7**(9), 1308–1318 (1989).

75. C. K. Kao et al., Fiber Cable Technology. *IEEE J. Lightwave Technol.* **LT-2**(4), 479–487 (1984).

76. D. Gloge, Optical Fiber Packaging and Its Influence on Fiber Straightness and Loss. *Bell Syst. Tech. J.* **54**(2), 245–262 (1975).

77. J. W. Berthold et al., Design and Characterization of a High Temperature Fiber-Optic Pressure Transducer. *IEEE J. Lightwave Technol.* **LT-5**(7), 870–875 (1987).

78. K. A. Murphy et al., Extrinsic Fabry–Perot Optical Fiber Sensor. *Opt. Fiber Sens. Conf. Proc.*, pp. 193–196 (1992).

79. F. Romero-Borja et al., Improved Sensitivity in Sensor Fibers by the Proper Source/Coating Selection. *Appl. Opt.* **29**(31), 4729–4732 (1990).

80. J. C. Bocquet et al., Optical Sensors Embedded in Composite Materials. *Proc. SPIE—Int. Soc. Opt. Eng.* **1588**, 210–217 (1992).

81. C-L. Chen, Excitation of Higher Order Modes in Optical Fibers with Parabolic Index Profile. *Appl. Opt.* **27**(11), 2353–2356 (1988).

82. G. J. Cannell et al., Flexible Networks Employing Nonintrusive Single Mode Optical Fiber Taps. *Electron. Lett.* **24**(25), 1534–1536 (1988).

83. J. A. Aberson and I. A. White, Local Power Injection into Single-Mode Fibers using Periodic Microbends: A Low-Stress High-Efficiency Injector. *OFC/IOOC Proc.*, p. 107 (1987).

84. J. M. Anderson et al., Lightwave Splicing and Connector Technology. *AT&T Tech. J.* **66**(1), 45–64 (1987).

85. J. N. Blake et al., Fiber-Optic Modal Coupler Using Periodic Microbending. *Opt. Lett.* **11**(3), 177–179 (1986).

86. J. N. Blake et al., Analysis of Intermodal Coupling in a Two-Mode Fiber with Periodic Microbends. *Opt. Lett.* **12**(4), 281–283 (1987).

87. D. S. Czaplak et al., Microbend Fiber-Optic Phase Shifter. *IEEE J. Lightwave Technol.* **LT-4**(1), 50–53 (1986).

88. J. T. McHenry and S. F. Midkiff, Hybrid Sensing and Communication Fiber Optic Multicomputer Networks for Smart Structure Applications. *Smart Mater. Struct.* **1**, 146–155 (1992).

89. Fibertek Inc., *Fiber Opt. News* **12**(35), 7 (1992).

90. D. J. Bak, Optical Fiber Senses Pressure. *Des. News*, March (1986).

91. B. W. Prahl, Pressure/Tactile Sensing with Intrinsic Fiber-Optic Sensors. *Sensors*, August, pp. 48–52 (1986).

92. R. Wolf and H. J. Miesseler, Monitoring of Prestressed Concrete Structures with Optical Fiber Sensors. *1st European Conf. on Smart Structures and Materials*, pp. 23–29 (1992).

13

Fluorescence Optrode Sensors for Composite Processing Control and Smart Structure Applications

RAM L. LEVY and SCOTT D. SCHWAB
McDonnell Douglas Aerospace–East
Advanced Materials Technology
St. Louis, Missouri

13.1. INTRODUCTION

The utilization of polymeric matrix composites in aircraft structures provides extremely valuable and well-documented advantages. A central aspect of polymeric composites, however, is the fact that their mechanical properties and durability are strongly dependent on the starting material characteristics, on the specific set of conditions occurring during their fabrication process [1], and on the environment induced changes during use [2]. Therefore, it is of great importance to fabricate such composites under optimal conditions and also that the extent of changes in their properties, induced by the service environment, is monitored and taken into account in determining the system's performance.

It is now widely recognized that the potential of polymeric composites for aerospace structures can be greatly expanded by integration of embedded sensors in such structures, thus creating "smart" or "intelligent" materials which can monitor the forces acting on them or their "state of health" [3]. The implementation of this emerging concept provides an opportunity to address both the problem of optimization and control of the manufacturing process and the problem of monitoring the changes induced during service.

A variety of embedded fiber optic sensor concepts which monitor the physical state of the composite (i.e., temperature, stress loads, strain, acoustic emissions, etc.) have appeared in the literature [4]. However, embedded fiber optic sensors capable of monitoring the aging of the resin or the sorbed water content of the polymeric matrix have not been described.

Fiber Optic Smart Structures, Edited by Eric Udd.
ISBN 0-471-55448-0 © 1995 John Wiley & Sons, Inc.

In this chapter we describe a novel embedded fiber-optic sensor concept of dual role capability; monitoring critical curing/processing parameters and subsequently monitoring the changes induced by exposure to certain elements of the service environment. This sensor is called fluorescence optrode cure sensor (FOCS) when used to monitor curing and fluorescence optrode sensor (FOS) when used to monitor environment-induced changes in cured composites.

13.2. EXPERIMENTAL

13.2.1. Sensor Hardware

A relatively simple fiber optic fluorometer that utilizes bandpass filters and dichroic mirrors to separate the excitation and emission beams was constructed and used as the basis for the first-generation FOCS shown in Fig. 13.1. The FOCS consists of a light source (20-W tungsten-halogen lamp), an excitation filter that provides a broad (390 to 410 nm) spectral band, a condensing lens, an optical waveguide consisting of fiber with a core diameter of either 200- or 400-μm glass-clad, glass core, as well as three dichroic mirrors, an emission filter, and two sensitive photodetectors. The distal end of the fiber optical waveguide (optrode) is embedded in the composite as shown in Fig. 13.2. A second-generation FOCS system capable of monitoring the changes in both fluorescence intensity and the wavelength of maximum emission, which

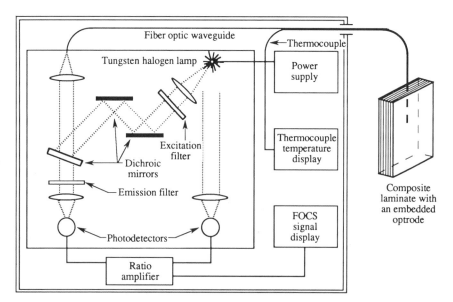

Figure 13.1. Functional schematic of optical and electronic components of the first-generation FOCS/FOS and a sensor waveguide embedded in a composite laminate.

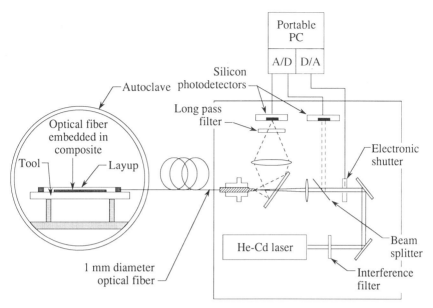

Figure 13.2. Laser-based FOS/FOCS for embedded sensor applications.

is described elsewhere [5], was used for monitoring the processing of PEEK laminates.

An improved version of this sensor, designed specifically for embedded sensor applications, is shown in Fig. 13.2. This version of the FOS/FOCS utilizes a He–Cd laser, emitting at 441 nm, as the source of excitation and permits operation with smaller-diameter fibers (100 to 250-μm). It also includes a shutter that minimizes the exposure of the resin matrix to the excitation light.

13.2.2. Embedding FO Waveguides and Curing of Laminates

During laminate layup, a 200-μm (or larger)-core-diameter optical fiber (Polymicro Technologies, Phoenix, Arizona) was embedded between two prepreg layers so that it was parallel with the carbon fibers of both layers as shown in Fig. 13.3. The remaining prepreg layers were laid up in a 0–90 arrangement. The embedded fiber was connected to the FOCS instrument console as shown in Fig. 13.1. A carbon–epoxy laminate (Hercules AS4/3501-6) containing an embedded waveguide was cured in a laboratory microautoclave which accurately reproduces the vacuum, pressure, temperature, and bleed control conditions existing in larger factor autoclaves. The laminate was cured under a standard cure cycle, which is given in Fig. 13.4. Carbon–PEEK (AS-4/APC02, ICI) thermoplastic composites (26 ply) with embedded 200-μm core fiber were fabricated at 360°C using a 50-ton press (Wabash Metal Product, Inc.).

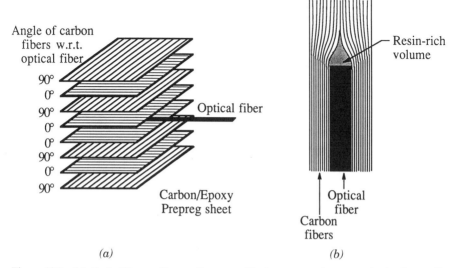

Figure 13.3. (a) Embedding a fiber optic waveguide in a cross-ply composite laminate; (b) configuration of the optrode sensor tip when embedded in a composite laminate.

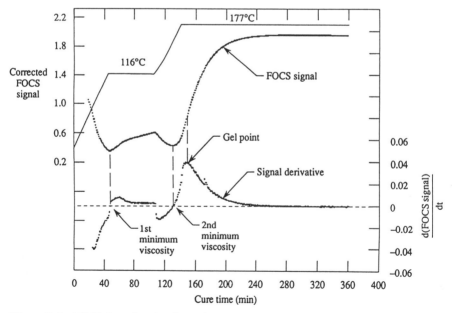

Figure 13.4. FOCS intensity signal monitoring the autoclave curing with a carbon–epoxy laminate (HERCULES AS4/3501-6).

13.2.3. Sorption/Desportion of Water

The completely cured laminate containing the embedded fiber optic waveguide was exposed to 100% relative humidity (RH) at 44°C for 1200 h, resulting in gradual sorption of water by the epoxy matrix. Subsequently, the laminate was exposed to 0% RH at 44°C for 1300 h, resulting in desorption of some of the sorbed water. The sensor was turned on 30 min prior to each measurement of the signal intensity.

13.2.4. Physical Aging

Concentrations ranging from 5 to 40 ppm of a compound that exhibits viscosity/free-volume-dependent fluorescence, 4-N,N-dimethylamino-4'-nitrostilbene (DMANS, Aldrich Chemical Co.) were dissolved in diglycidyl ether of bisphenol A (DGEBA) epoxy obtained from Dow Chemical Co. The DMANS containing epoxy was cured with a stoichiometric amount of a mixture of 1,6-diaminohexane and N,N-dimethyl-1,6-diaminohexane. These samples were cured at 50°C in silicone rubber molds according to the method described by Fanter [6]. To ensure that all chemical reaction had been completed, each sample was then postcured at 150°C for 3 h. The T_g value of these samples was found to be 47°C and the physical aging was conducted at 35°C (i.e., $T_g - 12$). Measurements of the changes in the fluorescence intensities were made with a SLM Model 4800 spectrofluorometer equipped with a temperature-controlled sample chamber.

13.3. RESULTS AND DISCUSSION

13.3.1. Principles of Operation

The FOCS is based on the combination of the viscosity/free-volume dependence of the resin fluorescence, fiber optic fluorometry, and a suitable tool-mounted optrode-laminate (OL) interface [5, 7–9]. The fluorescence quantum yields (Φ_f) of certain compounds exhibit a strong dependence on the viscosity or the free volume of the medium. Such viscosity-dependent fluorescence (VDF) is exhibited by molecules that can undergo nonradiative decay from the excited state by intramolecular twisting or torsional motions [10, 11]. In ordinary low-viscosity solvents these compounds exhibit low fluorescence. When such compounds are dispersed in a polymerizing medium [12–14] such as in the case of epoxy during curing [15], the motions of the probe molecules become progressively more inhibited by the increasing viscosity of the polymer, thereby leading to a proportional increase in fluorescence. Forster and Hoffman [11] have shown that the fluorescence quantum yield, Φ_f, of a viscosity-dependent fluorescent compound is linked to the viscosity of the medium

(η) according to the following expression:

$$\Phi_f = C\eta^{2/3}$$

where C is a constant for the specific VDF compound. This relationship applies only for a certain range of viscosities. However, similar expressions that contain different exponents can be empirically derived for the various ranges of viscosity or free volume.

Our experiments on the use of VDF probes for monitoring epoxy-cure kinetics [15, 16] led to the observation of VDF behavior in tetraglycidyl-diamondiphenylmethane (TGDDM), the main monomer constituent of the Ciba-Geigy MY720 resin used for fabrication of carbon–epoxy composites, as well as in the MY720 itself. Thus the changes in the resin viscosity and the degree of cure can be monitored without addition of a probe [15]. The VDF behavior of the MY720 resin, however, is attributed to an unidentified impurity which exhibits VDF behavior [8].

13.3.2. Cure Monitoring with an Embedded FOCS

Polymeric composites based on epoxy resins (thermosets) and carbon fibers are the most widely used type of composites in aerospace applications. The mechanical properties of such carbon–epoxy composites are, however, strongly dependent on the chemorheological events taking place during the autoclave cure stage [1, 19]. The autoclave curing of carbon–epoxy laminates is the most critical stage in the fabrication of composite structures [1, 17, 18, 19]. Extensive efforts are currently in progress to transform the autoclave curing process from a skilled craft to a science-based operation resulting in lower cost and higher-quality composites. Computer modeling, cure simulation, real-time control, and artificial intelligence expert systems are being developed toward this end. However, reliable cure sensors suitable for use in manufacturing are needed to fully implement the advances made in these projects. The need for suitable cure-monitoring sensors has been well recognized, and during the last decade, considerable effort has been made to develop such sensors [20]. Most of the approaches proposed and developed for cure monitoring are based either on the dielectrometric and conductometric [21] techniques or on the acoustic/ultrasonic techniques [22]. These approaches, however, have hitherto not gained acceptance as cure sensors for integration in composite manufacturing operations. Furthermore, these sensors cannot be embedded in the structure to monitor the curing in the center of a composite, which is particularly important when curing thick laminates. Therefore, both the tool-mounted and embedded FOCS developed in our laboratory represent a promising alternative to the existing techniques [5, 7–9].

The FOCS signal (from tool-mounted FOCS) as a function of cure time, recorded during curing of a Hercules AS4/3501-6 laminate, is shown in Fig. 13.4 along with the temperature–time profile of the cure cycle used. The observed changes in the FOCS signal follows the changes known to occur in the resin viscosity and distinctly indicate the two points of minimum viscosity as well as the rate of cure. Similar results were obtained with an embedded FOCS. The gel point is reached approximately 150 min into the cure cycle and manifests itself as a change of slope in the FOCS signal. After 200 min, the signal gradually levels corresponding to the greatly reduced rate of cure. Similar results were obtained with an embedded FOCS. These observations indicate that the FOCS can successfully monitor the chemorheological changes occurring in the laminate during the entire cure cycle.

13.3.3. Monitoring Sorbed Water

The service environment of most resin matrix composites includes intermittent or continuous exposure to atmospheric moisture. Under such conditions, most polymers, particularly epoxy resins, sorb water, which induces substantial changes in the polymer physical properties [2]. The main effect of sorbed water on polymeric matrices is plasticization, that is lowering of the glass transition temperature (T_g), thereby dramatically reducing the modulus of the polymer at elevated temperatures [23, 24]. In addition, sorption of water causes swelling, which can induce changes in residual stresses and possibly cause microcracks [25]. Furthermore, when carbon–epoxy composites containing significant quantities of sorbed water are suddenly heated to temperatures above 100°C, steam-induced delamination can occur.

The application of embedded fluorescence optrode sensor (FOS) for monitoring the sorbed water/organic solvents content in polymeric composites

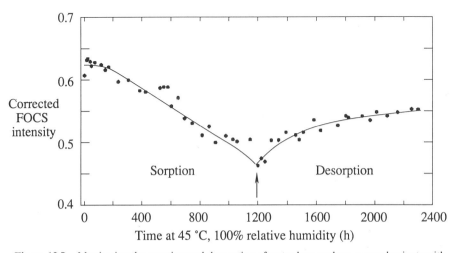

Figure 13.5. Monitoring the sorption and desorption of water by a carbon–epoxy laminate with an embedded FOCS.

was developed in response to this need. It is based on our earlier observation [16, 26] that the fluoresence of cured epoxy resins decreases in proportion to sorbed water and on combining this phenomenon with the ability of an embedded FOS to monitor the changes in the epoxy fluorescence.

The sensor signal intensity versus time of exposure to 100% RH at 44°C is shown in Fig. 13.5. During the 1200 h of exposure of the cured composite to 100% RH, the sensor signal monotonically decreases in response to the increase in sorbed water. During the subsequent 1300 h of exposure of the composite to 0% RH at 44°C (i.e., during desorption), the sensor signal increases to a level below the signal level observed from the initial dry specimen because (as expected) not all the sorbed water has been desorbed. Such a response clearly indicates the ability of the embedded FOCS to monitor the sorption and desorption of water by a carbon–epoxy composition. The multiple effect of sorbed water on the properties of epoxy resins and its potential impact on the composite's performance emphasizes the need for such sensors.

13.3.4. Physical Aging of Composite Matrices

The long-term properties of amorphous glasses such as epoxy resins change as a function of time at temperatures below their glass transition temperature (T_g) [27, 28]. This process is known as physical aging [27]. It is defined as the slow loss of excess free volume occurring as the material approaches equilibrium. The reduction of free volume diminishes the segmental mobility of the polymer chains, which leads to embrittlement of the material [28]. Thus the physical and mechanical properties of a polymeric matrix are time dependent and can change substantially from their initial values. Matrix properties strongly influence the overall properties of composite materials, and Kong [28] has shown that the ultimate tensile strength, strain to break, toughness and water sorption of carbon–epoxy composites decrease after physical aging. Physical aging, as opposed to chemical aging (e.g., oxidative degradation), is thermoreversible. If the polymer is heated above T_g and followed by a rapid quench to below T_g, the free volume lost during physical aging is restored [27, 28].

Many polymeric materials used in aerospace systems operate in a service-temperature envelope which includes at least part of their physical aging range. Therefore, the ability to monitor the aging process is important for understanding the long-term performance and stage of health of these polymers (i.e., the extent to which their initial properties have changed due to physical aging).

The potential of viscosity or free-volume dependent fluorescence (FVDF) probes to monitor the physical aging process of epoxy resins and its thermoreversibility was demonstrated by subjecting a cured epoxy containing a FVDF probe to physical aging at 35°C ($T_g - 12°C$) and monitoring the fluorescence intensity as function of time. The fluorescence intensity was found to increase in proportion to the decrease in the free volume induced by the

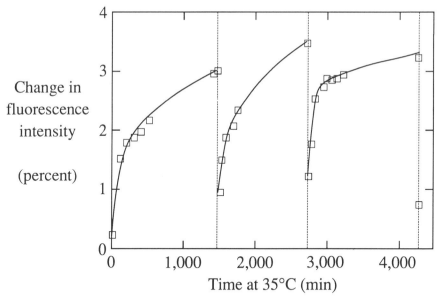

Change in
fluorescence
intensity

(percent)

Time at 35°C (min)

Figure 13.6. Monitoring three cycles of physical aging of a cured epoxy containing a fluorescence probe.

physical aging process as shown in Fig. 13.6 for the first 1400 min of aging. Subsequently, the epoxy specimen was heated briefly to a temperature above T_g and then cooled rapidly to the sub-T_g temperature of 35°C to restart the physical aging. As shown in Fig. 13.6, the fluorescence intensity decreases in response to the increase in the free volume. During the next 1300 min at 35°C the specimen again undergoes physical aging, which is monitored by a gradual increase in the fluorescence intensity. A third cycle of brief above-T_g heating followed sub-T_g physical aging is also shown in Fig. 13.6. Monitoring of the physical aging of a FVDF probe containing epoxy was achieved with a laboratory spectrofluorometer. However, we believe that similar results could also be obtained with an embedded FOS.

13.3.5. Monitoring the Processing of Thermoplastic

The processing of high-performance thermoplastics was also monitored with an embedded FOS. The fluorescence excitation maximum of poly(aryletheretherketone) (PEEK) occurs at about 390 nm and its emission maximum is at about 435 nm. Figure 13.7 shows a fluorescence intensity profile obtained with a 200 μm-core-diameter fiber embedded in the center of a 26-ply carbon–PEEK laminate. Clearly indicated in the profile are T_g along with the melting and recrystallization points, thus demonstrating the potential of the sensor to monitor the processing of carbon–PEEK composites.

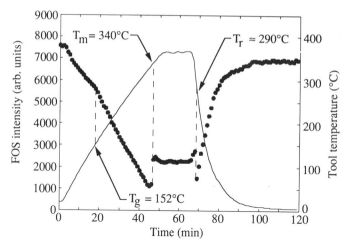

Figure 13.7. FOS intensity (•) and tool temperature (—) profiles during the processing of a carbon–PEEK laminate. Sensor was embedded in center of the 26-ply laminate. T_g, T_m, and T_r indicate the glass transition, melting, and recrystallization temperatures, respectively.

13.4. CONCLUSIONS

The combination of viscosity/free-volume-dependent fluorescence with fiber optic fluorometry produces a new family of sensors: fluorescence optrodes. When such sensors are embedded in the composite laminate they can fulfill a dual role: monitor the composite curing or processing conditions during fabrication, and subsequently monitor sorption of atmospheric moisture or organic solvents by the resin matrix during service. Preliminary experiments indicate that the fluorescence optrode sensor can in principle also monitor the physical aging of the epoxy matrix in a carbon–epoxy composite.

REFERENCES

1. A. C. Loos and G. S. Springer, *J. Compos. Mater.* **17**, 135 (1983).
2. C. E. Browning, Ph.D. Dissertation, University of Dayton, Dayton, OH (1977).
3. C. J. Mazur, G. P. Sendeckyi, and D. M. Stevens, *Proc. SPIE — Int. Soc. Opt. Eng.* **986**, 19 (1989).
4. E. Udd, *Laser Focus (Littleton, Mass.)* xx, 135 (1988).
5. R. L. Levy and S. D. Schwab, *SPE Proc.* **35**, 1530 (1989).
6. D. L. Fanter, *Rev. Sci. Instrum.* **49**, 1005 (1978).
7. R. L. Levy, *Polym. Mater. Sci. Eng.* **54**, 321 (1986).
8. R. L. Levy and S. D. Schwab, *ACS Symp. Ser.* **367**, 113 (1988).
9. S. D. Schwab and R. L. Levy, *Polym. Mater. Sci. Eng.* **59**, 591 (1988).

10. G. Oster and Y. J. Nishijima, *J. Am. Chem. Soc.* **78**, 1581 (1956).

11. T. Forster and G. Hoffman, *Z. Phys. Chem.* [N.S.] **75**, 63 (1971).

12. R. O. Loutfy, *Macromolecules* **14**, 270 (1981).

13. R. O. Loutfy, *J. Polym. Sci., Polym. Phys. Ed.* **20**, 825 (1982).

14. R. O. Loutfy, *Pure Appl. Chem.* **58**, 1239 (1986).

15. R. L. Levy and D. P. Ames, in *Adhesive Chemistry: Developments and Trends* (L. Lee, ed), p. 245. Plenum, New York, 1984.

16. R. L. Levy and D. P. Ames, *Polym. Mater. Sci. Eng.* **53**, 176 (1985).

17. S. D. Schwab and R. L. Levy, *Polym. Mater. Sci. Eng.* **59**, 596 (1988).

18. S. D. Schwab and R. L. Levy, *Adv. Chem. Ser., C.* (in press).

19. Y. A. Tajima, *Polym. Compos.* **3**, 168 (1982).

20. G. A. George, *Mater. Forum* **9**(4), 224 (1986).

21. S. Senturia, *Adv. Polym. Sci.* **80**, 227 (1978).

22. A. Lindrose, *Exp. Mech.* **18**, 227 (1978).

23. P. Moy and F. E. Karasz, *Polym. Eng. Sci.* **20**(4), 315 (1980).

24. C. Carafagna, A. Apicella, and L. Nicolais, *J. Appl. Polym. Sci.* **27**, 105 (1982).

25. M. Blikstad, P. W. Sjoblom, and T. R. Johannesson, *J. Compos. Mater.* **18**, 32 (1984).

26. R. L. Levy, *Polym. Mater. Sci. Eng.* **50**, 124 (1984).

27. L. C. E. Struik, *Physical Aging in Amorphous Polymers and Other Materials.* Elsevier, Amsterdam, 1978.

28. E. S.-W. Kong, *Adv. Polym. Sci.* **80**, 120 (1986).

14

Distributed Optical Fiber Sensors

JOHN P. DAKIN
Optoelectronics Research Centre
University of Southampton
Southampton, United Kingdom

14.1. INTRODUCTION

The highest state of the art in optical sensing is achieved with optical fiber distributed sensors. Such sensors permit the measurement of a desired parameter as a function of length along the fiber. This is clearly of particular advantage for applications such as "smart" skins, as a sensor can measure the variation of, for example, temperature over significant areas of the outer layer of vehicles.

There are three main criteria that must be satisfied to achieve a distributed sensor. First, it is necessary to construct (or select) a fiber that will modify the propagation of light in a way which can be relied upon to be dependent on the parameter to be measured. Second, one must be able to detect the changes in transmission (or light scattering) arising from the parameter to be measured. Third, it is necessary to locate the region of the fiber where the change in propagation occurs, to achieve the desired spatial distribution. We begin the chapter with a discussion of the basic methods that can be employed. These methods will then be expanded upon in later sections, and a number of examples of promising and practical sensors and sensor concepts will be described. The range of applications is very large, but potential applications in the area of smart skin and smart structure sensing will be given some emphasis, although a broader presentation of the technology will be given.

14.2. BASIC CONCEPTS OF DISTRIBUTED SENSORS

A distributed sensor consists of a continuous length of fiber, usually with no taps or branches along its length. It is therefore necessary to determine the location of any measurand-induced change in transmission or scattering properties by

Fiber Optic Smart Structures, Edited by Eric Udd.
ISBN 0-471-55448-0 © 1995 John Wiley & Sons, Inc.

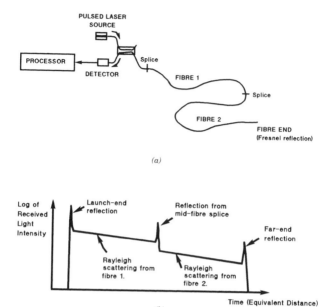

Figure 14.1. Concept of the basic optical time-domain reflectometer. (*a*) Basic optical arrangement of optical time domain reflectometer (OTDR). (*b*) Intensity versus time, OTDR return.

taking advantage of the propagation delays of light traveling in the fiber. Such propagation delays allow differences in the time of arrival of light, traveling in different modes of propagation, to be related to distances along the fiber.

The greatest differences in propagation delay occur when signals are traveling in opposite directions in the fiber. The simplest example of this is the optical time-domain reflectometer (OTDR), where a pulsed signal is transmitted into one end of the fiber, and returning backscattered signals are recovered from the same fiber end (Fig. 14.1). The concept is a guided-wave optical variant of the radar-location principle, where distance, z, is related to the two-way propagation delay, $2t$, by the simple formula

$$z = tV_g \tag{14.1}$$

where V_g is the group velocity of light in the fiber. A particular attraction of the backscatter methods for monitoring vehicle skins and structures is that it is only necessary to access one end of the fiber, thus simplifying the connection problem.

Rather than rely on weak backscattered light, it is possible to use counter propagating light and arrange for a nonlinear optical interaction to occur when the beams cross. Thus a continuous "probe" beam can be modulated by a "pump" pulse traveling in the opposite direction (Fig. 14.2). Changes in the probe intensity with time can again be related to changes with position along

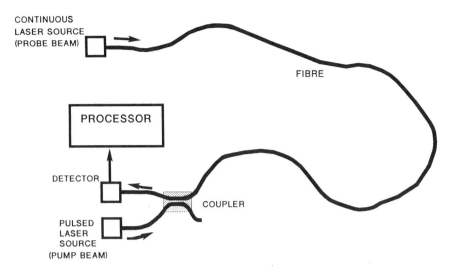

Figure 14.2. Schematic of distributed sensor concept with counterpropagating pump and probe beam.

the fiber. A third method of location involves signals that travel only in the forward direction.

To determine distance in this case, one must use a fiber that will support at least two forward-traveling modes of propagation. Light in these modes must travel at different velocities. The influence to be monitored must cause some conversion of light energy between these modes, such that for the remainder of the fiber length, the mode-converted light travels in a different mode to the remainder of the original light (Fig. 14.3). It is necessary to detect the small changes in propagation delay that arise from the intermodal difference in propagation velocity over the latter section. These changes are so small that it is generally necessary to use some form of interferometric approach to detect them. Having outlined the basic methods, each is now described in more detail and variants of each of the basic concepts are discussed.

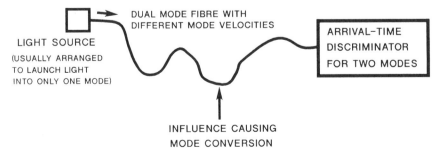

Figure 14.3. Basic concept of distributed sensing using forward-traveling waves only.

14.3. BACKSCATTERING METHODS, INCLUDING OTDR

14.3.1. General Concept

The OTDR, first reported in (1), has been established as a standard item of
fiber optic test gear for many years. Its main application is for fault finding and
attenuation monitoring in optical networks. As described above, the OTDR
relies on backscattering (or back reflection) of light that has been launched into
a fiber from an amplitude-modulated (usually, pulsed) source. The light is
Rayleigh scattered from refractive index fluctuations in the core of the fiber, or
may be reflected from discontinuities, such as connectors, splices, and fiber
breaks. The Rayleigh scattering component, which is detected returning from
the arrangement shown in Fig. 14.1, is given by (2)

$$P(z) = \tfrac{1}{2}S(z)\alpha_s(z)V_g \exp\left\{-\int_o^{z'}[\alpha_f(z') + \alpha_b(z')]dz'\right\} \qquad (14.2)$$

where $P(z)$ is the detected backscattered power as a function of the distance z
of the scattering point along the fiber, $S(z)$ the captured fraction of scattered
light coupled into backward-traveling modes in the fiber, $\alpha_s(z)$ the scattering
coefficient of the fiber, V_g the group velocity of light in the fiber, and α_f and α_b
the total attenuation coefficients of the fiber in the forward and backward
directions, respectively. (Usually, α_f and α_b will have the same value: i.e.,
$\alpha_f = \alpha_b = \alpha$). As stated in Section 14.1, the distance, z, is related to the two-way
time of flight, $2t$, by the relation

$$z = tV_g$$

When the attenuation coefficients and the capture factor are constant, the
expression for the detected backscattered power, P, has an exponential time
dependence of the form

$$P(t) = A_1 \exp(-B_1 t) \qquad (14.3)$$

where A_1 and B_1 are constants. This is, of course, the situation for a uniform
fiber.

The OTDR can sense changes in the total attenuation coefficient, α, if the
scattering coefficient, α_s, and the capture fraction, S, are constant. Under these
conditions

$$P(z) = A_2 \exp\left[-\int_o^z \alpha(z') \, dz'\right] \qquad (14.4)$$

where A_2 is a constant. The rate of change (differential with respect to time) of
the detected signal is proportional to the attenuation coefficient. Alternatively,

it can sense changes in scattering coefficient α_s if α and S are constant:

$$P(z) = A_3\alpha_s(z) \exp(-B_3z) \tag{14.5}$$

where A_3 and B_3 are constants. [Clearly, α will normally change with changes in α_s, but if the value of the integral $\int_o^z \alpha(z')\,dz'$ is small, the error will be small.]

Many parameters can cause (or be arranged to cause) variations in the attenuation or scattering coefficient of a fiber. These are discussed in the following section.

14.3.2. Distributed Sensors Based on Monitoring Attenuation Variations with the OTDR

14.3.2.1. Microbend Sensors for Mechanical Sensing. Any distortion of an optical fiber from its ideal cylindrical shape gives rise to an increase in attenuation. Gradual bends, having a radius above a few centimeters generally result in a much smaller loss than very small and sharp bends. Highly localized bends or kinks of this type, known as microbends, cause significant losses in both monomode and multimode fiber. Generally, in monomode fiber, the loss due to a given bend is fairly amenable to theoretical analysis (3). For this type

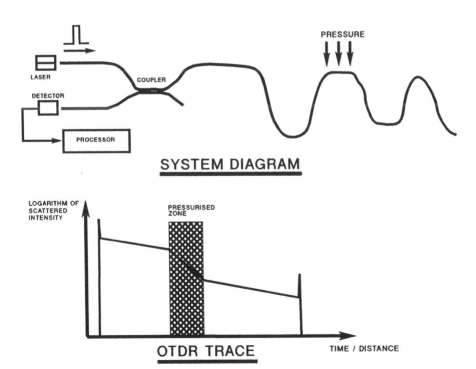

Figure 14.4. Concept of distributed sensing using a pressure-sensitive cable.

of fiber it is only necessary to determine the loss of energy from the fundamental mode. Thus monomode fibers are potentially capable of quantitative determination of the extent of bending of a fiber simply from a measurement of attenuation, provided that the shape of the bend deformation is known.

Multimode fibers, on the other hand, are less predictable in their behavior. A particular problem is that the attenuation due to a given bend is a complex function of the manner in which power is distributed between the multiplicity of waveguide modes at the entrance to the bend. This mode excitation is a function not just of the launching conditions, but also of the entire topology of the fiber (including any other microbend sensors) prior to the measurement region. Thus multimode microbend sensors will generally have some potential value for use as simple qualitative sensor, but are less likely to make practical quantitative types.

To construct a microbend sensing system it is necessary to arrange for the parameter to be measured to cause microbending of a cable that is monitored with an OTDR system (Fig. 14.4). A convenient form of sensing cable is one of a type originally designed by Harmer, which is now a commercial product ("Hergalite" pressure-sensitive cable, Herga Ltd., Bury St. Edmunds, UK; see also Fig. 14.5). This cable contains an inner communications fiber with a polymer fiber wound spirally round it. These fibers are then sheathed within a close-fitting outer tube. When compressed, the outer tube squashes against the spiral polymer fiber, which in turn deforms the inner optical fiber to give it a periodic, alternating, lateral displacement. Multimode fibers suffer particularly high microbending loss if the spatial period of the distortion matches the pitch of the zigzag path taken by the highest-order modes in the straight fiber.

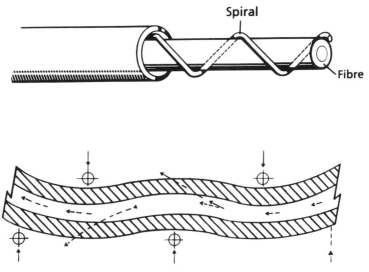

Figure 14.5. Principle of the microbend-sensing fiber cable (Hergalite).

Figure 14.6. Pressure-sensing mat for machine guarding. (Courtesy of Herga Electric Ltd., U.K.)

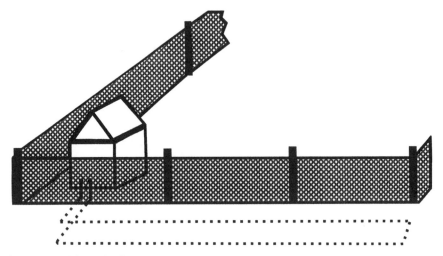

Figure 14.7.   Schematic of application of fiber optic intruder detector system buried in ground.

The spiral-wound fiber cable forms a distributed sensing element that is capable of many qualitative sensing tasks. For example, it can sense the pressure of a footstep and can hence form a security barrier for safety reasons (e.g., to switch off a dangerous machine or other manufacturing process if a person approaches: Fig. 14.6) or to protect against unauthorized intruders (Fig. 14.7). There are many sensing systems of this type, where it is sufficient for a simple transmission monitor merely to detect the total line-averaged effects of disturbance along a single section. However, the cable is capable of locating a disturbance along a continuous length if it is coupled to an OTDR system.

In addition to detection of foot pressure, the microbend sensor has been considered as a sensor for the monitoring of civil engineering structures (4). Clearly, if it can be made in a reliable form, it has considerable potential for use as a strain sensor for smart structures. If the strain in a structure can reliably be converted to a periodic ripple in a fiber, the structure can be monitored qualitatively. However, if multimode fiber is used in the sensor, the performance is likely to be rather variable, because the response will be dependent on the mode conversion occurring in earlier sections. In addition, the polymer elements that transfer the strain to the inner fiber can creep at room temperature and will flow more readily at high temperatures. Much work remains to be done, therefore, to develop versions suitable for high temperatures such as may arise in aerospace applications.

14.3.2.2. Radiation Sensing. The attenuation of an optical fiber increases when exposed to ionizing radiation. To cause attenuation, the radiation must be able to penetrate to the core region of the fiber and must have sufficient energy to cause structural change in the glass core. Three types of radiation

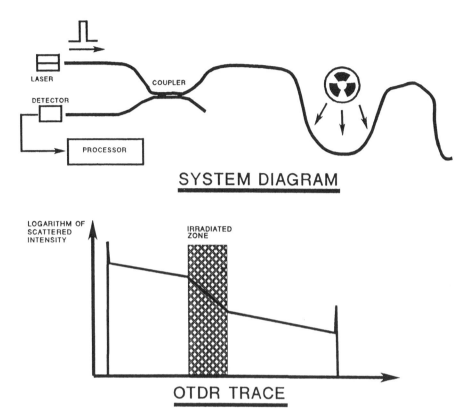

Figure 14.8. Concept of the distributed radiation dosimeter. [From Gaebler and Braunig (5).].

affect attenuation: neutrons, γ-rays, and x-rays. (The latter, also being photons, are essentially the same as γ-rays in nature, but they differ in the way in which they are generated and are usually of lower energy.) Gaebler and Braunig (5) were the first workers to recognize the potential of an optical fiber for distributed radiation sensing and to demonstrate the method experimentally using an OTDR. The basic concept of distributed radiation sensing is shown in Fig. 14.8. The irradiated section of fiber has a higher loss, and hence the gradient of the OTDR trace changes for positions corresponding to this region.

The distributed radiation sensor has particular attractions for detecting the presence of a single localized area of irradiation, as the line-integrated loss is reasonably low and the probe signal is not unduly attenuated. The sensor reported in reference 5 was probably less well suited for quantitative radiation dosimetry, as the sensitivity of normal germania-doped silica fiber is poor, and the response can exhibit a strong dose-rate dependence. The measurement can suffer significant error due to room-temperature annealing, which will slowly reduce the induced loss.

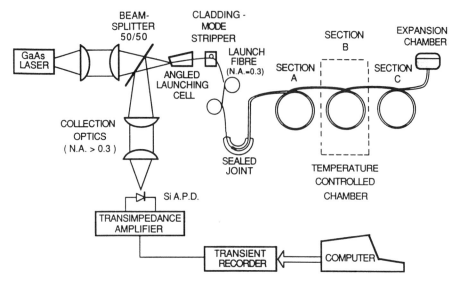

Figure 14.9. Rayleigh scattering temperature profiler using liquid-filled fiber. [From Hartog and Payne (6).]

14.3.2.3. Temperature Sensing. The first distributed temperature sensor, by Hartog and Payne, was based on a liquid-filled fiber (6) (Fig. 14.9). This fiber is simply a silica glass tube filled with a higher-refractive-index low-absorption liquid, which acts as the light-guiding core of the waveguide. The scattering loss coefficient, α_s, of the liquid depends on the density fluctuations caused by thermodynamic molecular motion, and therefore shows a strong temperature dependence.

The thermal variations are not significant in glass fibers, as the scattering is caused by "frozen-in" density fluctuations, formed as the glass was cooled from the melt. The scattering variations in the liquid-filled fiber are directly observable from an OTDR trace (Fig. 14.10), due to the dependence of the return signal on the scattering loss coefficient, α_s, in Section 14.3.1. This sensor successfully demonstrated the ability to measure temperature distribution for the first time, but the use of a liquid-filled fiber renders it impractical for a wide variety of applications, particularly for more severe aerospace environments. We shall therefore consider a means of using a solid glass fiber with an OTDR system. (Other methods using glass fiber are discussed in later sections.)

Optical fibers doped with rare-earth ions such as Nd^{3+} and Ho^{3+} show strong absorption peaks in the visible and near-infrared regions of the spectrum. Many of these absorbing transitions occur between electronic levels having a temperature-dependent occupancy, and therefore several absorption peaks have a significant thermal dependence (i.e., the coefficient α is temperature dependent). The OTDR can probe the absorption as a function of length, and hence determine the variation of temperature along the fiber (7). As

BACKSCATTER SIGNAL

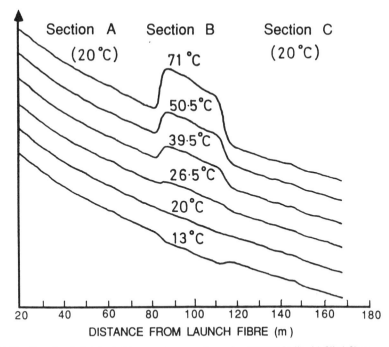

Figure 14.10. Results of distributed temperature sensing using OTDR in liquid-filled fiber.

already mentioned, a disadvantage of using variations in an absorbing fiber to sense a parameter is that both the probe and return backscatter signals are rapidly attenuated by this same absorption. The method is nonetheless attractive for systems requiring only a few measurement elements along the length of the fiber. It is also an excellent method when a fiber is used which has a low attenuation in normal use, yet which increases if sections become either too hot or too cold. This is clearly useful for methods requiring the detection and location of hot spots or cold spots, particularly if these are likely to occur only over a short length of the fiber.

The absorption loss along the length of a holmium-doped fiber, as measured at 665 nm, is shown in Fig. 14.11 (8). This fiber was cooled to $-196°C$ in liquid nitrogen for much of its length, but the central region was allowed to increase in temperature in stages, eventually being heated to $+40$ and $+90°C$. The attenuation profile shows the potential value of such a sensor for detecting loss of cryogenic coolant. It could, for example, also be used to detect the effects of solar heating on an otherwise-cold spacecraft surface. Many other combinations of rare earth dopants, host fiber, and interrogation wavelengths are possible, and there may therefore prove to be suitable candidates for sensors for the detection of a wide range of over- and underheating and cooling conditions.

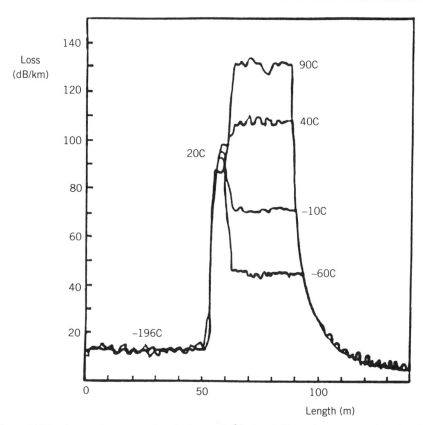

Figure 14.11. Attenuation versus length in an Ho^{3+}-doped fiber when a central section is exposed to varying temperature (outer sections are held at $-196°C$). Loss was measured with an OTDR at 665 nm. [From Cowle et al. (8).] Reprinted with permission.

In addition to fibers using special dopants, there is one type of commonly available fiber that has a significant temperature coefficient of attenuation. This is the polymer-clad-silica (PCS) fiber using a silicone cladding. This fiber has a low attenuation around room temperature and above but suffers severe attenuation when cooled to below $-50°C$ or so. This effect has been exploited as a practical sensor for detecting leakage in cryogenic liquids that has been launched as a commercial product (9). The attenuation occurs due to increase in the refractive index of the cooled cladding, which leads to eventual loss of guidance when it approaches the index of the silica core.

There are a large number of potential applications of distributed temperature sensors in smart skins and structures. One is the monitoring of surface heating due to air friction, solar radiation, or fire. Skin temperature is an important parameter in aircraft, rockets, and spacecraft. Of course, detection of cold spots may also have application in space vehicles, for example, to detect loss of vehicular heat due to breakdown of insulation or, as mentioned above, to detect leakage of cryogenic fluids, volatile rocket fuel, and so on. Fiber temperature sensors may also be used to monitor the temperature of thermally excited actuators such as shape-memory metal actuators, which change their length when electrically heated.

14.3.2.4. Chemical Sensing Using Absorbing Coatings. Distributed chemical sensors can be constructed by coating an optical fiber with indicator chemicals. As mentioned, a pure silica fiber can be coated with polymers of lower refractive index, such as silicone resins. These resins can be impregnated with the indicator before coating the fiber and before polymerization. The guided modes in such a fiber have a small, yet significant evanescent field penetration into the lower index polymer cladding, and hence the loss is dependent on the absorption of the dye. A schematic of the arrangement for distributed sensing is shown in Fig. 14.12. The chemical to be sensed can diffuse into the cladding, modify the absorption of the dye, and hence change the attenuation of the fiber. The first experimental results for such a sensor were obtained by performing a simple transmission measurement on a single section of fiber. Results for ammonia detection, using the change in color of a pH indicator, have been reported (10). It is quite possible that by the time this review goes to print, there may be reports of distributed chemical sensors capable of resolving the location of the interactions.

The evanescent-field chemical sensor has a number of potential problems that need to be solved before reliable operation can be achieved. The main problems arise from the variations in evanescent field coupling due either to

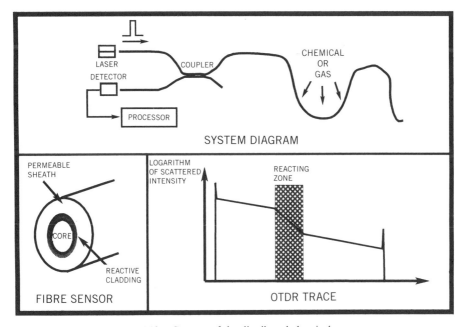

Figure 14.12. Concept of the distributed chemical sensor.

changes in the temperature of, or adsorption of chemicals (including water) in, the cladding material. In addition, any small bends or fluctuations in the diameter of the fiber will modify the evanescent field coupling. Finally, there is a need for care to ensure a sufficiently rapid diffusion time to ensure a fast enough response, and to choose materials that will display a reasonably reversible behavior, yet avoid any problems of the indicator becoming leached out.

Even if sophisticated distributed chemical sensors cannot be perfected, they may find application in a much simpler form in smart structures. It has already been shown possible to detect oil leaks using the increase in cladding refractive index when it becomes contaminated with the oil (11). This causes severe loss in the fiber and could therefore form a means of detecting leaks of fuel, lubricant, or hydraulic oil. For "smart skins" applications, this will be useful to detect leaks onto structures from containment vessels or transport piping.

14.3.3. Polarization Sensing
(Polarization Optical Time-Domain Reflectometry)

The polarization optical time domain reflectometry (POTDR) method (12) Fig. 14.13 is similar to OTDR, except that a pulse of polarized light is launched and the detector is arranged to be polarization sensitive by placing a polarizer before it. The method relies on the fact that Rayleigh and Rayleigh–Gans scattering in silica glass is polarized in the same direction as the incident light. Thus any changes in the polarization of the detected light result from changes during propagation over the two-way path to and from the scattering point. Changes in polarization of light from different points along the fiber result in differences in the detected signals at the appropriate delay times.

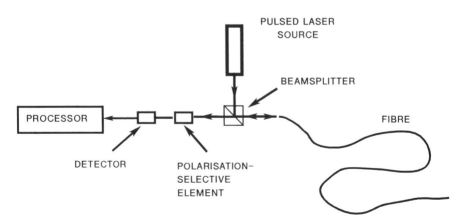

Figure 14.13. Basic optical arrangement of polarization optical time-domain reflectometer (POTDR). [From Rogers (12).]

The POTDR method was first suggested by Rogers (12), who pointed out its potential for distributed measurements of magnetic field (via Faraday rotation), electric field (via the Kerr quadratic electrooptic effect), lateral pressure (via the elastooptic effect), and temperature (via the temperature dependence, of the elastooptic effect). The first experimental measurements were reported by Hartog et al. (13), who used the technique for distributed measurement of the intrinsic birefringence of a monomode fiber. Kim and Choi (14) measured the birefringence induced by the bending of a wound fiber. Ross (15) carried out the first measurement of a variable external field using the Faraday rotation of polarization, caused by the magnetic field environment of the fiber. A comprehensive theoretical treatment of the POTDR method has been presented in the specialist paper of Rogers (16).

The POTDR technique appears to be attractive for the measurement of a large number of parameters. However, as with many potentially useful sensing methods, its main drawback is the variety of parameters to which it can respond. Spurious sensitivity to strain and vibration are particularly troublesome. In addition, POTDR requires the use of monomode fibers, which can, when used with narrow-linewidth laser sources, give rise to particular problems from coherent addition of light returning from multiple Rayleigh backscattering centers (17). Significant recent research is being directed toward solving the vibration problems, as the current monitor has strong commercial potential. More recent POTDR work has been carried out on birefringent fiber. The beat-length variations of such fiber are temperature dependent. Temperature distribution can be determined by monitoring the frequency of fluctuations of the detected signal when light is launched into both modes and the returning signal is passed through a polarizer (18). To reduce the frequency of the fluctuations to a convenient range, a relatively low birefringent fiber is necessary. This method is still in the early research stage, however.

14.3.4. Enhancement of the OTDR Using Active Fiber Components

We shall now consider how the performance of the basic OTDR can be enhanced using a variety of recently developed active fiber components.

14.3.4.1. Enhancement of the OTDR Using a Fiber Amplifier. An optical fiber amplifier can be inserted within the optical circuit of the OTDR, as shown in Fig. 14.14. The position of the amplifier is such that both the outgoing pulse from the laser source, and the weak backscattered signal from the fiber, are amplified in the same device. This is capable of having a dramatic effect on the performance of the system. At the time of writing, such systems have not yet found significant use in sensing systems. However, their potential is considerable (8), justifying their inclusion as an important part of this review.

The outgoing laser pulse power could, in principle, be amplified at least 1000 times in a typical erbium fiber amplifier. (This has typically $>30\,dB$ gain at 1500 nm, and recent laboratory results have shown up to 50 dB). In a

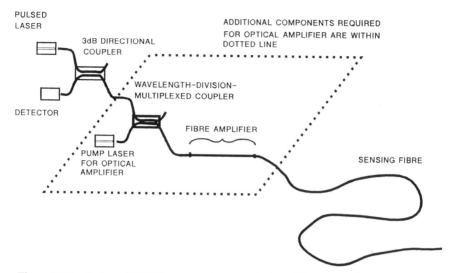

Figure 14.14. Enhanced OTDR system using an optical amplifier. [From Cowle et al. (8).] Reprinted with permission.

low-duty-cycle pulsed mode of operation, the small-signal gain can be approached. For example, a semiconductor laser source of 20 mW power can be boosted to launch 10 W into the fiber, even after allowing for a 3-dB loss in the 3-dB coupler. For short laser pulses of higher input power, much higher output powers of several hundred watts are possible from the amplifier. The peak output power for low-duty-cycle short pulses can be very high as a result of the high energy storage possible in the device. In practice, the maximum power output is more likely to be limited by other considerations, such as the onset of stimulated Raman scattering (particularly if monomode fibers are measured), and additional constraints such as limitation of eye hazard.

The effect on the returning backscattered signal is also highly significant. First, the amplifier provides gain before the lossy 3-dB directional coupler, and hence can provide an immediate 3-dB optical power improvement in the detectable signal level. However, for high-bandwidth signals, the performance of the detector, in combination with an optical fiber preamplifier, generally has a detection limit far superior to that with the detector alone. This advantage becomes particularly marked if the detector is equipped with a narrowband filter, included to remove any amplified spontaneous emission (ASE) that falls outside the passband necessary to amplify the fast-changing optical signal (19). The ASE suppressing filter must have a broad enough bandwidth to include any spurious frequency fluctuations in the source laser (including the modulation sidebands corresponding to the desired spatial resolution to be covered).

The advantages of the optical preamplifier are most marked when the required detection bandwidth is high. In high-bandwidth receivers, the thermal noise of conventional detection systems tends to rise rapidly, particularly above 200 MHz bandwidth. In addition, for the optical preamplifier system, the design of an appropriate optical filter to remove out-of-band ASE becomes progressively easier. An example of the detection limit improvement that can be achieved with an erbium fiber amplifier for a received signal bandwidth of 1 GHz is given in the paper by Laming et al. (20). At 1 GHz, an improvement of 7 dB in receiver sensitivity is possible. In an OTDR system, when the 3-dB gain due to amplification before the 3-dB coupler is included, a total potential receiver sensitivity improvement of 10 dB is anticipated. At frequencies above 1 GHz, as might be necessary to achieve ultrahigh-resolution measurements (a few centimeters), the potential improvement factor is much greater.

If one multiplies all the potential improvement factors available with an amplifier of moderate (30 dB) gain, there is a possible improvement factor of up to 10,000 times (40 dB), permitting a fiber sensor to operate with an additional 20-dB one-way loss. However, care must be taken to avoid problems of spontaneous oscillation due to optical feedback (caused, for example, by reflections from the source laser and detector, or from coherent Rayleigh backscatter in the sensing fiber). The maximum possible improvement may therefore be less than this in practice. In addition, care will have to be taken to ensure that the gain of the amplifier does not change significantly during the period when the backscatter signal is returning (or if it does, to compensate for gain changes that occur). Provided that the launched pulse does not seriously deplete the excited-state population in the amplifier (by virtue of its short duration), there is not likely to be a serious population change problem with the much weaker backscattered signals, as even the time-averaged signal power will be much less than that of the launched laser signal.

14.3.4.2. Use of High-Power Q-Switched Fiber Laser Sources. High-peak-power, short-duration pulsed sources are required in a number of distributed sensors, to provide good spatial resolution and signal-to-noise ratios. Q-switched neodymium-doped fiber laser are attractive sources for this application, by virtue of their ability to produce short-duration pulses using relatively low power laser diode pump sources. Neodymium-doped silica fibers can be fabricated to give 0.3 to -0.5 dB gain (at 1.06 μm for each milliwatt of launched pump power (at 810 nm). Such pump light can easily be obtained from AlGaAs laser diodes. The use of high-concentration (> 1000 ppm) Nd^{3+}-doped fiber permits short fiber lengths to be used, thus minimizing the photon lifetime in the cavity. This is an important requirement for the generation of short-duration Q-switched pulses.

A typical cavity configuration for a 1.06-μm Nd^{3+}-doped fiber laser would include a mirror with a high reflectivity at 1.06 μm, a doped fiber (length < 10 cm), an intracavity lens, an acoustooptic or electrooptic modulator, and a partially transmitting output coupling mirror. Pump light at 810 nm from a laser diode is launched through the input-end mirror, which has a high reflectivity at 1.06 μm but good transmission at 810 nm. Figure 14.15 shows a typical Q-switched Nd fiber laser output pulse with a peak power of about 600 W and a pulse duration of 5 ns. Such pulse characteristics would enable a

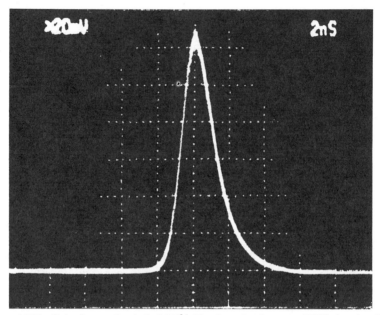

Figure 14.15. Pulse from a Q-switched Nd^{3+} fiber laser (peak power 600 W, duration 5 ns).

spatial resolution of 0.5 m to be obtained in a distributed sensor system. For these reported results, a 100-mW laser diode (SDL5411) was used as the pump source. For Nd^{3+}-doped fibers, the pulse characteristics remain unchanged up to a repetition rate of 400 Hz, above which the peak power lessons and the pulse duration increases. It should be emphasized that Q-switched fiber laser sources are more than just a research component, having already been incorporated into a commercial distributed temperature sensor (DTS commercial distributed temperature sensor, York Sensors, Chandlers Ford, UK).

14.3.5. Distributed Temperature Sensing Using Raman Scattering, Brillouin Scattering, and Fluorescence

All the backscattering methods described so far have relied on elastic scattering processes. In these processes the scattered light is at the same wavelength as the incident light. Rayleigh scattering and Fresnel reflections are both examples of elastic scattering. There are several commonly occurring physical processes that cause a wavelength change in the scattered light. In quantum theory, this implies either a loss or gain in energy by an incident photon, when creating a scattered photon. (It is this change of energy that explains the use of the term *inelastic*, by analogy with energy loss in inelastic mechanical systems.) A spectral plot of several inelastic processes is shown in Fig. 14.16. Use of these for distributed sensors is described below.

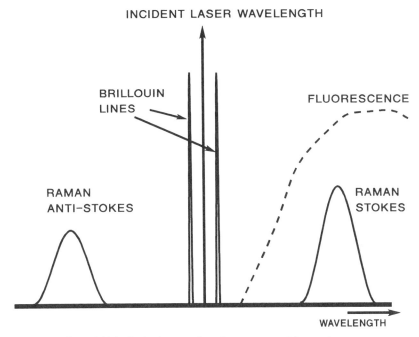

Figure 14.16. Inelastic scattering phenomena used for sensing.

14.3.5.1. Distributed Temperature Sensing Using Spontaneous Raman Scattering. If the spectral variation of backscattering from a germania-doped silica fiber (21) is examined (Fig. 14.17), it may be seen that there is a strong central line, due primarily to elastic Rayleigh (or Rayleigh–Gans) scattering. This central line also contains lines due to Brillouin scattering, which are much closer in frequency to the incident frequency, and hence more difficult to resolve. At each side of the central line, however, there are sidelobes due to Raman scattering, which are shifted in frequency to a much greater degree. These Raman lines may be used to detect temperature profiles in conventional (vitreous) communications fiber, using an OTDR system with additional filters to select out the Raman light (22). From standard texts on Raman scattering, the ratio, R, of anti-Stokes (higher-frequency-band) intensity to Stokes (lower-frequency-band) intensity at wavelengths λ_{a-s} and λ_s, respectively (separated at equal frequency shift, Δv, from the central line) is given by the relationship

$$R(T) = \left(\frac{\lambda_s}{\lambda_{a-s}}\right)^4 \exp\left(-\frac{hc\,\Delta v}{KT}\right) \qquad (14.6)$$

where h is Planck's constant, c is the velocity of light in vacuo, k is Boltzmann's constant, and T is the absolute temperature.

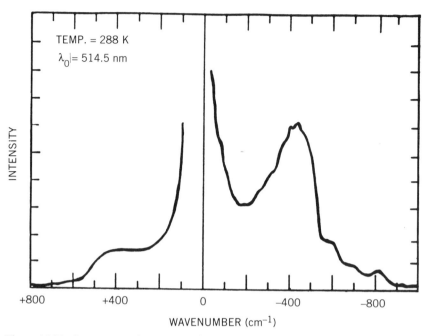

Figure 14.17. Raman scattering spectrum from a conventional germania-doped silica fiber.

Thus a measurement of the ratio of Stokes and anti-Stokes backscattered light in a fiber should provide an absolute indication of the temperature of the medium, irrespective of the light intensity, the launch conditions, the fiber geometry, and even the material composition of the fiber. In practice, however, a small correction has to be made for variations in fiber attenuation between the different Stokes and anti-Stokes wavelengths. The Raman technique appears to have only one significant practical drawback; that of a relatively weak return signal. Usually, the anti-Stokes Raman-scattered signal is between 20 and 30 dB weaker than the Rayleigh signal. To avoid an excessive signal averaging time, measurements have been taken using pulsed lasers which are capable of providing a relatively high launched power. In the first experimental demonstration of the method (22)], a pulsed argon-ion laser was used in conjunction with Corning telecommunication-grade 50- to 125-μm GRIN fiber.

The experimental arrangement is similar to that of the conventional OTDR except that a wavelength-selective directional coupler is used, which allows the launch laser to be coupled into the fiber but directs backscattered light at Raman wavelengths onto different detectors to receive Stokes and anti-Stokes bands, respectively.

More recent experimental results (23, 24) and first-generation commercially developed systems (York Sensors DTSII) have used a pulsed semiconductor

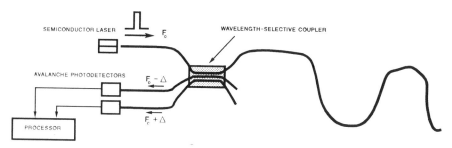

Figure 14.18. System diagram for Raman distributed temperature sensor, using semiconductor source and detector.

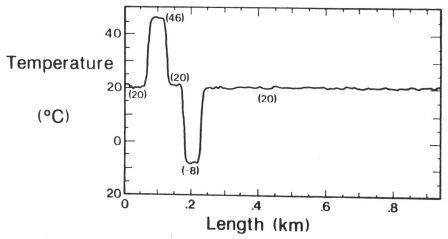

Figure 14.19. Typical temperature versus distance display for Raman DTS system.

laser and avalanche photodiode detectors, greatly reducing the size and power requirements (Fig. 14.18). Such first-generation systems have been capable of measurement of up to several hundred separate resolution elements over a few kilometers of fiber. A measurement accuracy of better than $\pm 1°C$ is possible with a resolution of a few meters. A typical form of temperature versus distance plot from a Raman DTS is shown in Fig. 14.19. A later-generation system (York Sensors DTS 80) has an option of using Q-switched fiber laser sources and the range is extended to over 10 km, with a resolution of about 1 m.

The Raman DTS is now well-established as a practical sensor and has a low cross-sensitivity to other parameters such as strain, pressure, variations in type of fiber and cable, etc. It has applications for monitoring electrical machines, cables and transformers, location of fires, and for sensing industrial plant. It is clearly attractive as a temperature sensor for smart skins (see applications discussed earlier in Section 14.3.2.3), being capable of measuring temperature at many thousands of points with a single instrument and fiber cable.

14.3.5.2. Sensing Using Brillouin Scattering. In addition to Rayleigh scattering and Raman scattering, glasses also exhibit Brillouin scattering. In a classical model, the latter form of scattering can be considered to be diffraction of light from the refractive index variations arising from acoustic waves. The light can be considered to be Doppler shifted as a result of the movement of the acoustic wave. The Doppler shift is characteristic of the acoustic velocity in the glass, which is a function of both temperature and pressure. In the more accurate quantum model, the light is more correctly considered to be scattered as a result of particle interactions, the incident photons being scattered from acoustic phonons, with an appropriate energy change in the scattered photon. This energy change in the photon represents a frequency change equivalent to the Doppler shift expected from the less-correct classical model. The frequency shift with Brillouin scattering is very small ($\approx 12\,\text{GHz}$) so unlike Raman scattering, it is difficult to select out Brillouin lines with conventional optical filters. However, the intensity of Brillouin scattering is at least an order of magnitude higher than that of Raman signals, so it is an attractive possibility for sensing. There have been various research attempts to detect these lines, the majority by mixing (or heterodyning) optical signals scattered from the fiber with light that is frequency shifted from the original laser source (i.e., the source used to excite the Brillouin scattering (25); Fig. 14.20).

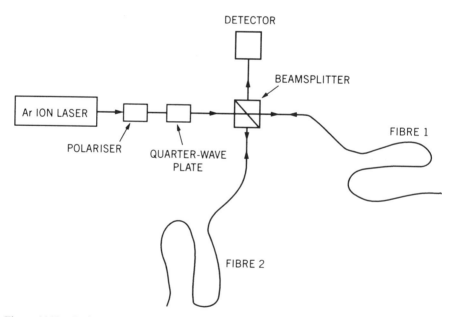

Figure 14.20. Basic optical arrangement for mixing of stimulated Brillouin scattering signals to create a heterodyne or difference frequency. [From Culverhouse et al. (25).]

The frequency-shifted light, required as a reference for the mixer, could be derived conveniently using a Brillouin fiber laser with the original laser as a pump source. (A Brillouin fiber laser is simply a fiber with reflective end mirrors, pumped optically at a level sufficient to cause lasing action via the phenomena of stimulated Brillouin scattering.) This laser light is shifted, essentially to the same extent as the spontaneous Brillouin scatttering in the measurement fiber. As a slightly different frequency is required for heterodyning, it should be tuned to a slightly different frequency, for example, by forming the Brillouin laser in a fiber composed of a different material to the measurement fiber.

The potential attraction of Brillouin scattering as a means of sensing is that the response to the measurand, such as a change in temperature or strain in the fiber, is manifest as a change in frequency of the scattered light. (These parameters change the acoustic velocity and hence the Brillouin "Doppler" shift.) By optical heterodyning, the frequency shifts can be down-converted to a more convenient electronic frequency and are then in a particularly convenient form for accurate logging. Unfortunately, the Brillouin method has yet to be developed into a practical sensor. However, recent research is starting to show very promising results for temperature and strain sensing in optical fibers (26). Apart from the relatively recent conception of the method, a further significant factor is the additional complexity of the optical system compared to either the conventional OTDR or its Raman version. In addition, the limited distance resolution of the Brillouin system presents a significant drawback. The Brillouin linewidth effectively restricts the distance interval to several tens of meters and will remain a problem unless satisfactory solutions are found.

14.3.5.3. Time-Domain Fluorescence Monitoring. The reemission spectrum of many fluorescent materials exhibits a significant temperature variation. It has been proposed that to construct a sensor, an optical arrangement similar to that used for Raman OTDR could be constructed, using a laser source as before, but with the detector filters now selected to examine regions of the fluorescent decay spectrum of the fiber (27) (Fig. 14.21). To produce a distributed temperature sensor, wavelengths that exhibit the maximum possible temperature variation should be used. The potential attraction of the method, first proposed by Dakin (27), is that the fluorescent quantum efficiency may be many orders of magnitude higher than that for Raman scattering, and higher doping levels can be used to greatly enhance the signals in short distributed sensor systems. However, there remains a problem with the availability of suitable fibers.

Silica-based optical fibers, having appropriate doping for high fluorescent efficiency, have been prepared using rare-earth doping (28). Unfortunately, these are likely to have limited distance resolution in a fluorescent OTDR system, due to the long fluorescent lifetimes. Reduction in the lifetimes of the

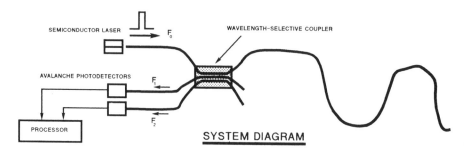

Figure 14.21.   System diagram for distributed sensing using fluorescent OTDR.

excited states of fluorescent dopants in glasses may generally be achieved by increased coupling to nonradiative processes, but this also reduces the fluorescent efficiency. Polymer fibers are believed by the present author (29) to offer more promise in short-distance systems, as these may be doped with organic dye materials having an excellent combination of high quantum efficiency ($\approx 50\%$ or better) and short fluorescent lifetimes (on the order of only a few nanoseconds). Of course, polymer fibers will have only a limited temperature range, a feature that could be a major problem for many applications.

A theoretical paper (30) has made a comparison of the performance expected from distributed temperature sensors based on the various techniques of temperature-dependent absorption, scattering, Raman scattering, and fluorescence. Although fluorescent doping necessarily increases the loss in optical fibers, it was found that this would, for short-distance operation, be more than compensated by the much higher light levels expected with fluorescence. If suitable glass fibers can eventually be produced, it is likely to be an attractive future method for distributed thermometry.

14.3.6.   Backscattering Systems Using Other Modulation Methods

The main methods discussed above have used a pulsed source, an approach used in first-generation electronic radar systems (31). This approach has the disadvantage that a low-duty-cycle signal is transmitted, leading to limitations in the mean power level. We now discuss three means by which the duty cycle of the transmitted signal can be increased. Each of these methods has been applied in sophisticated modern radar systems before finding use in experimental optical sensors.

14.3.6.1.   Methods Using a Pseudo-random Encoded Source Modulation.   The first method is that of encoded amplitude modulation of the source, using an orthogonal binary code sequence (32). There are numerous binary codes that can be used for this modulation. The basic requirement is that the code have a poor correlation with itself (i.e., an autocorrelation function close to zero) for

all conditions, except for the one where precisely time-synchronized sequences are compared. Under the latter condition, the autocorrelation function has a maximum value. The effect of using such a code is to increase the duty cycle of the source to 50%, compared to typical values of 0.1% or less for the pulsed system.

To decode the backscatter signal with such a code, it is necessary to correlate the returning signal with a sample of the code sequence transmitted. The signature can be recovered by sweeping the relative delay between the reference code and the detected signal. Alternatively, a parallel processor system can obtain correlation signals in a series of multiplier cells, each cell corresponding to a measurement range determined by delay of the reference signal applied to it.

Provided that the averaging systems are of similar type, the coded system gives a signal-to-noise ratio improvement on the order of $\sqrt{N/2}$, where N is the number of discrete, resolvable range cells. With finite-length codes and imperfect waveforms, care must be taken not to introduce additional artifacts in the correlation signal which fail to correspond to real features in the backscattered signal. It has been found that the use of complementary code sequences can lead to significant improvements in signal distortion in encoded OTDR systems (33). It should be emphasized that the encoded signal approach is applicable to both conventional OTDR (32) and Raman OTDR (34).

14.3.6.2. Optical Frequency-Domain Reflectometry. A second method of increasing the duty cycle is to "chirp" the optical source. This involves applying a periodic frequency modulation to the source, which should be a device such as a laser that has a reasonably narrow instantaneous bandwidth. The preferred frequency-modulating waveform has a sawtooth variation with time, with a fast flyback. This method is essentially similar to the frequency-modulated carrier wave (FMCW) technique used in radar systems. If an optical frequency-domain reflectometry (OFDR) system is operated in backscattering mode, in a continuous monomode fiber, the beat signal produced at the detector increases in frequency, in direct proportion to the distance from which the light is retroscattered (35, 36). If the beat signal detected is examined with a conventional electronic spectrum analyzer, the power in each frequency interval represents the level of scattered light received from a short section of fiber, situated at a distance corresponding to the frequency offset observed. The minimum theoretical range resolution, ΔR, possible, assuming a perfect linearly chirped source (and a sufficient signal-to-noise ratio) is given by Sorin (37) as

$$\Delta R = \frac{V_g}{2\Delta f}$$

where V_g is the velocity of light in the fiber and Δf is the peak-to-peak frequency deviation of the optical source.

As the frequency-slew rate of current-ramp-driven semiconductor laser diodes may be very high (100 GHz/s is easily achievable), and the frequency resolution of commercial electronic spectrum analysers is a few hertz or less, the technique has a far superior distance resolution capability than that of OTDR methods. The equation above predicts a typical distance resolution of the order of 1 mm, taking into account the reduced velocity of light in the fiber. Kingsley and Davies (38) have suggested using the technique to perform distributed measurements in integrated optical waveguide circuitry.

Unfortunately, a major potential problem with OFDR is caused by uncertainties in the coherence function of the source. The coherence function modulates the received spectrum and therefore distorts any spatial variation of scattering that it is desired to observe. Some recent research has been directed toward converting this problem into a virtue and actually uses the changes in the coherence to determine changes in the scatter function (39). Although several publications have reported OFDR results for fiber attenuation monitoring, none have so far used the method for distributed sensing of parameters external to the fiber.

14.3.6.3. Subcarrier Frequency-Domain Reflectometry. The third method of encoding is to amplitude modulate the source with a chirped electronic signal (40), called subcarrier frequency-domain reflectometry (SCFDR). This is essentially the same as the OFDR method except that the subcarrier is modulated rather than the source. This gives considerably more control of the source characteristics, as chirping of an electronic oscillator with a sawtooth modulation function is somewhat easier to achieve in practice. In addition, there is no longer a requirement for the source to have a long coherence length, as, unlike the OFDR method, no optical interference is necessary. Of course, the maximum frequency sweep and frequency-slew rate possible with the electronic subcarrier method is generally less than that with an optical source, so the range resolution will generally be less. Despite several interesting research papers, the OFDR and SCFDR methods have, unlike the OTDR, not found significant application in commercial instruments.

14.4. TRANSMISSIVE DISTRIBUTED SENSING SYSTEMS

We consider next sensing systems where only transmitted light is monitored. The advantage is that the signal strengths are much greater than those for weak backscattered signals. The disadvantage is that there is a much smaller difference in propagation delay between modes of propagation. As mentioned in Section 14.1, it is necessary to have the signals carried in two modes so that propagation delay differences can be derived and the location of measurand-induced changes can be computed. We now consider various methods that have been reported. All are still at the research stage, none having yet been taken to the stage of commercial instrumentation.

14.4.1. Transmissive Frequency-Modulated Carrier Wave Method for Disturbance Location

The optical frequency-modulated carrier wave (FMCW) method may be used to locate points where mode coupling in a fiber has occurred, provided that the fiber is capable of supporting two modes having significantly different phase velocities. The frequency of the source is chirped, as with the OFDR method, and the chirped signal is fed into one mode of a two-moded fiber. Any external disturbance capable of cross-coupling energy from the initially excited single mode will produce a heterodyne or beat signal on a detector situated at the far end of the fiber. This beat signal is at a frequency equal to the difference between the optical frequencies of the signals incident on the detector. Because of the frequency-ramped nature of the source, the difference frequency detected is proportional to the distance from the source, at which mode coupling to the second mode has taken place. This approach, first suggested by Franks et al. (41), is depicted in Fig. 14.22. Their particular implementation used a birefringent fiber, with the transmitted signal being launched into only one of the two principal polarization modes. The disturbance to be monitored was a transverse pressure applied to the fiber, which caused coupling of part of the propagating energy into the orthogonal polarization mode.

A convenient attribute of the technique is that the relatively close velocity matching between the polarization modes, even in high-birefringence fiber, allows the FMCW technique to be used over lengths far in excess of the coherence length of the source. Two potential difficulties exist with the approach, however. First, mechanical strains of certain critical magnitudes may cause coupling of power from one polarization mode to the other and then completely back again, resulting in no net energy transfer and hence no beat signal. Second, disturbances causing strain in a direction aligned exactly with either of the polarization axes of the fiber will cause no mode coupling. Otherwise, except for these somewhat unlikely conditions, the technique appears to a simple and elegant method of locating the position of disturbances

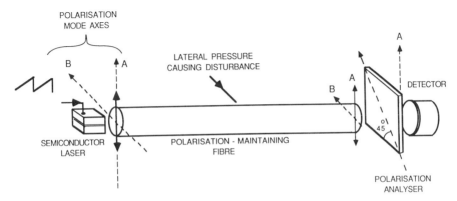

Figure 14.22. Transmission FMCW disturbance location sensor. [From Franks et al. (41).] Reprinted with permission.

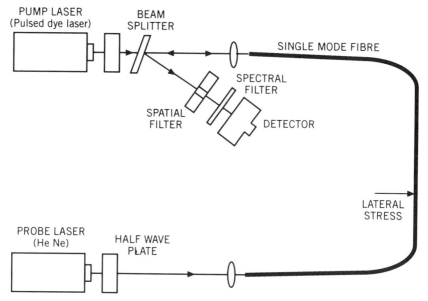

Figure 14.23. Sensor system with counterpropagating probe and pump-pulse signals, based on Raman amplification. [From Rogers (42).]

on a continuous fiber. Attractive applications include the detection and location of sources of noise or vibration: for example, for fracture location in materials or for intruder location (Fig. 14.7).

14.4.2. Distributed Sensing Using an Amplifying Fiber, with Counterpropagating Probe and Pump-Pulse Signals

If an optical signal from a steady continuous-wave (CW) source is transmitted through a fiber to a detection system, the power level received will depend on the total attenuation in the fiber. If the fiber is capable of amplification, the results can be more interesting. For example, if an intense optical pulse is transmitted in the optical fiber, in the direction opposite to the CW signal, the signal detected will be affected by any nonlinear gain processes that may be created by the effects of the pump.

The first example of such a system was reported by Farries and Rogers (42). This monomode fiber system used a reverse-traveling pulse from a Nd: YAG-pumped dye laser at 617 nm to provide Raman gain in a fiber. The gain was monitored using a continuous 633-nm forward-traveling signal from a helium–neon laser source. The Raman gain is strongly sensitive to the relative polarization of pump and probe signals. The arrangement of Fig. 14.23, in a fiber of low intrinsic birefringence, is capable of detection and location of lateral stresses, as these cause polarization mode conversion, which modifies the Raman gain.

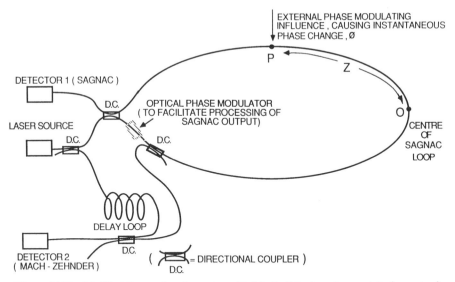

Figure 14.24. Modified Sagnac interferometer with Mach–Zehnder reference interferometer for disturbance location. [From Dakin et al. (44).] Reprinted with permission.

The technique has considerable academic interest, as it was the first of a new class of sensors. However, it suffered from a number of practical disadvantages because it uses inconvenient optical sources and is likely to be critically dependent on the pump power level of the dye laser source and to suffer from undesired polarization drift under the influence of normal environmental conditions on other sections of the fiber.

14.4.3. Disturbance Location Using Sagnac Interferometers

In its simplest form, the fiber optic Sagnac interferometer consists of a monomode fiber loop and a directional coupler.* The arrangement allows the launching of counter-propagating beams into the loop, from a single source, and the detection of the superimposed waves returning to a detector via an exit port of the same coupler (Fig. 14.24). Such arrangements have been studied extensively for use in optical fiber gyroscope systems, although additional components are usually inserted to produce a reciprocal configuration.

A major source of phase error in optical gyroscopes (43) may occur if rapid changes in optical path length due to thermal or mechanical effects on the

---

*For accurate low-drift operation of gyroscopes, reciprocity of counter-propagating beams is necessary. To ensure this, an additional single-polarization-mode filter is necessary, along with additional coupler to permit its inclusion into the system (reference). The sensor experiments described here would also benefit from such an arrangement, although the preliminary results reported were, for convenience, carried out with the simpler Sagnac loop.

fibers are allowed to occur at the position P away from the geometrical center O of the fiber length used to produce the coil. The reason for the error is that the two counter-propagating beams encounter the varying path-length changes at different moments in time, and therefore will suffer different phase changes. The difference in phase change is proportional to the product of two factors: (1) the rate of change, $d\phi/dt$, of the phase of the optical signal induced at point P by the external influence, and (2) the distance, z, between point P and coil center O.

Although the effect is a drawback for gyroscopes, the method has been researched as a means of locating disturbance in a fiber (44). In the first instance, to prove the concept, thermal changes were chosen as a suitable time-varying influence, although location of mechanical disturbance will probably have more potential applications. To calculate the distance, z, from the imbalance in the Sagnac interferometer (which, as has already been described, is proportional to the product of z and $d\phi/dt$), the method requires knowledge of the rate of change, $d\phi/dt$, induced by the influence. This quantity cannot readily be measured using only the Sagnac loop. However, an additional Mach–Zehnder interferometer may be introduced by adding an additional fixed-delay fiber link from the source and combining the output of this with a signal extracted from one of the counter-propagating beams in the Sagnac loop (Fig. 14.24). The output from the Mach–Zehnder interferometer gives an output proportional to change, ϕ, and differentiation of this phase output (conveniently performed in these measurements by a frequency-counting

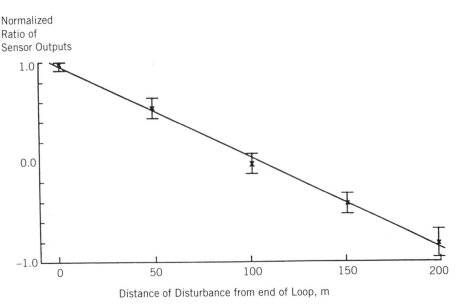

Figure 14.25. Response of Sagnac location system to thermal disturbances located at different points in the Sagnac loop.

system, which monitors the number of interference fringes passed through per second at the detector) yields the required rate of change, $d\phi/dt$. Simple division of the Sagnac phase offset by $d\phi/dt$ finally gives the desired distance, Z, of the point of disturbance, P, from the center point, O. Experimental results for location of the thermal disturbance are shown in Fig. 14.25.

With a suitable nonreciprocal Sagnac configuration and accurate inter-ferometric processing, the technique shows promise for accurate location of quite modest disturbance levels. The location capability of such a system has been analyzed in Dakin et al. (45). More recent theoretical proposals have been made by Udd to improve the configuration of the Sagnac disturbance-sensing system (46). These modifications are intended to allow the sensor to be used with broader-band sources and reduce the effects of back-reflected light to the source.

14.4.4. Transmissive Location Systems Using White-Light Interferometry

A recent method of location of disturbance in a two-moded fiber is based on white-light interferometry [47]. If a broadband source such as an LED is directed along two independent monomode paths and then recombined, interference fringes can only be observed when the paths are almost equal. For path differences much greater than the coherence length of the source, the superposition of multiple, differently phased fringes from each individual frequency component of the broadband source will lead to a total loss of visible interference. For a typical LED, the coherence length is on the order of only $30\,\mu m$, so quite small intervals in optical path length can be resolved.

A two-moded fiber having significantly different mode velocities does not have to be very long to generate path differences that are sufficient to prevent visible interference fringes from a broadband light source when the mode outputs are combined at the end of the fiber. For a high-birefringence fiber with 2-mm beat length, a length greater than 100 mm will suffice to lose coherence. However, if a separate twin-path interferometer is placed at the exit end and the optical path-length difference matched to that in the fiber, two equal-length interfering paths to the detector can be generated. When the external interferometer brings the system close to equal-optical-path-length conditions, intensity fluctuations can be observed as the path is changed (Fig. 14.26) (47). If the external interferometer is a free-space type, such as a Michelson arrangement, quite small differences in this external path can balance much longer lengths of fiber, as the latter will generally have only a small difference in propagation velocity between modes.

To form a disturbance location system, broadband light is launched into only one mode of a high-birefringence fiber. Pressure along the length can couple part of the propagating energy into the other mode. From an observa-tion of the time at which visible fringes are observed, a scanned-path-difference external interferometer can now determine the distance between the in-fiber coupling point and the output end of the fiber.

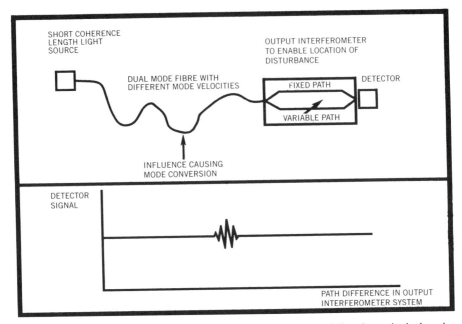

Figure 14.26. Illustration of white-light interferometry for location of disturbance in dual-mode fibers. [From Kotrotsios and Parriaux (47).]

14.5. CONCLUSIONS

The current status of distributed fiber sensors has been reviewed. The field is still one of rapid development, but commercial versions of the distributed temperature sensor, of the type first reported in reference 22, are now becoming available from several industrial sources. It is likely that practical systems will eventually be developed to cover other important application areas, such as the distributed measurement of mechanical strain, chemical concentration, and electrical and magnetic fields. There is still a considerable challenge in the technology and many areas where new innovations are required.

REFERENCES

1. M. K. Barnoski and S. N. Jensen, Fiber Waveguides: A Novel Technique for Investigating Attenuation Characteristics. *Appl. Opt.* **15**, 2112–2115 (1976).

2. P. Healey, Instrumentation Principles for OTDR. *J. Phys. E*, **19**, 334–341 (1986).

3. E. Marcuse, Microending Losses of Single-Mode, Step-Index and Multimode Parabolic-Index Fibers. *Bell. Syst. Tech. J.* **55**, 937–955 (1976).

4. A. J. A. Brunzma, P. Van Zuylen, C. W. Lamberts, and A. J. T. de Krijger, Fibre Optic Strain Measurement for Structural Integrity Monitoring. *Proc. Opt. Fiber Sens. Int. Conf.*, Stuttgart, *1984*, pp. 399–402. VDE, Berlin, 1984.

5. W. Gaebler and D. Braunig, Application of Optical Fibre Waveguides in Radiation Dosimetry. *Proc. Int. Conf. Opt. Fiber Sens., 1st*, London, *1983*, pp. 185–189 (1983).

6. A. H. Hartog and D. N. Payne, Fibre Optic Temperature Distribution Sensor. *Proc. IEE Colloq. Opt. Fiber Sens.*, London, *1982* (1982).

7. E. Theocharous, Differential Absorption Distributed Thermometer. *Proc. Int. Conf. Opt. Fiber Sens., 1st*, London, *1983*, pp. 10–12 (1983).

8. G. J. Cowle, J. P. Dakin, P. R. Morkel, T. P. Newson, C. N. Pannell, D. N. Payne, and J. E. Townsend, Optical Fibre Sources, Amplifiers and Special Fibres for Application in Multiplexed and Distributed Sensor Systems. *Proc. SPIE—Int. Soc. Opt. Eng.* **1586**, 130–145 (1992).

9. D. Pinchbeck, Optical Fibres for Cryogenic Leak Detection. In *Electronics in Oil and Gas.* Cahners Exhibits, London, 1985.

10. L. L. Blyer, Jr., J. A. Ferrara, and J. P. MacChesney, A Plastic-Clad Silica Fiber Chemical Sensor for Ammonia. *Proc. Opt. Fiber Sens. Int. Conf.*, New Orleans, pp. 369–372 *1988*, (1988).

11. H. Yoshikaw, M. Watanabe, and Y. Ohno, Distributed Oil Sensor Using Eccentrically Clad Fiber. *Proc. Opt. Fiber Sens.*, Tokyo *1988* (1988).

12. A. Rogers, Polarisation Optical Time Domain Reflectometry. *Electron. Lett.* **16**, 489–490 (1980).

13. A. H. Hartog, D. N. Payne and A. J. Conduit, POTDR: Experimental Results and Application to Loss and Birefringement Measurements in Single Mode Fibres. *Proc. 6th European Conf. on Optical Communication,* York, *1980*, post deadline paper (1980).

14. B. Y. Kim and S. S. Choi, Backscattering Measurements of Bending-Induced Birefringence in Single Mode Fibres. *Electron. Lett.* **17**, 193–195 (1981).

15. J. N. Ross, Measurement of Magnetic Field by POTDR. *Electron. Lett.* **17**, 596–597 (1981).

16. A. J. Rogers, POTDR: A Technique for the Measurement of Field Distributions. *Appl. Opt.* **20**, 1060–1074 (1981).

17. P. Healey, Fading in Heterodyne OTDR. *Electron. Lett.* **20**, 30–32 (1984).

18. A. J. Rogers, Distributed Optical Fibre Sensors for the Measurement of Pressure, Strain and Temperature. *Phys. Rep.* **169**, 99–143 (1988).

19. P. P. Smyth, R. Wyatt, A. Fidler, P. Eardley, A. Sayles, K. Blyth, and S. Craig-Ryan, 152 Photons per Bit Detection at 2.5 G Bit/s Using an Erbium Fibre Preamplifier. *Proc. European Conf. on Opt. Com. '90*, Amsterdam, *1990*, pp. 91–94 (1990).

20. R. I. Laming, P. R. Morkel, D. N. Payne, and L. Reekie, Noise in Erbium-Doped Amplifiers. *Proc. European Conf. on Opt. Com. '88*, Brighton, UK, *1988*, pp. 54–57 (1988).

21. Reference for Raman Spectrum Measured by CERL Research Labs, Leatherhead UK, in joint publication with CERL (Ref. 22).

22. J. P. Dakin, D. J. Pratt, G. W. Bibby, and J. N. Ross, Distributed Anti-Stokes Ratio Thermometry. *Proc. Int. Conf. on Opt. Fiber Sens. 3rd*, San Diego, *1985*, post deadline paper (1985).

23. J. P. Dakin, D. J. Pratt, G. W. Bibby, and J. N. Ross, Distributed Optical Fibre Raman Temperature Sensor Using a Semiconductor Light Source and Detector. *Electron. Lett.* **21**, 569–570 (1985).

24. A. H. Hartog, A. P. Leach, and M. P. Gold, Distributed Temperature Sensing in Solid-Core Fibres. *Electron. Lett.* **21**, 1061–1062 (1985).

25. D. Culverhouse, F. Farahi, C. N. Pannell, and D. A. Jackson, Exploitation of Stimulated Brillouin Scattering as a Sensing Mechanism for Distributed Temperature Sensors and as a Means of Realising a Tunable Microwave Generator. *Proc. Opt. Fiber Sens.*, Paris, *1989*, pp. 552–559 (1989).

26. T. Horiguchi, T. Kurashima, and Y. Koyamada, Measurement of Temperature and Strain Distribution by Brillouin Frequency Shift in Silica Optical Fibers. *Proc. SPIE—Int. Soc. Opt. Eng.* Vol. 1797 (1992).

27. J. P. Dakin, U.K. Pat. Appl. GB2,156,513A (1985).

28. S. B. Poole, D. N. Payne, and M. E. Fermann, Fabrication of Low-Loss Optical Fibres Containing Rare-Earth Ions. *Electron. Lett.* **21**, 737–738 (1985).

29. J. P. Dakin, Multiplexed and Distributed Optical Fibre Sensor Systems. *J. Phys. E.* **20**, 954–967 (1987).

30. J. P. Dakin and D. J. Pratt, Fibre-Optic Distributed Temperature Measurement—a Comparative Study of Techniques. *Proc. IEE Colloq. Distrib. Opt. Fibre Sens.*, London, *1986*, IEE Dig. 1986/74, pp. 10/1–10/4 (1986).

31. M. I. Skolnik, *1970 Radar Handbook*. McGraw-Hill, New York, 1970.

32. A. S. Subdo, An Optical Time Domain Reflectometer with Low Power InGaAsP Diode Lasers. *IEEE J. Lightwave Technol.* **LT-1**, 616–618 (1983).

33. S. A. Newton, A New Technique in Optical Time Domain Reflectometry. *Optoelectron. Mag.* **4**, 21–33 (1988).

34. J. K. A. Everard, Novel Signal Processing Techniques for Enhanced Optical Time Domain Reflectometry Sensors. *Proc. SPIE—Int. Soc. Opt. Eng.* **798** 42–46 (1987).

35. D. Uttam and B. Culshaw, Optical FM Applied to Coherent Interferometric Sensors. *Proc. IEEE Colloq. Opt. Fibre Sens.*, London, *1982*, IEE Dig. No. 1982/60 (1982).

36. S. A. Al Chalabi, B. Culshaw, D. E. N. Davies, I. P. Giles, and D. Uttam, Multiplexed Optical Fibre Interferometers: An Analysis Based on Radar Systems. *Proc. Inst. Electr. Eng.* **132**, 150–156 (1985).

37. W. V. Sorin, High Resolution Optical Fiber Reflectometry Techniques. *Proc. SPIE—Int. Soc. Opt. Eng.* **1797**, (1993).

38. S. A. Kingsley and D. E. N. Davies, OFDR Diagnostics for Fibre and Integrated-Optic Systems. *Electron. Lett.* **21**, 434–435 (1985).

39. K. Hotate and O. Kamatani, Optical Coherence Domain Reflectometry by Synthesis of Coherence Function. *Proc. SPIE—Int. Soc. Opt. Eng.* **1586**, 32–45 (1991).

40. H. Ghafoori-Shiraz and T. Okoshi, Fault Location in Optical Fibers Using Optical-Frequency-Domain Reflectometry. *IEEE J. Lightwave Technol.* **LT-4**, 316–322 (1986).

41. R. B. Franks, W. Torruellas, and R. C. Youngquist, Birefringent Stress Location Sensor. *Proc. SPIE—Int. Soc. Opt. Eng.* **586**, (1986).

42. M. C. Farries and A. J. Rogers, Distributed Sensing Using Stimulated Raman Interaction in a Monomode Optical Fibre. *Proc. Int. Conf. Opt. Fibre Sens. 2nd.*, Stuttgart, *1984*, pp. 121–132 (1984).

43. D. M. Shupe, Thermally-Induced Nonreciprocity in the Fiber-Optic Interferometer. *Appl. Opt.* **19**, 654–655 (1980).

44. J. P. Dakin, D. A. Pearce, C. A. Wade, and A. Strong, A Novel Distributed Optical Fibre Sensing System, Enabling Location of Disturbances in a Sagnac Loop Interferometer. *Proc. SPIE—Int. Soc. Opt. Eng.* 838, *Paper* 18 (1987).

45. J. P. Dakin, D. A. Pearce, A. Strong, and C. A. Wade, A Novel Distributed Optical Fibre Sensing System, Enabling Location of Disturbances in a Sagnac Loop Interferometer. *Proc. European Conf. on Opt. Com./Local Area Networks Int. Conf.,* Amsterdam, *1988* (1988).

46. E. Udd, Sagnac Distributed Sensor Concepts. *Proc. SPIE—Int. Soc. Opt. Eng.* **1586,** 46–52 (1992).

47. G. Katrotsios and O. Parriaux, White Light Interferometry for Distributed Sensing on Dual Mode Fibers. *Proc. Opt. Fiber Soc.,* Paris, *1989,* pp. 568–574 (1989).

15

Fiber Optic Sensor Multiplexing Techniques

ALAN D. KERSEY
Naval Research Laboratory
Washington, D.C.

15.1. INTRODUCTION

The application areas for fiber optic sensors include high-sensitivity military sensor systems, industrial process control sensors, chemical sensing, environmental monitoring, and smart structural and material sensing. In each of these areas, the ability to multiplex sensors in passive networks can be advantageous with regard to a number of system aspects, including reduced component costs, lower fiber count in telemetry cables, ease of E/O interfacing, and overall system immunity to electromagnetic interference (EMI). The development of efficient multiplexing techniques enhances the competitiveness of fiber sensors compared with conventional technologies in most application areas. Although these issues strongly affect the potential use of fiber sensors in smart materials and structures applications, such as those discussed in this book, the need for multiplexing in this case is of even more fundamental importance to facilitate the efficient interrogation of tens or perhaps hundreds of sensors that may be required to be distributed over a complex smart structure. The handling of such large flows of sensor information is one of the key areas in the development of advanced responsive structures. In this way, multiplexing and the concomitant organizing of sensor outputs onto common fiber links can be viewed as the first stage of data manipulation and reduction, which is important for the management and analysis of large sensor arrays.

A range of techniques for the multiplexing of fiber sensors have been developed, ranging from simple serial arrays of sensors based on optical time-domain reflectometry (OTDR) processing concepts to highly sophisticated interferometric fiber sensors using time- or frequency-division concepts. Figure 15.1 shows the most commonly utilized forms of basic topologies, or network architectures, for implementing multiplexed or fiber sensor arrays.

Fiber Optic Smart Structures, Edited by Eric Udd.
ISBN 0-471-55448-0 © 1995 John Wiley & Sons, Inc.

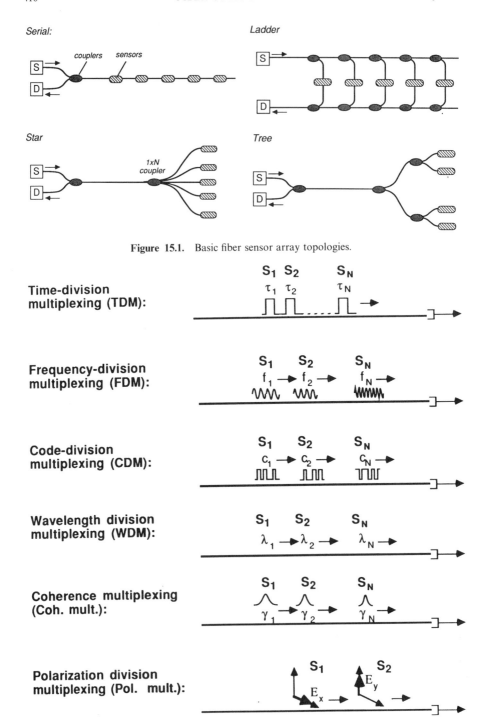

Figure 15.1. Basic fiber sensor array topologies.

Figure 15.2. Addressing techniques for sensor multiplexing.

These basic topologies form the foundation of the multiplexing. In addition to these physical fiber wiring diagrams of Fig. 15.1, a method for differentially encoding the sensor signals is required to allow separate sensors in the array to be addressed. These methods include time-, frequency-, code-, wavelength-, and polarization-division multiplexing. In addition, the optical coherence properties of a source can be used to encode the sensor signals. Figure 15.2 shows a graphical representation of these multiplexing techniques. In the following sections we describe many of these multiplexing concepts and their application to smart structures systems. In many cases, the sensors described are not of the type often utilized in smart structure systems, but the techniques are generic to many fiber sensor types.

15.2. SERIAL POINT SENSOR (QUASI-DISTRIBUTED) NETWORKS

The simplest form of multiplexed sensor involves the serial concatenation of point or quasi-point fiber sensors in a linear array. This type of system can be interrogated using optical time-domain reflectometry (OTDR) signal processing (Chapter 14) and is an extension of fully distributed fiber sensing techniques to the interrogation of a finite number of discrete sensors. Figure 15.3 shows such an implementation of a quasi-DFOS (QDFOS) system. As described in Chapter 14, the principle of operation of the basic OTDR-based

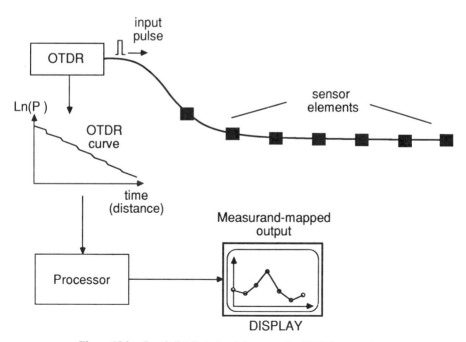

Figure 15.3. Quasi-distributed serial array using OTDR processing.

distributed sensor system relies on the detection of Rayleigh backscatter in the fiber. For a homogeneous fiber in a uniform environment, OTDR signal processing yields a steady exponential decay of the backscattered intensity from a pulse of light launched into the fiber. The logarithm of the backscattered optical signal yields a linear dependence with time, corresponding to the scattering distance along the fiber. The slope of this OTDR curve depends on the loss of the fiber. A splice or other fiber discontinuity in loss appears as a step in the OTDR curve, and the loss at the location can be calibrated against the intrinsic loss in the fiber. In intrinsic distributed sensors, this approach can be used to assess the loss introduced by a sensor element at any point along the fiber or continuously along the fiber. To achieve high spatial resolution, a short-duration optical pulse in the OTDR system is required. Unfortunately, shorter input pulses provide lower launched optical energies into the fiber and the signal-to-noise ratio of the output is thus reduced. Often, this results in the requirement for significant averaging of the data recorded for multiple input pulses using OTDR interrogation. In quasi-distributed sensing, the sensing is not continuous, but sensors can be positioned at certain finite number of predefined locations along the fiber path. Due to the fact that the sensor locations are defined, very short pulse OTDR operation is not necessary, and consequently, improved signal-to-noise performance can be achieved. Various loss-based sensing methods have been used to implement quasi-distributed sensor systems. Figure 15.4a–d shows some examples, including the use of modified fiber sections with sensitized optical properties spliced into a long fiber at certain intervals to provide localized variations in the loss, backscatter intensity, polarization, and fluorescence intensity. Alternatively, discrete non-fiber sensor elements that vary in transmittance or reflectance with the measurand field can be incorporated into the fiber line. Such an arrangement for distributed temperature sensing was demonstrated at an early stage in the development of QDFOS technology (32). This system used ruby glass sensor elements, the attenuation of which increase with temperature for light of wavelength about 600 to 620 nm (absorption edge shift rate about $1.2 \, \text{Å} \,^{\circ}\text{C}^{-1}$). An OTDR system operating at an wavelength close to the approximately 600-nm absorption edge was used to determine the loss at each sensor element, and a second wavelength removed from the absorption edge was used to provide a temperature-independent reference output. Other materials, such as semiconductors are also suitable for this approach, as are fibers doped with certain elements (e.g., hollium, neodymium). The major limitations of this system and similar approaches is the fact that the attenuation is accumulative; the light levels at the most distal sensor thus depend on the measurand at each sensor along the fiber. This places demanding requirements on the dynamic range of the detection system and limits the number of sensors that can be used in a practical system. This limitation can be relieved to some degree by adopting a sensor element with a variable reflectivity (8), as compared to a variable loss in the basic quasi-distributed OTDR system. In this case, the sensor reflectivity, which may vary in proportion to the measurand field over

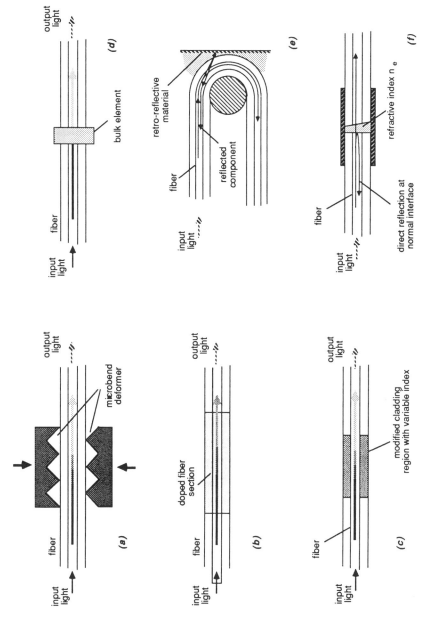

Figure 15.4. Sensor elements for quasi-distributed sensing based on attenuation (a–d) and reflective (e, f) sensing mechanisms.

the range zero to several percent, and the output of the system comprises a series of pulses, each proportional to the reflectivity at each sensor location. Figure 15.4*e* and *f* show two examples of transducers using this type of approach. The peak reflectivity of a sensor in an arrangement depends on the number of sensors to be addressed in the array but is required to be relatively low to minimize the intensity of pulses arising from multiple reflection in the system.

15.3. INTENSITY-SENSOR-BASED NETWORKS

The term *intensity-based sensor* is used to describe a generic class of sensors that depend on monitoring changes in some characteristic related to the intensity detected at the sensor output. Examples include sensors based on attenuation, reflectance, fluorescence signal, and modal modulation. In most cases, intensity-based sensors utilize multimode fibers and low-coherence optical sources [i.e., light-emitting diodes (LEDs), multimode lasers, etc.]. A number of different types of branching networks have been investigated for use with intensity-based sensors, particularly those based on simple concepts such as measurand-induced attenuation. In the following sections several examples of multiplexing techniques for intensity-based sensor arrays are examined. We also consider the effect of the optical power budget on the signal-to-noise ratio of a multiplexed system.

15.3.1. Time-Division Multiplexing

One of the most straightforward techniques for multiplexing of fiber sensors, and indeed, the first passive discrete-sensor network proposed by Nelson et al. (28), uses time division to address a number of fiber sensors arranged in a network with different delays from the source and detector. For such a system, a short-duration pulse of light input to the network produces series of distinct pulses at the output. These pulses represent time samples of the sensor outputs interleaved in time sequence, as depicted in Fig 15.5, which shows both ladder and star topologies. The required duration of the input pulse is determined by the effective differential optical delay (τ) between the fiber paths to the sensor elements, and repetitive pulsing of the system allows each sensor to be addressed by simple time-selective gating of the detector.

One consideration in these types of array topologies is the balancing of the optical power returned for each sensor in the system. For the star arrangement, adequate power branching can be accomplished by setting the splitting ratio of the 1:N splitter at $1/N$ for all output ports (provided that the losses in the delays coils are not excessively large). In the case of the ladder array, however, the splitting ratio of the couplers in the input and output fiber distribution and receiving paths (or input and output fiber buses) have to be selected to balance

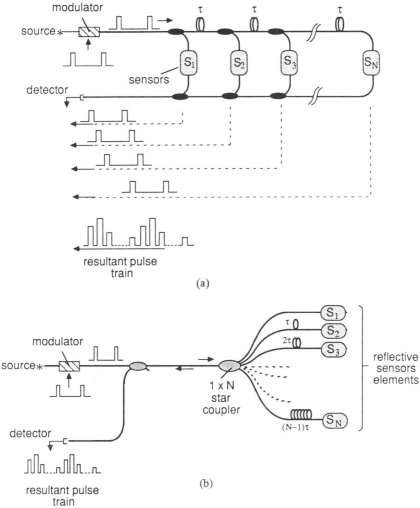

Figure 15.5. Time-division-multiplexed intensity sensor arrays using (a) a ladder and (b) a star topology.

the returned power levels. For a system with very low excess loss, a simple formula for determining the splitting ratios of the fiber couplers in the network can be derived which can suffice for balancing the returned power from small arrays (four or fewer sensors). This is obtained by equating the power received from the first sensor with that from the jth sensor. The power received from the first sensor is proportional to

$$P_1 = \eta_1 k_1^2 \tag{15.1}$$

where k_1 is the coupling ratio of the first coupler in the input and output fiber buses and η_1 is the transmittance of sensor 1. The optical power returned from the jth sensor element is given by

$$P_j = \eta_j[(1 - k_1)^2(1 - k_2)^2 \cdots (1 - k_{j-1})^2]k_j^2 = \eta_j k_j^2 \prod_1^{j-1} (1 - k_{j-1})^2$$

$$(15.2)$$

where $k_{1\ldots j}$ are the coupling ratio of the first through jth couplers in the input and output buses, η_j corresponds to the transmittance of the jth sensor, and the Π symbol represents the product series over the limits indicated. Equating the power levels from the first and jth sensors, the following equality is obtained:

$$\eta_1 k_1^2 = \eta_j k_j^2 \prod_1^{j-1} (1 - k_j)^2 \qquad (15.3)$$

Normalizing for the transmission factors of the sensors (i.e., assuming that $\eta_1 \approx \eta_j \approx 1$), the solution to this equality is

$$k_j = \frac{1}{N + 1 - j} \qquad (15.4)$$

Consequently, for an array of N sensors, the optical throughput for a single sensor (i.e., effective transmission factor from the source to detector through any given sensor path, normalized for the sensor transmission factors $\eta_{1,2,\ldots,N}$) is equal to

$$T_{s,x} = \frac{1}{N^2} \qquad (15.5)$$

(it is noted that this result is also the result obtained in the case of the star topology). This strong dependence of T_s on array size results in a precipitous falloff in the effective optical throughput of the sensors in arrays with large numbers of sensors. Decreasing T_s results in lower optical signal per sensor and thus poorer signal-to-noise performance. This represents one of the principal limitations to the number of sensors that can be multiplexed using passive fiber networks.

In practice, due to the fact that the light coupled to and from the more distal sensors in a ladder topology has to pass through more couplers and splicers, the accumulated system losses can be high enough to cause appreciable attenuation of the optical signals from these sensors. The result is a gradual decrease in received power with sensor position if the coupler splitting ratios are chosen according to Eq. (15.4). Figure 15.6 shows this affect for a 16-sensor system with different degrees of loss in the couplers and fiber splices in the fiber input and output buses. As can be seen, even only modest levels of excess loss can cause a large disparity in the levels of optical signal obtained from each sensor in the system. The losses can be accounted for by adjusting the coupler splitting ratios slightly to balance the effect of the loss. Figure 15.7 shows graphically the resultant change in splitting ratio required to maintain bal-

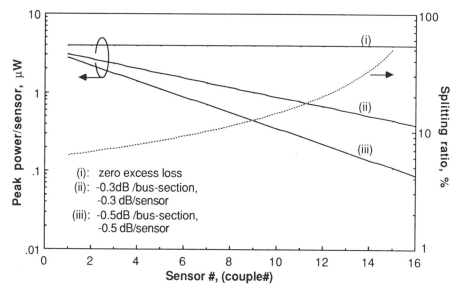

Figure 15.6. Coupler ratios for a 16-sensor ladder array and return power/sensor with loss.

anced optical outputs for two different cases of loss. As can be seen, the power returned from each sensor can be balanced by appropriate tailoring of the coupler splitting ratios. This reduces the power returned from the initial sensors in the array while boosting that returned from the more distal sensors. The effect of such losses on the optical power received from each sensor in

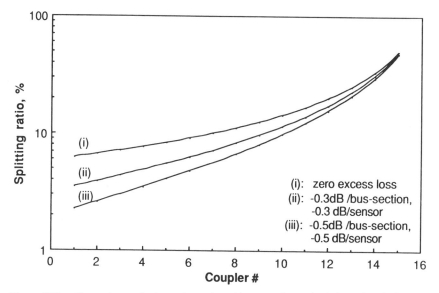

Figure 15.7. Change in required coupler ratios with system losses for balanced optical returns.

larger array is shown in Fig. 15.8. Here the optical power coupled from source to detector for a power-balanced array is shown for different excess losses for arrays up to 100 sensor elements: Curves (ii) and (iii) highlight the deviation in the dependence of T_s on N from the ideal case (i.e., zero loss) for the two different loss examples considered in Figs. 15.6 and 15.7. As can be seen, the optical throughput/sensor depends strongly on the level of loss in the fiber buses when the total number of sensors N exceeds 10. This figure, however, does not reflect the further reduction in average power received per sensor due to the pulsed operation of the source in TDM systems: For an array of N sensors, the optimum duty cycle for the source is $1/N$, which even for an ideal zero-loss system results in an overall $1/N^3$ dependence between the received power and array size. Figure 15.9 shows the result of Fig. 15.8 factoring in the duty cycle. Clearly, for ladder arrays of more than 30 sensors the average optical power level that can be achieved from each sensor in an array with practical losses and reasonable input powers (ca. a few milliwatts) are so small that the system noise performance will typically be compromised to some degree. The signal-to-noise performance for such an array system depends on the received optical power, and three principal noise-limiting regimes can be identified: (1) source intensity noise (RIN) at high received powers, (2) shot noise due to weak power levels at intermediate powers, and (3) detector noise. In general, the power budget considerations for other topologies, such as the star or tree arrangement, are comparable to those described above for the ladder topology.

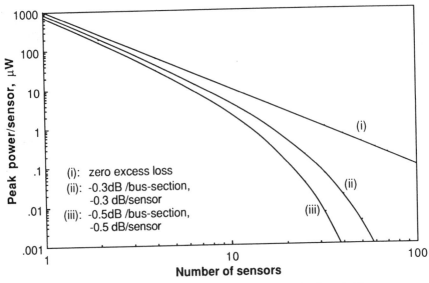

Figure 15.8. Returned optical power/sensor with array size for a ladder array.

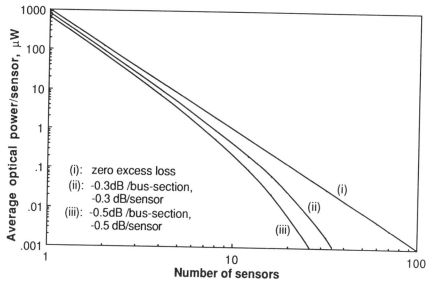

Figure 15.9. Average power/sensor using TDM.

The type of sensors that can be multiplexed using TDM approaches include microbending loss elements, modified cladding sensors, variable reflectivity devices, fiber–fiber coupling via gratings/masks, and in-line absorption sensors, such as gas absorption and doped fiber sections.

15.3.2. Frequency-Division Multiplexing

Figure 15.10 shows a basic approach to frequency-division multiplexing of output optical signals from a network of fiber sensors. Here an array of N sensors are fed by separate input fibers which carry light modulated at subcarrier frequencies $f_1, f_2, \ldots, f_N$, and the outputs of the sensors are combined onto a single output fiber bus and fed to a detector. The subcarrier modulation frequencies allow the sensor outputs to be separated at the detector electrical output by appropriate frequency band filtering or synchronous detection techniques. Although this simple approach provides for the combining of several sensor signals onto a single output fiber, the scheme requires multiple sources and input fibers.

A number of novel concepts for frequency-domain subcarrier-based multiplexing schemes for intensity sensors that operate from a common source and input fiber bus have been developed. One approach utilizes the radar-based frequency-modulated continuous-wave (FMCW) technique to provide frequency-division-based discrimination between the sensor signal returned from an array of intensity-based sensors (22). As shown in Fig. 15.11, in this case a source modulated in intensity by a chirped radio-frequency (RF) signal is used to interrogate a number of intensity sensors arranged in a ladder (or tree or

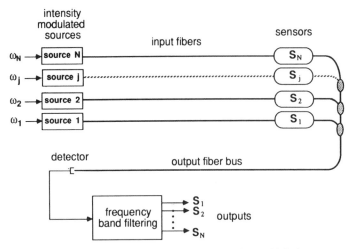

Figure 15.10. Frequency-domain addressing of fiber sensors using multiple frequency-modulated sources.

star) topology, with delays incorporated between the various paths. Consequently, the detected signal comprises a series of RF-modulated signals from the sensors in the array. For the ladder array considered, the optical power levels returned by each sensor can be balanced as in the TDM case discussed in the preceding section. The detector output is electrically mixed with a

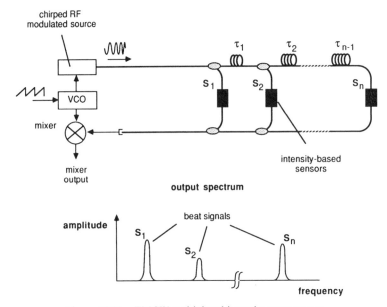

Figure 15.11. FMCW multiplexed intensity sensor array.

reference chirp signal, which produces a beat frequency associated with each sensor element, the frequencies of the beat signals depending on the round-trip optical delays from the source to detector for each sensor. Figure 15.12 shows the principle of operation for a single sensor path: In (a), the modulation frequency of the source is chirped over a frequency excursion Δf, with a period T, and the optical signal obtained from one of the sensors, with round-trip optical delay ΔT_j, is mixed with a reference chirp signal (b). Due to the delay in the sensor signal, this mixing process produces a beat note at a frequency equal to the instantaneous difference between that of the modulation of sensor signal and the reference chirp [shown in (c)]. This can be expressed in the form

$$f_j = \Delta f \frac{\Delta T_j}{T} \tag{15.6}$$

The use of different delays for each sensor provides for the generation of different beat frequency components associated with each sensor, and thus allows for frequency demultiplexing of the outputs using band filtering. A system comprising three sensors has been demonstrated using this approach.

In a variation on this approach (25), the individual sensor information is carried not by separate beat frequencies, but by the phase and amplitude of an

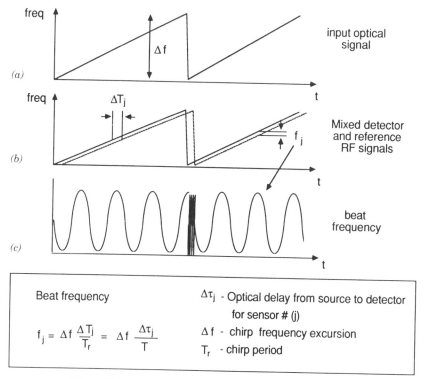

Figure 15.12. Principle of operation of the FMCW technique.

RF subcarrier amplitude modulation of source light returned from the array sensor elements. Figure 15.13 illustrates this concept. Due to the differing delays between the source and detector for each sensor path, the resultant detector signal has a magnitude and phase which are (for a given set of sensor transmission factors) dependent on the source modulation frequency. Modulation of the source at a number of discrete (nonharmonic) frequencies provides sufficient information to allow the status of each sensor to be interpreted. A system comprising three sensors has been demonstrated using this technique and showed reasonably good crosstalk performance (ca. $-40\,\mathrm{dB}$).

Another technique for the multiplexing of fiber sensors which is based on subcarrier signal processing utilizes a series of fiber transversal filters. A fiber transversal filter consists of two fibers of unequal length connected in parallel, as shown in Fig. 15.14a. In response to an RF intensity-modulated source, the recombined light at the output of such a filter exhibits a series of minima when the differential delay in the two fiber paths corresponds to a half-integral number of cycles of the modulation frequency. The normalized frequency response of a single sensor is given by (33)

$$g_i(f) = |\cos(\pi\,\Delta\tau_i\,f)| \tag{15.7}$$

where f is the frequency of the modulation and $\Delta\tau_i$ is the differential delay of sensor i. To track the null frequency, the null-tracking technique described by Wade et al. (34) can be used. For a linear array of sensors the frequency response of the combination is given by

$$G(f) = \prod g_i(f) \tag{15.8}$$

the product of the frequency response functions of the individual elements in the array (35). Figure 15.15 shows the experimentally observed response curve

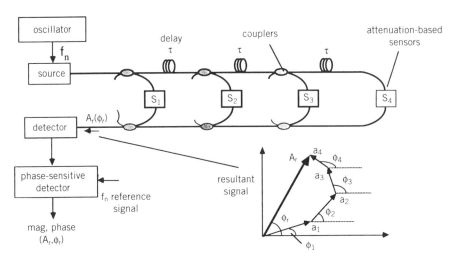

Figure 15.13. Frequency-phase-delay-based multiplexed intensity sensor array.

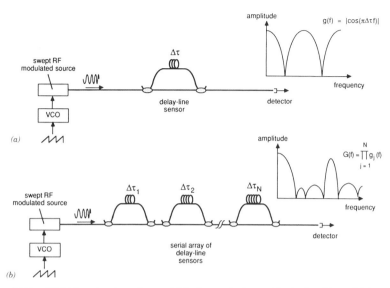

Figure 15.14. FDM-based subcarrier multiplexing technique: (*a*) single, (*b*) multiple sensor system.

for a cascade of three fiber filters. Here curves (*a*), (*b*), and (*c*) show the frequency response from 0 to 20 MHz for each of three sensors of fiber delays, $\Delta\tau_i = 0.294\,\mu s$, $\Delta\tau_i = 0.227\,\mu s$, and $\Delta\tau_i = 0.112\,\mu s$, independently operated, whereas the observed frequency response of the three-sensor network is shown in curve (*d*) of Fig. 15.15 from 0 to 20 MHz and in curve (*e*) from 10 to 16 MHz. This system has been used to multiplex three temperature sensors (23).

15.3.3. Code-Division Multiplexing

Code-division multiplexing is based on the use of signal-encoding, typically binary, sequences which possess noiselike properties. This property effectively spreads the sensor information over a bandwidth much larger than the sensor information alone. Codes with such properties, such as maximal-length (or *m*-sequence) codes and Gold codes, have been examined extensively for use in communications systems (9). Such spread-spectrum (SS) and code-division-multiplexed (CDM) techniques have been applied to a variety of communications applications, including optical fiber systems (30). In many respects, the use of code-division multiplexing to encode sensor outputs is closely related to frequency-division concepts: In FDM systems, the sensor outputs are encoded at different subcarrier frequencies, whereas with code-division multiplexing, the noiselike sequences encode the sensor outputs. In the FDM case, the sensor information is thus carried as sidebands of the subcarrier frequencies, but in CDM the sensor signals are spread over the wide bandwidth of the code sequence. In both cases, however, the sensor signals can be recovered by synchronous detection (i.e., mixing of the received detector signal with a

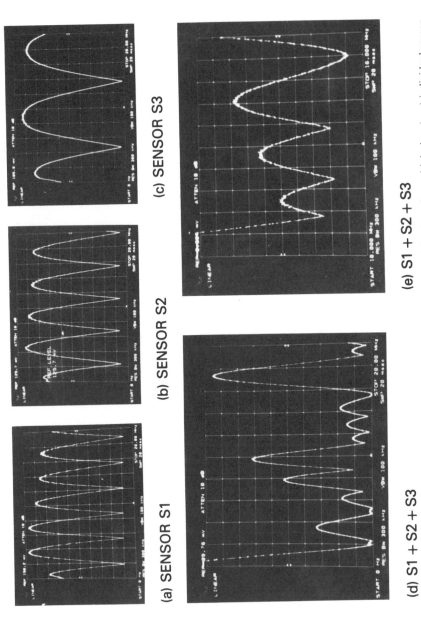

(a) SENSOR S1

(b) SENSOR S2

(c) SENSOR S3

(d) S1 + S2 + S3

(e) S1 + S2 + S3

Figure 15.15. Experimentally observed three-sensor frequency response with subcarrier multiplexing: (*a–c*) individual sensor elements with delays: 0.294, 0.227, and 0.112 μs; (*d, e*) response of the three sensors in series over 0 to 20 MHz and 16 to 20 MHz.

reference subcarrier or code), which returns the sensor signal to the baseband. This type of signal processing has been previously investigated for optical time-domain reflectometry (OTDR)-based sensing (10), and more recently, has been proposed and tested as a means for multiplexing interferometric sensors (1). In a CDM system, the interrogating optical source is modulated using a pseudo-random bit sequence (PRBS), and correlation is used to provide synchronous detection to identify specific sensor positions. A delay equal to an integer multiple of the bit (or "chip") period separate the sensors. The received signals from the array are then encoded by delayed versions of the PRBS, and correlation techniques can be used to extract the individual signals. Figure 15.16 shows schematically the principle of operation of the CDM approach applied to an N-sensor array in a ladder topology. The PRBS-coded input optical signal is fed to each of the N sensors, delayed by a multiple, n_j, of the bit period T, where j denotes a specific sensor ($1 \leqslant j \leqslant N$). The total output signal comprises the intensity sum of the overlapping delayed PBS sequences (each modified by the appropriate sensor transfer function). This results in a complex up–down staircase-like function at the optical detector which can be decoded using synchronous correlation detection involving multiplication of the received signal with an appropriately delayed reference PRBS code.

As mentioned, although a variety of pseudo-random codes can be used, the basic maximal-length (m-sequence) code of length $2^m - 1$ bits, which can be generated by a simple m-stage binary shift register with feedback. Typically, when used in communications systems, the isolation between channels multiplexed onto a common link (i.e., microwave link, for example) is determined by the finite length of the code: According to the theory of the autocorrelation of m-sequence codes, an isolation between an interfering asynchronously coded channel and signal synchronously code channels equal to $1:(2^m - 1)^{-1}$ can be achieved. This result is generally invoked to define the performance of a SS communications system with regard of its resistance to jamming from an

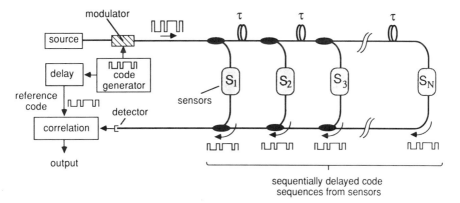

Figure 15.16. Basic concept involved in code-division multiplexing.

unwanted transmission, compared to the wanted signal. In the case of a multiplexed sensor array, however, all of the received sensor signals are encoded with the PRBS code, but with different delays of multiples of the bit period T. In this case, by judicious choice of the signal processing/correlation electronics employed, however, it can be shown that high isolation between the sensor channels can be obtained provided that the code length is simply longer than the number of sensor elements in the array. This arises due to the modified autocorrelation of unipolar m-sequence code and a bipolar version of the code (17). In this case, a correlation function that has a value 2^{m-1} for a synchronously aligned code but a value of zero for any asynchronous alignment of the codes is obtained. This type of modified signal processing provides for very low crosstalk between sensors, provided that the code length is greater than the number of sensors in the array. This greatly eases the bandwidth constraints that can be introduced using very long codes.

Clearly, the CDM technique can also be considered a variant of conventional TDM; however, in general, the CDM method may provide advantages in terms of power budget over TDM systems, as stronger optical signals are produced at the output. This does not directly affect the signal-to-noise performance due to shot noise but can improve the limit due to detector noise. Experimentally, the code-division multiplexing technique has been applied only to interferometric sensors (17), but it is equally applicable to intensity sensor networks.

15.3.4. Wavelength-Division Multiplexing

Wavelength-division multiplexing has been evaluated experimentally for use in fiber communications systems for many years (36). The technique provides for greatly increased channel capacity in coherent and noncoherent-based communications systems. The usefulness of the approach in local area networks has also been demonstrated. The use of this technique in sensor application has not, however, received as much practical attention. Figure 15.17 shows an example of the type of topology possible using WDM. In comparison to other topologies considered for TDM and FDM systems, where the input light is divided into a number of sensor channels in a generally wavelength-independent fashion, in the WDM approach the input light to the array is divided into a number of wavelength bands that are directed selectively toward specific sensor elements using wavelength-dependent splitter/recombiners. The scheme, which is essentially applicable to both intensity and interferometric sensor types, is theoretically the most efficient technique possible, as all the light from a particular source could in principle be directed to a particular sensor element and then onto a corresponding photodetector with minimal excess loss. Apart from the use of multiple sources of different wavelengths, a variety of optical sources provide wide spectral outputs that can be sliced for use in WDM multiplexed systems including LEDs, ELEDS, SLDs, and broadband rare-earth doped-fiber sources. The reason for the lack of practical demonstrations

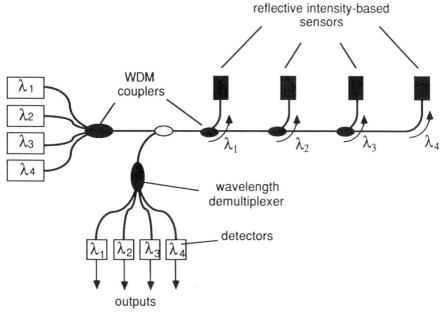

Figure 15.17. Wavelength-division-multiplexed sensor array topology.

of WDM arrays is due to the limited availability of wavelength-selective couplers (splitters and recombiners) which are required to implement the technique. This, combined with the complexity of the WDM fiber components (e.g., $N \times N$ star and $1 \times N$ tree couplers) needed to build systems based on a number of sensors and the limited wavelength selectivity of such devices are the major drawbacks of the approach. Consequently, apart from the obvious use of WDM techniques in fiber Bragg grating systems, which are described in the following section, wavelength-division multiplexing of large numbers of discrete intensity-based (or interferometric) sensors utilizing common servicing fibers may not prove to be feasible in terms of both cost and performance.

15.3.5. Polarization-Division Multiplexing

A further parameter of an optical system that can be used to encode optical sensor signals is that of polarization. In this case, sensor signals are encoded on orthogonal components of polarization of the input source light. At the detector, the received light is resolved into the two component polarization states, which can be separately detected to demultiplex the sensor outputs. The principal limitation of this approach is that only two channels of information (sensor signals) can be encoded onto a signal carrier. Variations of this basic approach are, however, possible utilizing subcarrier modulation of the state of polarization. Such schemes are strictly hybrids of PDM and FDM processing.

15.4. FIBER BRAGG GRATING–BASED SENSORS

Intracore fiber Bragg grating (FBG) sensors have attracted considerable
interest over the past few years because of their intrinsic nature and
wavelength-encoded operation. The gratings are holographically written into
Ge-doped fiber by side exposure to ultraviolet (UV) interference patterns (24).
Other means for producing such gratings also exist, and other fiber dopants
may be used to improve efficiency or to alter the required writing wavelength.
These sensors will prove to be useful in a variety of applications, particularly
in smart materials and structures, where the intrinsic nature of the FBG
facilitates embedment into advanced composites, or other materials, to allow
real-time evaluation of load, strain, temperature, vibration, and so on. Figure
15.18 shows the generic sensing concept involved for a single sensor element.
The fiber Bragg grating (FBG) sensor is illuminated using a broadband source
(BBS), such as an edge-emitting LED, superluminescent diode, or superfluores-
cent fiber source. The narrow-wavelength component reflected by the sensor is
determined by the Bragg wavelength:

$$\lambda_B = 2n\Lambda \tag{15.9}$$

where n is the effective index of the core and Λ is the period in the index
modulation of the core induced by the UV exposure. Figure 15.19 shows a
typical FBG transmission characteristic of a device in the 1550-nm region.
Measurand-induced perturbation of the grating sensor changes the wavelength
reflected by the FBG element, which can be detected and related to the
measurand field (e.g., strain) at the sensor position (21). The wavelength-
encoded nature of the output has a number of distinct advantages over other
direct intensity-based sensing schemes—most important, the self-referencing
nature of the output; the sensed information is encoded directly into

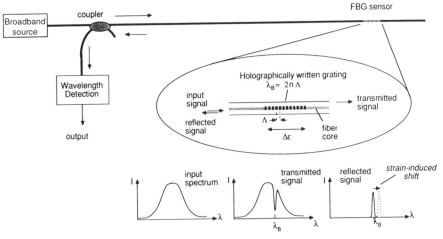

Figure 15.18.   Fiber Bragg grating sensor wavelength-encoding operation.

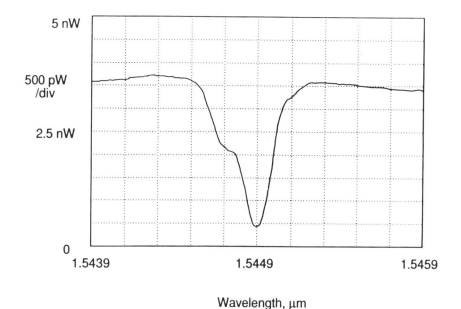

Figure 15.19. Typical transmission spectrum of a FBG element.

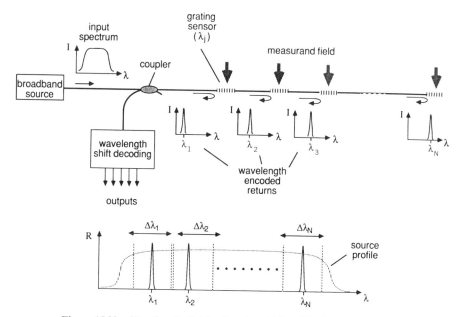

Figure 15.20. Wavelength-division-based multiplexed FBG sensor system.

wavelength, which is an absolute parameter and does not depend on the total light levels, losses in the connecting fibers and couplers, or source power. The dependence of the (normalized) shift in Bragg wave-length with fiber strain, ε, is $(1/\lambda_B)(d\lambda_B/d\varepsilon) \approx 0.74 \times 10^{-6}\,\mu\text{strain}^{-1}$, where 1 μstrain is a strain of 1 part in 10^6, and a temperature dependence $(1/\lambda_B)(d\lambda_B/dT)$ of $\approx 8.2 \times 10^{-6}\,^\circ\text{C}^{-1}$ at room temperature to around $11.2 \times 10^{-6}\,^\circ\text{C}^{-1}$ at temperatures above 300°C (27).

FBG elements are ideal for multiplexed networks and a variety of configurations have been proposed (27). Figure 15.20 shows a generalized concept for multiplexing based on wavelength-division addressing. Here the gratings are assigned a particular wavelength range, or domain, for operations that do not overlap. The Bragg wavelengths of the individual grating can thus be determined by illuminating the system with a broadband source and using an optical spectrum analyzer (spectrometer) to analyze the return signal. Figure 15.21 shows the type of response observed in transmission using this serial array of FBG sensor elements. This simplest approach is practical for only a limited number of devices, due simply to the fact that the bandwidth of sources are limited and can thus only accommodate a specific number of grating operational wavelength bands.

A means of overcoming this limitation is to adopt a combination of multiplexing techniques. For example, time-division multiplexing (TDM) can be combined with the inherent wavelength-division multiplexing (WDM) capability of the grating sensors. Figure 15.22 shows a concept using such a

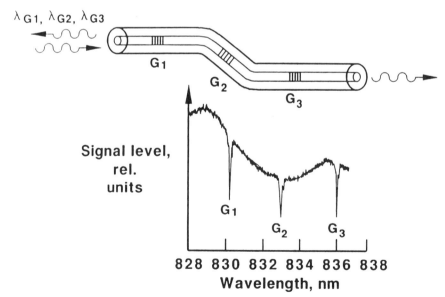

Figure 15.21. Typical transmission spectrum of a multiple FBG system. [From Morey et al. (26).]

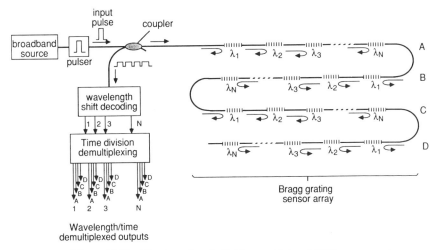

Figure 15.22. Time/wavelength-division-multiplexed FBG array.

TDM and WDM for addressing a large number of elements. This type of signal processing may allow, a large number of grating sensors to be interrogated in a serial array comprising time-division addressed strings (A to D) of gratings sensors spaced in wavelength ($\lambda_1 \rightarrow \lambda_N$). The development of this type of array is of particular interest for use in strain mapping in complex structures.

15.5. INTERFEROMETRIC SENSOR MULTIPLEXING

Interferometric fiber sensors offer extremely high sensitivity to weak measurand fields and have been widely researched for use in a range of application areas, including acoustic pressure, acceleration, and magnetic and electric fields (14). A number of different multiplexing topologies have been devised and tested by research groups working in this field (20). Early work in this area concentrated on demonstrating the principle of operation of various multiplexing approaches, such as time-division (TDM), frequency-division (FDM), and coherence multiplexing (Coh.M) using a relatively low number of sensors. These early experiments were necessary to demonstrate that interferometric devices could be multiplexed without incurring serious excess loss, crosstalk, or degradation in sensitivity. In more recent years arrays with up to 10 sensors (16) multiplexed on a common input–output fiber pair have been reported and represent the first demonstrations of significant multiplexing gain achieved in practical systems. Additionally, a system utilizing a hybrid TDM/WDM approach has been demonstrated with 14 sensor elements supported on a single input–output fiber pair (15). Systems have also been taken beyond the laboratory environment: An array comprising a total of 8 networked sensors based on a FDM scheme was tested at sea in 1990 (7).

A wide variety of interferometric sensor types, including the Mach–Zehnder, Michelson, fiber Fabry–Perot, ring resonators, and others, are used in this field. Most typically, the former two configurations are used for the detection of physical fields such as acoustic pressure waves. Although these configurations can be used in structural sensing, interferometric sensor elements more suitable for sensing in materials or structures include two-mode sensors, and intrinsic and extrinsic Fabry–Perot sensors. In many cases, these interferometric elements are suitable for use in the array topologies developed for use with interferometric sensors described in the following sections, where we discuss the various array topologies and comment on some of the reported experimental demonstration systems.

15.5.1. Frequency-Division Multiplexing

The multiplexing of sensors based on the FMCW concept (11) described in Section 15.2 was one of the earliest approaches demonstrated for use with interferometric sensors. In comparison with the approach used for intensity-based sensors, in the interferometric approach the optical frequency of the light is chirped and coupled to an array of unbalanced interferometric sensors. The interferometers can be arranged in a serial topology, as shown in Fig. 15.23 (a parallel topology is also possible). An unbalanced interferometer is sensitive

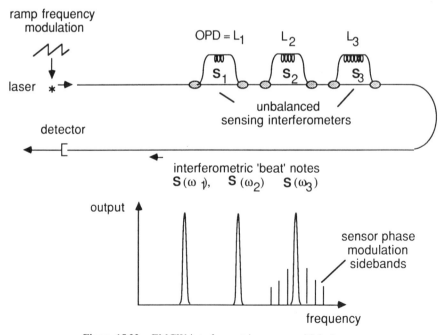

Figure 15.23. FMCW interferometric sensor multiplexing.

to changes in the frequency of the input light according to

$$\Delta\phi = \frac{2\pi n \Delta L}{c} \Delta v \qquad (15.10)$$

where $\Delta\phi$ is the change in output phase shift of the interferometer arising due to a shift Δv in the optical frequency of the input light, ΔL the fiber-length difference between the interferometer arms, n the effective fiber index for the guided mode, and c the free-space velocity of light. Due to this inherent dependence between the interferometric phase shift of an unbalanced sensor on the input optical frequency, a beat frequency is generated at each interferometer output by the use of the frequency chirped optical source. The beat frequency of the jth sensor depends on the frequency excursion, Δv_0, of the chirp, the chirp period, T_0, and the interferometer optical path difference (OPD $= n \Delta L_j$) according to the relationship

$$f_j = \frac{n \Delta L_j}{T_0 c} \Delta v_0 \qquad (15.11)$$

Assigning a different OPD to each interferometer allows the beat frequencies associated with each sensor element to be distinct and thus separable using band filtering. One major problem that arises with this type of multiplexing technique is cross-terms due to unwanted interferometric components arising differentially or additively between sensors, or ghost interferometers arising from connecting fiber paths in conjuction with interferometers. These cross-terms lead to sensor-to-sensor interference, or crosstalk, which is a problem in most applications where the full capability of an interferometric sensor, in terms of the detection sensitivity and dynamic range, are important. These stray components can be minimized using certain topologies, but cause significant design complexity for an array involving an appreciable number of sensors elements. A three-sensor multiplexed system based on FMCW multiplexing was demonstrated by Sakai (31) utilizing a variant FMCW approach using sine-wave modulation. Due to problems associated with high phase noise and poor crosstalk performance, little further work has conducted on this approach.

A preferred FDM approach utilizes the spatial and frequency-domain separation of sensor signals shown in Fig. 15.24. In Fig. 15.24a, the outputs from K sensor elements all powered from a common source are spatially multiplexed onto separate fibers. In Fig. 15.24b, the outputs from J sensors, which are independently illuminated by separate sources, are combined onto a single output fiber. The latter topology was described earlier for use with intensity-based sensors, where the source outputs are intensity modulated at different carrier frequencies. In the case of interferometric sensors, the use of phase-generated-carrier (PGC) interrogation (5), with each laser optical frequency modulated at a different frequency, and unbalanced interferometers allows the sensor outputs to be separated using synchronous detection or band filtering of the phase carrier signals generated by each interferometer. Combin-

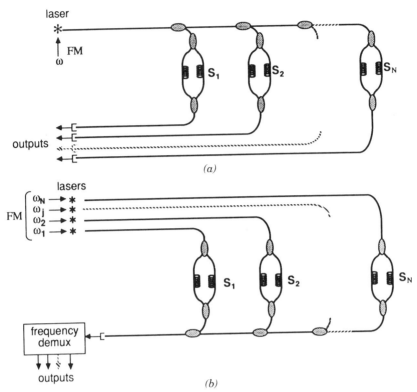

Figure 15.24. (*a*) Spatial and (*b*) frequency-domain addressing of interferometers.

ing the techniques shown in Fig. 15.24 allows a matrix-type array (6) to be configured, which contains $N = J \times K$ sensor elements, as represented schematically in Fig. 15.25. This system is somewhat unique in that the operation of the remote PGC interrogation (demodulation) scheme automatically provides both the demodulation and demultiplexing functions, provided that the sources are modulated at different carrier frequencies. Figure 15.26 shows a practical implementation of this type of array for a 3×3 (nine-sensor) system. This type of array has been shown to be capable of providing good phase detection sensitivity and low crosstalk for systems involving up to eight sensor outputs combined onto a single output fiber. This type of array is the most highly developed topology demonstrated to date; an array comprising 48 acoustic sensors was demonstrated successfully in a sea test conducted in 1990 (7).

15.5.2. Coherence Multiplexing

The basic principle of the coherence-multiplexing concept for interferometric sensors relies on the use of white-light or low-coherence differential inter-

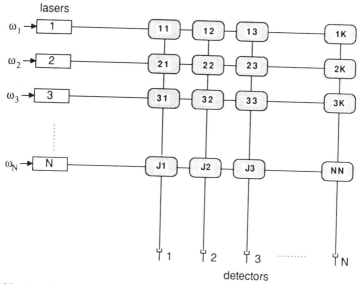

lasers

detectors

Figure 15.25. $J \times K$ matrix array configuration based on the spatial and frequency-domain multiplexing concepts of Fig. 15.24.

ferometric optical techniques. The basis of these concepts is the use of a sensing interferometer that has an optical path imbalance between the arms which is greater than the coherence length of the source in conjunction with a second matched compensating or receiver interferometer, as shown in Fig. 15.27. Due to the short coherence of the input light, no interferometric signal is observed at the output of the sensor interferometer directly. However, when the light is coupled to a second receiving interferometer which has a path imbalance closely matched (to within the source coherent length) to that of the sensor, an interference signal is generated at the receiver interferometer output. This output interferometric signal is generated by the two optical paths which experience nearly equal optical delays through the two interferometer arrangement [i.e., short(s)–long(r) and long(s)–short(r) combination of paths through

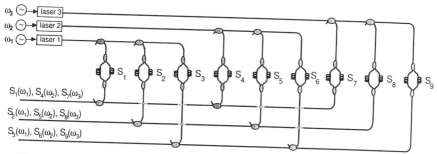

Figure 15.26. Practical implementation of the $J \times K$ (3 × 3) FDM matrix array system.

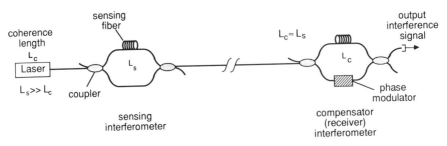

Figure 15.27. Principle of white-light interferometry.

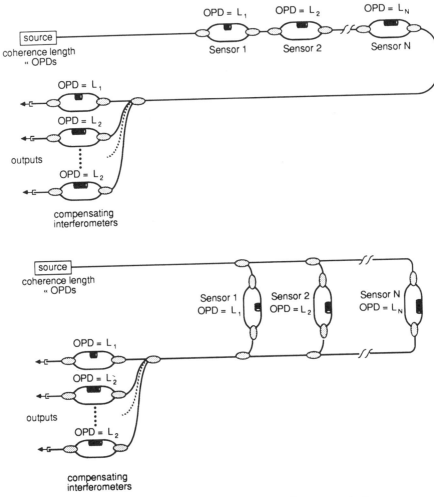

Figure 15.28. Coherence-multiplexed interferometric array topologies.

the sensor(s) and receiver(r) interferometers, respectively]. The phase shift of the resultant interferometric signal produced is dependent differentially on the sensor and receiver phase shifts. The term *path-matched differential interferometry* (PMDI) has also been coined to describe this approach. Due to the fact that the receiver interferometer has to be matched to that of the sensor to within the coherence length of the source, coherence addressing is possible, where a mutiple of unbalanced sensor interferometers of differing OPDs are matched to respective receiver interferometers. Figure 15.28 shows possible array topologies for serial and parallel sensor configurations.

Although there was initially significant interest in the coherence multiplexing of interferometric sensors (2), problems associated with crosstalk, excess phase noise, and poor power budget have limited the practical use of this approach with interferometric sensors. Nevertheless, interest in the use of this approach remains for other less demanding applications: for example, for use with interferometric sensors configured to detected quasi-static (dc) measurands using white-light interferometry (29). Coherence addressing and multiplexing of polarimetric sensors have also been demonstrated (12).

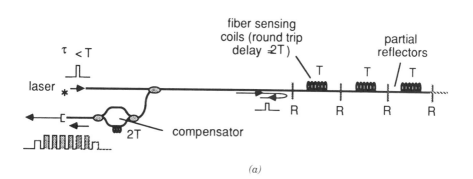

(a)

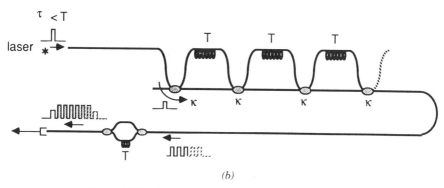

(b)

Figure 15.29. Time-division-multiplexed serial arrays.

15.5.3. Time-Division Multiplexing

A number of multiplexing topologies based on time-division addressing have been developed and tested. Array structures based on serial and parallel topologies, of the types shown in Fig. 15.29 and 15.30, respectively, are possible. The system of Fig. 15.29a, which is referred to as a reflectometric sensor array, was the first interferometric TDM array to be demonstrated (4, 13). This configuration utilized in-line partially reflective fiber splices, or in-fiber reflectors, between fiber sensing coils each of length L, to form in-line interferometric elements, which can be interrogated with pulsed operation of the source and a compensating interferometer of delay equal to the round-trip delay ($T_d = 2L/nc$) between reflectors. Provided that the width of the input pulse, τ, is less than T_d, an interferometric signal from each element in the array can be generated in time sequence at the compensator output. The initial demonstration of this concept utilized a differential delay heterodyne interrogation approach, where two pulses of differing optical frequencies and separated in time by T_d were launched into the array such that at the output pulses reflected from consecutive partial reflectors in the array overlapped to produce heterodyne beat signals associated with each sensing element, without the need for a compensating interferometer. An array of six acoustic sensors based on this approach has been field tested (13).

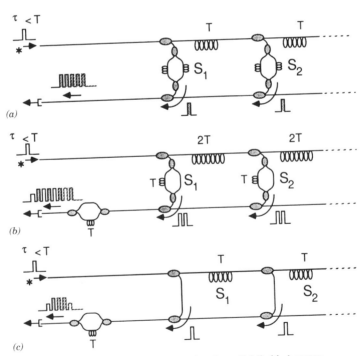

Figure 15.30. Time-division-multiplexed parallel (ladder) arrays.

This type of array can be implemented using FBG reflector elements. The basic topology is as shown in Fig. 15.31. Here fiber Bragg gratings with relatively low (<1%) reflectivities are written into a fiber at points spaced by a distance L, to form the sensor elements. The Bragg wavelengths of the FBG reflectors are required to be aligned with the source. This may cause a system limitation due to thermal/mechanically induced shifts in the Bragg wavelength, which would result in a reduction of the effective reflectivity at the source wavelength; however, the use of very short FBG elements with wide optical bandwidths (ca. 1 nm) should alleviate this problem to a large degree.

The second array configuration shown in Fig. 15.29b, shows a schematic of an array topology which is the transmissive analog of the reflectometric system. Here the sensors are formed by a cascade of Mach–Zehnder-like interferometric blocks, formed using very-low-coupling-ratio fiber couplers (typically $\leqslant 1\%$). The input light is launched into a fiber that couples the light to a series of sensor coils. Between each sensor, a small fraction of the light is tapped off to an output fiber bus. The delay between successive pulses coupled to the output bus is determined by the path difference between the longer and shorter paths in each MZ section. The same form of dual pulse operation used with the reflectometric system can be used to interrogate this form of tapped serial array (TSA).

The reflectometric array and TSA topologies give rise to multiple pulse interactions, which lead to crosstalk between the sensor elements. Figure 15.32 shows an example of the optical paths in the reflectometric array system that give rise to this type of effect. Analysis of this crosstalk shows that time-averaged sensor–sensor crosstalk levels can be <30 dB for an array of 25 sensors using partial reflectors with 0.25% reflectivity (19).

Other work in TDM systems has concentrated on ladder-array configurations (3, 18) of the form shown in Fig. 15.30. A 10-element array based on the

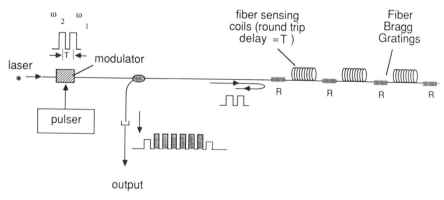

Figure 15.31. Reflectometric array utilizing FBG reflectors.

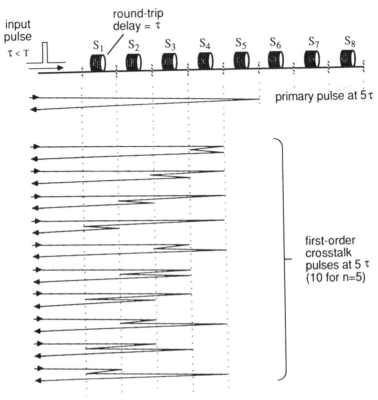

Figure 15.32. Origin of the multicoupling crosstalk paths in the reflectometric array.

topology of Fig. 15.30*a* has been demonstrated successfully. This topology does not lead to direct optical crosstalk between the sensor elements, and phase detection sensitivities comparable to those obtained in single sensor systems have been achieved. In the demonstration of this approach, interferometers which are slightly unbalanced to allow for passive demodulation via frequency modulated laser-based phase-generated carrier or synthetic heterodyne techniques are used. Naturally, Mach–Zehnder arrays are not the only possible system topologies that can be used. Arrays based on Michelson interferometers have, for instance, been proposed and demonstrated. Figure 15.33 shows a ladder topology based on Michelson interferometers. Additionally, star and tree topologies are, of course, also possible.

15.5.4. Other Techniques

The power budget of interferometric sensor arrays is determined by the same types of considerations as discussed in Section 15.2. Due to the high sensitivity of interferometric devices, however, the reduction in received optical power/

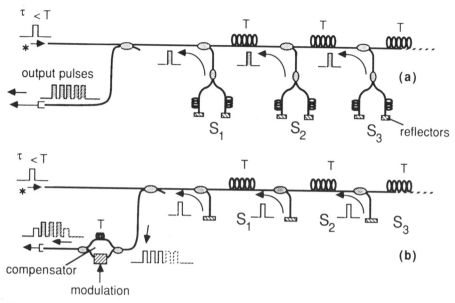

Figure 15.33. Time-division-multiplexed serial arrays based on Michelson interferometer sensor elements.

sensor can result in a degradation in sensitivity for only modest array sizes (i.e., ca. 20 sensors). A means of improving the multiplexing capability is to utilize a system based on a hybrid of addressing approaches. A possible means for this is combining time- or frequency-division addressing with wavelength-division multiplexing (WDM). As discussed earlier, due to the complexity of the components, (i.e., tree- or star-type WDM couplers) required to selectively tap certain wavelengths from a fiber bus to sensors and recombine them onto a single output fiber, the WDM approach has received little experimental attention. Furthermore, the crosstalk between sensors is determined by the degree of wavelength isolation that can be achieved with the WDM couplers, which is typically only about 15 to 20 dB. Consequently, although WDM may not prove to be feasible as the primary multiplexing basis of a large number of sensors, combining WDM concepts with time- or frequency-division addressing has the potential to allow a severalfold improvement in the number of multiplexed sensors in an array. This type of approach has the capability of extending the multiplexing capability of an interferometric sensor network to possibly 100 sensors per input–output fiber. Initial experimental demonstrations of this technique have been carried out (15). A variety of other approaches to the multiplexing of sensors or sensor transducers using interferometric techniques have been described in the literature, and this is an area that continues to receive strong research interest.

15.6. SUMMARY

A wide range of enabling techniques for implementing multiplexed fiber sensors networks have been developed and demonstrated. These techniques encompass a diverse range of addressing approaches which provide compatibility with practically all forms of fiber sensors developed to date. A key area of interest in the development of multiplexing techniques has been for high-performance interferometric fiber sensor arrays based on the use of Mach–Zehnder and Michelson interferometer configurations. Although these configurations have been developed primarily for high-sensitivity acoustic and magnetic sensor systems, their use in smart structural sensing is also conceivable. Other interferometers more typically associated with embedded sensor systems, however, such as intrinsic and extrinsic Fabry–Perot and two-mode configurations, can also be supported using many of the topologies outlined in this chapter.

REFERENCES

1. H. S. Al-Raweshidy and D. Uttamchandani, Spread Spectrum Technique for Passive Multiplexing of Interferometric Fiber Optic Sensors. *SPIE Proc. Int. Soc. Opt. Eng.* **1314**, 342 (1990).

2. J. L. Brooks et al., Coherence Multiplexing of Fiber Optic Interferometric Sensors. *IEEE J. Lightwave Technol.* **LT-3**, 1062 (1985).

3. J. L. Brooks et al., Fiber Optic Interferometric Sensor Arrays with Freedom from Source Induced Phase Noise. *Opt. Lett.* **11**, 473 (1986).

4. J. P. Dakin, C. A. Wade, and M. L. Henning, Novel Optical Fibre Hydrophone Array Using a Single Laser Source and Detector. *Electron. Lett.* **20**, 53 (1984).

5. A. Dandridge, A. B. Tveten, and T. G. Giallorenzi, Homodyne Demodulation Scheme for Fiber-Optic Sensor Using Phase Generated Carrier. *IEEE. J. Quantum Electron.* **QE-18**, 647 (1982).

6. A. Dandridge, A. B. Tveten, A. D. Kersey, and A. M. Yurek, Multiplexing of Interferometric Sensors Using Phase Generated Carrier Techniques. *IEEE J. Lightwave Technol.* **LT-5**, 947 (1987).

7. A. Dandridge et al., AOTA Tow Test Results. *Proc. AFCEA/DoD Conf. Fiber Opt. '90*, McLean, 1990, p. 104 (1990).

8. F. X. Desforges et al., Progress in OTDR Optical Fiber Sensor Networks. *Proc. SPIE—Int. Soc. Opt. Eng.* **718**, 225 (1986).

9. R. C. Dixon, *Spread Spectrum Systems.* Wiley, New York, 1984.

10. J. K. A. Everard, Novel Signal Processing Techniques for Enhanced OTDR Sensors. *Proc. SPIE—Int. Soc. Opt. Eng.* **798**, 42 (1987).

11. I. P. Giles, D. Uttam, B. Culshaw, and D. E. N. Davies, Coherent Optical-Fiber Sensors with Modulated Laser Sources. *Electron. Lett.* **19**, 14 (1983).

12. V. Gusmeroli et al., A Coherence Multiplexed Quasi-distributed Polarimetric Sensor Suitable or Structural Monitoring. *Proc. Opt. Fiber Sens.*, Paris, *1989*, p. 513 (1989).

13. M. L. Henning and C. Lamb, At-Sea Deployment of a Multiplexed Fiber Optic Hydrophone Array. *Proc. Int. Conf. Opt. Fiber Sensors, 5th*, New Orleans, *1988*, OSA Tech. Dig., p. 84 (1988).

14. A. D. Kersey, Recent Progress in Interferometric Fiber Sensor Technology. *Proc. SPIE—Int. Soc. Opt. Eng.* **1367**, 2 (1991).

15. A. D. Kersey, Demonstration of a Hybrid Time/Wavelength Division Multiplexed Interferometric Fiber Sensor Array. *Electron. Lett.* **2**, 554 (1991).

16. A. D. Kersey and A. Dandridge, Multiplexed Mach–Zehnder Ladder Array with Ten Sensor Elements. *Electron. Lett.* **25**, 1298 (1989).

17. A. D. Kersey and A. Dandridge, Low Crosstalk Code Division Multiplexed Interferometric Array. *Electron. Lett.* **28**, 351 (1992).

18. A. D. Kersey, A. Dandridge, and A. B. Tveten, Multiplexing of Interferometric Fiber Sensors Using Time Division Addressing and Phase Generated Carrier Demodulation. *Opt. Lett.* **12**, 775 (1987).

19. A. D. Kersey et al., Analysis of Intrinsic Crosstalk in Tapped Serial and Fabry–Perot Interferometric Fiber Sensor Arrays. *Proc. SPIE—Int. Soc. Opt. Eng.* **985**, 113 (1988).

20. A. D. Kersey, K. L. Dorsey, and A. Dandridge, Transmissive Serial Interferometric Fiber Sensor Array. *IEEE J. Lightwave Technol.* **LT-7**, 846 (1989).

21. A. D. Kersey et al., High Resolution Fiber Grating Based Strain Sensor with Interferometric Wavelength Shift Detection. *Electron. Lett.* **28**, 236 (1992).

22. K. I. Mallalieu et al., FMCW of Optical Source Envelope Modulation for Passive Multiplexing of Frequency-Based Fiber-Optic Sensors. *Electron. Lett.* **22**, 809 (1986).

23. M. J. Marrone et al., Quasi-distributed Fiber Optic Sensor System with Subcarrier Filtering. *Proc. Opt. Fiber Sens.*, Paris, *1989*, p. 519 (1989).

24. G. Meltz et al., Formation of Bragg Gratings in Optical Fiber by a Transverse Holographic Method. *Opt. Lett.* **14**, p.823 (1988).

25. J. Mlodzianowski et al., A Simple Frequency Domain Multiplexing System for Optical Point Sensors. *IEEE J. Lightwave Technol.* **LT-5**, 1002 (1987).

26. W. W. Morey et al., Bragg-Grating Temperature and Strain Sensors. *Proc. Opt. Fiber Sens.*, Paris, *1989*, p. 526 (1989).

27. W. W. Morey et al., Multiplexing Fiber Bragg Grating Sensors. *Proc. SPIE—Int. Soc. Opt. Eng.* **1586**, 216 (1992).

28. A. R. Nelson et al., Passive Multiplexing System for Fiber-Optic Sensors. *Appl. Opt.* **19**, 2917 (1980).

29. D. O'Connell, A. D. Kersey, A. Dandridge, and C. A. Wade, Coherence Multiplexed Fiber Optic Temperature Sensor Using a Wavelength Dithered Source. *Opt. Fiber Com.*, Houston, *1989*, p. 145 (1989).

30. P. R. Prucnal et al., Spread Spectrum Fiber Optic Local Area Network Using Optical Processing. *IEEE J. Lightwave Technol.* **LT-4**, 547 (1986).

31. I. Sakai et al., Multiplexing of Optical Fiber Sensors Using a Frequency-Modulated Source and Gated Output. *IEEE J. Lightwave Technol.* **LT-5**, 932 (1987).

32. E. Theochorous, Distributed Sensors Based on Differential Absorption. *Proc. IEE Colloq. Distrib. Opt. Fiber Sens.*, London, *1986*, Dig. 1986/74, paper 13 (1986).

33. C. A. Wade et al., Optical Fiber Displacement Sensor Based on Electrical Subcarrier Interferometry Using a Mach–Zehnder Configuration. *Proc. SPIE—Int. Soc. Opt. Eng.* **586**, 223 (1985).
34. C. A. Wade, A. D. Kersey, and A. Dandridge, Temperature Sensor Based on a Fiber-Optic Differential Delay RF Filter. *Electron. Lett.* **24**, 1305 (1988).
35. C. A. Wade, M. J. Marrone, A. D. Kersey, and A. Dandridge, Multiplexing of Sensors Based on Fiber-Optic Differential Delay RF Filters. *Electron. Lett.* **24**, 1557 (1988).
36. G. Winzer, Wavelength Multiplexing Components—A Review of Single-Mode Devices and Their Applications. *IEEE J. Lightwave Technol.* **LT-2**, 369–378 (1984).

16

Neural Network Processing for Fiber Optic Sensors and Smart Systems

BARRY G. GROSSMAN and MICHAEL H. THURSBY
Florida Institute of Technology
Center for Fiberoptic Sensors and Smart Structures
Melbourne, Florida

16.1. INTRODUCTION AND OVERVIEW

Smart structures systems require certain key functions for proper system operation. A block diagram of a generic smart system is shown in Fig. 16.1. The sensor system senses measurands, such as temperature or strain, and outputs a signal to the processing system. The processing system extracts the measurand information and performs signal processing and/or control functions that are required to obtain the desired output for subsequent systems. Warning systems, and actuators that can modify the system characteristics in response to a stimulus, are examples. Fiber optic sensors are popular for use in composite materials for a number of reasons. Because of their small diameter they can readily be embedded between the plies of the structure during manufacturing with minimal degradation of the structural strength. They can measure a large variety of parameters, including strain, temperature, and vibration, and can monitor large areas of the structure (e.g., multiplexed point sensors or distributed sensors).

With a single sensor and actuator, simple processors such as high-speed digital microprocessors can be most effective from both cost and performance standpoints. For large systems, however, where hundreds or thousands of sensors and actuators may be used, real-time operation requires higher computing speeds. Artificial neural processors have a parallel computing architecture, and when implemented in hardware can process multiple inputs and multiple outputs in nanoseconds. Additionally, artificial neural networks have some rudimentary properties of the human brain and are able to learn from experimental data, and adapt. They can learn to process data in one way,

Fiber Optic Smart Structures, Edited by Eric Udd.
ISBN 0-471-55448-0 © 1995 John Wiley & Sons, Inc.

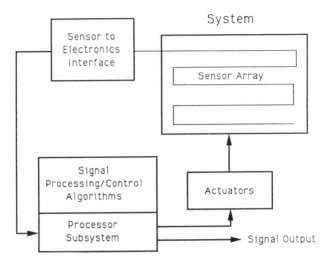

Figure 16.1.   Smart system block diagram.

and when the conditions change, the processing can adapt to the new conditions. Thus a processor trained to control a structure on the ground can modify its algorithms when the environment changes, such as in space.

The chapter is divided into two sections. In the first section we provide a tutorial explanation of what artificial neural networks are, what components make them up, and what gives them the ability to learn correct responses and adapt to varying situations. A few popular network architectures and learning algorithms are discussed to illustrate the diversity in neural network types and operation. In the second section, on applications, we provide some examples of how neural networks can be used in smart systems. The emphasis of this section is on recent work at Florida Institute of Technology. Both sections are intended to provide a basis for further study and investigation and to provide insight into the many possibilities of neural networks for smart system applications.

16.2.   INTRODUCTION TO NEURAL PROCESSORS

Neural processors have the capability to identify images, control systems, and process signals in a manner that no other computational system can. It is these accomplishments, together with the potential benefits of the neural paradigm (i.e., generalization, abstraction, and learning) that make them an ideal candidate for signal processing in smart systems. The history of artificial neural processors dates back nearly 50 years. Neural network research activity can be grouped into three periods, as shown in Table 16.1 (1, 2). Early effort by McCulloch and Pitts, Hebb, and Rosenblatt in the 1940s through 1960s established fundamental concepts. Researchers, including Widrow, Hoff,

Table 16.1   Historic Perspective of Neural Networks

Early work, 1940–1960: fundamental concepts	McCulloch and Pitts	Boolean logic
	Hebb	Synaptic learning rule
	Farley and Clark	First simulation
	Rosenblatt	Perceptron
	Steinbuch, Taylor	Associative memories
Transition, 1960–1980: theoretical foundations	Widrow and Hoff	LMS algorithm
	Albus	Cerebellum model (CMAC)
	Anderson	Linear associative memory
	Von Der Malsburg	Competitive learning
	Fukushima	Neocognitron
	Grossberg	ART, BCS
Resurgence, 1980–present: theory, biology, computers	Kohonen	Feature map
	Feldman and Ballard	Connectionist models
	Hopfield	Associative memory theory
	Reilly, Cooper	RCE classifier
	Hinton and Sejnowski	Boltzmann machine
	Rumelhart	Back-propagation
	Rumelhart and McClelland	PDP books
	Edelman, Reeke	Darwin III

Grossberg, and Fukushima, contributed a variety of mathematical theories between the 1960s and 1980s. Since the early 1980s, activities in all areas of neural networks have increased dramatically, fueled by several events, including a realization of the deficiencies of the single-layer perceptron and the need to extend the theoretical work to multilayer systems. A revolution in computer technology, which produced powerful and comparatively inexpensive computing devices and diagnostic tools, enabled further work on neural network mathematical theories and produced simultaneous breakthroughs in the understanding of neurobiological processes.

Artificial neural processors are made up of neurons, which are the basic computational element of the more complex artificial neural network. Individual neurons perform summation (of their inputs) and have an output that is a result of passing this sum through an output function, usually a nonlinearity. Figure 16.2 shows this schematically. The artificial neural network (ANN) processes information in a straightforward manner. The data to be processed are presented to the input of the network in analog or digital form. The input layer distributes this information through connections called weights to the next layer, where they are summed and a nonlinear function applied. The results are input to the next layer, where a similar process is carried out. The output of the network is provided by the output of the individual neurons in the last processing layer. The network transfer function is defined by the weights (multiplicative factors) associated with each of the inputs to the neurons and the architecture of the interconnection of these neurons. The

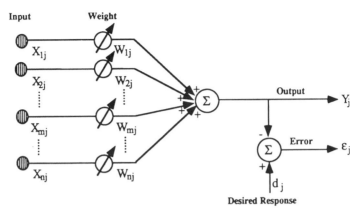

Figure 16.2. General form of a linear network.

artificial neural network provides a method of defining a parallel architecture for processing signals using simple computational elements called neurons.

A typical artificial neural network architecture, the multilayer perceptron, is shown in Fig. 16.3. It is composed of a number of simple processors that act like summing gates. These processors, also called nodes or artificial neurons, are represented by circles. The output of each node is connected to many other nodes in the following layer through weighting values, which allow the amount of output signal from each node to be varied. The last layer of nodes generate

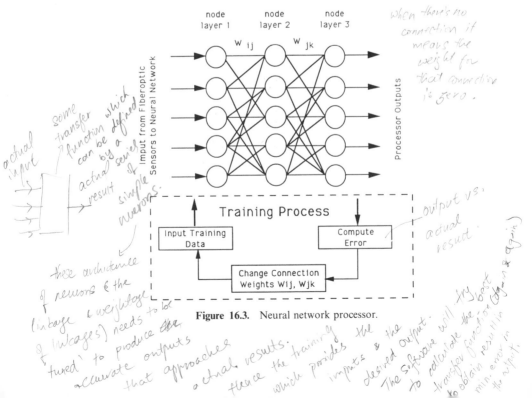

Figure 16.3. Neural network processor.

[handwritten margin note, top:] for a set of input, there must be only one certain desired output. Or else the system to be monitored is essentially in a state of chaos!

the system outputs. The inputs may be the outputs of a sensor array, for example, one input for each sensor, and three outputs may be the x, y, and z spatial location of a damaged area of the structure. A fourth output may be used to signal a warning. *[margin: Yes.]*

The training process requires the adjustment of the weight based on the response of the network to a set of known inputs; this is shown in Fig. 16.1. The weights are set initially to small random values. Training data are input to the network. At the output layer, the output values calculated are compared to the desired value. Using one of a number of learning algorithms the weights are modified and the procedure is repeated for another set of training inputs. This process is repeated hundreds or thousands of times until the weights have been adjusted to produce the minimum error over the complete set of training data. At this point the weights can be fixed and the network used to process *[handwritten: i.e. totally diff sets of input & outputs]* similar data. Similarly, should conditions change, the network can continue to adapt, thus providing an adaptive scheme for processing. Although a new technology compared with conventional computers, it is obvious that artificial neural networks will eventually have a niche application area for smart structures. Neural networks are frequently referred to as an extension of artificial intelligence. There are, however, two major differences between them: architecture and training. Neural networks can be taught to respond to trained *[handwritten: (unknown input)]* inputs and can then make decisions on untrained input (generalization). Artificial intelligence, however, does not have this kind of humanlike capability. The ability of an ANN to solve problems (converge to a solution) is as much dependent on the correct posing of the problem as it is on the architecture chosen for the network. Neuroscientists have also been developing neural networks to better understand biological nervous systems. Similarly, mathematicians, engineers, and computer scientists have been studying biological nervous systems to help develop new artificial neural networks to solve real-world physical problems.

[right margin handwritten: only if the desired output is known, so as to allow the error to be calculated & the corrections returned. ANN will interpolate (best match) or extrapolate (not always good results) unless the desired output is just scaled up.]

In the following section we look first at several different ANN architectures that can be used for processing. We discuss the application of ANNs in signal processing and control. The usefulness of ANNs will be demonstrated in several examples using ANNs for processing in fiber optic systems for smart structures applications.

16.2.1. Brief Survey of Artificial Neural Processor Architectures

In the following paragraphs, the basics of biological and artificial neurons are presented. Figure 16.4 illustrates typical biological neurons and shows two neurons in synaptic contact (1, 3). The parts of the neuron are described below. The *soma* or nerve cell, which is the central portion of the neuron, provides the energy for operation. The *axonhilloc* is attached to the soma and is electrically active, producing the action potential (electrical impulse used by the nervous system to communicate between neurons). The *axon* is the means of transmitting the information from one neuron to another (transmission line). The

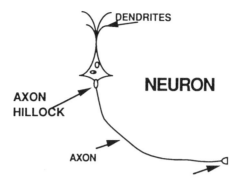

Figure 16.4. Schematic of biological neuron.

synapse is the portion that facilitates communications between the axon of one cell and the dendrites of another. The *dendrite* receives inputs from other neurons by means of a specialized contact area known as the synaptic button. From this structure and the electrophysiology of the neuron, we can construct electrical analogies of this system which are called artificial neural networks (ANNs). Figure 16.5 depicts the simplest configuration. Here neurons become processing elements, axons and dendrites become wires, and synapses become variable resistors carrying weighted input that represent data from other processing elements. Figure 16.6 shows several different types of nonlinearities that can be used as the activation function.

An ANN in its untrained state is not capable of solving problems. It must be trained to perform the desired function. Despite current research successes, ANNs are not destined to replace conventional general-purpose computers. They are, however, useful as specialized processors and preprocessors for solving problems that traditional systems have found intractable. Neural networks are inefficient compared to microprocessors at performing simple math functions and implementing common logic. Neural networks have been able to solve many problems that were previously unsolvable using conventional digital techniques.

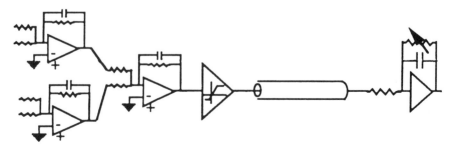

Figure 16.5. Artificial neuron.

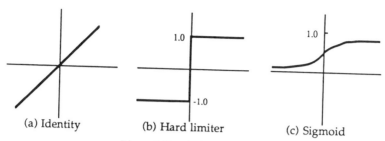

Figure 16.6. Activation functions.

16.2.1.1. Typical Neural Networks and Training. Many types of neural networks have been developed (4–7), including associative networks, clustering/classifying networks, and self-organizing networks. Their names imply the types of problems for which they are best suited. It is important that the proper network be employed to solve a given type of problem. More than 30 different models have been developed in the field of neural networks (8). They can be classified as single-layer and multilayer neural networks according to their architectures. A single-layer network has only one connection weight set between input units and output units. A layer is said to be fully connected if each output node is fed from each input node. A multilayer neural network is a network with one or more hidden layers between the input and output layers. This network can solve more complicated problems than can a single-layer network, but it may take more time to train.

There are numerous learning algorithms used to train neural networks (9). Some of the most common are Hebb's rule, competitive learning, and the least mean square (LMS) algorithm. The selection of a learning algorithm is driven by the network structure and by the function the network is to perform. For instance, the competitive learning algorithm is employed to solve classification problems. Hebb's rule is one of the simplest methods of training. During Hebbian learning, the weights are determined by forming the outer product of the input and output. This method is simple because the mapping between the input and output is straightforward and much easier to visualize than most other methods.

A very popular learning algorithm is the least mean square (LMS) algorithm, or more specifically, the delta rule (Widrow–Hoff learning rule) or back-propagation. This algorithm, as the name implies, minimizes the mean-squared error of the network. This algorithm is effective in applications where the desired output is known for a given input. In the following section the general LMS algorithm, as applied to neural network, is discussed.

16.2.1.2. Network Architectures. Several extensive reviews of neural architectures exist in the literature (10). We will not duplicate them here but will describe several networks that have use in the field of smart structures. In particular, we discuss the multilayer perceptron architecture, the Kohonen network, and the Jordan recursive network.

Perceptron and Backpropagation Algorithm. The perceptron is one of the earliest mathematical models and was developed by Rosenblatt to mimic the human nervous system (10). This network can learn to recognize simple patterns. The perceptron forms two decision regions separated by a hyperplane; a single node computes the weighted sum of its input elements and passes the result through a hard-limiting nonlinearity such that the output is either $+1$ or -1. Connection weights and thresholds can be fixed or adapted using a number of different algorithms. The perceptron convergence procedure introduced by Rosenblatt is:

Step 1. Initialize weights and threshold to small random nonzero values.

Step 2. Present new input and desired output.

Step 3. Calculate actual output.

$$y(t) = f_h\left[\sum_{i=0}^{N-1} w_i(t)x_i(t) - \theta\right] \tag{16.1}$$

Step 4. Adapt weights.

$$w_i(t+1) = w_i(t) + \propto[d(t) - y(t)]x_i(t) \tag{16.2}$$

Step 5. Repeat by going to step 2.

In this equation α is a positive gain fraction less than 1 and $d(t)$ is the desired output ($+1$ or -1) for the current input. $W_i(t)$ is the weight from input i at time t and θ is the threshold in the output node. Weights are unchanged if the correct decision is made (10). Perceptrons can differentiate patterns only if the patterns are linearly separable.

Multilayer Perceptron. A more complex structure is required when classes cannot be separated by a single hyperplane. Multilayer perceptrons are feedforward nets with one or more layers of nodes between the input and output nodes. These additional layers contain hidden units or nodes that are not connected directly to both the input and output nodes [10]. A typical two-layer perceptron is shown in Fig. 16.7 (11).

The capabilities of the multilayer perceptron stems from the nonlinearities. If nodes were linear, a single-layer net with appropriately chosen weights could exactly duplicate those calculations performed by any multilayer net. The multilayer perceptron can be trained by the back-propagation algorithm. It is named for the procedure used to pass the error signal back from the output to previous layers. This is a generalization of the LMS algorithm that uses a

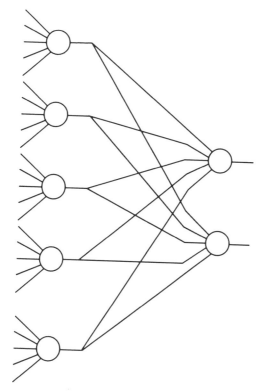

Figure 16.7. Multilayer feedforward network.

gradient search technique to minimize the mean-squared error between the actual output and the desired output. Since the back-propagation algorithm requires taking the derivatives of the activation function, a continuous, differentiable nonlinear function is needed. Frequently, a sigmoid nonlinearity expressed by Eq. (16.3) is chosen:

$$F(\alpha) = \frac{1}{1 + e^{-(\alpha - \theta)}} \tag{16.3}$$

where θ is the threshold value of the nonlinear function.

The back-propagation algorithm is described below [10].

Step 1. Initialize weights and offsets. Set all weights and node offsets to small random values.

Step 2. Present input and desired outputs.

Step 3. Calculate actual outputs. Apply the sigmoid nonlinearity from above and formulas as in Eq. (16.3).

Step 4. Adapt weights. Start at the output nodes and work back to the first hidden layer.

$$W_{ij}(t + 1) = W_{ij}(t) + \alpha\delta_j x_i \tag{16.4}$$

where W_{ij} is the weight from hidden node i or from an input to node j at time t, x_i' the derivative of the activation function with respect to a change in the net input to the neuron i, α the learning rate, and δ_j an error term for node j.

If node j is an output node, the error term is determined to be

$$\delta_j = O_j(1 - O_j)(D_j - O_j) \tag{16.5}$$

where D_j and O_j are the desired and actual outputs of node j, respectively. If node j is an internal hidden node, then

$$\delta_j = x_j'(1 - x_j')\sum_k \delta_k W_{jk} \tag{16.6}$$

where k is over all nodes in the layers above node j.

Step 5. Repeat by going to step 2.

The training steps described above can be discontinued when the error is acceptably small for each of the training vector pairs. The utility of the back propagation algorithm stems from the computational power of two- or three-layer perceptrons with one or two hidden layers. These networks can form any desired decision region as described by Lippmann (10).

Despite the many successful applications of back-propagation, the learning process of the network may involve uncertain training and long training times, especially for complex problems. Back-propagation employs a type of gradient descent; that is, it follows the slope of the error surface downward, constantly adjusting the weights to provide a minimum error. It is not, however, guaranteed to find the global minimum. The network can get trapped in a local minimum. If a local minimum is reached, the error of the network may be unacceptably high and thus leads to a form of instability (11).

Least Mean Square Algorithm. The least mean square (LMS) algorithm (12–16) has been widely used for signal processing applications because of its relatively simple implementation procedures and ease of computation. If the input vector and the desired response are known for each exemplar, the LMS algorithm is usually the best choice for signal processing. A primary concern with the LMS algorithm is its ability to converge to the optimum weight vector. Therefore, the LMS algorithm is trained by comparing the output with the desired response to obtain an error, e, and then adjusting the weight vector to minimize this error. The error is propagated back through the network to adjust the weights. This back-propagation of error gives rise to the name *back-propagation* which is used interchangeably with *LMS* and *delta rule* (*Widrow–Hoff learning rule*). We refer the reader to reference 13, where a detailed derivation of LMS algorithm can be found.

To summarize, to find the optimal weight matrix W, the weights are adjusted in the direction of the negative gradient in accordance with

$$W_{j+1} = W_j - \mu \hat{V}_j \qquad (16.7)$$

Here a new weight matrix (jth + 1) is computed by subtracting a proportional amount of the gradient from the jth or previous weight matrix. The proportionality μ regulates the speed and stability of network and is called the *learning rate* of the network. Equation (16.7) can be rewritten as follows:

$$W_{j+1} = W_j + 2\mu\varepsilon_j X_j \qquad (16.8)$$

Equation (16.8) is the essence of the LMS algorithm. The case where a sigmoid is used as the activation function is derived in the following paragraphs.

LMS Algorithm with Sigmoid Activation Function. There are a large number of activation functions (squashing functions) that can be used in neural networks. The sigmoid nonlinearity is the most popular. This nonlinear activation function has a profound effect on the error response. We can define delta (δ_j) as follows:

$$\delta_j = \varepsilon_j f'(S_j) \qquad (16.9)$$

Equation (16.9) is called the *delta rule*. An adjusted weight matrix is found from Eq. (16.9).

$$W_{ji}(\text{new}) = W_{ji}(\text{old}) + \mu\delta_j X_i \qquad (16.10)$$

In Eq. (16.10) the term δ_j can be computed for each processing element independently and simultaneously. Once the delta are computed, each of the weights is updated from Eq. (16.10). Hence the LMS algorithm is inherently parallel.

LMS Algorithm (Delta Rule)

Step 1. Compute the sum of the weighted input signals.

$$S_j = W_j^T X = \sum_{i=1}^{L} W_{ji} X_i \qquad (16.11)$$

Step 2. Evaluate the activation function to compute the output signal.

$$Y_j = f(S_j) = f(W_j^T X) \tag{16.12}$$

Step 3. Compute the adjustment factor (delta) for the unit's weights.

$$\delta_j = \varepsilon_j f'(S_j) \tag{16.13}$$

Step 4. Update the value of the unit's weights.

$$W_{ji} = W_{ji} + \mu \delta_j X_i \qquad (i = 1, \dots, L) \tag{16.14}$$

So far, the algorithm describes how weights are adjusted in a single-layer perceptron network to achieve a LMS error solution. In the multilayer case, the desired response at a hidden layer for a given input is not specified. We need to know how the weights of the hidden layers are to be adjusted. Generally, a network's desired response is the output of an ideal network. The error at the network output must be distributed throughout the hidden layers. The method of propagating the error connection from output units back to the hidden units is called *back-propagation*.

Figure 16.7 illustrates a simple multilayer feedforward neural network [17]. An input signal is transmitted through the network in a forward direction, applying the appropriate weights and activation functions. At the output layer, the difference (error) between the desired response and actual output is computed. The computed error, $e(\delta_k)$, is propagated backward to all units of the hidden layer. In a similar manner, the error is propagated from the hidden layer to the input layer of the network. The weights at the output layer are adjusted as described using Eq. (16.10), where the input to the layer is taken as the output of the previous layer.

This training procedure repeats itself until the global error is decreased to an acceptable value. An essential component of the algorithm is the iterative method described that propagates the error terms, backward, from units in the output layer to units in hidden layers. The use of multilayer perceptrons trained with back-propagation can be complex and difficult to implement. However, it has been used successfully in many applications, including classifying spoken vowels and nonlinear signal processing (5–7).

16.2.2. Kohonen Neural Net Classifier

The cortex of the brain is organized in anatomical regions according to function. The sensory areas are also represented in a topological map by function. Kohonen constructed an artificial system organized in a similar topological manner, the self-organizing feature map. The Kohonen neural network is an important architecture. Its learning process can be either supervised or unsupervised, depending on the application. For pattern classifi-

1.8 · With known inputs
       or unknown inputs

cation purposes, a supervised Kohonen neural network is preferable. We now describe unsupervised and supervised learning algorithms.

16.2.2.1. Network Structure. In the brain one finds specificity associated with spatial location of a neural response (18). One important organizing principle of sensory pathways in the brain is that the placement of neurons is orderly and usually reflects some physical characteristic of the external stimulus being sensed (10). There are many kinds of maps or images of sensory experiences in the brain (19). These individual areas which exhibit a logical ordering of their functionality are referred to as *ordered feature maps* (14). Mapping is said to be *ordered* if the topological relations of the images and the patterns are similar, where an image is regarded as the unit with the maximum response to a particular input pattern (18).

Kohonen's algorithm produces self-organizing feature maps similar to those occurring in the brain. Unlike the back-propagation network, the Kohonen network is not a hierarchical system but consists of a single layer of neurons. These neurons are, however, highly interconnected within the layer; each Kohonen layer neuron receives the entire input pattern of the network but also numerous inputs from the other neurons within the layer via the neighborhood concept, as shown in Fig. 16.8 (10, 11).

16.2.2.2. Competition in Learning. In Kohonen learning, each neuron computes how close its weight vector is to the input vector. The neuron with the largest dot product (closest) is declared the winner in the computation. In Kohonen networks, the philosophy of "winner-take-all" is used. Only the winner is permitted to provide output, and only the winner and its neighbors are permitted to adjust their weights. The algorithm that forms feature maps requires a neighborhood to be defined around each node as shown in Fig. 16.8. This neighborhood decreases in size slowly with time.

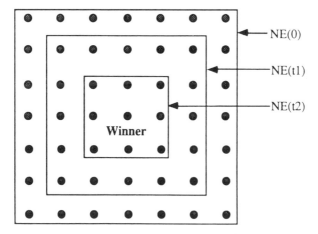

Figure 16.8. Topological neighborhoods at different times.

Each neuron in the layer produces an output which can be used as an excitatory or inhibitory input to the neighboring neurons. Inhibitory inputs tend to reduce the activation of the neurons and thus reduce their output. In most, the effect is inhibitory, but in the immediate neighborhood of the neuron (i.e., physically closest to the excited neuron) it is actually excitatory. This results in every neuron trying to fire and if that is not possible, trying to help its neighbors fire by exciting them. Thus, if any of the neuron's neighbors wins, it will still be allowed to change its weights. The degree of lateral interaction is usually described as having the form of a Mexican hat, thus is called the *Mexican hat function* (Fig. 16.9). The exact size of the neighborhood — the width of the Mexican hat function — is not limited to immediate neighbors of the winner. The size of the neighborhood may vary during the training process. Typically, it starts large and is slowly reduced to include only the winner and possibly its immediate neighbors (10, 11, 19).

16.2.2.3. Kohonen Learning Rule. The learning rule for the Kohonen network can be either supervised or unsupervised. The Kohonen network maps the input vectors to its output neurons described by the distribution of clusters (18). The clusters can be considered as regions in the neural network where the network is most sensitive to corresponding external inputs. In other words, the cluster is centered around the local maximum of the input excitation (18). When the learning is unsupervised the network is called a *Kohonen self-organizing feature map* (KSOFM).

16.2.2.4. Unsupervised Learning. The unsupervised learning starts by randomizing all weights. Each input is applied to the network in sequence, repeating all inputs as many times as necessary to define a winner for each input. The location of the maximum is regarded as the best match between the input and weight vectors. A widely applied matching criterion is the Euclidean distance; thus the processing element with minimum Euclidean distance is defined to be the winner given by Eq. (16.15) (18). Unsupervised networks are

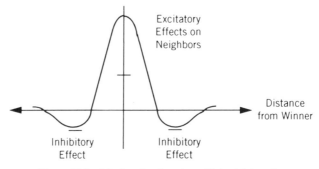

Figure 16.9. Mexican hat function of lateral interaction.

most suitable for statistical analysis having naturally formed clusters.

$$\|X - W_{\text{winner}}\| = \min\|X - W\| \tag{16.15}$$

Next, the weights of the winner and specified neighbors are adapted as follows, leaving the others alone:

$$W_{\text{NE}}(t) = W_{\text{NE}}(t) + \alpha(t)[X(t) - W_{\text{NE}}(t)] \tag{16.16}$$

where $W_{\text{NE}}(t)$ represents the weights of neurons within neighborhood at time t and $\alpha(t)$ is the learning rate. The learning rate $[0 < \alpha(t) < 1]$ is usually a slowly decreasing function of time. This learning rule causes the weight vector to be moved closer and closer to the input vector. After sufficient training, the weights will be organized such that topologically close nodes are sensitive to inputs that are similar [10]. The trained Kohonen network offers a feature map on which several isolated regions are formed, representing specific clusters (11).

16.2.2.5.   Supervised Learning. Supervised Kohonen networks are used as pattern classifiers having artificially defined clusters. A prerequisite of using a supervised net is that the class affiliation of the input patterns must be known a priori, or that desired clusters be declared empirically. Each desired cluster can have a single neuron, or several neurons. The size of the neural network is related to the number of separable input patterns and the accuracy desired (20). Unlike KSOFM, the supervised nets, called *learning vector quantizers*, require no neighborhoods because clusters for the input patterns have already been defined.

During training, with initial small random weight values, each neuron in the network has to compete to respond to a particular input pattern by comparing the Euclidean distance between its weight and input vectors, as in the unsupervised case, Eq. (16.15). In the LVQ1 model, only the winner gets to update weights, depending on its current location in the network. If the winner is located within the desired cluster, the winner's weight vector will be moved closer to that input vector by adding a fractional difference of the input and weight vectors.

$$W_{\text{winner}}(t + 1) = W_{\text{winner}}(t) + a(t)[X(t) - W_{\text{winner}}(t)] \tag{16.17}$$

If the winner is outside a desired cluster, the winner's weight vector will be moved away from that input vector by subtracting the same fractional difference; that is,

$$W_{\text{winner}}(t + 1) = W_{\text{winner}}(t) - a(t)[X(t) - W_{\text{winner}}(t)] \tag{16.18}$$

In both cases, all loser's weight vectors remain unchanged. It is not guaranteed that the winner will be inside its desired cluster when a training pattern is presented since the weight vector of the desired cluster still may not be closest to that input vector. On the other hand, it is possible that a matched winner may miss its desired cluster because the weight vector of that cluster may be changed by other unmatched winners.

From a network viewpoint, the training process can be considered as the winner's movement around the entire network. After successful training, each winner settles down inside its desired cluster, and the weight vectors over the entire network are optimized. The weight values represent the desired input–output relationship. When a trained pattern or a similar untrained pattern appears, the network generates the category with the best match. The main drawback of the algorithm is that the probability of any particular input initially matching its desired cluster is $1/N$, where N is the number of classes to be classified. Thus the larger the number of desired clusters, the smaller the general probability will be (20–23) and the longer the training required.

16.2.2.6. M-LVQ1, Modified Learning Rule for the LVQ1. With the LVQ1 learning rule, a long training process is frequently involved, especially for a large number of input groups. The modified learning rule for the LVQ1 network introduced by Grossman et al. in a fiber optic sensor/structure application (22) reduces the required training time significantly. The new learning rule overcomes the problem of training time by increasing the probability of any particular pattern matching is desired cluster. They demonstrated the effectiveness of the new learning rule in the application of optical neural networks.

This new learning rule is identical to the original one when a winning neuron of a particular input matches its desired cluster, but is modified when the winning neuron of an input does not match. When a mismatch occurs, the new learning rule not only adjusts the winner's weight vector farther from that input but also updates the weight vector of the desired cluster by moving it closer to that input vector. For the mismatched case the modified learning rule is

$$W_{\text{winner}}(t + 1) = W_{\text{winner}}(t) - a(t)[X(t) - W_{\text{winner}}(t)] \qquad (16.19)$$

$$W_{\text{d.c.}}(t + 1) = W_{\text{d.c.}}(t) - a(t)[X(t) - W_{\text{d.c.}}(t)] \qquad (16.20)$$

where the subscript d.c. refers to the desired cluster. The operation of the modified learning rule can be viewed as a "push–pull" process. Whenever the unmatched case occurs, the winner is pushed away from its current location by reducing the local sensitivity to that particular pattern. At the same time, the winner is also pulled toward its desired cluster by increasing the local sensitivity of the desired cluster to that pattern. Mathematically, the learning rule increases the general probability of any particular input pattern matching its desired cluster. Physically, the winner's movement around the network controlled by M-LVQ1 is directly guided, whereas it is indirectly guided under LVQ1. The result is that the new learning rule converges more rapidly (22, 23).

16.2.3. Iterative Networks

The identification of a system (i.e., the smart structure) is of primary concern in the design of a smart structure or material (24): The study of the stability properties of multilayer networks is of interest. The use of ANNs for system identification is reasonable. We describe a network for the prediction of a nonlinear system response. Simple networks have been trained to respond to a nonlinear system (25), which leads us to the recursive network as originally described by Jordan (26) for a similar task.

16.2.3.1. Jordan Recursive Network. Recurrent networks suggested by Jordan (26) can behave like a state machine. Thursby et al. (27) have applied the Jordan concept to the design of a network for antenna element control. The objective of this system is to control the frequency of operation of a conformal antenna. The recursive network concept is a simple one—the output of a network is fed back to the input to provide state history to the network. Figure 16.10 shows the basic structure of such a network. The ability of a network to process serial input data and to provide a serial output response is useful in control systems. The question of performance of the network centers on its time response, including measures like settling time, overshoot, and steady-state error. The training of a recursive network whose output is changing dynamically with time requires special considerations. The solution is best seen by modeling a time-evolving waveform such as the damped sinusoid, described by

$$Y(t) = e^{-\alpha t}\sin \omega t \qquad (16.21)$$

To train a network to respond to $Y(t)$ where the target value for the network is changing at each sample time and the desired response of the network is the value of the function at that time, we can adapt the weights to an average change as specified by the ensemble of errors calculated at each instance of time that the network is operating. The length of time that the network is allowed

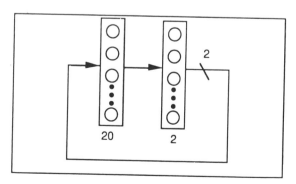

Figure 16.10. Recurrent network architecture.

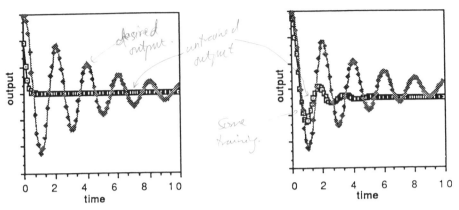

desired output

untrained output

some training.

Figure 16.11. Typical untrained response.

to evolve will be determined by the desired final results and the nature of the signal. Figure 16.11 shows the output signal for one typical untrained response and the response after some training. The final network was able to track the desired time response for all four initial condition types, although one was used to train the network.

One can consider the time evolution of the error signal to be the result of a series of identical networks concatenated to form a virtual network that has as many sets of layers as the number of iterations of the actual network, shown in Fig. 16.12. As Jordan points out, one must store the error signals for each of the iterations. After the time course of the signal is completed, then, we average the individual weight changes and modify the network accordingly.

16.2.3.2. Training Algorithm. We chose a multilayer perceptron with two layers. The layers consisted of two neurons in the input and output layers and 20 neurons in the hidden layer. The network was configured into a recurrent structure. This architecture was chosen to produce a time-varying response similar to the evolution of state in a dynamical system. The network structure is shown in Fig. 16.13.

How to decide how many neurons to use.

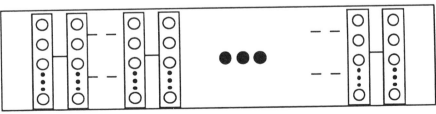

Figure 16.12. Unfolded virtual layers for backpropagation of error.

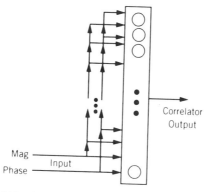

Figure 16.13. Schematic representation of the neural correlator.

16.3. APPLICATION OF NEURAL NETWORK PROCESSORS TO SMART SYSTEMS

In this section we provide examples of how neural networks can be used with fiber optic sensors and as a part of smart systems. The intent is to provide a basis for further study and investigation and to provide insight into the many possibilities of utilizing neural networks for smart system applications.

16.3.1. Smart Electronic System

We will use the recursive network concept, described earlier, to develop a smart electromagnetic structure (SEM). This structure is "smart" in that it integrates sensing elements (e.g., antennas), processing elements (neural networks), and control elements (diodes). Smart electromagnetic structures (SEMSs) can provide an adaptive electromagnetic (EM) environment for the structure on which they are mounted. With their sensing capabilities they can detect and modify the surrounding EM fields and therefore their far-field pattern. This ability will also allow the structure to change its radar signature (28). The ability to adapt comes from the closed-loop nature of the SEMSs. The speed of adaptation is determined by the speed of the loop, which in turn is determined by the speed of the computational elements. By embedding a control element in the structure of a single microstrip patch element (28, 29), its electrical characteristics can be changed. The inclusion of an ANN in the antenna model can be taught to adapt the magnitude and phase response of microstrip patch antennas to incoming signals and thus produce a SEMS. This we refer to as the neural antenna.

With an artificial neural network (ANN) as the processor, the SEMS loop can respond in approximately 15 gate delays, as shown later in this section. ANNs and their ability to model and control dynamical systems for smart structures, including sensors, actuators, and plants, are directly applicable to the SEMS concept (27).

16.3.1.1. Neural Net Antenna. The micropatch antenna has been the main-stay of conformal antennas for many years (30). The antenna has many advantages, including simplicity and size, and a few drawbacks, such as narrow bandwidth. The electrical characteristics of the antenna can be adjusted using control elements embedded in the patch itself (31). We describe research into the control of such patch antenna elements using a neural network (NN) in the feedback loop to enhance the operating characteristics of the patch. The advantages of a neural net over a classical processor and algorithmic control scheme are that the net has the potential to be taught by example rather than requiring the calculation of new control points for each new condition. The neural net has the advantage of being able to make the required determinations in near real time. The additional ability of the net to adapt to previously unknown inputs (generalization) and its fault tolerance makes the neural antenna an ideal candidate for flexible tactical antennas for the future. In addition, the antenna could be manufactured with a set of control devices placed at convenient points on the patch surface. The network could then perform the desired tuning after manufacture, thus reducing the required manufacturing tolerances. The combination of a simple neural network with a microstrip patch antenna shown in Fig. 16.14 has the potential to enhance the characteristics of the patch antenna. An example of this is center-frequency tuning, described below.

16.3.1.2. Neural Network Model. Our first goal for training the smart an-tenna structure was to provide high-speed frequency tracking of an unknown incoming signal. The network has two layers, 14 neurons in the input layer and 20 neurons in the hidden layer. One neuron is in the last layer. Each neuron uses a sigmoidal function, allowing numerical outputs from $+1$ and -1. One linear neuron with fixed weights and an integrating connection is used for the output. Two neurons in the input layer receive sensor information from the receiver; the rest of the input layer neurons receive delayed values in a shift register fashion from their neighbor. There are a total of seven pairs of neurons

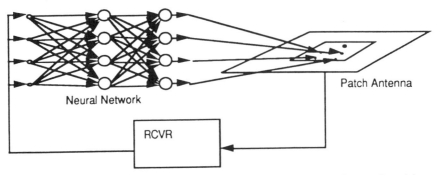

Figure 16.14. The microwave patch antenna with tuning points and a neural network to drive the points can be considered a smart antenna structure.

in the input layer. For training, a back-propagation algorithm was used to adapt the weights. Since the input layer has time-delayed input values, we change the weights after all input neurons have received an input from a test pattern.

16.3.1.3. Smart Antenna Concept. The neural network must provide the proper bias voltage for the tuning element to tune the antenna center frequency to coincide with the unknown incoming signal frequency. Figure 16.15 shows this relation for the tunable patch antenna. We wish to apply this technique to move the center frequency of the antenna from position (1) to (2) This concept was first described by Schaubert et al. (31). The method described requires the determination of the solution of a set of complex equations for each control point. If we can train a ANN to determine the desired driving function, we can save significant time in processing. First we must develop an algorithm to train the sequential neural network. We explain the training strategy in the next section.

16.3.1.4. Training Algorithm. The neural network was trained with a conventional back-propagation algorithm, modified to include a strategy for choosing the input training patterns and for calculating the error terms. Input values for the neural network come from the receiver. Target values were determined by subtracting accumulated voltage from the desired target voltage. We initially chose four different pairs of diode voltages as training patterns: initial bias values, desired bias values; (5 V, 25 V), (10 V, 20 V); (20 V, 10 V), and (25 V, 5 V). A step-by-step description of the training is given below.

Step 1. Initialize all input neurons, weights, and offsets. Set all input neurons, weights, and node offsets to random values.

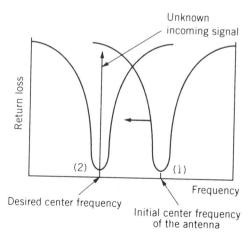

Figure 16.15. Relationship between the center frequency and unknown incoming signal frequency.

Step 2. Choose an input pattern at random. Set the receiver output to an initial bias voltage randomly.

Step 3. Apply the input and propagate it forward through the network.

Step 4. Determine the desired output. The time course of the system results in the bias voltage that corresponds to the final tuning frequency of the antenna desired. The input and desired pattern are changed at every iteration of the network. We choose a desired output as a value that makes the accumulated voltage equal to the target voltage.

$$T_i(t) = AV(t) - TV$$
$$= [O_i(t) + AV(t-1)] - TV \qquad (16.22)$$

where $AV(t)$ is the accumulated voltage at time t and TV is the target voltage used for a given initial starting voltage.

For a better understanding of how to decide the desired output, we show a graph of accumulated voltage versus time. Figure 16.16 shows the untrained network response for a starting input diode voltage of 20 V and target voltage of 10 V; the graph shows 30 iterations of the network (epochs). With a starting voltage of 20 V and the target of 10 V, the untrained network was well behaved for only 18 epochs. The error can be seen as the difference between the desired value (a constant 10 V) and the actual net output (e.g., at $t = 9$ the error is approximately zero). After 18 epochs the output saturates and no further change information is available. The desired response of the system would be for the output to go directly from 20 V to 10 V in one step. This would provide an ideal step response with no overshoot or ringing. For the epochs before the system saturates, the error values can be measured and used for back-propagation. Figure 16.17 shows the output after training.

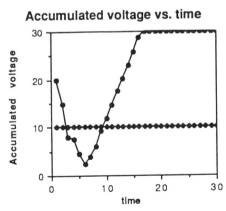

Figure 16.16. Plot of accumulated voltage and target voltage (10 V) versus time without training.

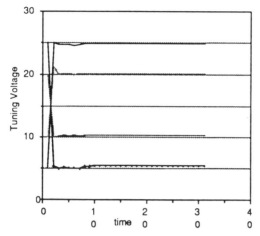

Figure 16.17. Results of training network to tune the antenna.

Step 5. Compute the change of the weights using the back-propagation rule. Weight changes are computed by the relation

$$\Delta W_{ji}(t + 1) = \eta \delta_j O_i + \alpha \Delta W_{ji}(t) \tag{16.23}$$

where h is the learning rate of the network and δ is given by

$$\delta_{pj} = \frac{\partial E_p}{\partial \mathrm{net}_{pj}} \tag{16.24}$$

and α is the momentum. E_p is the error function for the network. For further discussion of the learning algorithm, see Rummelhart et al. (32).

Step 6. Calculate the new input. Based on the tuning voltage determined by the network, the receiver and antenna system output is determined. For a detailed description of the model, see Thursby et al. (33).

Step 7. Repeat steps 3 through 6 during the response time. The response time is defined as the time from the start of the stimulus until the system has reached steady state or has caused the system to saturate.

Step 8. Update all weights. The weight change is computed after a complete response has been obtained. The weights are calculated as follows. For each epoch i the difference between the desired output and the actual output is used to determine the weight change for each epoch. After the weight changes for all epochs have been accumulated, the average weight change for each weight is calculated. The average weight change for the weight between neuron l and k for pattern i is given by

$$[\Delta W_{1k}^i] = \frac{1}{N} \sum_{i=1}^{N} \Delta W_{1k}^i \tag{16.25}$$

Change the weights with averaged weight changes using

$$W_{1k}^{new} = W_{1k}^{old} + [\Delta W_{1k}] \tag{16.26}$$

Repeat steps 2 through 8 until the error is reduced to an acceptable level. The error is defined as

$$E = \sum_{j} (T_j - O_j) \tag{16.27}$$

16.3.1.5. Results of Training. The system was trained to respond to four different starting and target frequencies. The final error was less than 0.05%. The network converged to similar error values for several different initial weight sets. Figure 16.17 shows the results of training the network to produce the desired response. Figure 16.18 shows the results of several tests of initial and final values that were not used for training. The ability of the antenna system to tune to the desired center frequency is demonstrated by the rapid movement from the starting voltage to the desired final voltage.

16.3.1.6. Further Test Results. We have investigated the response of the trained network from several points of view. First, the network response to different initial conditions for a fixed target should not change significantly over the operating range of the system. To test this we presented the system with several different initial conditions for a given target value. The response of the system to these conditions is shown in Fig. 16.19. Four different target voltages, 5 V, 10 V, 20 V, and 25 V, were chosen for this set of tests. For all starting voltages the system converge to the desired target voltage.

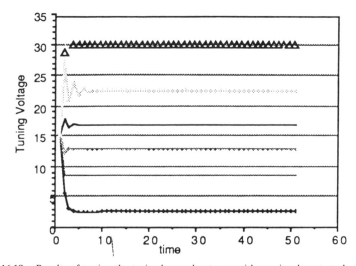

Figure 16.18. Results of testing the trained neural antenna with previously untested conditions.

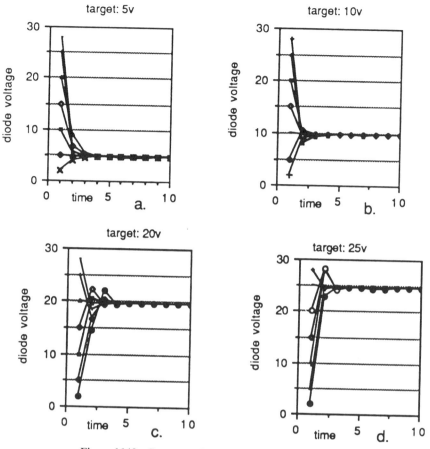

Figure 16.19. Response of the network with training patterns.

The second case of interest where the desired output is varied over the range of operation the system corresponds to the actual operating range of the embedded control device is shown in Fig. 16.20. The steady-state error of the network when compared with the desired tuning voltage is shown in Fig. 16.21, which is well contained and centers around the desired response. We have tested the system with a linearly changing input frequency for the unknown input signal. Both positive and negative sloped frequency characteristics have been tested. The response of the system to a positively increasing frequency signal is shown in Fig. 16.22. The greater the slope, the better the network is able to track the input.

16.3.1.7. Conclusions. We have demonstrated a smart antenna system controlled by a neural processor that can tune its operating frequency to optimize its response to an incoming signal in near real time. The network is relatively simple and contains only 36 neurons. The use of a fixed weight integrating

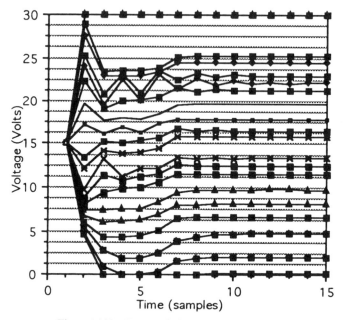

Figure 16.20. Test result for nontrained target voltage.

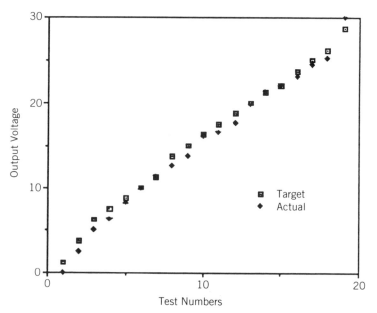

Figure 16.21. Error produced by the neural antenna system over the set of test cases shown in Fig. 16.20.

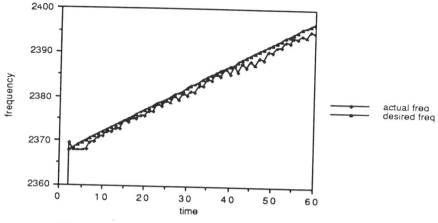

Figure 16.22. System response for linearly sweeping target frequency.

neuron was found to improve the response of the network. The network responded with less than 0.05% error to the training set and when tested with data previously not seen, the system responded with error less than 12%, as measured with respect to the desired amplitude. The system was able to respond to a dynamically varying velocity input signal with minimal steady-state error. Although this neural network was used to control the smart antenna, the concept of delayed input value and accumulated weight change can be applied to many control problem areas.

16.3.2. Structural Damage Assessment Employing a Kohonen Optical Neural Network Processor and an Embedded Fiber Optic Sensor Array

A two-dimensional damage assessment system for real-time monitoring of the structural health of a composite structure has been described and models tested (22). The system determines the occurrence of impact damage and indicates its spatial location on the surface of a structure. With impact damage, like other forms of damage in composite structures, the external appearance of the structure may appear normal while internal damage may be extensive, including cracking and delamination. These types of failures make in situ real-time nondestructive evaluation and health monitoring of such structures essential. Predicting catastrophic structural failures far enough in advance to prevent them and providing real-time structural health and damage monitoring is coming closer to reality with continued sensor/processing system development. Fiber optic sensors appear best suited for these applications for a number of reasons, including their small diameter, all dielectric nature, geometric versatility, sensitivity, and dynamic range.

16.3.2.1. System. The system, whose block diagram is shown in Fig. 16.23, combines a fiber optic strain sensor array and a Kohonen optical neural

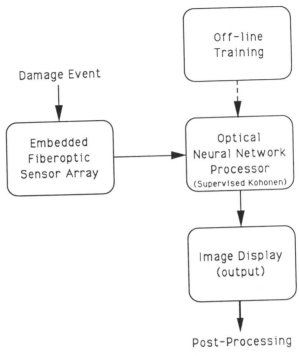

Figure 16.23.   System block diagram.

network processor made from passive components to determine damage. A two-dimensional fiber optic sensor array can be embedded into a composite structure during the manufacturing process and can subsequently be used to detect changes in the mechanical strain distribution caused by impact damage to the structure. The optical processor is a pretrained, supervised Kohonen neural network employing a modified LVQ1 training algorithm (23). This neural network architecture has the capability to indicate damage locations because it associates each input data set with a unique spatial location in the neural plane.

   The system optical schematic is shown in Fig. 16.24. Light from a laser is coupled through a 1 : 24 star coupler into a 12 × 12 fiber polarimetric strain sensor array embedded into a composite plate. Damage at a particular spatial location in the plate produces a strain distribution which results in a unique set of 24-sensor output light intensities. These sets of signals become the input to the neural network. The light from each optical fiber is projected through an array of transparencies whose transmissivities are equal to the neural network weight values, which are determined for the network during an off-line training process. The weighted light from all the fibers is projected into an image screen corresponding to the neural plane. The weighted light intensities add at the image screen, with the brightest area on the screen corresponding to the neuron that would "fire." The spatial location of this neuron corresponds

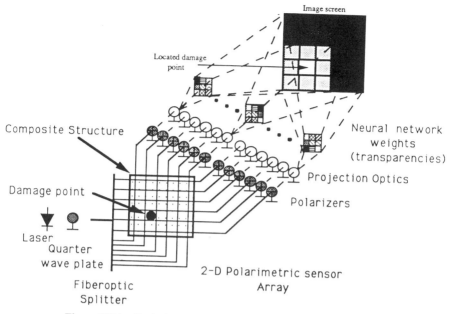

Figure 16.24. Optical two-dimensional damage assessment system.

to the spatial location of the damage location. Off-line training determines the appropriate weight values to ensure that damage occurring at a particular location on the composite causes a bright spot to be produced at the same relative location at the image (neural) plane. The image on the screen, if desired, can then be captured by a camera or CCD array and subsequently displayed on a video monitor after any desired postprocessing, such as thresholding, is performed.

16.3.2.2. Composite Structure. The structure analyzed is a square plate of epoxy–fiberglass composite, shown in Fig. 16.25, with dimensions $180 \times 180 \times 0.9$ in. To calculate the static strain distribution of the composite structure for various damage locations, the MSC/NASTRAN program was utilized. The plate was assumed to be a six-ply symmetrical laminate, with each ply 0.15 in thick and with lamina orientation of $-45/45/0/0/45/-45°$. One edge of the plate is attached to a fixed structure; thus displacements caused by the damage are not transferred onto the rigid structure. The other three edges are free edges forming a winglike structure.

The plate was divided into 1296 equal elements, each 5.0×5.0 in. Each element was connected to its immediate neighbor elements at their nodes. To reduce the amount of data and processing time, the QUAD4 type of element was selected for the analysis. For each specific damage location, the NASTRAN program computes mechanical displacement and strain values for

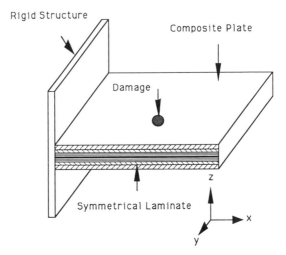

Figure 16.25. Composite structure.

each finite element in the composite structure. Thus the strain distribution is formed by combining all discrete strain values on the elements. Both x- and y-oriented strain values were calculated. The strain is maximum near the impact point and values decrease away from this point.

16.3.2.3. Fiber Optic Sensor System. In polarimetric sensors we interfere two optical beams propagating down a single-mode fiber with mutually orthogonal linear polarizations. High birefringence polarization-preserving fiber has two preferred polarization axes; light polarized parallel to each axis propagates with constant polarization but with different velocities. This type of fiber is designed to minimize the coupling of energy between the two polarizations. Strain on the fiber, due to damage of the structure, for example, results in a phase shift between the two polarizations. This then results in a change in the state of polarization of the emerging beam. By measuring the light output of the fiber through a polarizing analyzer oriented at an angle of 45° to these two axes, the strain-induced polarization change is converted into an intensity change. The output light intensity varies sinusoidally between zero and a maximum with varying strain.

Wih no damage-induced strain, the light output of the sensors is set to zero. The 24 sensors, configured into a 12 by 12 array, are aligned along the two coordinate directions, x and y. A unique strain distribution will be formed corresponding to a particular damage location; the strain sensor array will thus produce a unique set of 24 output intensities associated with this damage location. This results in a one-to-one correspondence between the damage location coordinates and the set of sensor array output intensities, which can be exploited by an artificial neural network processor.

16.3.2.4. Kohonen Neural Network. As shown in Fig. 16.26, the Kohonen architecture has a single-layer parallel structure. The neural network has the capability to classify outputs from the fiber sensors into desired clusters in the network (18, 34); the desired cluster has the same relative spatial location in the neural plane as the damage has in the composite plate. The output light intensities from the fiber sensors are processed in parallel by the Kohonen neural network. It determines the outputs almost instantaneously, the time delay being the time for light to propagate through the system (i.e., less than a nanosecond).

The neural network was simulated in software and trained off-line. After the strain distributions were calculated and the optical intensity patterns from the fiber sensor array determined for each damage location, a simulated neural network was trained to classify these patterns into the desired spatial clusters in the neural layer. The training data were the output light intensity values of the fiber sensors corresponding to the damage locations. The number of input elements of the neural network depends on the number of fiber sensors embedded in the composite structure, in this case 24. A single-layer neural

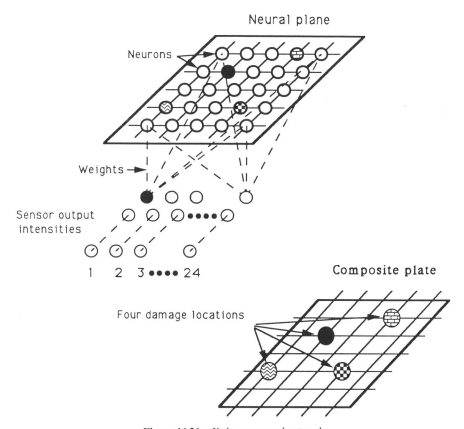

Figure 16.26. Kohonen neural network.

network architecture was employed according to the simulation results of the composite structure and the arrangement of fiber optic sensors described earlier. An 8×8 square neuron array corresponded topologically to the composite plate. Each neuron in the Kohonen layer represented a single damage point having the same relative two-dimensional position with respect to the neural network. Sixty-four sets of training data were prepared, corresponding to 64 selected damage locations, each of them represented by the 24 corresponding light intensity values from the embedded fiber optic sensor array. Thus the neural network processor had 24 inputs and a possible 64 outputs. The sets of input data were randomly fed into the neural network during the training process. At the beginning of training, no inputs matched their desired clusters because of random initial weights. After 900 iterations, the number of matched inputs grew rapidly; the training process was terminated when the complete set of inputs matched their desired clusters after 2048 iterations. After training, the resultant set of weights for the entire network determines the relationship between the light intensity patterns and the damage locations.

The optimized weight values can be implemented by use of transparencies having optical transmissivity values equal to the weight values. The bigger the weight value is, the larger the corresponding light transmissivity would be. The outputs from the 24 fiber sensors would then illuminate the entire image screen after passing through their weight transparencies. Each set of output intensities will produce a specific intensity distribution on the screen. The sum of the 24 individual intensity distributions will form a feature map on which one brightest point will occur. This brightest point is located in the same relative spatial position as the damage in the composite structure. In other words, for a particular image pattern from fiber sensor array, which is the optical representation of the damage location, there is a brightest point at the appropriate spatial position on the image screen. The brightest point is the damage location. The combined image on the screen, if desired, can be captured by a camera or CCD array and can then be displayed on a video monitor after any desired image processing, such as thresholding, is performed.

16.3.2.5. Simulation Results. The simulation resulted in 100% of the training inputs correctly classified into their desired clusters. The trained neural network was then tested by using untrained damage inputs. One hundred and twenty-eight sets of untrained inputs were used, which corresponded to 128 damage locations different from the trained inputs. The testing results demonstrated that 100% of the untrained inputs were classified correctly. The trained neural network processor was also tested with noisy light inputs to the processor. The noise, which can be considered as variations due to vibration or other sources, was varied from 0 to 21% with respect to the values used for training inputs. The results shown in Fig. 16.27 show that all trained damage inputs with up to 15% noise are associated correctly with their desired location, which means that the processor will operate as desired. With 18% and

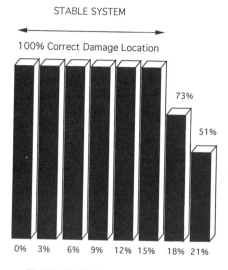

Figure 16.27. Effect of noise on classification accuracy.

21% noise, the accuracy degraded to 73% and 51% trained damage inputs classified into their desired locations. Thus the trained network processor appears to be extremely robust.

16.3.2.6. Future Work. Plans for further development include system breadboard development and multiple-damage assessment. In a multiple-damage situation, the multiple clusters of the neural network processor will indicate the damage locations for the strain distribution caused by multiple damage occurring simultaneously in the composite structure.

16.3.3. *N*-Mode Fiber Optic Sensor with Neural Network Processing

The spatial distribution of the output light intensity from standard multimode fibers, where hundreds or thousands of modes propagate, is a very complicated distribution determined by the interference between the modes. The pattern is very sensitive to strain and other measurands, such as temperature, when compared to the two-mode case, producing patterns that do not repeat over a large range of strain values. Multimode interferometric sensors have the potential for being low cost, due to the low cost of the fiber, and are very sensitive. Two disadvantages, however, are that it is not easy to process the "output signal" to make full use of the sensitivity and dynamic range and that they are very sensitive to microbend effects. Microbend effects result in a decrease in output intensity independent of the measurand, as well as a redistribution of energy between excited propagation modes, and from excited to unexcited modes.

One technique would be to store the intensity distribution data experimentally for each strain value. To determine an unknown strain value, one could use image processing techniques to correlate stored images with the one produced from the unknown strain, thus identifying the unknown strain value. Limitations of this technique are that a very-high-resolution imaging system is required and extensive computer processing time is required to do the correlation and interpolation. This results in an expensive signal-processing system where only static (versus dynamic strain) could be measured due to the long processing time required.

Another technique is to sample a small spatial area of the output and from this signal determine the measurand (35). One problem with this technique is that microbend effects will redistribute the energy between modes, resulting in an output distribution that varies with severity of microbend.

16.3.3.1. Sensor and Processing System. In our approach (36) a few-mode sensor scheme is utilized in which a small number of modes are allowed to propagate in a standard single-mode communications-grade fiber and which interfere at the output. At the output a multielement photodetector samples the interference pattern intensity distribution and converts it to electrical signals — one per element. These outputs are connected to an artifical neural network which has been trained to associate an experimentally obtained training set of intensity distributions with their corresponding strain values. When an unknown strain is measured, the neural network calculates the strain value and will interpolate or extrapolate as required between the training values. The advantages so obtained are a sensor that is sensitive, having large dynamic range and constructed with the use of a low-cost fiber.

Although the possibility of using more than the two lowest-order modes, the LP_{01} and LP_{11}, with regular single-mode fiber has been known (37, 38), operation was limited to two modes primarily because of the difficulty in processing multimode (three or more) sensor outputs. The two-mode sensor did not become feasible until a method using elliptical core fiber was discovered to prevent pattern rotation, since any rotation would preclude the use of a photodetector positioned at one of the lobes of the output pattern. With our technique for sensor processing, any number of modes (two or more) can be used since we are employing a form of pattern recognition to recognize the modal pattern associated with the phase delay between modes induced by the measurand upon the fiber. An added feature of this processing technique allows the processing of outputs with pattern rotation since the output pattern is sampled at a number of points rather than just one. This allows a fiber/source to be picked having an optimum number of modes for processing. The experimental system is shown in Fig. 16.28. A 632-nm helium–neon laser is coupled into an optical fiber designed for single-mode operation at 1300 nm, which has diameter 8.3 μm and numerical aperture of 0.12. This results in a normalized frequency parameter of 5.13, which produces a predominantly three-lobe pattern which oscillates and rotates with strain. The output light spatial distribution depends on the relative phase between the interfering modes,

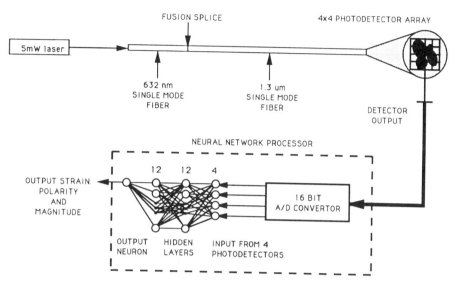

Figure 16.28. *N*-mode sensor system.

which is changed when the fiber is strained. A 632-nm single-mode fiber is fusion-spliced to the 1300-nm fiber input to ensure a stable launch angle. The 1300-nm fiber was epoxied onto a cantilever beam, and the beam was strained by static loading. The output was projected onto a 16-element (4 × 4) photodetector array. The values of the 16 photodetector output voltages are converted into a digital value by a 16-bit analog-to-digital converter. Four of these signals are then processed by a neural network processor on an Apple Macintosh computer.

16.3.3.2. Experimental Results. Figure 16.29 shows the output voltage variation for all 16 photodetectors as a function of strain. The range is from −2200 μstrain to +2200 μstrain. The 16 graphs, one for each photodetector, are arranged spatially in the figure as they are in the 4 × 4 array. There is a significant variation in output with applied strain between the various detectors. It is apparent from the figure that only a few of the photodetectors are required to determine the strain value. The optimum choice for which outputs to use for input to the neural network involves three considerations. The first is minimizing the number of detectors needed and thus the size of the neural network required. This will result in the simplest system and the minimum processing time and cost. Second, we wish to maximize the variation (i.e., minimize the correlation) between outputs, which will maximize the accuracy of the processor output. The outputs must be sufficiently different such that there are no two values of strain at which the outputs are the same. The third consideration is that the output signal modulation for the detectors used be maximum. This ensures that the uncertainty in the output is minimized over the strain range. This, too, will maximize the system accuracy.

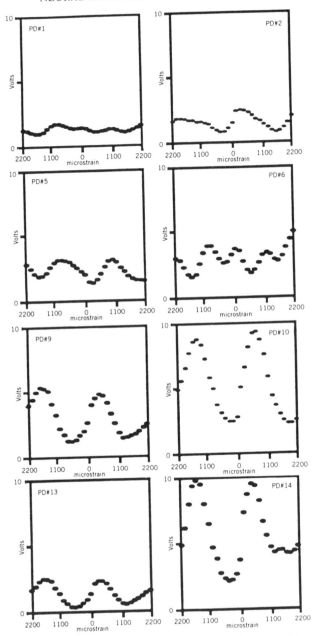

Figure 16.29. *N*-mode sensor voltage-strain characteristics.

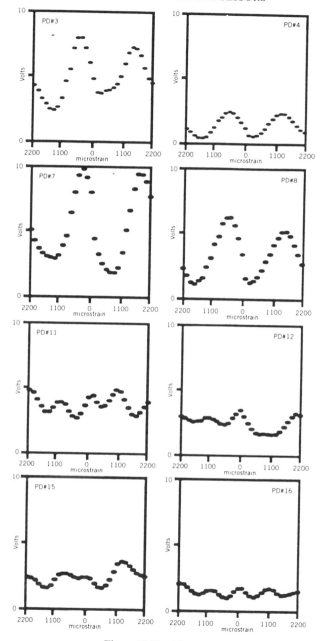

Figure 16.29. (*Continued*).

The outputs from four of the photodiodes were chosen to be input to the neural network. The neural network is a standard multilayer-perceptron architecture trained with the back-propagation rule. The network consists of the four inputs from the photodetectors routed to two layers of 12 neurons each and a final, output layer which provides the strain magnitude and polarity. Figure 16.30 shows the results of the training and testing of the neural network. The network was first trained with 52 sets of voltages, each set corresponding to the four detector outputs for each of 52 strain values spread over the $\pm 2200\,\mu$strain range. The training resulted in an rms error of less than 0.2%. The network was then presented with 33 testing sets that it had not been trained on. The network determined the corresponding strain with an rms error of 0.63%. Using $2200\,\mu$strain as a maximum, this corresponds to a $13.8\,\mu$strain maximum error. As can be seen from the graph, the actual error varies with strain value.

Our next step is to vary the number of neurons, training sets, and the training time to minimize the error. We also investigate the possibility of incorporating input–output lead insensitivity.

16.3.4. Smart Structure Incorporating an Artificial Neural Network, Fiber Optic Sensors, and Solid-State Actuators

Artificial neural networks have been shown to be useful in sensor processing and dynamic system modeling and control, as well as other areas of signal

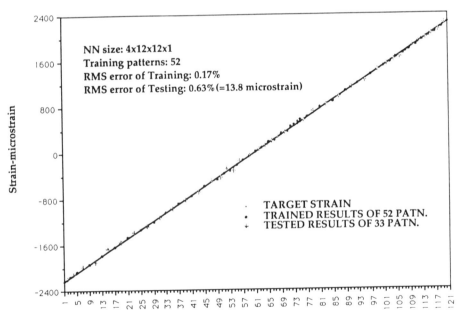

Training and Testing Vector Numbers

Figure 16.30. Neural network processor results.

analysis. Their application to the development of a real-time smart structure is indicated further by the extremely fast postlearning speed of the neural network. Fiber optic sensors possess fast response times, high sensitivities, and are well suited to the detection of structural changes. The integration of these two technologies can result in a highly adaptive smart structure. A smart structure applicable to artificial human joints and airfoil control surfaces has been modeled (39). The muscles in the model are shape-memory alloy strands, and the joint-position sensors of the neuromuscular system are modeled by fiber optic sensors. The control (neural) circuits are an artifical neural network. Our emphasis was on the suitability of each of these subsystems to the problem.

16.3.4.1. Smart Structure System. Biological concepts have long been applied successfully to human-made structures. Artificial honeycomb structures and fish-eye lenses are some examples of devices that have been copied from nature. Artificial smart structures can and should be patterned after biological models. The work of Brooks (40) demonstrated that arrays of single or dual neuron nets, when operated in concert, result in a complexity of performance that approximates that of simple invertebrates. The logical extension of this line of thinking is the application of such concepts to the smart structure model to approximate some of the neural circuits of, for example, the leg or arm systems of higher animals and humans.

The biological system we chose to model is the neuromuscular functional unit. Figure 16.31a depicts a representative neuromuscular unit (41). Figure 16.31b shows a mechanical analogy to the biological system. Figure 16.32 shows the final abstraction of the system to a cantilevered beam instrumented with a fiber optic strain sensor and shape memory alloy (SMA) actuators. The processor is an artificial neural network.

The system modeled and shown in Fig. 16.32 consists of an aluminum cantilever beam attached at its end to a rigid structure. A fiber optic sensor is

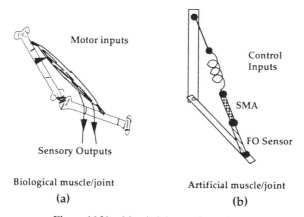

Figure 16.31. Muscle joint configurations.

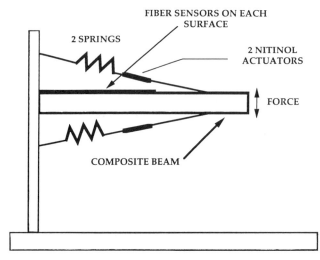

Figure 16.32. Smart beam configuration.

attached to one surface to monitor the strain and thus the displacement of the beam. There are spring-shape memory alloy actuator combinations on each side attached to the beam. The objective is to have the beam remain stationary regardless of the force exerted on the end of the beam by some external source. The operation of the system is as follows. As a force is applied to the end of the beam, a deflection occurs, which is detected by the fiber optic sensor system which determines the magnitude and polarity of the induced strain. At the same time, the spring on the side of the beam in the direction of the force compresses while the other spring is stretched. Without use of the actuators, the beam would deflect to an equilibrium position dependent on the geometries, materials, spring constants, and so on. Since we wish the beam to remain in its original position regardless of the force applied to the end, the shape memory actuator on the side opposite to the direction of the force must be actuated and caused to contract an amount sufficient to return the beam to its original position.

16.3.4.2. Fiber Optic Sensor System. The fiber optic sensor system used in the simulation is a polarimetric architecture similar to that discussed previously. Light polarized at 45° to the polarization axes of a birefringent fiber is input to the fiber. As the fiber is strained, the phase delay between the two polarizations at the fiber output varies, changing the polarization of the output light. After passing through a polarizer or polarizing fiber aligned at 45° to the fiber axes, this polarization modulation is converted into an intensity modulation. Figure 16.33 shows the usual sine-squared dependence of the output intensity to the induced strain on the beam. This output signal becomes the input to the processor.

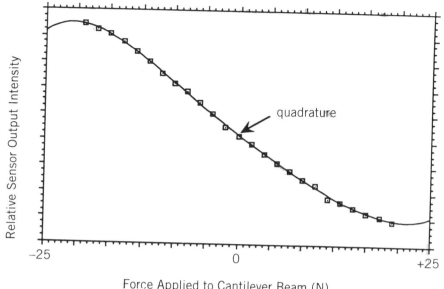

Figure 16.33. Output of sensor versus force.

16.3.4.3. Shape-Memory Alloy Actuators.

Shape-memory alloy strands were chosen as the actuators for the system. This class of materials has the property of changing crystallographic structure abruptly when heated or cooled past a temperature transition point determined by the material composition. This change in crystal structure results in a large reversible change in length, either positive or negative, depending on training. Although many compounds exhibit this effect, nickel–titanium alloys exhibit it to an unusually large degree and are the most commonly used. By varying the relative concentrations of the two elements when the material is manufactured, one can set the transition temperature to the desired value within the overall range. The effect is large and reversible and produces an actuator capable of large displacements and large forces. The major disadvantage is the millisecond or slower activation rate because thermal mechanisms are involved rather than direct electrical or optical. Typically, the actuators are controlled by electrical current, which heats the alloy due to ohmic heating. Optical energy for heating delivered by an embedded optical fiber has also been utilized successfully for this application (36).

16.3.4.4. Neural Network Simulation Results.

The neural network architecture employed for the signal processing was the multilayer perceptron with sigmoid activation functions, the operation of which was discussed earlier. The configuration is shown in Fig. 16.34. Output from the fiber optic sensor monitoring the strain in the beam is input to a two-layer neural network

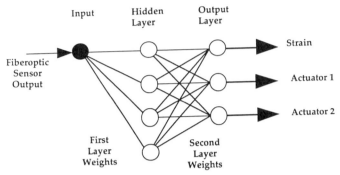

Figure 16.34. Neural network processor for smart beam.

simulated on a computer. There are four hidden neurons and three output neurons, one to output the strain experienced by the beam, and two others, one for each actuator. The outputs for the actuator represent the calculated change in length of each actuator/spring as force is applied to the beam.

The strain as a function of sensor output intensity should be an inverse sine-squared curve. The calculated and actual values are shown in Fig. 16.35. Over a strain range of $\pm 3000\,\mu$strain, the rms error was less than 2.8%. Most of the error occurred at the ends of the range; the error over $\pm 2500\,\mu$strain was approximately 1%. Figures 16.36 and 16.37 show the calculated length changes for actuators 1 and 2, respectively. Over a range of $\pm 2500\,\mu$strain the neural network calculated the correct values, as compared to theory, within

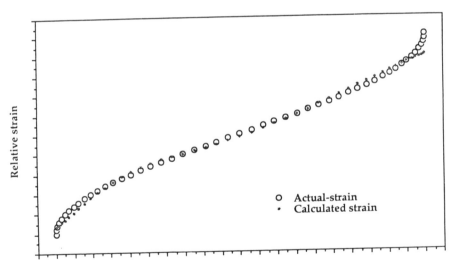

Relative fiberoptic sensor output intensity

Figure 16.35. Neural network results for strain measurement.

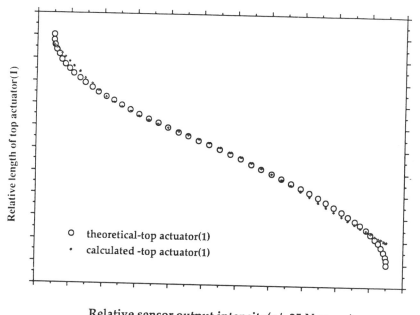

Figure 16.36. Neural network results for top actuator (1).

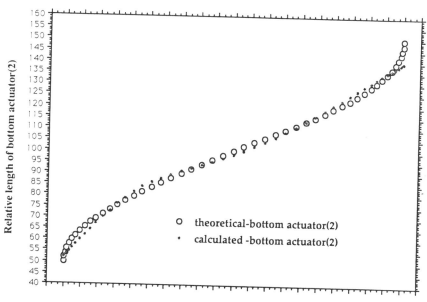

Figure 16.37. Neural network results for bottom actuator (2).

2%. The accuracies attained are determined in large part by the number of neurons in the hidden layer. To maximize processing speed, four neurons were used. For greater accuracy, more neurons could be added.

16.3.4.5.   Conclusions and Future Work.   The biological neuromuscular system is a reasonable model after which to pattern smart structures. Our work suggests that the mechanical model proposed for the neuromuscular system can be implemented with existing technology. The use of SMA as a muscle fiber is feasible. We have characterized the SMA response to optical stimulus (36), and we are presently incorporating single- and multiple-SMA strands into a separately controllable bundle. Fiber optic sensor data may be interpreted by the use of neural networks to analyze their optical output. We have demonstrated a neural network that analyzes the output of a polarimetric sensor. Neural networks can be trained to provide control functions for a shape-memory-alloy actuator muscle system. Our next step is to integrate these elements into a closed-loop neuromuscular system.

16.3.5.   Liquid Crystal Neural Network Implementation

One of the simplest tasks for a neural network processor is processing the output of a single fiber optic sensor as shown in Fig. 16.38. In this case the output of an interferometric sensor must be processed to determine the strain or temperature being sensed. The processor must map the light intensity output back to the measurand value. Using the neural network configuration shown in Fig. 16.38, a four-layer perceptron network with one neuron in the first layer, nine neurons in the second layer, four in the third, and one output

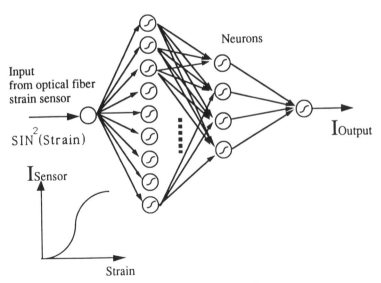

Figure 16.38.   Neural network processor for single sensor.

neuron, the performance of the network after training is shown in Fig. 16.39. The network correctly calculates the strain over ± 3000 μstrain with an rms error less than 0.1%. The true utility of neural networks, however, lies not in processing single inputs but in processing multiple inputs. One technique for producing a compact processor incorporates liquid crystal spatial light modulators.

16.3.5.1. Liquid Crystal Neuron Implementation. A key element of any neural network processor is the neuron itself. One activation function commonly used is the sigmoid function, which allows for an analog output that is required for most signal-processing applications. Twisted nematic liquid crystal displays (LCDs) used in optical computing applications and commercially as a display on laptop computers, hand-held calculators, digital watches, and so on, have an inherent "sigmoid-like" transfer function for their optical transmissivity with respect to applied voltage. In most applications, however, they are used in a binary (light or dark) display mode. Light modulation is achieved by sandwiching a birefringent liquid crystal material between a pair of polarizers. Transparent electrodes are provided so that an ac or pulsed dc voltage can be applied across the liquid crystal material.

The operation of the display is shown in Fig. 16.40. When there is no voltage applied to the LCD, the light propagating through the front polarizer is rotated 90°. Increasing the voltage applied to the LCD causes a decrease in the polarization rotation. If the relative orientation of the two polarizers is the

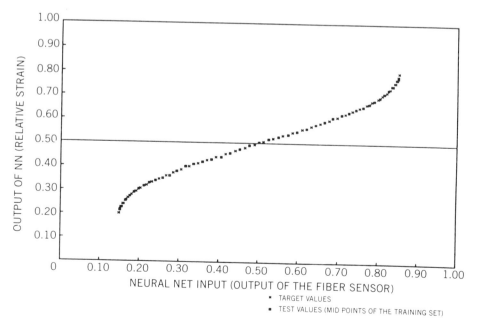

Figure 16.39. Neural network strain calculations.

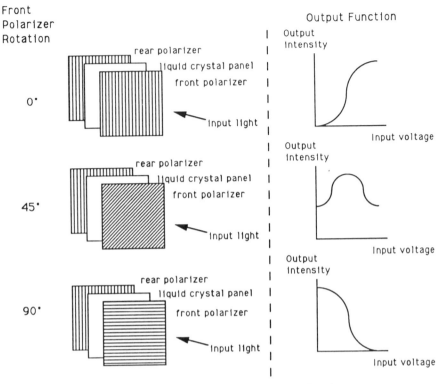

Figure 16.40. Liquid-crystal-display activation function.

same, as shown in the figure as 0° rotation, with no voltage applied, the light rotates 90° and is blocked by the rear polarizer, resulting in zero output light. As the voltage is increased, the light is rotated less, resulting in more light being transmitted. By rotating the two polarizers sandwiching the liquid crystal material, it is possible to obtain either an upgoing sigmoid, a down-going sigmoid, a bell-shaped, function, or an inverted bell-shaped transfer function. Figure 16.40 shows some possible polarization orientations and the corresponding transmission functions as the voltage of an applied 1-kHz square wave is increased. These characteristics can be used to implement a LCD-based neural network. The upgoing and downgoing curves approximate a sigmoid curve and can be used as activation functions for the neurons. Figure 16.41 shows an experimentally measured LCD transmission function along with the theoretical sigmoid curve fit.

16.3.5.2. Compact Liquid Crystal Display–Based Neural Network. With the area of a typical laptop-computer-size LCD screen, it is possible to implement hundreds or thousands of neurons. In the past, several LCD-based neural networks have been demonstrated (42, 43), but the LCDs were used primarily as weights or used in a binary mode unsuitable for analog signal processing. The use of the inherent optical transfer function of the LCD simplifies the

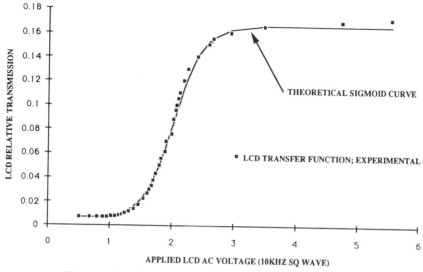

Figure 16.41. Measured liquid-crystal-display activation function.

required electronics, and if all the LCD neurons are driven directly (rather than in a time-multiplexed fashion), the dynamic range of the LCD transmissivity can be increased significantly relative to an active matrix LCD configuration.

In intensity-based optics, there is no simple way to represent negative numbers. Therefore, the neural network needs to be constructed in such a way that we can represent negative inputs and negative weights. This is why the downgoing sigmoid is important. An alternative to using different orientation polarizers for each type of LCD neuron is to place two liquid crystal panels between two polarizers of a set orientation. By doing this, each segment of a LCD cell can have either an upgoing or a downgoing sigmoid transfer function, depending on the applied voltage on the cells of the second LCD panel. In effect, the second LCD panel is used to rotate the polarization of light selectively from the first LCD cells by 90°; this effectively rotates the polarization of the selected cells by 90°.

By combining the LCD cells with a corresponding set of photoconductors and by driving them from a common ac voltage source, it is possible to implement an optical neuron where the photoconductors receive and sum up the weighted light inputs and the LCDs perform the sigmoid transfer function on the weighted sums. It is possible to implement a very compact neural network system by backlighting the LCD with a flat light source such as an electroluminescent (EL) panel and utilizing a weight matrix transparency and a layer of photoconductors to control the neural outputs. An exploded picture of a possible configuration is shown in Fig. 16.42. The inputs to the network from an array of fiber optic sensors illuminate the photoconductive cells on the first surface of the system. These signals are retransformed into the optical domain by using the photoconductors to modulate the voltage directly to the

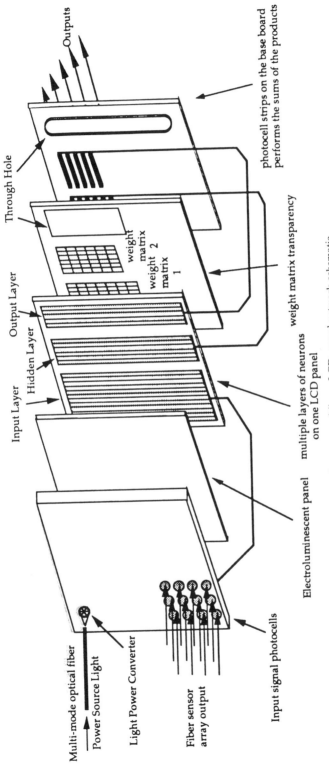

Figure 16.42. Compact multilayer LCD neural network schematic.

492

first set of LCD stripes (input layer). The ac voltage is supplied to each photoconductor so that the amount of light that shines on each photoconductor determines the transmissivity of the input-layer LCD stripes. The purpose of using the photoconductor–LCD set for the input layer of the neural network is so that the inputs are evenly distributed to the neurons in the next layer (hidden layer). The light from the EL panel propagates through the LCD stripes, through the weight matrix, and hits the photoconductor stripes on the back layer. These photoconductor stripes are orthogonally oriented with respect to the LCD stripes to allow matrix sums of products. The photoconductors control the voltage inputs to the second layer of neurons (the hidden layer), on the same LCD panel as for the first set of neurons. The light from the EL panel propagates through the hidden-layer LCD stripes and corresponding weight matrix and illuminates the second set of photoconductor stripes at the backplane. These photoconductors modulate the input voltage to the third set of LCD neurons (the output layers), again on the same LCD panel. The outputs from this third set of LCD stripes (neurons) are the outputs of the neural network. By avoiding the use of lenses or fibers to distribute the light output between layers of neurons, the configuration can be made very compact. Figure 16.43 shows the neural network architecture corresponding to Fig. 16.42.

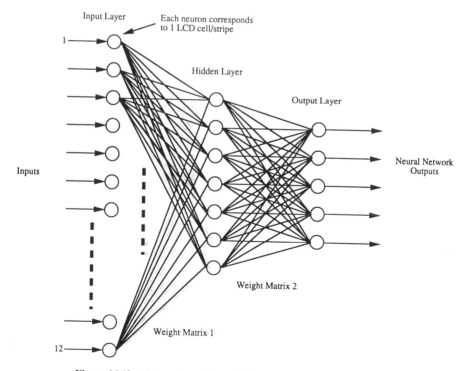

Figure 16.43. Compact multilayer LCD neural network architecture.

ACKNOWLEDGMENTS

Some of the work decribed in this chapter was supported under the U.S. Army Research Office Grant DAAL03-89-G-0085. For their exceptional dedication and contributions, we would also like to thank our students T. Alavie, W. Constandi, S. Doi, J. Franke, X. Gao, F. Gonzalez, H. Hou, Y. Lin, R. Nassar, and K. Yoo.

REFERENCES

1. DARPA Neural Network Study, *Theoretical Foundations*, pp. 17–36. Armed Forces Communications and Electronics Association (AFCEA) International Press, Fairfax, VA 1988,.
2. J. A. Anderson and E. Rosenfeld, eds., *Neurocomputing*. MIT Press, Cambridge, MA, 1989.
3. M. Thursby, K. Yoo, and B. Grossman, Neural Control of Smart Electromagnetic Structures, *Active Materials and Adaptive Structures Workshop*, Alexandria, VA, November (1991).
4. R. P. Lippmann and W. Y. Huang, Comparisons between Conventional and Neural Net Classifiers. *Conf. Neural Network, 1st, 1987*, pp. iv–485 (1987).
5. J. Alspector, Neural-Style Microsystems that Learn. *IEEE Commun. Mag.*, pp. 29–36 November (1989).
6. T. X. Brown, Neural Networks for Switching. *IEEE Commun. Mag.*, pp. 72–81 November (1989).
7. R. E. Ziemer, Application of Neural Networks to Pulse-Doppler Radar Systems for Moving Target Indication. *Int. J. Conf. Neural Networks (IJCNN)*, January, Vol. 2, (1990).
8. L. V. Fausett, Basic Concepts and Activation Functions. In *Fundamentals of Neural Networks: Architectures, Algorithms, Applications*, pp. 5–7 Prentice Hall, Englewood Cliffs, NJ (1991).
9. R. P. Lippmann, An Introduction to Computing with Neural Nets. *IEEE Trans. Acoust., Speech, Signal Process.* pp. 4–22 (1987).
10. R. P. Lippmann, An Introduction to Computing with Neural Nets. *IEEE Trans. Acoust., Speech, Signal Process.* pp. 18–20 (1987).
11. P. D. Wasserman, *Neural Computing: Theory and Practice*. Van Nostrand-Reinhold, New York, 19xx.
12. L. V. Fausett, *Fundamentals of Neural Networks: Architectures, Algorithms, Applications*, pp. 106–108, Prentice Hall, Englewood Cliffs, NJ (1994).
13. B. Widrow and S. D. Stearns, The LMS Algorithm. In *Adaptive Signal Processing*, pp. 99–114. Prentice Hall, Englewood Cliffs, NJ, 1985.
14. J. A. Freeman, D. M. Eskapuna, *Artificial Neural Systems Theory and Practice*. Addison-Wesley, Reading, MA, (1991).
15. DARPA Neural Network Study, *Single- and Multi-layer Perceptrons*, pp. 75–80. Armed Forces Communications and Electronics Association (AFCEA) International Press, Fairfax, VA, 1988.

16. R. H. Nielson, Back-Propagation Neural Network. In *Neurocomputing*, pp. 124–137. Addison Wesley, Reading, MA, 1989.

17. R. G. Palmer, Multi-layer Networks. In *Introduction to the Theory of Neural Computation*, John A. Hertz, Ed. pp. 115–162. Addison-Wesley, Reading, MA, 1991.

18. T. Kohonen, *Self-Organization and Associative Memory*, 3rd ed. Springer-Verlag, Berlin, 1989.

19. T. Kohonen, Self-Organized Formation of Topologically Correct Feature Maps. *Biol. Cybernet.* **43**, pp. 59–69 (1982).

20. B. Grossmann, M. Thursby, H. Hou, R. Nassar, and A. Ren, Neural Network Processing of Fiberoptic Sensors and Sensor Arrays. *Proc. SPIE — Int. Soc. Opt. Eng.* **1370**, pp. 205–216 (1990).

21. T. Kohonen, Improved Versions of Learning Vector Quantization. *Int. Jl. Conf. Neural Networks (IJCNN)* **1** (1990).

22. B. Grossman, M. Thursby, and X. Gao, Composite Damage Assessment Employing an Optical Neural Network Processor and an Embedded Fiberoptic Sensor Array. *Proc. SPIE — Int. Soc. Opt. Eng.*, **1588**, pp. 64–75 (1992).

23. B. Grossman, M. Thursby, and X. Gao, A Modified Learning Rule for the LVQ2 MODEL of the Kohonen Neural Network. *Neural Comput.* (1990).

24. K. S. Narendra and K. Parthasarathy, Identification and Control of Dynamical Systems Using Neural Networks. *IEEE Trans. Neural Networks* **1**(1), 4–27 (1990).

25. B. Widrow, A Broom Balancer Implementation Using Neural Networks... *Neural Inf. Process. Syst. Conf.*, Denver, Personal Communication 1988.

26. M. I. Jordan, Serial Order: A Parallel Distributed Processing Approach. *ICS Report 8604*, Institute for Cognitive Science University of California, San Diego, La Jolla, CA May (1986).

27. M. H. Thursby, B. Grossman, and K. Yoo, Neural Networks and the Control of Smart Systems. *US Army Res. Off. Workshop, Smart/Intell. Mater. Syst.*, pp. 242–251 1990.

28. S. Long, Reactively Loaded Patch Antennas. *Missile/Projectile/Airborne Test Instrum. Antenna Workshop*, pp. 457–490 Ga. Tech Res. Inst., *1987* (1987).

29. W. F. Richards, S. E. Davidson, and S. A. Long, Dual-Band Reactively Loaded Microstrip Antenna. *IEEE Trans. Antennas Propag.* **AP-33**, 556–560 (1985).

30. K. R. Carver and J. W. Mink, Microstrip Antenna Technology. *IEEE Trans. Antennas Propag.*, **AP-29**, 2–24 (1981).

31. D. H. Schaubert, F. G. Farrar, A. Sindoris, and S. T. Hayes, Microstrip Antennas with Frequency Agility and Polarization Diversity. *IEEE Trans. Antennas Propag.* **AP-29**, 118–123 (1981).

32. D. E. Rummelhart, G. E. Hinton and R. J. Williams, Learning Internal Representations by Error Propagation. In *Parallel Distributed Processing Explorations in the Microstructure of Cognition*, D. E. Rummelhart and J. L. McClelland, eds., pp. 318–362, Vol. 1, MIT Press, Cambridge MA, 1986.

33. M. H. Thursby, B. G. Grossman, and Z. Drici, Smart Electromagnetic Structures — A Neural Network Antenna. *SPIE Symp. Opt. Eng. Photon. Aerosp. Sens.* (1990).

34. T. Kohonen, *Some Practical Aspects of the Self-Organizing Maps. Int. Jl. Conf. Neural Networks (IJCNN), 1990* pp. II253–II256. (1990).

35. W. Spillman et al., Statistical-Mode Sensor for Fiberoptic Vibration Sensing Uses. *Appl. Opt.*, **28**, pp. 3166–3176 (1989).

36. B. Grossman, F. Caimi, T. Alavie, J., Franke, X. Gao, H. Hou, R. Nassar, W. Costandi, M. Thursby, Smart Structures and Fiberoptic Sensor Research at Florida Institute of Technology — 1990. *Invited Paper, SPIE OE/FIBERS 1990 Conf. Proc.*, **1370**, pp. 69–83, San Jose, CA, 1990.

37. B. D. Duncan, *Modal Interference Techniques for Strain Detection in Few Mode Optical Fibers.* Fiber and Electrooptics Research Center, Virginia Polytechnic Institute, Blacksburg, 1987.

38. M. R. Layton and J. L. Bucaro, Optical Fiber Acoustic Sensor Utilizing Mode–Mode Interference. *Appl. Opt.* **18**, 666–670 (1979).

39. M. Thursby, B. Grossman, T. Alavie, and K. Yoo, Smart Structures Incorporating Artificial Neural Networks, Fiber Optic Sensors, and Solid State Actuators. *Proc. SPIE — Int. Soc. Opt. Eng.* **1170**, 316–325 (1990).

40. R. Brooks, Robotics, Modularity, and Learning Neural Information Processing Systems Natural and Synthetic. Denver, Nov. 28–Dec. 1, 1988.

41. T. Alavie, B. Grossman, M. Thursby, and F. Ham, Processing of Few-Mode and Polarimetric Sensor Signals. *Proc. SPIE — Int. Soc. Opt. Eng.* **1169** (1989).

42. X. Yang, T. Lu, and T. S. Francis, Liquid Crystal Television Spatial Light Modulators. *Appl. Opt.* **29**, 5223–5225, (1990).

43. K. Liu and T-H. Hhao, Liquid Crystal Television Spatial Light Modulators. *Appl. Opt.* **28**, 4722–4780 (1989).

17

Actuators for Smart Structures

ZAFFIR CHAUDHRY AND CRAIG ROGERS
Center for Intelligent Material Systems and Structures
Virginia Polytechnic Institute and State University
Blacksburg, Virginia

17.1. INTRODUCTION

Materials that allow an intelligent or smart structure to adapt to its environment are known as actuators. These materials have the ability to change the shape, stiffness, position, natural frequency, damping, friction, fluid flow rate, and other mechanical characteristics of intelligent material systems in response to changes in temperature, electric field, or magnetic field. What distinguishes them from conventional hydraulic and electrical actuators, and makes them especially attractive for smart structures, is their ability to change their dimensions and properties without utilizing any moving parts. These actuator materials contract and expand just like the muscles in the human body. When integrated into a structure (either through embedding or through surface bonding), they apply localized strains and directly influence the extensional and bending responses of the structural elements. Because of the absence of mechanical parts, they can easily be integrated into base structures. Integration within structures ensures an overall force equilibrium between the forcing actuator and the deforming structure, and thus precludes any rigid body forces and torques. A brief description of the most common actuator materials, along with their typical properties and applications, follows.

17.2. PIEZOCERAMIC ACTUATORS

Piezoelectricity describes the phenomenon of generating an electric charge in a material when subjecting it to a mechanical stress (direct effect), and conversely, generating a mechanical strain in response to an applied electric field. Piezoelectric properties occur naturally in some crystalline materials and

Fiber Optic Smart Structures, Edited by Eric Udd.
ISBN 0-471-55448-0 © 1995 John Wiley & Sons, Inc.

can be induced in other polycrystalline materials. The distortion of the crystal domains produces the piezoelectric effect. The domains may be aligned (poled) by the application of a large field, usually at high temperature (see Fig. 17.1). Subsequent application of the electric field will produce additive strains in the local domains, which translate into a global strain in the material. The piezoelectric effect was discovered in 1880 by Pierre and Jacques Curie. The direct piezoelectric effect has been used for a long time in sensors such as accelerometers. Use of the converse effect, however, has until recently been restricted to ultrasonic transducers. Barium titanate, discovered in the 1940s, was the first widely used piezoceramic. Lead zirconate titanate (PZT), discovered in 1954 (4), has now largely superseded barium titanate because of its stronger piezoelectric effects. It is only recently that researchers in the area of structural control have taken notice of the very desirable features of piezoelectric actuators and have started using them for many structural control applications. Piezoceramics are compact, have a very good frequency response and can be easily incorporated into structural systems. Actuation strains on the order of 1000 μstrain have been reported for PZT material. Within the linear range they produce strains that are proportional to the applied electric field/voltage. These features make them very attractive for many structural control applications.

17.2.1. Constitutive Modeling of Piezoelectric Actuators

There are several approaches to modeling the constitutive behavior of piezoelectric materials. The first approach is that taken by material scientists.

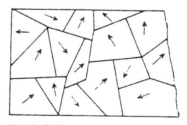

Unpolarised

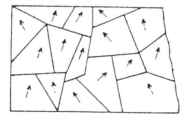

Polarised

Figure 17.1. Domain alignment in piezoceramics due to polarization.

In this model, the effects of the microstructure and the chemical composition of the material properties are studied. This approach is useful in improving the basic electromechanical properties of the piezoelectric materials; however, the parameters involved in such a model are not in a form where they can be used directly by structural engineers in structural equations.

The second approach is an electrical one, where an equivalent electrical circuit for the piezoelectric material is developed. This is useful and necessary to represent the effect of the piezoelectric material on the driving electrical circuit. This approach does not deal directly with the mechanically induced strains, which are the source of control in structural applications. Nevertheless, this modeling is important to structural control engineers because it is through this model that the electrical power required to drive the piezoelectric actuators/sensors bonded to the structure is computed.

The third approach is the thermodynamic approach, which is all encompassing and can deal with all forms of energy and their interconversion. Thermodynamics is concerned with restrictions on the responses of bodies arising from the energy balance and the entropy inequality. Without these restrictions, mathematically characterized ideal bodies may not represent physical bodies and their predicted response may not be real (3). This approach, although the best, can be unnecessarily complicated for use in the solution of typical structural problems.

The macromechanical approach is the fourth approach and the one that is the most popular with structural design engineers. It provides the relationship between the electrical and mechanical effects in a manner that can be incorporated into the existing constitutive relations of typical isotropic or orthotopic structural materials. The details of this approach are presented herewith.

17.2.1.1. Linear Constitutive Equations. For linear piezoelectric materials, the interaction between the electrical and mechanical variables can be described by linear relations of the form

$$S_i = S_{ij}^E T_j + d_{mi} E_m$$
$$D_m = d_{mi} T_i + \varepsilon_{mk}^T E_k$$

The mechanical variables in the foregoing equations are the stress T and the strain S, and the electrical variables are the electric field E and the electric displacement D. The first equation, describes the converse piezoelectric effect (i.e., a strain is produced in response to an applied electric field), and the second equation describes the direct effect. The stress and strain are second-order tensors, while the electric field and the electric displacement are first order. The superscripts T and E signify that these quantities are measured at zero stress and constant field, respectively, s is the compliance, d is the piezoelectric constant and ε is the permittivity.

Figure 17.2 shows the typical coordinate system used to represent a poled piezoelectric. The 3-axis is in the direction of the initial polarization. The 1- and 2-axes are arbitrary in the plane perpendicular to the poling direction. The 1- and 2-axes are arbitrary because a poled piezoelectric is transversely isotropic, being isotropic in the 1–2 plane. The equations above written explicitly in matrix form are

$$
\begin{bmatrix} S_1 \\ S_2 \\ S_3 \\ S_4 \\ S_5 \\ S_6 \\ D_1 \\ D_2 \\ D_3 \end{bmatrix} =
\begin{bmatrix}
s_{11}^E & s_{12}^E & s_{13}^E & 0 & 0 & 0 & 0 & 0 & d_{31} \\
s_{12}^E & s_{11}^E & s_{13}^E & 0 & 0 & 0 & 0 & 0 & d_{31} \\
s_{13}^E & s_{13}^E & s_{33}^E & 0 & 0 & 0 & 0 & 0 & d_{33} \\
0 & 0 & 0 & s_{55}^E & 0 & 0 & 0 & d_{15} & 0 \\
0 & 0 & 0 & 0 & s_{55}^E & 0 & d_{15} & 0 & 0 \\
0 & 0 & 0 & 0 & 0 & s_{66}^E & 0 & 0 & 0 \\
0 & 0 & 0 & 0 & d_{15} & 0 & \varepsilon_1^T & 0 & 0 \\
0 & 0 & 0 & d_{15} & 0 & 0 & 0 & \varepsilon_1^T & 0 \\
d_{31} & d_{31} & d_{33} & 0 & 0 & 0 & 0 & 0 & \varepsilon_3^T
\end{bmatrix}
\begin{bmatrix} T_1 \\ T_2 \\ T_3 \\ T_4 \\ T_5 \\ T_6 \\ E_1 \\ E_2 \\ E_3 \end{bmatrix}
$$

Because of the transverse isotropy and the peculiar electrical and electromechanical properties, there are 10 material constants in the constitutive equation above: five elastic constants, three piezoelectric constants, and two electric constants. The matrix is sparsely populated, and that is desirable for structural control.

The piezoelectric constants that are of most interest from a structural control standpoint are the d constants. These constants relate the strain developed in the material to the applied electric field; obviously, the highest value of these constants is desirable. The d_{33} constant relates the strain in the 3-direction to the field in the 3-direction; similarly, d_{31} relates the strain in the 1-direction to the electric field in the 3-direction. The electric field is voltage applied across the piezoelectric divided by its thickness. It is important to point

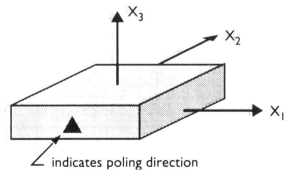

indicates poling direction

Figure 17.2. Typical coordinate system used to represent a poled piezoelectric.

out that d_{33} is usually a positive number and d_{31} is a negative number. This means that a positive field (i.e., a field applied in the poling direction) will produce a positive mechanical strain in the 3-direction and a negative strain in the 1-direction. The d_{15} constant relates the shear strain to the field in the 1-direction. To use this constant, it is necessary to apply electrodes in the 1-direction after the poling in the 3-direction. This constant is used when it is desirable to induce shear between two surfaces attached to either side of a piezoelectric actuator. It should be noted that the d constants alone do not determine the authority of the actuator; high stiffness is equally important and desirable.

17.2.2. Properties of PZT Materials

A properties table of typical PZT and PMN actuator materials is given in Table 17.1.

17.2.3. Design Considerations for PZT Actuators

The force and stroke of a PZT actuator are determined from the force displacement plot, shown in Fig. 17.3. When the actuator is fully constrained it develops maximum force (known as the blocking force), as indicated by point A. When there is no constraint, the actuator expands/contracts freely to its limit (point B, i.e., free induced strain Λ), and no force is developed. When the actuator is coupled to a system or structure, depending on the actuator/ structure stiffness ratio, the actuator develops a certain force and at the same time undergoes a certain stroke and the equilibrium state lies on line AB. In a dynamic application, since the structural impedance or dynamic stiffness changes with frequency, the relationship above does not hold, and an imped-ance model must be used to compute the actuator force and stroke (5).

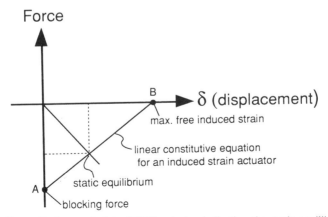

Figure 17.3. Force displacement plot of PZT actuator indicating the static equilibrium state when the actuator is attched to a structure.

Table 17.1 Piezoelectric Properties

Electromechanical Properties:	Barium Titanate				Lead Zirconate Titanate								Lead Titanate EC-97	Lead Magnesium Niobate EC-98
	EC-21	EC-31	EC-55	EC-57	EC-63	EC-64	EC-65	EC-66	EC-67	EC-69	EC-70	EC-76		
Physical properties														
Density ($\times 10^3$ kg/m³)	5.7	5.55	5.55	5.3	7.5	7.5	7.5	7.45	7.5	7.5	7.45	7.45	6.7	7.85
Young's modulus ($\times 10^{10}$ N/m²)	11.4	10.7	11.6	12.5	8.9	7.8	6.6	6.2	9.3	9.9	6.3	6.4	12.8	6.1
Curie temperature (°C)	130	115	115	140	320	320	350	270	300	300	220	190	240	170
Mechanical Q for a thin disk	1400	400	550	600	500	400	100	80	900	960	75	65	950	70
Electric properties at 25°C														
Dielectric constant at 1 kHz	1070	1170	1220	640	1250	1300	1725	2125	1100	1050	2750	3450	270	5500
Dissipation constant at 1 kHz low field	0.5	0.7	0.5	0.6	0.4	0.4	2.0	2.0	0.3	0.3	2.0	2.0	0.9	2.0
K_{31}	0.17	0.19	0.19	0.15	0.34	0.35	0.36	0.36	0.33	0.31	0.37	0.38	0.01	0.35
K_p	0.26	0.32	0.31	0.25	0.58	0.60	0.62	0.62	0.56	0.52	0.63	0.64	0.01	0.61
K_{33}	0.38	0.48	0.46	0.38	0.68	0.71	0.72	0.72	0.66	0.62	0.74	0.75	0.53	0.72
K_{15}	0.37	0.49	0.48	0.23	0.69	0.72	0.69	0.68	0.59	0.55	0.67	0.68	0.35	0.67
d_{31} ($\times 10^{-12}$ m/V)	-49	-59	-58	-32	-120	-127	-173	-198	-107	-95	-230	-262	-3	-312
d_{33} ($\times 10^{-12}$ m/V)	117	152	150	87	270	295	380	415	241	220	490	583	68	730
d_{15} ($\times 10^{-12}$ m/V)	191	248	245	154	475	506	584	626	362	330	670	730	67	825
g_{31} ($\times 10^{-3}$ V·m/N)	-5.2	-5.8	-5.6	-5.5	-10.3	-10.7	-11.5	-10.6	-10.9	-10.2	-9.8	-8.6	-1.7	-6.4
g_{33} ($\times 10^{-3}$ V·m/N)	12.4	14.8	14.3	16.2	24.1	25.0	25.0	23.0	24.8	23.7	20.9	19.1	32.0	15.6
g_{15} ($\times 10^{-3}$ V·m/N)	15.7	20.4	20.1	28.0	37.0	39.8	38.2	36.6	28.7	28.9	35.0	28.9	33.5	17.0
Elastic constant														
S_{11}^E ($\times 10^{-12}$ m²/n)	8.8	9.3	8.6	8.0	11.3	12.8	15.2	16.1	10.8	10.1	15.9	15.6	—	16.3
S_{22}^E ($\times 10^{-12}$ m²/n)	-2.7	-2.9	-2.6	-2.2	-3.7	-4.2	-5.3	-5.5	-3.6	-3.4	-5.4	-4.7	—	-5.6
S_{33}^E ($\times 10^{-12}$ m²/n)	10.0	9.7	9.8	9.3	14.3	15.0	18.3	17.7	13.7	13.5	18.0	19.8	7.7	21.1
Aging rate (% change per time decade)														
Dielectric constant	-4.5	-1.8	-1.6	-4.8	-4.1	-4.2	-0.8	-1.5	-3.2	-3.0	-2.1	-2.0	-0.3	-1.5
Coupling constant	-4.0	-2.0	-1.4	-5.0	-2.1	-2.1	-0.3	-0.4	-1.6	-1.4	-0.4	-0.4	-0.4	-0.4
Resonant frequency	0.85	0.4	0.6	0.8	1.0	1.0	0.2	0.3	0.8	0.75	0.4	0.3	0.05	0.4
Electric field dependence														
rms kV/m at 25°C	79	79	79	79	394	394	79	79	394	394	79	79	79	79
Dielectric constant (% increase)	1.4	6.0	3.0	0.4	18.0	18.0	12.0	12.3	3.8	3.2	12.6	14.0	1.5	22.5
Dissipation factor (%)	0.7	4.0	0.9	0.7	3.0	3.0	7.3	6.9	0.9	0.6	7.1	7.9	0.8	6.2

Values are nominal; actual production values vary $\pm 10\%$.

Source: EDO Corporation (2), **compliments of the EDO Corporation**

502

There are certain limitations and special characteristics of PZTs that should be kept in mind when considering a certain application. The linear constitutive law was examined above, but in actuality the field strain behavior is nonlinear (see Fig. 17.4). These materials also exhibit varied degrees of hysteresis. In addition, there are certain other constraints on the use of these materials such as depoling, failure stress, and creep. These issues and how they affect design are addressed in the following sections.

17.2.3.1. Hysteresis. Hysteresis is the phenomenon that occurs when, upon removal of the electric field, a certain mechanical strain remains in the piezoelectric material. The cause of the hysteresis is the energy dissipation due to internal sliding events in the polycrystalline piezoelectric body. The degree of hysteresis depends on the field level, the cycle time, and the material of the piezoelectric actuator. Hysterisis is often measured as a percentage of the total strain, and is typically in the range 0.1 to 10%. Since the energy dissipated in the hysteresis cycle appears in the form of heat, it is a concern in high-frequency operations, where it can lead to excessive temperatures.

17.2.3.2. Creep. *Creep* refers to plastic deformations that take place (under constant loads) over time. In piezoelectric materials, it is the prolonged application of electric field that causes the induced mechanical strain to grow with time. Creep is a concern when the piezoelectric actuator is constantly being subjected to high dc voltages. Creep and hysterisis behavior are illustrated in Fig. 17.5.

17.2.3.3 Depoling. When a high electric field is applied to the piezoelectric opposite to the poling direction, its piezoelectric properties may degrade or it

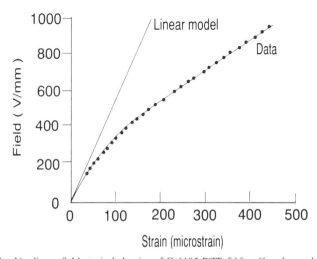

Figure 17.4. Nonlinear field-strain behavior of G-1195 PZT. [After Crawley and Anderson (1).]

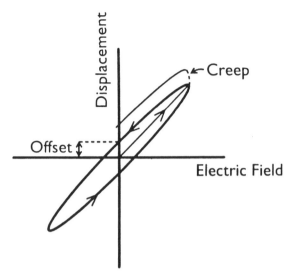

Figure 17.5.   Creep and hysterisis in piezoelectric material. (Courtesy of Piezo Systems Inc. (6).]

may get poled in the opposite direction. The field that causes depoling is called a coercive field. It is important to note that a high stress level can also cause depoling. For ac operation, the coercive field limit is higher because the duration of the applied field is shorter. The coercive field limit is generally available from the manufacturer; if not, it can easily be determined through a simple experiment.

17.2.3.4   Electrical Breakdown. There is a maximum electric field that can be applied in the poling direction before there is an electrical breakdown through the body of the piezoelectric or across its edges. Electrical breakdown destroys the piezoelectric properties. The electrical breakdown voltage is quite high and can typically be twice as much as the coercive field. Arcing can occur at the edges of the actuator at low field levels and should not be confused with electrical breakdown. This arcing is generally due to the debris left adhering to the edges after cutting. The arcing vaporizes the debris, thereby cleaning the actuator.

17.2.3.5   Curie Temperature. Curie temperature is the maximum operating temperature beyond which the material loses all its piezoelectric properties. The operating temperature should generally be much lower than the Curie temperature because at temperatures close to the Curie temperature, depoling is facilitated, aging and creep are accelerated, and the maximum safe mechanical stress also decreases. For typical PZT material, this temperature is about 350° C.

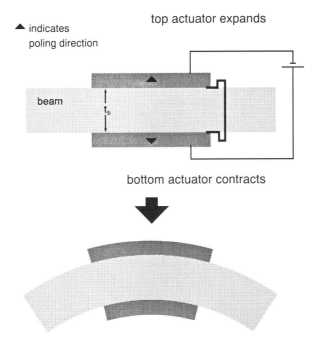

Figure 17.6. Surface-bonded PZT actuators used to induce bending.

17.2.4. Applications

The use of PZT actuators is widespread and can be divided into two broad categories: the linear actuators and the actuators used for structural control applications. In the first category, the PZT actuators, arranged in the form of stacks, are used in a fashion similar to shakers or conventional hydraulic or electrical actuators. Of course, the stroke is severely limited; therefore, in static applications they are used for micropositioning work. In structural control applications, the actuators are typically embedded within the structure and they apply localized strains that can be used directly to control structural deformations. In such a configuration, the actuation does not require a back-reaction and does not produce rigid-body modes. A typical arrangement is shown in Fig. 17.6; the two actuators on the upper and lower surfaces of the structure are actuated out of phase (the upper actuator expands and the bottom contracts), which causes bending.

17.3. ELECTROSTRICTIVE ACTUATORS

Another ferroelectric with interesting properties is the compound lead–magnesium–niobate, or PMN. This material exhibits a strong electrostrictive effect. Electrostriction involves a nonlinear electromechanical coupling for which the material develops a strain proportional to the square of the polarization. The ideal strain versus applied field curves for electrostrictive material is shown in Fig. 17.7. As shown, the strain is insensitive to the polarity of the field and due to the second-order effect is always extensional. Electrostrictors exhibit considerably less hysteresis, and the phenomenon of depoling does not exist. Free induced strains for PMN-PT (lead magnesium niobate doped with lead titanate) on the order of 1200 μstrain have been reported. The nonlinear electromechanical coupling of this material can be used to develop transducers with tunable sensitivities. Because the slope of the repeatable electric field-to-strain relationship varies, the transduction sensitivity, whether the ceramic is used as an actuator or a sensor, can be controlled by adjusting the dc bias electric field about which the electrostrictive transducer operates. Typical properties for PMN are included in Table 17.1 together with the properties of piezoceramic actuators. The disadvantage of the nonlinear (quadratic) response is the reduced actuator authority for ac operations. Furthermore, the electromechanical response is frequency dependent and is restricted to a rather small temperature range in which the dielectric permittivity is large. The deformation for low applied electric fields is small.

17.3.1. Constitutive Relations

The constitutive relations for electrostrictive actuators are written similarly to those for piezoelectric actuators but second-order terms are included:

$$S_{ij} = s^E_{ijkl} T_{kl} + d_{kij} E_k + M_{klij} E_k E_l$$

In this equation the first two terms are the same as those used to describe the piezoelectric constitutive law: Hook's law and the converse piezoelectric effect.

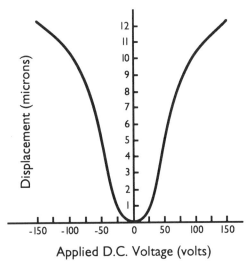

Figure 17.7. Voltage versus displacement for a 5 mm × 5 mm × 15 mm rectangular electrostrictive stack. [Courtesy of AVX Corporation (7).]

The third term represents the electrostriction effect. The components of M_{klij} are the electrostrictive coefficients. From the equation above, electrostriction appears as an electric field-dependent contribution to the linear piezoelectric effect. In fact, the two effects are separable because the piezoelectric effect is possible only in noncentrosymmetric materials, whereas the electrostrictive effects are not limited by symmetry and are present in all materials (8). In addition to the direct electrostrictive effect, the converse electrostrictive effect also exists.

17.3.2. Applications

Just like PZTs, PMN actuators can be used as linear actuators in a stack configuration (this utilizes the electrostrictive effect parallel to the direction of the applied field) or to induce bending in the bonded structure (this utilizes mechanical strain induced transverse to the direction of the applied field). Bonded electrostrictive actuators have been used successfully to control the shape of deformable mirrors. In the stack configuration, they have been used for impact dot-matrix printing. Advantages over conventional electromagnetic actuators include order-of-magnitude higher printing speeds, order-of-magnitude lower energy consumption, and reduced noise emissions. A tunable ultrasonic medical probe composed of electrostrictors embedded in a polymer has also been developed (9).

17.4. ELECTROCERAMIC COMPOSITE ACTUATORS

Electroceramic composite materials are formed by combining a piezoelectric ceramic and a nonpiezoelectric material, such as a polymer or air-filled voids. The properties of the piezocomposite are dependent on the properties of the constituent phases and the way in which they are connected. Through proper design, the following improvements over the constituent piezoceramic can be achieved (14):

- An acoustic impedance that can be chosen to improve mechanical coupling, leading to increased sensitivity and a broader bandwidth.
- Low transverse coupling.
- High flexibility; in some cases, the composite can be shaped to focus the acoustic beam.
- An electrical impedance that can be chosen to be compatible with the electronics of the system.
- Sensors and actuators that were not possible with conventional piezoelectric ceramics (examples include piezoelectric paint and epoxy sensors and ink-jet printers with channels in the actuator for the ink to flow through).

The properties of several types of composites are listed in Table 17.2.

Table 17.2 **Properties for Piezoelectric Composites**

	Density (g cm^{-3})	Dielectric Constant	g_h $(10^{-3}\,\text{V}\cdot\text{m N}^{-1})$	d_h $(10^{-12}\,\text{C N}^{-1})$	$d_h g_h$ $(10^{-15}\,\text{m}^2\,\text{N}^{-1})$
PZT ceramic	7.6	1800	2.5	40	100
Composite					
0–3 PbTiO$_3$–chloroprene rubber	—	40	100	35	3,500
1–3 PZT rods–epoxy	1.4	54	56	27	1,536
3–1 Perforated composite	2.6	650	30	170	5,100
3–2 Perforated composite	2.5	375	60	200	12,000
3–3 Replamine PZT/silicone	3.3	40	30	35.6	1,068

Source: Gallegro-Juarez (11).

17.4.1. Connectivity

The concept of connectivity was introduced by Newnham in 1978 to describe the way in which the individual phases are self-connected (13). For a two-phase composite, there are 10 possible connectivities: 0–0, 1–0, 2–0, 3–0, 1–1, 2–1, 3–1, 2–2, 3–2, and 3–3. Two of the most commonly used connectivity patterns are shown in Fig. 17.8. In the figure the arrows indicate the directions of connectivity.

17.4.2. Manufacturing Piezoelectric Composites

A great deal of research has been performed on manufacturing 1–3 piezoelectric composites, the most widely used connectivity pattern. The three ways of fabricating 1–3 composites with ceramic rods embedded in a polymer matrix, such as the one shown in Fig. 17.9. are (14):

- The dice-and-fill method, in which deep grooves are cut into a solid piece of ceramic and polymer is cast into the grooves.
- Align the piezoelectric rods and cast polymer in between them; then slice off the desired composite disk.

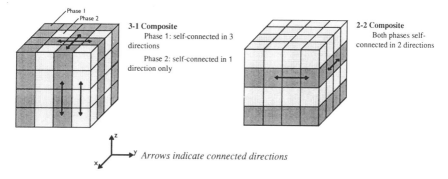

Figure 17.8. 3–1 and 2–2 connectivity patterns in two-phase electroceramic composites.

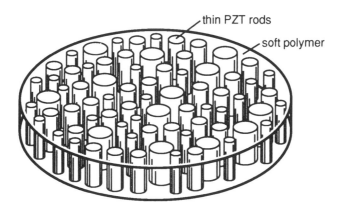

Figure 17.9. Schematic of a 1–3 electroceramic composite.

- A lamination process, in which alternate plates of piezoelectric and passive material are bonded together to form a layered structure. Slices of the layered structure are again alternated with passive layers and bonded into a final structure, from which plates of the desired composite are sliced.

To manufacture 1–3 composites consisting of a ceramic matrix with open spaces or spaces to be back-filled with compliant polymers or conducting electrodes, *fugitive phase processing* can be performed. In such a process, the fugitive phase is removed from the composite through thermal or chemical attack. Drop-on-demand ink jets are made using designed-space-forming technology developed at N.E.C. This combines tapecasting, photolithography, and a fugitive phase. A photosensitive polymer is exposed selectively to ultraviolet light, and the desired pattern remains on the carrier film following developing. This pattern, which will form the spaces in the ceramic, is transferred to the tape-cast green ceramic. During firing, it burns off and leaves the desired spaces in the ceramic. When electrical power is supplied to electrodes embedded in the ceramic, ink droplets are propelled through the nozzle (12).

17.4.3. Electroceramic Composite Properties

One-dimensional solutions for the piezoelectric properties of series and parallel connections were developed by Newnham to illustrate how the composite properties are affected by connectivity (13):

17.4.3.1. Series Connection. For the series connection in Fig. 17.10a, consisting of phase 1 (designated with a superscript 1) and phase 2 (designated with a superscript 2), with phase 1 volume fraction ^1v, piezoelectric coefficient $^1d_{33}$, and permittivity $^1\varepsilon_{33}$, and phase 2 volume fraction ^2v, piezoelectric coefficient $^2d_{33}$, and permittivity $^2\varepsilon_{33}$, the composite piezoelectric coefficient is:

$$\bar{d}_{33} = \frac{{}^1v\,{}^1d_{33}^2\varepsilon_{33}}{{}^1v\,{}^2\varepsilon_{33} + {}^2v\,{}^1\varepsilon_{33}}$$

Using the relationship

$$\bar{g}_{33} = \frac{\bar{d}_{33}}{\bar{\varepsilon}_{33}}$$

yields the composite piezoelectric voltage coefficient

$$\bar{g}_{33} = \frac{{}^1v\,{}^1d_{33}}{{}^1\varepsilon_{33}} + \frac{{}^2v\,{}^2d_{33}}{{}^2\varepsilon_{33}}$$

$$\bar{g}_{33} = {}^1v\,{}^1g_{33} + {}^2v\,{}^2g_{33}$$

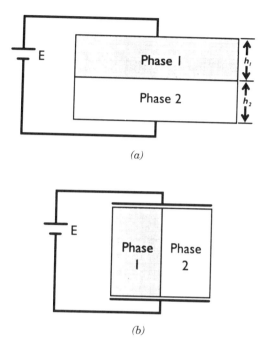

(a)

(b)

Figure 17.10. Series connection model (*a*) and parallel connection model (*b*).
Reprinted from Newnham et. al., (13). with kind permission from Elsevier Science Ltd., The Boulevard, Langford Lane, Kidlington OX5 1GB, UK.

For the series connection, even a very thin low-permittivity layer rapidly lowers the composite *d*-coefficient, but has little effect on the corresponding composite *g*-coefficient.

17.4.3.2. Parallel Connection. For the parallel connection shown in Fig. 17.10*b*, when transverse coupling is neglected, the composite piezoelectric coefficient is

$$\bar{d}_{33} = \frac{{}^1v\,{}^1d_{33}^2 s_{33} + {}^2v\,{}^2d_{33}^1 s_{33}}{{}^1v\,{}^2s_{33} + {}^2v\,{}^1s_{33}}$$

where ${}^1s_{33}$ and ${}^2s_{33}$ are the elastic compliances for stresses normal to the electrodes. The composite voltage coefficient is

$$\bar{g}_{33} = \frac{{}^1v\,{}^1d_{33}^2 s_{33} + {}^2v\,{}^2d_{33}^1 s_{33}}{({}^1v\,{}^2s_{33} + {}^2v\,{}^1s_{33})({}^1v\,{}^1\varepsilon_{33} + {}^2v\,{}^2\varepsilon_{33})}$$

Consider a compliant nonpiezoelectric material in parallel with a stiff piezoelectric material. In this case ${}^1d_{33} \gg {}^2d_{33} = 0$, ${}^1s_{33} \ll {}^2s_{33}$, and ${}^1\varepsilon_{33} \gg {}^2\varepsilon_{33}$, allowing g_{33} (an important parameter for hydrophone applications) for the composite to be much larger than that of either constituent.

When designing composite piezoelectric materials, it is important to remember that simplified models such as the one-dimensional analysis discussed previously cannot predict all of the aspects of real systems but can aid in design. As would be expected, the volume fraction of the piezoelectric phase strongly affects the mechanical and piezoelectric properties of the composite (10). Connectivity can affect the overall composite properties by orders of magnitude.

17.4.4. Applications

Piezoelectric composites are being used in a variety of applications and have been investigated for use in others, including:

- Hydrophone sensors (1–3, 0–3)
- Medical ultrasound probes (1–3)
- Nondestructive ultrasound probes (1–3)
- Ink-jet printers (1–3 with open spaces)
- Smart paints and epoxies (0–3)

17.5. POLYVINYLIDENE FLUORIDE

Polyvinylidene fluoride (PVDF), also abbreviated PVF_2, is a polymer with strong piezoelectric and pyroelectric properties. In the α phase, PVDF is not polarized and is used as a common electrical insulator among many other applications. To impart the piezoelectric properties, the α phase is converted to the β phase and polarized. Stretching α-phase material produces the β phase and metallization of the surface, and the application of a strong electric field provides the permanent polarization for the piezoelectric properties (Fig. 17.11). The flexibility of the PVDF overcomes some of the drawbacks associated with the brittle piezoelectric ceramics. As an actuator it provides less transducing power than its ceramic counterparts, but it produces higher voltage/electric field in response to a mechanical stress. Its easy formability, along with this property, makes it superior to ceramics as a sensor. The g-constants, which represent the voltage generated in response to a mechanical stress, are typically 10 to 20 times that of piezoceramics. PVDF film also produces an electric voltage in response to infrared light, and the pyroelectric coefficient defines this relationship.

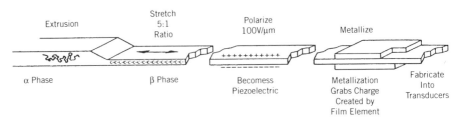

Figure 17.11. PVDF film manufacturing process. [Courtesy of Elf Atochem Sensors, Inc. (15).]

As a sensor PVDF behaves like a strain gage but does not require a conditioning power supply. The output signal is also typically greater than an amplified strain gage signal. This high sensitivity is due to the low thickness of the typical PVDF film (1 mil). Even a very small extensional force creates a large stress because of the small cross-sectional area. The constitutive relations for PVDF can be described as

$$\varepsilon_i = s_{ij}^{E,T} + d_{mi}^T + \alpha_i^E T$$

$$D_m = d_{mi}^T T_i + \varepsilon_{mn}^T E_n + D_m T$$

The symbols used in the equations above are the same as for PZTs, and α is the coefficient of thermal expansion. A typical set of properties for PVDF is given in Table 17.3.

Table 17.3   Properties of PVDF

Thickness, t	9, 28, 52, 110 × 10^{-6} m (μm)
Piezoelectric strain constant, d_{31}	23 × 10^{-12} $\dfrac{m/m}{V/m}$ or $\dfrac{C/m}{N/m}$
Piezoelectric strain constant, d_{33}	-33 × 10^{-6} $\dfrac{m/m}{V/m}$ or $\dfrac{C/m}{N/m}$
Piezoelectric stress constant, g_{31}	216 × 10^{-3} $\dfrac{V/m}{N/m^2}$ or $\dfrac{m/m}{C/m^2}$
Piezoelectric stress constant, g_{33}	-339 × 10^{-3} $\dfrac{V/m}{N/m^2}$ or $\dfrac{m/m}{C/m^2}$
Electromechanical, k_{31}	12% (at 1 kHz)
Coupling Factor, k_{33}	19% (at 1 kHz)
Capacitance, C	380 pF/cm^2 for 28-μm film $\varepsilon/\varepsilon_0 = 12$
Young's modulus, Y	2 × 10^9 N/m^2
Speed of sound, v_0	1.5–2.2 × 10^3 m/s (transverse-thickness)
Pyroelectric coefficient, p	-25 × 10^{-6} C/$m^2 \cdot$ K
Permittivity, ε	106–113 × 10^{-12} F/m
Relative permittivity, $\varepsilon/\varepsilon_0$	12–13
Mass density, P_m	1.78 × 10^3 kg/m^3
Volume Resistivity, P_e	10^{13} $\Omega \cdot$ m
Surface metallization, $R_\square$	2.0 $\Omega/\square$ for CuNi
Resistivity, R	0.1 $\Omega/\square$ for Ag ink
Loss tangent, tan δ_e	0.015–0.02 (at 10–10^4 Hz)
Compressive strength	60 × 10^6 N/m^2 (stretch axis)
Tensile strength	160–300 × 10^6 N/m^2 (transverse axis)
Temperature range	-40 to 80°C
Water absorption	<0.02% H_2O
Max operating voltage	750 V/mil = 30 V/μm
Breakdown voltage	2000 V/mil = 100 V/μm

Source: Elf Atochem Sensors, Inc. (15)

Because of its good sensor properties (i.e., the high g constant), light weight, flexibility, and toughness, PVDF is used in numerous sensor applications. PVDF is very commonly used in switching applications, where it develops a voltage in response to a pressure (it develops approximately 10 V with finger pressure). Keyboards make use of this property. One surface of the film is completely metallized, whereas the other surface is metallized only in areas corresponding to the keys and their associated signal-detection wires. The film is stretched under the metallic keyboard with recesses at various key locations. The finger pressure bends the film, producing a signal corresponding to the specific key, which is then used to perform appropriate tasks (16).

PVDF film is also used as for acoustic applications both as sensors and as actuators. It is used in compact microphones, earphones, and loudspeakers. Specially shaped PVDF sensors have also been used as modal sensors for sensing particular modes of vibration of structural elements such as beams, plates and cylinders. It has also been used in vibration control applications as an actuator, but of course as an actuator its authority is limited.

Many surveillance systems make use of the pyroelectric properties of the PVDF sensor. This application typically uses two closely coupled sensors. Much like the ambient-temperature compensation scheme used with strain gauge measurements, only one sensor is exposed to the infrared emissions, whereas both sensors respond equally to other stimuli, such as noise, vibration and global temperature changes. Signals from both sensors are combined to detect the infrared signal alone.

17.6. SHAPE-MEMORY ALLOYS

The first observation of the shape-memory effect (SME) was made in 1932 with gold–cadmium. The phase transformation associated with the shape-memory effect was later discovered in 1938 with brass. It was not until 1962 that Buehler et al. at the Naval Ordnance Laboratory (NOL) discovered a series of nickel–titanium alloys that demonstrated this shape-memory effect. The shape-memory alloy discovered by Buehler et al., (19) later named Nitinol, has been made commercially available ever since. Many materials are known to exhibit the shape-memory effect. They include the copper alloy systems of Cu–Zn, Cu–Zn–Al, Cu–Zn–Ga, Cu–Zn–Sn, Cu–Zn–Si, Cu–Al–Ni, Cu–Au–Zn and Cu–Sn, the alloys of Au–Cd, Ni–Al and Fe–Pt, and others. The most common of the shape-memory alloys or transformation metals is the Nickel–Titanium alloy, Nitinol.

The shape-memory effect (SME) can be described very basically as follows: an object in the low-temperature martensitic condition, when plastically deformed with the external stresses removed, will regain its original (memory) shape when heated. This transformation is illustrated in Fig. 17.12. This process, or phenomenon, is the result of a reverse martensitic transformation taking place during heating. The martensitic transformation, which is the

Shape Memory Effects

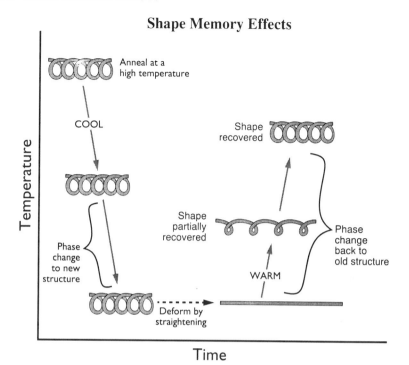

Figure 17.12. Shape-memory effect.

essence of the response of shape-memory alloys, may be illustrated simply by the change in martensite volume fraction with respect to temperature, as shown in Fig. 17.13. The four important transition temperatures are martensite finish (M_f), martensite start (M_s), austenite start (A_s), and austenite finish (A_f).

Nickel–titanium alloys of proper composition exhibit unique mechanical "memory" or restoration force characteristics. The shape-recovery performance of Nitinol is phenomenal. The material can be plastically deformed in its low-temperature martensite phase and then restored to the original configuration or shape by heating it above the characteristic transition temperature (the characteristic temperature can be varied from -50 to $166°C$). This unusual behavior is limited to NiTi alloys having near-equiatomic composition. Plastic strains of typically 6 to 8% may be recovered completely by resistively heating the material so as to transform it to its austenite phase. Restraining the material from regaining its memory shape can yield stresses of 100,000 psi, as shown in Fig. 17.14.

In addition to the special characteristics noted above, SMAs have some other special properties that are discussed briefly. The change in the Young's modulus of shape-memory alloys is very different from that of conventional metal materials. For most metals, the Young's modulus decreases as the temperature increases. However, the Young's modulus of shape-memory alloys increases within the phase transformation temperature range for the heating

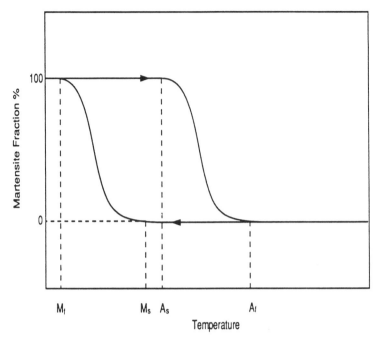

Figure 17.13. Martensite volume fraction versus temperature. [From Liang et al. (24).]

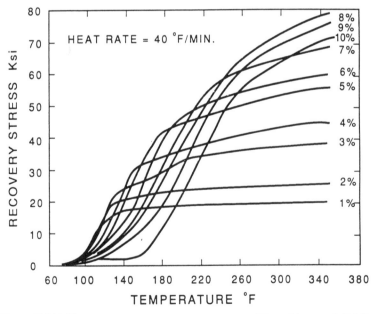

Figure 17.14. Shape-recovery stress versus temperature. [From Liang et al. (24).]

process. The Young's modulus of Nitinol increases three to four times from temperatures below M_f to temperatures above A_f. This is shown in Fig. 17.15.

Another important property of shape-memory alloys that has attracted a great deal of attention is the damping characteristics (i.e., the internal friction characteristics of shape-memory alloys). Figure 17.16. shows the change in the internal friction of $Ti_{50}Ni_{49}Fe_1$ with respect to temperature for different frequencies. The damping properties of shape-memory alloys may be exploited for passive and adaptive dynamic control applications.

Shape-memory alloys are different from other materials in many other aspects. The influence of the phase transformation on the electrical resistance is an example and is shown in Fig. 17.17. The distinction between the heating and cooling process demonstrates the effect of the phase transformation. Measuring the electrical resistance of shape-memory alloys is often used to determine the phase transition temperatures, as indicated in Fig. 17.17.

17.6.1 Properties for SMAs

The properties of nitinol are listed in Table 17.4.

17.6.2. Constitutive Relations of Shape-Memory Alloys

The mechanical behavior of shape-memory alloys is closely related to the microscopic martensitic phase transformation; the constitutive relations developed for ordinary materials such as Hooke's law and plastic flow theory are

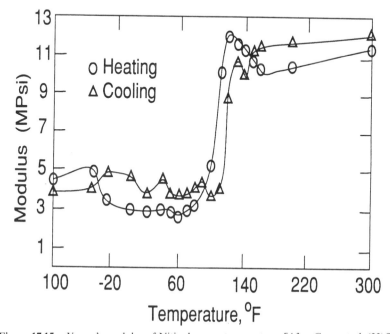

Figure 17.15. Young's modulus of Nitinol versus temperature. [After Cross et al. (22).]

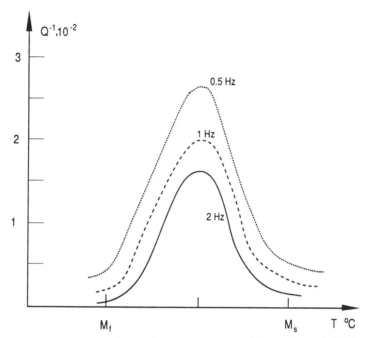

Figure 17.16.   Internal friction, Q^{-1} versus temperature. [(From Lin et al. (26).]

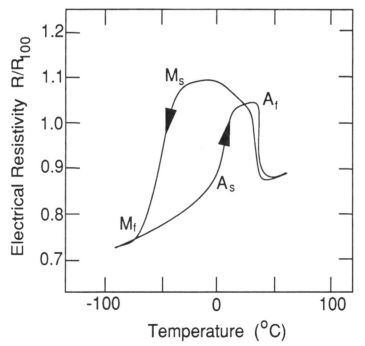

Figure 17.17.   Electric resistivity versus temperature. [After Funakubo (23).].
(© 1977, Gordon & Breach. Reprinted with permission.)

Table 17.4 Properties of NiTi Alloys

Physical	
Melting point	1300°C
Density	6.45 g/cm^3; 0.233 lb/in^3
Electrical resistivity	Austensite ~ 100 $\mu\Omega \cdot$ cm
	Martensite ~ 70 $\mu\Omega \cdot$ cm
Thermal conductivity	Austensite 0.18 W/cm $\cdot$ °C
	Martensite 0.085 W/cm $\cdot$ °C
Corrosion resistance	Similar to 300 series stainless steel or Ti alloys
Mechanical	
Young's modulus: Austensite	$\sim 12 \times 10^6$ psi
Martensite	~ 4–6×10^6 psi
Yield strength: Austensite	28–100×10^3 psi
Martensite	~ 10–20×10^3 psi
Ultimate strength	$\sim 130 \times 10^3$ psi
Elongation at failure	20–40%
Transformation	
Transformation temperature	-200–110°C
Latent heat of transformation	40 cal/g-atom
Shape-memory strain	8.5% maximum

From Liang et al. (24).

not directly applicable to shape-memory alloys. Therefore, new constitutive relations, which take into consideration the phase transformation behavior of SMA, have been developed. Several constitutive relations have been developed for shape-memory alloys over the last 20 years (21, 27, 30). While each is unique, none of the constitutive relations has a strong experimental justification, and each possesses distinct limitations (24). There are two approaches to developing material constitutive relations. One is the macroscopic phenomenological method, that requires a significant amount of experimental work; the other is the microscopic or physical method which derives material constitutive relations based on fundamental physical concepts. Combining both approaches takes advantage of each method and yields a more accurate constitutive model capable of predicting and describing the material behaviors of SMA.

17.6.2.1. Modeling of SMA with Internal Variables. Tanaka, and Sato et al. (30, 29a, 29b, and 29c) developed a model based on the concept of the free-energy driving force. Tanaka's model considers a one-dimensional metallic material of length L that is undergoing either martensitic transformation or its reverse transformation. The state variables for the material are strain, temperature, and extent of phase transformation, ξ, which is defined as the martensite fraction. The general state variable is defined as

$$\Lambda \equiv (\bar{\varepsilon},\ T,\ \xi)$$

The Helmholtz free energy is a function of the state variable Λ. The general constitutive relations can then be derived from the first and second laws of thermodynamics as

$$\bar{\sigma} = \rho_0 \frac{\partial \Phi}{\partial \bar{\varepsilon}} = \sigma(\bar{\varepsilon},\ T,\ \xi)$$

The stress is a function of the martensite fraction, an internal variable. From the equation above, the rate form of the mechanical constitutive equation is obtained as

$$\bar{\sigma} = \frac{\partial \sigma}{\partial \bar{\varepsilon}}\, \bar{\varepsilon} + \frac{\partial \sigma}{\partial T}\, \dot{T} + \frac{\partial \sigma}{\partial \xi}\, \dot{\xi} = D\bar{\varepsilon} + \Theta\dot{T} + \Omega\dot{\xi}$$

where D is Young's modulus, Θ the thermoelastic tensor, and Ω the transformation tensor, a metallurgical quantity that represents the change of strain during phase transformation. The material properties derived from thermomechanics are given as

$$D = \rho_0 \frac{\partial^2 \Phi}{\partial \bar{\varepsilon}^2}$$

$$\Theta = \rho_0 \frac{\partial^2 \Phi}{\partial \bar{\varepsilon}\, \partial T}$$

$$\Omega = \rho_0 \frac{\partial^2 \Phi}{\partial \bar{\varepsilon}\, \partial \xi}$$

If the expression of free energy is known, the minimization of the free energy may determine the equilibrium states of phases (i.e., the relation of the martensite fraction with the applied stress and temperature can be determined). However, instead of making the effort to find the free-energy expression, the martensite fraction, ξ, is assumed to be an exponential function of stress and temperature based on the study of transformation kinetics in Tanaka's model. These functions are

$$\xi_{M \to A} = \exp[A_a(T - A_s) + B_a\sigma]$$

$$\xi_{A \to M} = 1 - \exp[A_m(T - M_s) + B_m\sigma]$$

where A_a, A_m, B_a, and B_m are material constants in terms of the transition temperatures, A_s, A_f, M_s, and M_f, and so on.

Based on Tanaka's work, Liang and Rogers (25) modified and extended this model to predict and describe quantitatively the behavior of shape-memory alloys. The above equation integrated with respect to time yields the constitutive equation

$$\sigma - \sigma_0 = D(\varepsilon - \varepsilon_0) + \theta(T - T_0) + \Omega(\xi - \xi_0)$$

A cosine model that better models the phase transformation was proposed

$$\xi_{A \to M} = \tfrac{1}{2}\{\cos[a_a(T - A_s)] + 1\}$$
$$\xi_{M \leftarrow A} = \tfrac{1}{2}\{\cos[a_M(T - M_f)] + 1\}$$

where the two material constants are determined from

$$a_A = \frac{\pi}{A_f - A_s}$$

$$a_M = \frac{\pi}{M_s - M_f}$$

The three transition temperatures, M_f, M_s, and A_s are linearly dependent on the applied stress, as seen in Fig. 17.18. The A_f line is also assumed to be straight and parallel to the others, as shown in Fig. 17.18. To express the influence of stress on the transition temperatures, the following two material constants corresponding to the slope are introduced:

$$C_M = \tan \alpha$$
$$C_A = \tan \beta$$

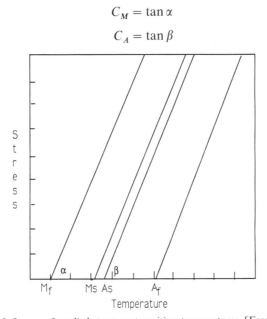

Figure 17.18. Influence of applied stress on transition temperatures. [From Liang et al. (24).]

where α and β are shown in Fig. 17.18. This effect of stress on the martensitic fraction versus temperature equations is introduced as follows:

$$\xi_{M\to A} = \frac{\xi_M}{2}\left\{\cos[a_a(T-A_s)+b_A\sigma]+1\right\}$$

$$\xi_{M\leftarrow A} = \frac{1-\xi_A}{2}\left\{\cos[a_M(T-M_f)+b_M\sigma]+1\right\}+\frac{1+\xi_a}{2}$$

where the two new material constants are

$$b_A = -\frac{a_A}{C_A}$$

$$b_M = -\frac{a_M}{C_M}$$

The constitutive equation, together with the martensitic volume fraction equations, provides the information necessary to solve all problems involving linear stiffness structures coupled to an SMA actuator.

17.6.3. Applications

Shape-memory alloys have been used in many fields and applications over the past 30 years. The first major industrial application of a shape-memory alloy was a cryogenic pipe fitting device developed by Raychem Corporation in 1969. Perhaps the best known example of this material is the prototype heat engine developed in the mid-1970's (18). Because of its biocompatibility and superior resistance to corrosion, Nitinol has been used in the medical field as a bone plate (20), an artificial joint (27a), and in dental applications (17). The major industrial application for this material has been as force actuators and robot controls (Funakubo, 1987) (23).

17.6.3.1. SMA Hybrid Composites. Rogers and Robertshaw (28) suggested the embedment of SMA fibers into composites (known today as SMA hybrid composites) to adaptively control the performance of the composites or composite structures. The class of the materials referred to as SMA hybrid composites are simply composite materials that have shape-memory alloy fibers (or film) embedded in them. The host composite material properties can be changed or internal forces applied to it by activation of the embedded SMA wires (activation refers to changing the phase of the SMA; this is usually done through resistive heating). One of the many possible configurations of SMA hybrid composites is one in which the shape-memory alloy fibers are embedded in a material off the neutral axis on both sides of the structure in agonist–antagonist pairs, as shown in Fig. 17.19. SMA fibers can be embedded in a variety of matrix materials, such as graphite–epoxy, glass–epoxy, thermoplastic materials, and other moldable or formable materials.

Two active structural modification techniques have been developed for SMA hybrid composites: one is called active properties tuning (APT), another

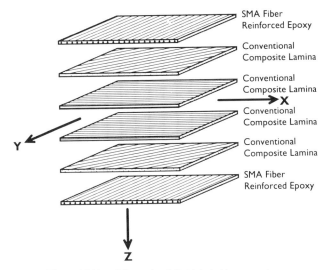

Figure 17.19. Schematic of SMA hybrid composite.

is active strain energy tuning (ASET). In active properties tuning, the embedded fibers are actuated by resistive heating (i.e., passing an electrical current through the fibers). Once the temperature of the fibers exceeds the transition temperature, the Young's modulus of the SMA fibers may be increased in a controlled manner from 4 Mpsi to 12 Mpsi. Active strain energy tuning uses the same mechanisms as described above for active property tuning but adds another important control parameter, the restoring stress associated with embedded inelastically elongated SMA fiber actuators trying to regain their original shape by contracting.

Using either active properties tuning or active strain energy tuning can result in significant control of the structural response of a composite structure. The difference between the two concepts is that APT changes the structural response by varying the stiffness of the structure alone, while ASET depends primarily on the recovery force of the plastically elongated fibers to change the structural response. Using active strain energy tuning in almost all cases results in more versatility of control than is possible in APT and can give a wide range of deflection control. The ability of ASET in active vibration and structural acoustic control of SMA hybrid composites has been demonstrated both theoretically and experimentally.

17.7. MAGNETOSTRICTIVE ACTUATORS

Magnetostrictive materials expand in the presence of a magnetic field, as their magnetic domains align with the field. The magnetostrictive effect was first discovered in nickel in 1840 and was later found to be present in cobalt, iron, and their alloys. However, magnetostriction in these materials was relatively small, on the order of 50 μstrain. Strains on the order of 10^{-2} were observed in the rare-earth elements terbium (Tb) and dysprosium (Dy) at very low temperatures (below 180 K). Scientists working with the Ames Laboratory and the Naval Ordnance Laboratory (now the Naval Surface Weapons Center) developed Tb-based compounds, such as the commercially available Ter-fenol-D ($Tb_{.3}Dy_{.7}Fe_{1.9}$), which have magnetostrictions of up to 2000 μstrain at temperatures up to 80°C and higher.

17.7.1. Behavior of Terfenol-D

Figure 17.20 shows magnetostriction and magnetization of Terfenol-D over a large range of temperatures and magnetic fields at a compressive stress of 13.3

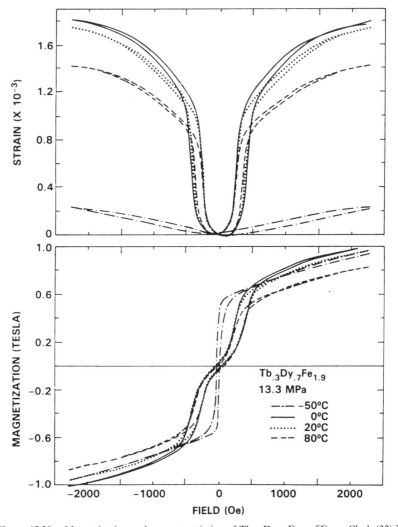

Figure 17.20. Magnetization and magnetostriction of $Tb_{0.3}Dy_{0.7}Fe_{1.9}$. [From Clark (32).]

MPa [from Clark, (32)]. Note that at very low temperatures ($-50°C$), only small magnetostriction occurs and the magnetization curve resembles that of a typical magnetic material. At temperatures of 0°C and above, a "jump" in the magnetostriction curve occurs at small magnetic fields as internally stored magnetic energy is converted to elastic energy. The change in strain in this narrow region is roughly half that of the overall change in strain at very high magnetic fields.

Note that magnetostriction is dependent on applied stress, as shown in Figure 17.21. When there is no applied compressive stress, the "jump" in the magnetostriction does not occur, while above a certain compressive stress, magnetostriction decreases. The reason for this unusual behavior is that applied stress changes the alignment of the magnetic domains within the material. Magnetostriction is a result of magnetic domain rotation; the more rotation that occurs, the larger the observed strain. With sufficient compressive stress (7.6 MPa) and no applied magnetic field, a large percentage of the magnetic domains are oriented perpendicular to the direction of applied stress, allowing larger magnetostrictions to occur, as the domains must rotate larger angles to align with the applied field (in the direction of the applied stress). Note that as the stress is increased further (see the 18.9 MPa curve), larger magnetic fields are required to produce the same strains, as the material performs more mechanical work when pushing against a larger stress (32).

A simple two-dimensional model to explain magnetostriction in Terfenol-D [developed by Clark, (32)], is shown in Fig. 17.22. In the first stage of

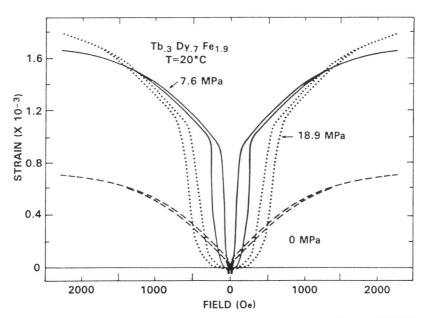

Figure 17.21. Magnetostriction of $Tb_{0.3}Dy_{0.7}Fe_{1.9}$ at compressive strsses of 0, 7.6, and 18.9 Mpa. [From Clark (32).]

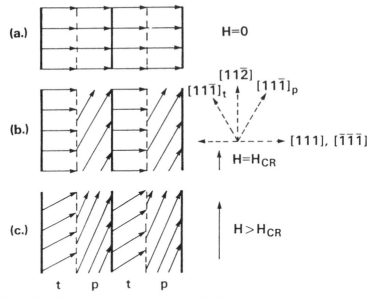

Figure 17.22. Model of the magnetization process for [112] twinned Terfenol-D single crystal: for no magnetic field (*a*), at critical magnetic field causing parent phase to "jump" (*b*), and above critical magnetic field when twin slowly rotates toward the [112] direction (*c*). [From Clark (32).]

magnetostriction, the domains are oriented perpendicular to the applied compressive stress at an initial magnetic field of zero. Note that the material consists of alternating parent and twin phases. Magnetostriction does not occur until a critical magnetic field is reached, when the parent phase "jumps" to a new direction and large strains are observed. Magnetostriction continues to increase as the applied field is increased beyond this critical value, as both the parent and twin rotate toward the direction of the applied field. This model accurately predicts the observed magnetostrictions.

17.7.2.   Properties of Terfenol-D

The properties of Terfenol-D are given in Tables 17.5 and 17.6

17.7.3.   Constitutive Relations

The following actuator and sensor equations describe the constitutive behavior of magnetostrictive materials (31):

$$\varepsilon = \frac{\sigma}{E} + d_m H$$

$$B = d_m \sigma + \mu^T H$$

Table 17.5 Nominal Properties of ETREMA Terfenol-D

Property	Units	Value
Density, ρ	kg/m^3	9.25×10^3
Young's modulus		
$\quad Y^H$	N/m^2	$2.5–3.5 \times 10^{10}$
$\quad Y^B$	N/m^2	$5.0–7.0 \times 10^{10}$
Sound speed		
$\quad c^H$	m/s	1.72×10^3
$\quad c^B$	m/s	2.45×10^3
Permeability		
$\quad \mu^T$	Tm/A	$9.2 \times 4\pi \times 10^{-7}$
$\quad \mu^S$	Tm/A	$4.5 \times 4\pi \times 10^{-7}$
Constant, d	m/A	1.50×10^{-8}
Constant, g	1/T	1.28×10^{-3}
Coupling, k	—	$0.70–0.75$
Resistivity, ρ_e	$\Omega \cdot$m	60×10^{-8}
Impedance		
$\quad \rho c^H$	rayls	1.57×10^7
$\quad \rho c^B$	rayls	2.27×10^7
Frequency constant		
$\quad f^H 1,$	Hz$\cdot$m	0.845×10^3
$\quad f^B 1,$	Hz$\cdot$m	1.255×10^3

Source: Etrema Products, Inc. (33).

Table 17.6 ETREMA Terfenol-D Thermal, Mechanical, and Magnetic Properties

Property	MKS Units	English Units
Bulk modulus	9×10^{10} N/m^2	13.1×10^6 lb/in^2
Tensile strength	28 MPa	4.1×10^3 lb/in^2
Compressive strength	700 MPa	101.5×10^3 lb/in^2
Curie temperature	380°C	716°F
Thermal expansion	12×10^{-6}/°C	
Thermal conductivity	0.1 W/cm$\cdot$°C (T_b)	
	1.0 W/cm$\cdot$°C (F_e)	
Magnetization	1.0 T	
Magnetostriction	1500–2000 ppm	
Energy density	$14–25 \times 10^3$ J/m^3	

Source: Etrema Products, Inc. (33).

where ε is mechanical strain, E is the Young's modulus ($= 26.5$ Gpa for Terfenol-D), σ is the mechanical stress, d_m is the magnetostrictive coefficient ($= 20 \times 10^{-9}$ m/A, H the magnetic field intensity, B the magnetic flux density, and μ^T the permeability ($= 11.56 \times 10^{-6}$ Tm/A). The magnetic field intensity H in a rod of length L is related to the current in the surrounding coil (with n turns per inch) through the following equation:

$$H = nI$$

A typical arrangement of a linear magnetostrictive actuator is shown in Fig. 17.23.

17.7.4. Applications and Limitations

To date, the commercial use of magnetostrictive actuators is limited. Their high stroke and force characteristics make them very good candidates for applications such as vibration and acoustic control, which are being actively investigated. The magnetostrictive actuators do not easily lend themselves to being embedded in a structure because of the difficulty of delivering a magnetic field within a structure. Also, magnetostriction is temperature-sensitive, making it difficult to design actuators for stable operation over a wide temperature range. One of the major limitations is the ohmic heating losses due to the solenoidal coil.

17.8. ELECTRORHEOLOGICAL FLUIDS

Electrorheological (ER) fluids experience reversible changes in rheological properties (apparent viscosity, plasticity, and elasticity) when subjected to electric fields. These fluids contain micron-sized (1 to $100\,\mu$m) dielectric particles suspended in a nonconductive base medium. The base medium is usually a silicone or mineral oil, although many different types of fluids can be

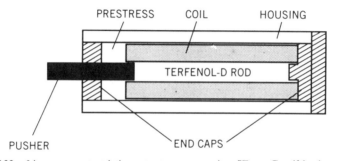

Figure 17.23. Linear magnetostrictive actuator cross section. [From Goodfriend et al. (34).]

used. The dielectric properties of the particles and the base medium must be different to ensure that the application of electric field will polarize the particles. This causes the particles to align and produce the unique characteristics. ER fluids have been the focus of an increasing amount of research because of the great promise they show for use in valves, mounts, clutches, brakes, and dampers.

Although researchers agree that the electrorheological effect is a result of polarization of the suspended particles, the mechanism for the observed particle chain formation has not been agreed upon. Some feel that it is a result of induced polarization of the layers surrounding the particles, while others feel it results from a bridging of the particles by water. A third explanation claims that chain formation is merely the result of interaction of interparticle coulombic forces. In any case, particle chains are formed when ER fluids are exposed to electric fields, as shown in Fig. 17.24, and these chains affect the mechanical behavior of the material.

Curves depicting idealized constitutive shear behavior of ER fluids are shown in Fig. 17.25. The two areas of the shear stress–shear strain curves evident in this figure are the preyield and postyield stages. Behavior in both regions is dependent on the applied electric field. For small strains in the

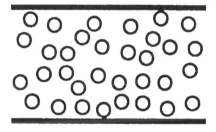

No electric field

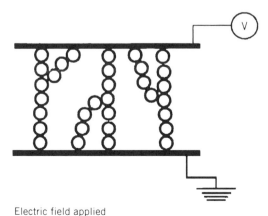

Electric field applied

Figure 17.24. Particle chain formation in ER fluid.

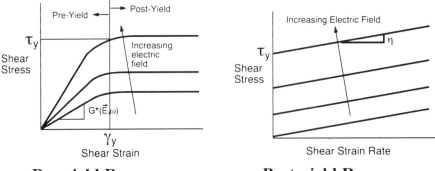

Pre-yield Response   **Post-yield Response**

Figure 17.25. Idealized constitutive shear behavior of ER fluids. [From Coulter et al. (35).]

preyield region, shear stress increases linearly with shear strain, with the curve having slope G^*, the complex shear modulus of the fluid, as shown in Figure 17.25. For structural applications where the ER fluid does not flow, preyield complex shear modulus is controlled to change the overall properties of the structure.

In the postyield region, behavior is often idealized using the Bingham plastic approximation,

$$\tau = \tau_y + \eta\dot{\gamma}$$

which is plotted in Fig. 17.25. In this equation τ_y is the dynamic yield stress, which is highly dependent on the electric field; η is the plastic viscosity, which is slightly dependent on electric field and is often assumed to be constant; and $\dot{\gamma}$ is the shear rate.

17.8.1. Properties and Requirements of ER Fluids

The following properties of ER fluids must be known to properly design an ER device (36):

- Shear stress versus shear strain rate data over the appropriate range of applied electric fields
- Dynamic and static yield stress values (if Bingham plastic behavior is observed) as a function of applied electric field
- Zero-field viscosity
- Response-time estimate
- Current density as a function of shear rate and electric field
- Preyield complex shear modulus as a function of applied electric field and frequency

However, the properties reported in patents and literature are often "vague, obtained by different test methodologies, inconsistent in definition, and varying in units.... This lack of uniformity in reporting material properties has hindered the development of feasible device designs and the commercialization of ER materials" (36). The minimum properties that ER materials should exhibit to be useful in a broad range of applications are listed in Table 17.7. The actual properties of several ER fluids developed by Lord Corporation are shown in Fig. 17.26. Note that ER fluids that exhibit the recommended

Table 17.7 Minimum Properties Needed by an ER Material to Be Utilized in a Variety of Applications

Property Description	Suggested Value
Dynamic yield stress at 4 kV/mm	$>3\,kPa$
Current density at 4 kV/mm	$<10\,\mu A/cm^2$
Zero-field viscosity	0.1–0.3 Pa-s
Operating temperature range	-40–$200°C$
Dielectric breakdown strength	$>5\,kV/mm$
Particle size	$\sim10\,\mu m$
Stability	Low sedimentation
	No dynamic separation
	No electrophoresis
	No chemical changes
	Low volatility
Miscellaneous properties	Nonabrasive
	Nontoxic
	Noncorrosive
	Nonflammable

Source: Weiss (37).

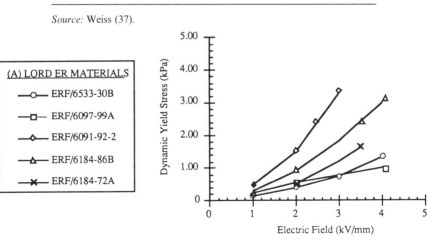

Figure 17.26. Properties of ER fluids developed at Lord Corporation. [From Weiss et al. (36).]

dynamic yield stress, zero-field viscosity, and temperature stability are currently available.

17.8.2. Applications of ER Fluids

The two general types of applications of ER fluids are illustrated in Fig. 17.27. In the first case, known as the sliding-plate mode (Fig. 17.27a), the material is stationary and the wall/electrodes of the fixture move relative to each other. In the second case, the flow (or valve) mode (Fig. 17.27b), the fluid flows through stationary walls/electrodes.

The derivation of the force–velocity relationship for the sliding plate mode is contained in one of the references (35). The force has two components: one from the yield stress of the ER fluid and the other from Newtonian behavior. Expressions for the flow rates and pressure drops that occur in the flow (valve) mode are also available. In the fixed-plate configuration the pressure drop, like the force in the previous configuration, has contributions from both yielding

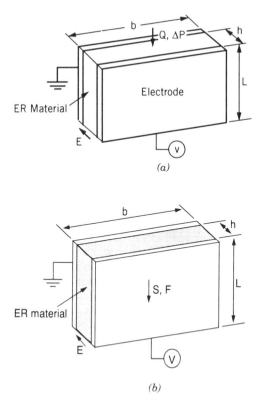

Figure 17.27. (a) Fixed electrode and (b) sliding electrode configuration used in ER devices. [From Coulter et al. (35).]

and viscous behavior. In addition to achieving the required mechanical behavior, the volume of ER fluid needed and the electrical power consumption should also be minimized when designing an ER device.

Examples of "early" ER devices are the clutch shown in Fig. 17.28 and the damper shown in Fig. 17.29. In 1988, the use of ER fluids was introduced in a patent application by Carlson, Coulter, and Duclos. (The patent was granted in 1990.) They proposed filling beams with ER fluid, which would change the natural frequency of the structure in response to an electric field. Sketches of two configurations of adaptive beams are shown in Fig. 17.30. It is expected that as ER fluids are improved and their behavior is better understood, the number of ER devices, which have simple designs that eliminate moving parts and greatly improve response time, will grow.

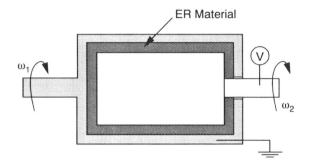

Figure 17.28. Sliding plate ER controllable clutch configurations. [From Coulter et al. (35).]

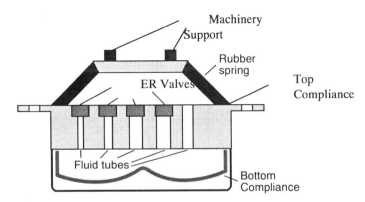

Figure 17.29. Fixed-plate ER damper configurations. [From Coulter et al. (35).]

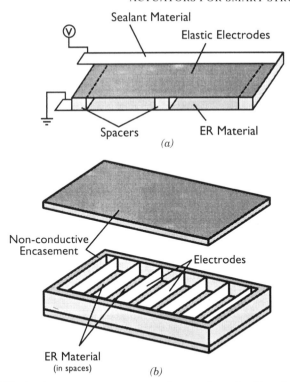

Figure 17.30. Basic ER adaptive structures components: shear configuration (*a*) and extensional configuration (*b*). [From Coulter et al. (35).]

REFERENCES

Piezoelectric Actuators

1. E. F. Crawley and E. H. Anderson, Detailed Models of Piezoceramic Actuation of Beams. *J. Intell. Mater. Syst. Struct.* **1**, 4–25 (1990).

2. EDO Corporation, *Piezoelectric Ceramics — Material Specifications, Typical Applications.* Salt Lake City, UT.

3. A. C. Eringen, *Continuum Physics.* Academic Press, New York, 1976.

4. B. Jaffe, R. S. Roth, and S. Marzullo, Piezoelectric Properties of Lead Zirconate–Lead Titanate Solid Solution Ceramics. *J. Appl. Phys.* **25**, 809–810 (1954).

5. C. Liang, F. P. Sun, and C. A. Rigers, An Impedance Method for Dynamic Analysis of Active Material Systems. *ASME J. Vib. Acoust.* **116**(1), 120–128 (1994).

6. Piezo Systems, Inc., *Piezoelectric Motor/Actuator Kit Manual.* Piezo Syst., Inc., Cambridge, MA.

Electrostrictive Actuators

7. AVX Corporation, *Electrostrictive Actuators.* AVX Corp., Myrtle Beach, SC.

8. D. Damjannovic and R. E. Newnham, Electrostrictive and Piezoelectric Materials for Actuator Applications. *J. Intell. Mater. Syst. Struct.* **3** (1992) 190–208.

9. K. Uchino, Electrostrictive Actuators: Materials and Applications. *Ceram. Bull.* **65**(4), 647–652 (1986).

Electroceramic Composites

A very extensive bibliography on piezocomposites is contained in the paper by Smith (14).

10. H. Banno and K. Ogura, Piezoelectric Properties of 0-3 Composite of Polymer and Ceramic Powder Mixture of PZT and $PbTiO_3$. *Jpn. J. Appl. Phys.* **30**(9B), Sept. 2247–2249 (1991).

11. J. A. Gallego-Juarez, Review Article: Piezoelectric Ceramics and Ultrasonic Transducers. *Phys. Rev. E.* **22**, 804–416 (1989).

12. R. E. Newnham, The Golden Age of Electroceramics. *Adv. Ceram. Mater.* **3**(1), 12–16 (1988).

13. R. E. Newnham, D. P. Skinner, and L. E. Cross, Connectivity and Piezoelectric-Pyroelectric Composites. *Mater. Res. Bull.* **13**, 525–536 (1978).

14. W. A. Smith, The Role of Piezocomposites in Ultrasonic Transducers. *Proc. 1989 IEEE Ultrason. Symp.,* pp. 755–766 (1989).

PVDF

15. Elf Atochem Sensors, Inc., *Standard and Custom Piezo Film Components.* Elf Atochem Sens., Inc., Valley Force, PA.

16. A. J. Lovinger, Ferroelectric Polymers. *Science,* **220**, 1115–1121 (1983).

Shape Memory Alloy

17. G. F. Andreason, and R. E. Morrow, *Am. J. Orthod.,* **73**(2), pp. 142–151 (1978).

18. R. Bank, *Shape Memory Effects in Alloys,* p. 537. Plenum, New York, 1975.

19. W. J. Buehler and R. C. Wiley, Nickle-Based Alloys. U.S. Pat. 3,174,851 (1965).

20. L. S. Castleman, S. M. Motzkin, F. P. Alicandri, and V. L. Bonawit, *J. Biomed. Mater.* (10) pp. 695–731 (1976).

21. J. S. Cory, Nitinol Thermodynamic State Surfaces. *J. Energy* **2**, No. 5 (1978).

22. W. B. Cross, A. H. Kariotis, and F. J. Stimler, *Nitinol Characterization Study. NASA. [contract Rep.] CR* **NASA CR-1433**, (1970).

23. H. Funakubo, *Shape Memory Alloys.* Gordon & Breach, New York, 1987.

24. C. Liang, The Constitutive Modeling of Shape Memory Alloys. Ph.D. Dissertation, Virginia Polytechnic Institute and State University, Blacksburg (1990).

25. C. Liang and C. A. Rogers, A One-Dimensional Thermomechanical Constitutive Relation of Shape Memory Materials. *J. Intell. Mater. Syst. Struct.* **1**(2), 207–234 (1990).

26. Z. C. Lin, K. F. Liang, J. X. Zhang, and W. G. Zhang, Resistance of Phase Interface Motion in the Process of Pre-Martensitic Transformation of Ti50Ni39Fe1 and Ti47.9Ni50.3 Alloys, *Proc. of MRS Int'l Mtg. on Adv. Mats.,* Japan, June (1988).

27. I. Müller, A Model for a Body with Shape Memory. *Arch. Ration. Mech. Anal.* **70**, 61–77 (1979).

27a. Ohnishi, Hamaguchi, Nabeshima, Miyagi, Hamada, Suzuki, and Shikida, *Proc. 3rd Conf. of the Japanese Soc. for Biomaterials*, p. 121 (1982).

28. C. A. Rogers and H. H. Robertshaw, Shape Memory Alloy Reinforced Composites. *Eng. Sci. Repr.* **ESP25.88027** (1988).

29a. K. Tanaka and R. Iwasaki, A Phenomenological Theory of Transformation Superplasticity, *Engineering Fracture Mechanics*, **21**(4), pp. 709–720 (1985).

29b. K. Tanaka, A Thermomechanical Sketch of Shape Memory Effect: One-Dimensional Tensile Behavior. *Res. Mech.* **18**, 251–263 (1986).

29c. Y. Sato and K. Tanaka, Estimation of Energy Dissipation in Alloys due to Stress-induced Martensitic Transformation, *Res. Mechanica*, **23**, 381–393, Elsevier Applied Science Publishers Ltd., England (1988).

30. K. Tanaka and S. Nagaki, A Thermomechanical Description of Materials with Internal Variables in the Process of Phase Transformation. *Ing. Arch.* **51**, 287–299 (1982).

Magnetostrictive Actuators

31. J. L. Butler, *Application Manual for the Design of ETREMA Terfenol-D Magnetostrictive Transducers*, Tech. Lit. Etrema Products, Inc., Subsidiary of EDGE Technologies, Ames, IA, 1988.

32. A. E. Clark, High Power Rare Earth Magnetostrictive Materials. In *Proc. Recent Advances in Adaptive and Sensory Materials*, pp. 387–397. Technomic Publishing, Lancaster, PA, 1992. C. A. Rogers and R. C. Rogers Eds.

33. Etrema Products, Inc., *Application Manual for the Design of Etrema Terfenol-D Magnetostrictive Transducers*. Etrema Prod. Inc.,. Ames, IA.

34. M. J. Goodfriend, K. M. Shoop, and O. D. McMasters (1992). Characteristics of the Magnetostrictive Alloy Terfenol-D Produced for the Manufacture of Devices. In *Recent Advances in Adaptive and Sensory Materials*, pp. 448–456. Technomic Publishing, 1992. C. A. Rogers and R. C. Rogers Eds.

Electrorheological Fluids

35. J. P. Coulter, K. D. Weiss, and J. D. Carlson, (1992). Electrorheological Materials and Their Usage in Intelligent Material Systems and Structures. Part II. Applications. In *Recent Advances in Adaptive and Sensory Materials and Their Applications*, pp. 507–523. Technomic Publishing, 1992. C. A. Rogers and R. C. Rogers, Eds.

36. K. D. Weiss, J. P. Coulter, and J. D. Carlson, Electrorheological Materials and Their Usage in Intelligent Material Systems and Structures. Part I. Mechanisms, Formulations, and Properties. In *Recent Advances in Adaptive and Sensory Materials and Their Applications*, pp. 605–621. Technomic Publishing, Lancaster, PA, 1992. C. A. Rogers and R. C. Rogers Eds.

18

High-Temperature Optical Fiber Sensors

RICHARD O. CLAUS, KENT A. MURPHY, ANBO WANG,
and RUSSELL G. MAY
Fiber & Electro-Optics Research Center
Virginia Polytechnic Institute and State University
Blacksburg, Virginia

18.1. INTRODUCTION

Optical fiber systems have been developed during the past 25 years for primary applications in long-distance, high-speed digital information communication. Optical fiber sensors have been developed over the past 15 years for applications in aerospace and hydrospace materials, civil structures, industrial process control, consumer products and biomedical systems. Optical fibers may be used as sensors for the measurement of environmental parameters such as strain, temperature, vibration, chemical concentrations, electromagnetic fields, and more. Their advantages for measurements include (1) immunity to electromagnetic interference, (2) avoidance of ground loops, (3) wide variety of environmental measurands, (4) excellent resolution, and (5) avoidance of sparks for applications in explosive environments. Additionally, optical fibers may allow effective long-term sensor operation at temperatures of approximately 900°C for silica waveguides and above 1700°C for sapphire waveguides. This chapter specifically concerns the last of these advantages, the use of optical fiber sensors at high temperatures at which other conventional sensors either do not perform as well as they do at low temperatures or at which they degrade unacceptably.

All sensor system elements degrade at very high temperatures due to the gradual or catastrophic destruction of the sensing elements themselves, typically the burning or melting of the elements. The upper operational temperatures of piezoelectric devices, for example, are limited by the Curie temperature of the piezoelectric material, resistance foil strain gauges by the melting temperature of the foil and connecting wire lead materials, and platinum gauges similarly, but by the higher melting temperatures of bulk platinum material or platinum films.

Fiber Optic Smart Structures, Edited by Eric Udd.
ISBN 0-471-55448-0 © 1995 John Wiley & Sons, Inc.

Optical fibers, and sensors fabricated from those fibers, are limited in temperature due to the individual temperature limitations of the materials that form the fibers and the sensors. The fiber material, as well as the coatings, the sensor housings, and the associated sensor system elements, all have specific maximum operating temperatures, and together these temperatures determine the way in which optical fiber sensors must be implemented in high-temperature environments.

In this chapter we consider the temperature limitations of optical fiber sensor elements, review methods that have been used to increase the upper operational temperatures of these elements, give special consideration to related material properties of both the fiber and the structural elements to be characterized using the sensors, and describe several examples of sensors that have been demonstrated up to temperatures exceeding 1700°C. These sensor methods include (1) multiwavelength pyrometers, (2) intrinsic blackbody radiators, (3) fluorescent decay time/temperature probes, (4) protected silica fiber sensors operated at temperatures approaching the softening temperature of silica, and (5) sapphire fiber sensors. Emphasis is given to the latter sapphire fiber devices because they alone offer the operation of the usual range of optical fiber sensor alternatives at temperatures well above the range of most conventional silica fiber or wire-based devices.

18.2. TEMPERATURE LIMITATIONS OF OPTICAL FIBERS AND SENSORS

Optical fiber sensors typically are fabricated using a number of individual components, and each component has an associated temperature limitation. Such components include the optical fiber material itself, the coatings on the fiber, the fiber sensor element housing materials, if any, and the passive connection and active optoelectronic circuit devices required for operation of the entire sensor system.

Optical fibers are typically manufactured from polymer, glass, or single-crystal ceramic materials. Polymer fibers are not intended for operation at high temperatures, for they have upper temperature limits of at most several hundred degrees Celsius. The silica in silica-based fibers has a softening temperature of approximately 1000°C, although the properties of the coating materials applied to the fiber usually determine its high-temperature limit. Conventional acrylate coatings, acceptable for many terrestrial communication applications, degrade rapidly above temperatures as low as 150°C. Polyimide coating materials may extend this range to perhaps as much as 400 to 500°C, but these still do not allow effective long-term operation at temperatures of particular interest for some applications. Metal coatings such as gold (bulk melting temperature 1053°C), platinum (bulk melting temperature 1755°C), and iron (bulk melting temperature 1530°C) may permit silica fiber operation up to the softening or melting temperature of the silica itself, as do ceramic coatings. It should be noted that the effective melting temperatures of bulk materials given here are generally higher than the melting temperatures of the materials in the form of thin coatings.

In addition to the temperature limitations of the fibers and the coatings on the fibers, the limitations of the sensor housings and other individual elements need to be considered. Sensor element housings need to be constructed from materials that do not degrade at high temperature, materials such as ceramics or high-melting temperature metals. Most passive and active optoelectronic support elements in optical fiber sensor systems cannot be heated to temperatures much above 150°C, so they must be isolated in remote and relatively cool instrumentation compartments. These combined limitations suggest that a high-temperature optical fiber sensor system would probably consist of a high-temperature probe end connected via a path of fibers of decreasing temperature performance to input and output electronics and optics in a relatively cool zone.

18.3. METHODS TO INCREASE TEMPERATURE LIMITATIONS OF FIBER SENSOR ELEMENTS

Several methods have been attempted to raise the maximum temperatures that various fiber sensor elements can withstand. In this section we briefly review some of those methods.

18.3.1. Optical Fiber

The straightforward way to increase the upper temperature limit of optical fiber waveguide is to use materials for the waveguides and coatings which can withstand higher temperatures. For operation at temperatures below approximately 100°C, conventional acylate coatings are serviceable in that they do not degrade mechanically, thereby possibly allowing microbending of the glass waveguide or exposure of the silica surface. For operation up to or slightly in excess of approximately 400 to 500°C, polyimide coatings may be employed. Polyimide coatings in particular are currently used on fibers and on fiber sensors which are embedded in polymer-based composite smart materials and structures because the processing temperatures of those materials are in the range 250 to 350°C. Different commercially available polyimide fiber coatings display considerably different degrees of adhesion within such composite laminates, as measured using single-fiber pullout tests.

For operation at temperatures above the approximate 400°C upper temperature limit for polyimide coatings, metal fiber coatings are suitable. Because different metals melt at different temperatures, the coating metals must be chosen with the specific applications in mind. Aluminum, for example, has a melting temperature of approximately 650°C, well less than the roughly 1000°C softening temperature of silica. Gold coatings, however, have a melting temperature of about 1060°C, thus would potentially allow a full temperature range of operation for a gold-coated silica fiber. Such gold-coated fiber has been used successfully in applications requiring the attachment of optical fiber sensors to the surfaces of metal alloys where the attachment is performed at high processing temperatures and the metals are subsequently held at high temperature during measurements (1). Here a major concern is the amount of time the silica fiber as well as the fiber coating remain at temperature during processing, since the material processing temperatures are typically higher than the in-service operational temperature of the resulting component.

Another limitation on the high-temperature performance of silica fiber waveguides is the diffusion of dopants between the core and cladding regions, a process that gradually destroys the waveguiding properties of the structure. For example, by altering the basic glass chemistry of the fiber and using fluorine rather than germanium dopant materials, the softening temperature of the doped silica may be increased by tens of degrees Celsius, while the low-temperature waveguiding properties are not altered. Thus far, however, this general approach has not created silica glass structures capable of substantially higher operational temperatures.

18.3.2. Sensor System Elements

Fiber sensor housings and components typically cannot be designed for operation at excessively high temperatures, and thus must be actively cooled by thermally conducting gases or liquids, by connection to a cold metal plate having high thermal conductivity, by recessing into a cooler region of the host environment for operation, or by a combination of all of these methods. One exception to this may be the implementation of very-high-temperature ceramic single-crystal optical elements which may be integrated at the end of lower-temperature probes fibers while exposed to the external high-temperature environment. Examples here include the use of sapphire or ruby windows.

18.3.3. Sapphire Optical Fiber

Sapphire optical fibers may also be used to extend typical optical fiber sensor operation temperatures much higher than those achievable using silica fibers. Sapphire optical fiber waveguides typically are manufactured as single-crystal or multicrystalline rods or hollow tubes using either edge-defined film growth (EFG) or crystalline draw methods. Typical sapphire fibers have outer diameters of 50 μm or more, a uniform index of refraction of approximately 1.76 at room temperature, lengths of several tens of centimeters, and attenuation on the order of 1 dB or more per meter. Such fibers are fabricated mainly as structural support fibers for use in either advanced high-temperature metal matrix and ceramic matrix structural composite materials, and in low-transmissivity and relatively thin sapphire windows and plates. Structural support fibers in general are not manufactured for their optical transmission or sensing characteristics and may have attenuation on the order of 10 dB per meter or much more. Major influences on the attenuation are diameter variations as a function of distance along the fiber and material inclusions caused in part by the metal dies used in the EFG process. Improved EFG-fabricated sapphire fibers and single-crystal fibers have much lower optical attenuation and have been used in all of the sensor applications described below.

Larger-diameter sapphire fiber "rods" may be used either as extended transmissive windows into hot regions, or as elements in interferometric sensor systems, as described below. Obviously, the major limitation of rods over fibers for such applications is their inability to be bent due to a smaller bending radius determined by (1) mechanical load to failure, and (2) geometric optical path considerations. Because the melting temperature of single-crystal c-axis sapphire is approximately 2040°C, such rod-based sensors may operate at corresponding high temperatures. It should be noted again that the surface chemistry of the sapphire at elevated temperatures differs from that of the bulk material, and degradation of the edges of sapphire rod fibers has been noted at temperatures as low as 1700°C.

18.4. SPECIAL CONSIDERATIONS FOR HIGH-TEMPERATURE FIBER SENSORS

Several special considerations need to be given to optical fiber sensor systems that operate at temperatures at or above 1000°C. The first are the coefficients of thermal expansion (CTE) of the fibers and of the mechanical supporting elements that surround the fibers. If the CTEs of the various materials differ substantially, large excursions in temperature will produce large interfacial strains, which may lead to cracking or spalling. Thermal shock is of similar concern, for the effects of such interface stresses may be increased if the rate of temperature rise is high and the limited thermal conductivity of the materials does not or cannot allow temperature conduction fast enough to prevent large stress accumulation that must be relieved by interface or internal material fracture.

Second, although sapphire is relatively inert chemically at temperatures below 1500°C, some reactions may occur at higher temperatures, so the exposed surfaces of the fibers need to be protected from reactive materials. Such protective coatings on sapphire fiber rods, for example, have been shown to be beneficial for measurements in coal-fired combustion systems which produce high-alkali-content fly ash as a by-product (2). Protective zirconia and other coatings have been employed in these cases to mitigate surface reactions between the ash and the sapphire fiber elements (3).

Third, for quantitative measurements of displacement-related phenomena in and on materials, including the quantitative determination of strain and ultrasonic wave propagation, the sensor elements must be mechanically affixed to the material and not free to expand by temperature-induced CTE effects alone. Thus these fiber sensor geometries are subject to the same type of thermal and strain cross-sensitivity that plagues low-temperature systems, although they are enhanced due to the large differential temperature involved. To overcome this problem, temperature and strain typically need to be monitored independently (2) or multiparameter measurement techniques implemented.

18.5. HIGH-TEMPERATURE OPTICAL FIBER SENSORS

In this section we briefly review high-temperature optical fiber sensor systems and performance.

18.5.1. Blackbody Radiation Sensors

A general class of high-temperature fiber sensors uses glass or crystalline fibers as windows to gather emitted blackbody or effectively filtered graybody radiation from inside hot regions. By observing the ratio of output signal powers in two or more relatively narrow wavelength ranges, the temperature of the region may be determined from Planck's radiation law and the emissivity of the target. Representative spectral curves are shown in Fig. 18.1. Here the requirements on the optical fibers are those of (1) operation at temperatures

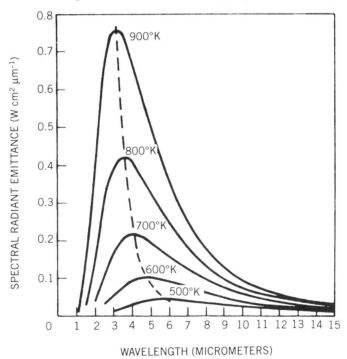

Figure 18.1.   Blackbody radiation curves. By observing the spectral content of radiation emitted from a blackbody source, the temperature of the source may be determined.

high enough to survive the environment around the viewing port, and (2) good optical transmission in the infrared portion of the spectrum. High-temperature operation is typically achieved using sapphire rod materials or windows, or by actively cooling a lower-temperature viewing rod material. Spectral transmission requirements typically limit the range of useful fibers to sapphire-, fluoride-, or chalcogenide-based fibers.

A similar but slightly different sapphire rod–based high-temperature sensor was demonstrated by Dils (4). In this sensor, a specific metal with a high melting temperature was deposited onto the end of the probe fiber rod to itself serve as the blackbody radiator, and then overcoated with a thin layer of protective sapphire. The measured temperature information was extracted by taking the ratio of two specifically selected wavelengths emitted by the platinum film. Figure 18.2 shows a diagram of a commercially available sensor that is based in part upon this concept.

18.5.2. Fluorescent Decay Temperature Sensors

A second method for the monitoring of temperature, albeit most effectively at temperatures only up to several hundred degrees Celsius, where the effects of internal blackbody radiation are low, involves the use of fluorescent materials which are deposited onto the end of a high-temperature and transmissive fiber material (5). The arrangement of the phosphor tip of such a sensor element is shown in Fig. 18.3. When an optical pulse at the correct wavelength impinges on the material, it emits a fluorescent signal at a different wavelength with a time decay delay which is a function of temperature. If the fiber is a sapphire rod waveguide, and the coating material TiO_2, temperature resolution on the order of approximately 1°C is possible.

18.5.3. Silica Fiber Sensors Operated at High Temperature

Certainly, silica fiber sensor elements and systems can be operated at ambient temperatures that do not cause material degradation. A number of silica fiber–based sensors may also be operated conveniently in temperature regions

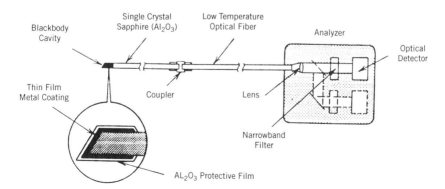

Figure 18.2. Geometry of blackbody emission-based temperature sensor.

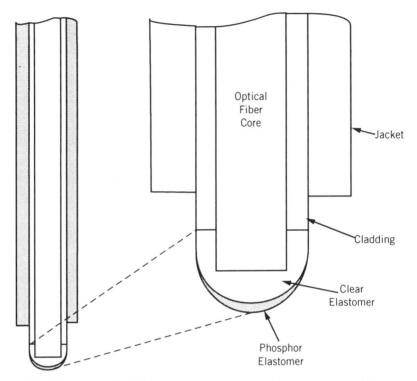

Figure 18.3. Geometry of tip of fiber temperature sensor based on measurement of flouorescent lifetime decay.

well in excess of 1000°C with proper thermal management and active cooling. Figure 18.4, for example, shows the profile of a silica fiber-based acoustic pressure transducer cooled by flowing water; such a device has been used for the measurement of skin friction on the surface of a wind tunnel combustor (6) combustion chamber. The sensors operated in this way may typically be those in which changes in the index of refraction of the glass fiber are relatively large and may mask the desired sensor information by cross-sensitivity effects, as indicated above. For that reason, sensors such as extrinsic Fabry–Perot interferometric systems in which the gauge length of the sensor is formed from air rather than glass appear particularly attractive.

18.5.4. Sapphire Optical Fiber Sensors

In this section we discuss the use of sapphire optical fiber-based sensors of temperature, displacement, strain, and ultrasonic waves.

18.5.4.1. Intensity-Based Sapphire Fiber Sensors. Simple intensity-based sapphire fiber sensors have been proposed and evaluated. For example, operation may be based on the temperature dependence of the critical angle for total internal reflection (TIR). To illustrate this particular concept, consider the sensor diagram in Fig. 18.5. Applying Snell's law at the interface between the

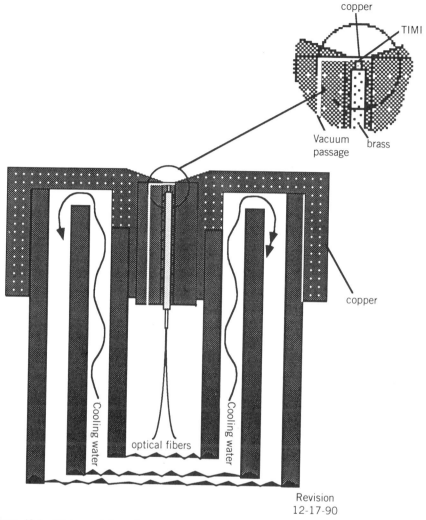

Figure 18.4. Water-cooled sensing head of silica fiber-based temperature-insensitive Michelson interferometer transducer and experimental measurement of pressure in a combustor at 1400°C.

inclined sapphire end face and the assumed surrounding air, we have

$$n \sin \alpha = \sin \frac{\pi}{2}$$

where n is the sapphire index of refraction, which is temperature dependent, and α is the critical angle corresponding to total internal reflection. Therefore, this equation can be simplified to

$$\alpha = \sin^{-1} \frac{1}{n(T)}$$

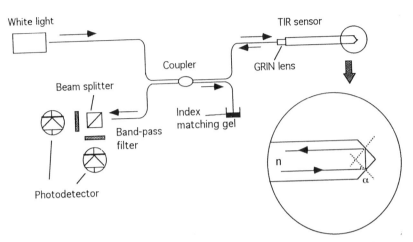

Figure 18.5. TIR-based sapphire rod temperature sensor geometry. Temperature changes vary the back-reflected light at the end of the sapphire rod.

where T is the temperature at the sensor head. The metal coating on the end surface serves as a mirror to reflect all the totally internal reflection light back to the GRIN lens and photodetector. Changes in temperature will alter the refractive index $n(T)$, thus changing the critical angle α for total internally reflected signal, and this will subsequently change the optical power received by the photodetector

Due to the difficulty in polishing sapphire, quartz glass may be used instead of sapphire rod to demonstrate the feasibility of this concept. For example, for a quartz glass rod ($n = 1.452$), the inclination angles with respect to the longitudinal axis of the mirror and the total internal reflection surfaces should be $66.6°$ and $43.5°$, respectively. Preliminary results of an experiment in which the sensor was heated from room temperature to $1030°C$ are shown in Fig. 18.6. The output signal was small but repeatable.

18.5.4.2. Extrinsic Fabry–Perot Sapphire Fiber Sensors. A family of sapphire fiber-based extrinsic Fabry–Perot interferometric (EFPI) sensor devices has been developed and demonstrated for environmental measurements as indicated in this section.

Measurement of Temperature and Displacement. Sapphire fiber–based systems have been used first to measure both displacement and temperature. For the measurement of the displacement of the surface of a material specimen, the flat end of the sapphire rod is placed close and parallel to the surface to be monitored. Here the Fabry–Perot etalon is formed by an air gap between the end of the sapphire rod and the reflective surface of the specimen (7). Figure 18.7 shows the output electrical signal for a displacement of an aluminum surface 1.0 mm away from the end of the sapphire rod. The top trace shows all

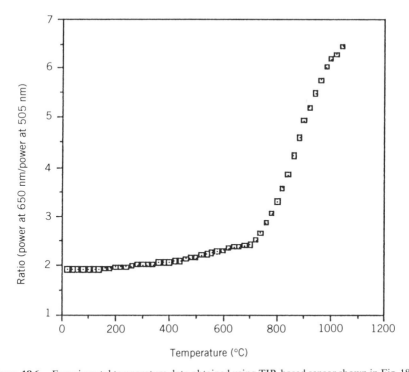

Figure 18.6. Experimental temperature data obtained using TIR-based sensor shown in Fig. 18.5.

of the fringes associated with this displacement; the bottom trace shows an expanded view of several fringes.

For the measurement of temperature, the entire sapphire rod may be inserted into a temperature-controlled chamber. In this case, the etalon is formed by the interface between the GRIN lens and the sapphire rod, and the opposite end surface of the rod. By slowly increasing the temperature in the chamber so that the temperature of the entire rod remains approximately uniform, a simple model may be used to relate the measured fringe shifts at the

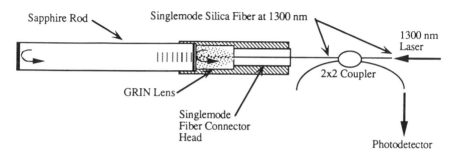

Figure 18.7. Geometry of external Fabry–Perot etalon (top) and interference fringes caused by reflector displacement (bottom).

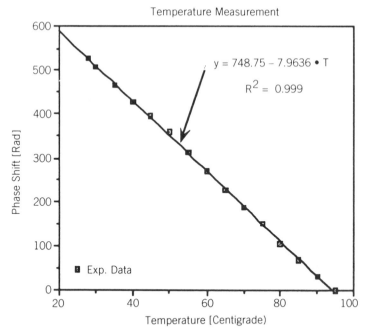

Figure 18.8. Temperature data obtained by monitoring interferometric fringes from sapphire sensor.

photodetector to the temperature of the rod. Specifically, fringe shifts are due to the combination of both changes in the length of the rod, due to the coefficient of thermal expansion of sapphire (measured to be approximately $5.3 \times 10^{-6}°C^{-1}$) and the value of dn/dT for sapphire (measured for the rod material used to be approximately $16 \times 10^{-6}°C^{-1}$ at $25°C$).

Measurement of Ultrasonic Waves at High Temperature. Next, the sapphire sensor rod and a second sapphire rod used solely as a reflective target were supported inside a sapphire alignment tube as shown in Fig. 18.9. The inner diameter of the tube is slightly larger than the outer diameter of the fiber rods. If the rods are attached to the tube, the distance between the attachment points

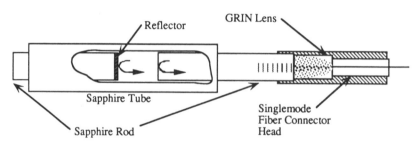

Figure 18.9. Extrinsic sapphire rod–based Fabry–Perot interferometric sensor geometry.

defines the gauge length of the sensor element and suggests its use as an optical fiber strain gauge. Here, the Fabry–Perot etalon is formed by the air gap between the end of the rod and the facing surface of the target rod. A simple displacement of the target reflector rod by a distance of 0.0012 mm produced the set of fringes shown in Fig. 18.10. Here the top trace includes all the fringes detected; the bottom trace is an expanded view of the first few fringes.

Finally, the sapphire fiber rod sensor was supported above the surface of a material specimen as shown in Fig. 18.11 to measure the out-of-plane particle displacements associated with ultrasonic surface elastic waves propagating on the surface. If the density and moduli of the material are known, this component information may be used to determine the entire surface-wave displacement field. For this geometry, the etalon again is formed by the gap between the end of the sapphire rod and the facing surface.

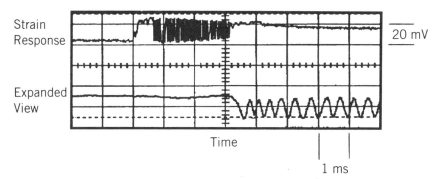

Figure 18.10. Interference fringes measured at output of sensor element shown in Fig. 18.8 due to applied stress at room temperature.

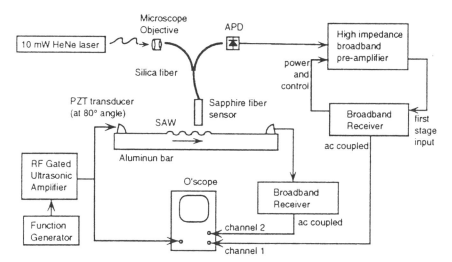

Figure 18.11. Extrinsic etalon geometry for measurement of ultrasonic surface waves on a material at high temperature.

Figure 18.12 shows the photodetector output for a 2.25-MHz wave generated on the surface using a conventional piezoelectric transducer. Here, the
averaging effect is observed due to the relative size of the probing beam and
the acoustic wavelength of the surface wave. The ultrasonic transducer produces a surface elastic wave having a maximum out-of-plane displacement on
the order of 10 Å, so the resulting maximum optical phase change corresponds
to much less than one fringe, and the output signal remains in the linear range
of the system.

18.5.4.3. Birefringence-Balanced Polarimetric Sapphire Fiber Sensor. For a
birefringence-modulated fiber sensor in which the input and output polarizers
are set parallel to one another and the slow axis of the birefringent element is
oriented at 45° with respect to the input polarizer, the output intensity of the
sensor is given by

$$I = I_0 \cos^2\left(\pi \Delta n \frac{L}{\lambda} + \frac{\phi_0}{2}\right)$$

$$= \frac{I_0}{2}\left[1 + \cos\left(2\pi \Delta n \frac{L}{\lambda} + \phi_0\right)\right] \tag{18.1}$$

where Δn is the birefringence of the sensing element, L the propagation length
of the light in the sensing element, ϕ_0 the initial phase delay, and λ the
operating wavelength in vacuum (8). Since Δn is a function of the measurand
due to a certain effect, such as the Kerr, Pockels, or photoelastic effects, the
output intensity of the sensor is a function of the measurand. To obtain the
maximum sensitivity, ϕ_0 is usually chosen to be $N\pi + \pi/2$, where N is an
integer. Note that we have assumed that the operating light is monochromatic
in Eq. (18.1). It has been reported that for a multimode fiber network, such as
that for an extrinsic multimode fiber sensor, a serious modal noise, caused by
the spatial filtering effect, is observed when coherent light is launched into this
system. One of the effective ways of reducing this noise is to use a low-

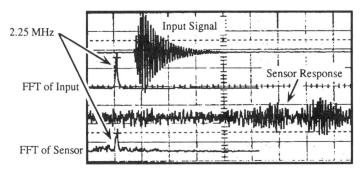

Figure 18.12.   Ultrasonic surface wave detected using sensor geometry shown in Fig. 18.11.

coherence light source. In multimode fiber communication networks, the dispersion and hence the information transmission rate will be limited by such a low-coherence light source. However, in most multimode fiber sensor applications, the bandwidth is not a problem. Light-emitting diodes (LEDs) are the most commonly used light sources in multimode fiber sensors.

LED light sources may also be used for these types of sensors. In this case, the spectral width of the light source should be considered. The intensity-spectrum distributions of most LEDs can be approximately expressed by a Gaussian function: that is,

$$Q^2(\omega) = \frac{2P_0}{\sqrt{\pi}\,\Delta\omega} \exp\left[-\frac{(\omega - \omega_0)^2}{\Delta\omega^2}\right] \qquad (18.2)$$

where $\Delta\omega$ is the half spectral width, ω_0 the central angular frequency of the source, and P_0 the total output power of the source. Then Eq. (18.1) can be rewritten as

$$I = \frac{AP_0}{\sqrt{\pi}\,\Delta\omega} \int_0^\infty \left[1 + \cos\left(\frac{\Delta n\,L}{c}\omega\right)\right] \exp\left(-\frac{(\omega - \omega_0)^2}{\Delta\omega^2}\right) d\omega \qquad (18.3)$$

where the constant A includes the effects of the fiber losses. Note that we have ignored the wavelength dependence of the fiber losses within the spectral width of the light source. Assuming that $\Delta\omega \ll \omega_0$ and evaluating the integral in Eq. (18.3), we obtain

$$I = \frac{AP_0}{2}\left\{1 + \exp\left[-\frac{(\Delta n\,L\Delta\omega)^2}{4c^2}\right]\cos\frac{\Delta n\,L\omega_0}{c}\right\} \qquad (18.4)$$

It is noted that when $\Delta(nL)\Delta\omega/c \ll 1$, $\exp[-(\Delta nL\Delta\omega)^2/(4c^2)]$ approaches unity and Eq. (18.4) becomes Eq. (18.1). When $\Delta(nL)\Delta\omega/c \gg 1$, the sensor output is nearly a constant with respect to ΔnL. We define the fringe contrast to be

$$\text{fringe contrast} = \exp\left[-\left(\frac{\Delta nL\Delta\omega}{2c}\right)^2\right] \qquad (18.5)$$

The value of the fringe contrast is between 0 and 1. The intensity-spectrum profile of the LED used to obtain experimental results described below is shown in Fig. 18.13; $\Delta\omega$ of this LED was measured to be 46 nm. Based on Eq. (18.5), the dependence of the fringe contrast on the ΔnL is plotted in Fig. 18.14. From this figure it is noted that the value of ΔnL should be controlled within the order of about 10 μm to observe a clear signal variation in the sensor output. The birefringence of the sapphire single crystal used in our experiment is 0.008. Thus 10 μm of ΔnL corresponds to 1.25 mm. For a contact high-temperature measurement, this length limitation is not acceptable.

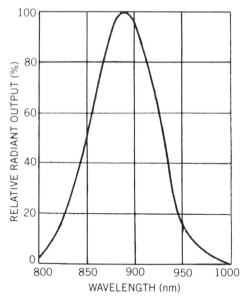

Figure 18.13.   Intensity spectrum distribution profile of the LED.

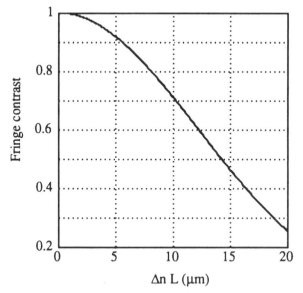

Figure 18.14.   Dependence of the fringe contrast on ΔnL.

To solve the length-limitation problem posed above, birefringence-balanced polarimetric fiber sensors based on the inherent properties of sapphire crystal have been developed. The schematic of the birefringence-balanced polarimetric multimode fiber sensor is shown in Fig. 18.15. Here the light from an LED is

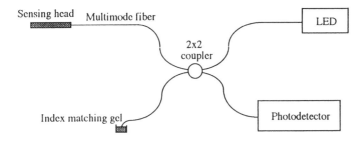

Enlarged view of the sensing head

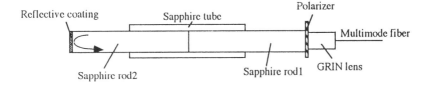

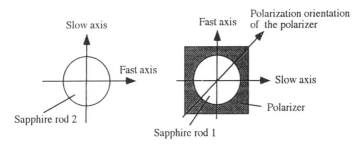

Figure 18.15. Schematic of the birefringence-balanced polarimetric fiber optic sensor based on sapphire single-crystal materials.

launched into a multimode fiber and propagates to the sensing head. The light out of the fiber is collimated by a quarter-pitch GRIN lens and passes through a polarizer (P). The polarization orientation of the polarizer was oriented at 45° with respect to the fast or slow axis of the sapphire rod adjacent to the polarizer. This linearly polarized light is launched into a sapphire rod and propagates along the longitudinal axis of this rod. To effectively shorten the length of the wave media, another sapphire rod of the same type and size is attached end to end with the first in such a way that the fast axis of the first sapphire rod is aligned with the slow axis of the second. Hence the fast component of the electric field in the first rod becomes the slow component in the second rod, and vice versa. At the free end face of the second sapphire rod, the incident light is reflected and travels back in the same rod to the polarizer.

The effective optical path difference between the fast and slow axes is given by

$$(\Delta nL)_{\text{effective}} = 2(\Delta n_1 L_1 - \Delta n_2 L_2) \qquad (18.6)$$

Assuming that $\Delta n_1 = \Delta n_2$, Eq. (18.6) becomes

$$(\Delta nL)_{\text{effective}} = 2\Delta n(L_1 - L_2) \qquad (18.7)$$

Both Δn and L depend on temperature. If the difference between the lengths of the two sapphire rods can be controlled within 0.1 mm, it may be possible to observe dramatic intensity changes in the sensor output under temperature perturbations of either sapphire rod.

To demonstrate the performance of this device, light from a pigtailed LED at 890 nm was launched into a multimode silica fiber with dimensions of 100 to 140 mm and a numerical aperture of 0.28. Two single-crystal sapphire rods, manufactured by Saphikon, Inc., with 3.23-mm diameters and lengths of 37.9 mm were connected end to end. To increase the reflection from the free end face of the second sapphire rod, a chromium film, with a melting temperature of approximately 1870°C, was deposited onto this end face. A 2 × 2 fused biconical taper directional coupler was used to separate the input and output light of the sensor. To demonstrate temperature measurements up to 1500°C, an oxypropane torch was used to directly heat the second sapphire rod. A length of about 10 mm of the second sapphire rod was heated, and a high-temperature thermocouple (type B) attached to the heated sapphire rod surface was used simultaneously to monitor the temperature. The experimental results are presented in Fig. 18.16. The sensitivity of the sensor was derived

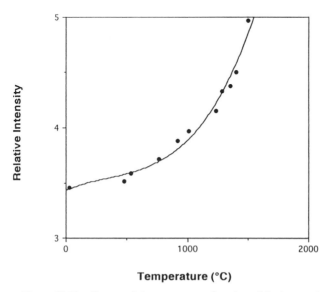

Figure 18.16. Output of the sensors as a function of the temperature.

from 3σ (σ is the standard deviation in the sensor output at room temperature). A sensitivity of 5°C was obtained with a dynamic range of 25 to 1500°C.

18.5.4.4. Sapphire Fiber–Based Intrinsic Interferometric Sensor. Sapphire fiber–based intrinsic Fabry–Perot interferometric sensor devices have also been implemented, as shown in Fig. 18.17 (9). Here a length of multimode sapphire fiber is connected by fusion splicing to a single-mode silica fiber to form the Fabry–Perot cavity. A laser beam is launched into the single-mode silica fiber and propagates to the sapphire fiber. Due to the difference in the refractive index between the silica and sapphire fibers, part of the incident laser power is reflected. The light transmitted into the sapphire fiber excites propagation modes. Since the numerical aperture of the single-mode fiber is very small, primarily low-order modes in the sapphire fiber are excited. The propagating light in the sapphire fiber is reflected at the free end face in air. Without reflection coatings, this end face provides about 7% reflectance (the refractive index of sapphire is approximately 1.7), caused by internal Fresnel reflection. At the silica–sapphire fiber splice, a fraction of this reflected light from the free end face of the sapphire fiber is recoupled into the silica single-mode fiber. The interference of the two reflections, from the silica–sapphire fiber splice and the free end face of the sapphire fiber, respectively, gives rise to the interference fringe output. Since the optical phase of the light traveling in this sapphire fiber is highly sensitive to temperature and longitudinal strain, this sensor is useful for the measurement of these parameters.

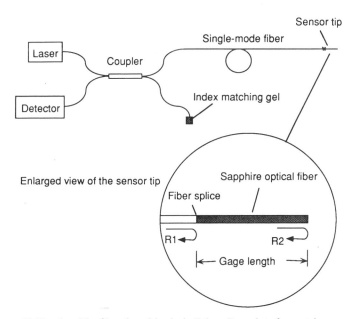

Figure 18.17. Sapphire fiber–based intrinsic Fabry–Perot interferometric sensor.

The sapphire fiber used in the experimental device described below had a relatively large diameter (much larger than the diameter of the single-mode fiber core), with no cladding. Due to this large size, the propagation constant of the fundamental mode in the sapphire fiber should be very close to the propagation constant in bulk sapphire. Thus the sensor interference fringe output can be written approximately as

$$\phi \cong 2nL \tag{18.8}$$

where n is the refractive index of sapphire and L is the length of the sapphire fiber. Then we obtain an expression for the phase difference responsible for the interference fringe variations of the sensor output as

$$\Delta\phi \cong \frac{4\pi}{\lambda_0}(n\Delta L + L\Delta n) \tag{18.9}$$

where λ_0 is the source wavelength in vacuum.

The schematic of the experimental setup is shown in Fig. 18.17. In the experiment, the light from a pigtailed semiconductor laser at 1300 nm was launched into a single-mode silica fiber with dimensions of 9 to 125 mm and a numerical aperture of 0.13. A 2×2 fused biconical taper bidirectional coupler was used to separate the input and output light of the sensor. A 31-mm length of uncoated structural-grade sapphire fiber with a diameter of 125 mm was connected to the lead-in single-mode silica fiber.

The sapphire fiber, which has a melting point of 2053°C, was fabricated by Saphikon, Inc. of Milford, New Hampshire, by an edge-defined fiber growth (EFG) technique. In the experiment it was found that the fringe contrast of the sensor output was dependent on the quality of the silica fiber to the sapphire fiber splice. Specifically, the fringe contrast is a function of the relative position between the silica and sapphire fibers at the splice. To optimize the sensor performance, it is necessary to study this effect. To change the relative position between the fibers precisely, both the silica and sapphire fibers were mounted onto two micropositioners. To eliminate the additional Fabry–Perot cavity formed by the air-glass interfaces at the fiber splice, an index-matching liquid with a refractive index of 1.548 was used to fill the gap between the two fibers.

This setup is similar to that of the sensor that was used for high-temperature measurements. In that sensor, a thin layer of glass is sandwiched between the fiber end faces at the fiber splice, and the refractive index of the glass is 1.547. To determine the dependence of the fringe contrast of the sensor output on the relative position between the silica and sapphire fibers at the fiber splice, the temperature of the sapphire fiber was changed slightly to obtain a several fringe shift in the sensor output for each relative position. Figures 18.18 and 18.19 show the normalized fringe contrast as functions of the lateral offset and the longitudial gap between the end faces of the two fibers at the fiber splice. It is noted that the lateral offset and the longitudinal gap should be controlled within about 10 and 25 μm, respectively.

In the experiment, the splice was achieved using auminosilicate glass

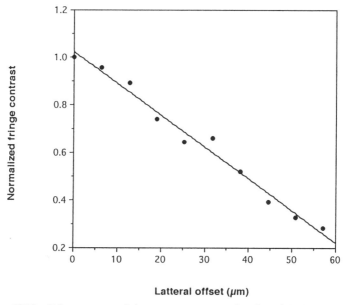

Figure 18.18. Fringe contrast of the sensor output as a function of the lateral offset.

deposited on the sapphire fiber end and a cleaved silica single-mode fiber was then spliced to the glass-covered end of the sapphire fiber. The difference between this splice and the one reported previously (4) is that the end faces of

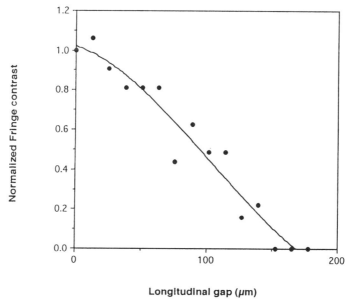

Figure 18.19. Fringe contrast of the sensor output as a function of the longitudinal gap separation.

the fibers are closer in the glass "melt" in the present splice. To achieve more reflection power at the free end face of the sapphire fiber coupled into the single-mode fiber, the single-mode fiber was positioned as close to the center of the sapphire fiber end face as possible.

The lateral offset was controlled to within 10 μm at the fiber splice. Also, the thickness of the glass layer between the sapphire and single-mode fiber end faces was controlled to within about 20 μm. Since the softening point of the aluminosilica glass is 908°C, the silica-to-sapphire fiber splice may be operated at temperatures up to approximately 800°C. The free end face of the sapphire fiber was polished to be perpendicular to the axis of the fiber. Due to this optimal fabrication of the sensor, an undistorted interference fringe output of the sensor was observed without any coatings at the free end face of the sapphire fiber. The fringe contrast was measured to be 0.41. Since no film was deposited onto the free end face of the sapphire fiber, the problem of oxidation of the film at high temperature has been solved and the sensor may thus be capable of being operated at temperatures up to 2000°C.

This intrinsic interferometric fiber sensor was demonstrated for high-temperature measurement. To obtain a high temperature up to 1500°C, an oxypropane torch was used to heat the sapphire fiber directly. About 5 mm of the sapphire fiber was heated, and a conventional high-temperature thermocouple (type B) was located very close to the heated sapphire fiber region to monitor the temperature simultaneously.

The experimental results are presented in Fig. 18.20. A linear output of

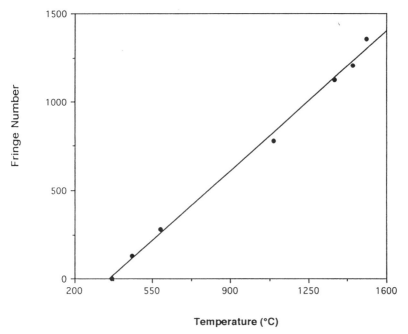

Figure 18.20. Sensor output as a function of temperature.

phase change as a function of temperature was obtained for the measurement range 256 to 1510°C. The noise level in the sensor output was measured to be 23 mV and the amplitude of the sensor output fringe was 72 mV. Based on the sensor temperature coefficient of 6.66 rad/°C, as shown in Fig. 18.20 at quadrature point operation the resolution of the sensor was determined to be about 0.1°C for a signal-to-noise ratio of 3 dB. It is believed that the noise in the sensor output was caused primarily by random laser phase changes, induced by the backward reflection. This phase noise may be reduced by inserting an optical isolator.

18.6.   SPECIAL CONSIDERATIONS FOR HIGH-TEMPERATURE COMPONENTS

In this section we discuss briefly special considerations that must be given to optical fiber sensor system components and processes for the detection of material properties at high temperature.

18.6.1.   Sapphire-to-Silica Fiber Splices

Sapphire-to-silica fiber splices have been implemented to allow the connection of high-temperature sapphire sensor fibers to lower-temperature silica fiber waveguides. Such splices have been fabricated using four procedures and then evaluated for optical losses and mechanical strength (10). The procedures have been diffusion-based splices, direct arc-fusion splicing of the two fibers, use of an intermediate jumper of aluminosilicate glass, and the deposition of fused silica on the sapphire fiber, followed by subsequent splicing.

The approach employed in the preparation of diffusion-bonded splices has been to polish the ends of both a sapphire and a silica fiber, butt the fibers together, and heat the interface for an extended period to a high temperature less than the working temperature for the fused silica. It was hoped that diffusion of aluminum or silicon ions across the interface would promote a mechanical bond between the two fibers. A scanning electron microscope (SEM) with electron dispersive x-ray (EDX) spectroscopy was used to examine the interface for evidence of ion diffusion.

A fused silica rod (approximately 96% pure, 2 mm in diameter) and a sapphire rod (single crystal, 1.8 mm in diameter) were held in contact, and the interface gradually heated in air using an oxypropane torch, while the temperature of the interface was carefully monitored using a type B thermocouple. The interface was heated to temperatures exceeding 1400°C but less than 1700°C for 90 min. During these experiments, a small but observable bond between the silica rod and the sapphire rod was apparent. The physical bonding between the rods was found to be very weak and was easily given to fracture at the interface. This may be due to the difference in the CTEs of silica and sapphire, which would affect the interface during the cooling process.

Photomicrographs of the sapphire rod end face were obtained after the diffusion experiment using the SEM. The photomicrographs showed that there is a definite interaction across the interface, characterized by an intricate network of short, pin-shaped microstructures. These results suggest a high probability of mullite formation at the interface. Comparison with the photomicrographs of a reference piece of mullite suggests that mullite formation does occur at the interface.

Alternative and lower-loss approaches have included the CVD deposition of a thin layer of silica on the outer diameter and end face of the sapphire prior to fusion splicing. Upon splicing, some of this silica tends to melt and form around the sapphire fiber end, forming a semimechanical connection rather than a true fusion splice at the interface. Improved splices have also been manufactured using intermediate aluminosilicate glass jumper fibers designed to partially match the properties of the silica on one side of the splice and of sapphire on the other side (Table 18.1). Such splices have been fabricated using drawn jumper inserted into sapphire holder tubes prior to fusion. Resulting losses are less than 1 dB, and mechanical strength is equal to or greater than that of the silica fiber itself.

18.6.2. Fiber Attachment Techniques at High Temperature

Finally, some discussion is needed concerning the attachment of high-temperature fiber elements to host materials for operation at high temperature. Again, the likely mismatch in CTE between the fiber and the host may give rise to large interfacial stresses, and effective methods to overcome these stresses are needed. Several methods have been proposed and demonstrated.

First, fibers may be attached at high temperatures using ceramic adhesives which are a mixture of silica, alumina, and binder (2). Although such adhesives

Table 18.1 Properties of Sapphire, Fused Silica, and Aluminosilicate Glass

	Sapphire	Fused Silica	Aluminosilicate Glass
Structure	Rhombohedral crystal	Vitreous	Vitreous
Melting temperature	2054°C	—	—
Softening point		1580°C	908°C
Refractive index	1.768 (slow) 1.760 (fast)	1.458	1.547
CTE (25°C)	$5.3 \times 10^{-6} \, °C^{-1}$ $\parallel$ to c-axis $4.5 \times 10^{-6} \, °C^{-1}$ $\perp$ to c-axis	$0.5 \times 10^{-6} \, °C^{-1}$	$4.6 \times 10^{-6} °C$

have a tendency to peel and spall after thermal cycling, if the temperature differential is low enough, they may provide enough adhesion to overcome the interface stresses caused by CTE mismatch.

Second, flame spray or plasma spray coating methods may be used to provide material contact. Although such application methods are more cumbersome than simple adhesive attachment, preliminary tests have demonstrated significantly improved fiber adhesion after repeated thermal cycling at temperatures up to 1200°C on metal matrix and ceramic matrix composite material specimens (1).

Finally, silica and sapphire fibers may be embedded directly within high-temperature materials during processing. Major concerns here again are the CTE mismatch between fiber and host and the buildup of stresses at the fiber–host interface. Recent work in the development of multilayer CTE grading coatings seems to reduce this problem partially by increasing the distance over which the stress is applied within the material.

18.7. SUMMARY

Optical fiber sensors may be extended for use at high temperatures by cooling the fibers and the sensors, or by using specialized high-temperature materials. Sapphire optical fiber waveguides, in particular, appear to be advantageous for measurements at temperatures between 1000 and 2000°C.

REFERENCES

1. S. E. Baldin, E. Nowakowski, H. Smith, E. J. Frieble, M. Putnam, R. Rogowski, L. Melvin, R. Claus, T. Tran, and M. Holben, Cooperative Implementation of a High Temperature Acoustic Sensor. *Proc. SPIE O/E Fibers Conf.* Boston, MA, **1588**, pp. 125–131, Sept. (1991).

2. K. A. Murphy, C. Koob, M. Miller, S. Feth, and R. O. Claus, Optical Fiber-Based Sensing of Strain and Temperature at High Temperature. *Proc. Review. of Progress in Quantitative NDE*, **10B**, pp. 1231–1239, (Brunswick, ME), July 1991.

3. S. Desu, R. O. Claus, R. Raheem, and K. A. Murphy, High temperature sapphire optical sensor fiber coatings. *SPIE OE/Aerospace Sensing*, (Orlando, FL), April (1990).

4. R. R. Dils, U.S. Patent # 4,750,139

5. F. D. Tilstra, A Fluorescense-Based Fiber Optic Temperature Sensor for Aerospace Applications. *SPIE OE/Fibers*, Boston, **1589**, pp. 32–37, Sept. (1991).

6. R. F. Hellbaum, M. F. Gunther, K. A. Murphy, and R. O. Claus, Some Nonconventional Concepts for Measurement of Acoustic Pressure Levels in Hostile Environments. *Proc. Sixth Annual Sensors Expo*, Chicago, IL, pp. 108–118, October (1991).

7. K. A. Murphy, G. Z. Wang, B. R. Fogg, A. M. Vengsarkar, and R. O. Claus, Sapphire Fiber Interferometer for Microdisplacements at High Temperature. *SPIE OE/Fibers*, Boston, MA, **1588**, pp. 117–1231 (1991).

8. A. Wang, G. Wang, K. A. Murphy, and R. O. Claus, Birefringence-Balanced Polarization-Modulation Sapphire Optical Fiber Sensor. *Optics Letters*, **17**(19), pp. 1391–1393 (1992).

9. A. Wang, S. Gollapudi, R. G. May, K. A. Murphy, and R. O. Claus, Advances in Sapphire-Fiber-Based Intrinsic Interferometric Sensors. *Optics Letters*, November **17**(21), pp. 1544–1546 (1992).

10. R. G. May, S. Gollapudi, G. Wang, and R. O. Claus, Silica-to-Sapphire Splices and Applications in High Temperature Materials Evaluation. *Proc. Fiber Optic Sensor-Based Smart Materials and Structures Workshop*, Blacksburg, VA, April, pp. 107–110 (1992).

19

Interferometric Optical Fiber Sensors for Ultrasonic Wave Measurement

RICHARD O. CLAUS, V. S. SUDARSHANAM, and KENT A. MURPHY
Fiber & Electro-Optics Research Center
Virginia Polytechnic Institute and State University
Blacksburg, Virginia

19.1. INTRODUCTION

Optical fiber systems have been developed during the past twenty-five years for primary applications in long-distance, high-speed digital information communication. Optical fibers may also be applied to the measurement of environmental parameters such as strain, temperature, vibration, chemical concentrations, and electromagnetic fields. In this chapter we consider specifically the use of a class of fiber sensors for the analysis of ultrasonic waves in and on materials.

19.1.1. Optical Fibers and Materials

Optical fibers are typically cylindrical and fabricated from polymer, glass, or ceramic materials. They consist of central core regions of material surrounded by concentric cladding regions. Typical outer dimensions of fiber claddings are more than 100 μm; core dimensions are smaller and their diameters are in part determined by the desired waveguiding properties of the fiber. The materials in the core and the cladding of the fiber are designed to have slightly different indices of refraction. The index of the core is required to be slightly greater than the index of the cladding, so a relatively large family of light rays incident at the core–cladding interface is repeatedly reflected back into the core of the fiber upon multiple reflections and contributes effectively to light propagation along the length of the fiber.

Light-ray propagation in the fiber may also be considered by solving the electromagnetic waveguide equations in the core and cladding regions of the

Fiber Optic Smart Structures, Edited by Eric Udd.
ISBN 0-471-55448-0 © 1995 John Wiley & Sons, Inc.

fiber, equating solutions at the core–cladding interface, and identifying proper solutions (1). Each solution to the propagation equation is termed a different waveguide mode and is specified by its spatial periodicity in the radial, azimuthial, and fiber axis dimensions. The number of modes supported by a fiber may be controlled by varying the difference between the indices of refraction of the core and cladding, the outer diameter of the core, and the optical wavelength of operation. For large numbers of modes, the square of the normalized propagation constant of the fiber, the V-number, given by

$$V = \frac{2\pi a}{\lambda} (n_1^2 - n_2^2)^{1/2}$$

is approximately proportional to the number of modes in the fiber. Here a is the radius of the core of the fiber, λ the optical source wavelength in free space, and n_1 and n_2 the indices of refraction of the core and cladding materials, respectively. The fiber waveguide parameters in the V-number also control the apparent field of view or numerical aperature (NȦ) of a fiber given by

$$NA = (n_1^2 - n_2^2)^{1/2}$$
$$= n_0 \sin \theta_{max}$$

where n_0 is the index of the material surrounding the end of the fiber and θ_{max} is the angle between the front input surface of the fiber and the wave propagation vector corresponding to the ray that enters the fiber at the steepest angle and propagates down the guide by total internal reflection, as shown in Fig. 19.1.

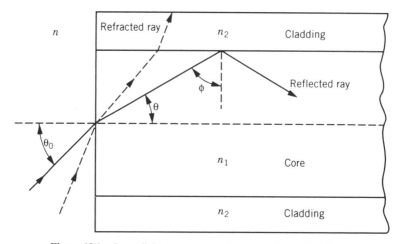

Figure 19.1. Input light ray geometry for step-index optical fiber.

Fibers are termed single-mode, two-mode, few-mode, or multimode depending on the number of waveguide-mode solutions that exist for the particular choice of material, waveguide and source parameters (2). For V-numbers below approximately 2.4, a number corresponding to the first zero crossing of the radial field component basis function, a fiber is single mode; that is, a single linearly polarized mode optical field solution exists for the guide. If the fiber parameters are altered to increase the V-number to between 2.4 and 3.8, a second linearly polarized mode field solution exists, this one degenerate for circular-core fibers. The degeneracy may be removed by distorting the shape of the fiber core into an ellipse, resulting in two-mode elliptical-core fiber (3–5). Still further increases in V-number lead to a few-mode and multimode fibers.

Typical single-mode fiber core diameters, for both an optical wavelength of operation of approximately 1 μm and conventional silica fiber waveguide materials, are on the order of 10 μm or less. Multimode fiber typically has a core diameter of 50 or 62.5 μm and supports up to thousands of discrete optical modes.

Finally, the materials that form the optical fiber and the coating material that surrounds the waveguiding region of the fiber determine both the thermal and mechanical performance of the fiber and the sensors that can be fabricated using the fiber. Most optical fibers are fabricated from silica, which has a very low coefficient of thermal expansion—on the order of $10^{-7}\,^{\circ}C^{-1}$, an index of refraction of 1.458 for the core and 1.456 for the cladding at room temperature and approximately a 1-μm wavelength, and a softening temperature of approximately 1000°C, depending on the levels and types of impurities used to dope the glass index (6).

19.1.2. Optical Fiber Sensors

Optical fiber sensors have been developed over the past 15 years for applications in materials and structural analysis, electromagnetic field and chemical concentration detection, and biomedical applications (7). Their advantages for measurements include (1) inherent immunity to electromagnetic interference, (2) avoidance of ground loops, (3) wide variety of measurands, (4) excellent resolution, (5) operation at temperatures up to approximately 800°C for silica fibers and above 1700°C for sapphire fibers, and (6) avoidance of sparks for applications in explosive environments.

Optical fiber sensors may be categorized into their method of operation, the optical field observable that is analyzed, or the environmental perturbation that is detected. Broadly, such sensors may be distinguished as either extrinsic or intrinsic devices. Light that propagates in extrinsic fiber sensors exits the fiber at some point, interacts with the environment to be analyzed, and reenters the input fiber or another fiber connected to detection optoelectronics. Intrinsic fiber sensors depend on the modulation of one or more of the optical field observables associated with the field, which propagates in the fiber itself to allow the measurement of external environmental effects.

More specifically, optical fiber sensors may be classified according to the environmental properties they are used to measure or the optical properties which are key to the performance of the measurement. Thus optical fiber sensors for the detection of temperature, pressure, strain, vibration chemical concentrations, and many other phenomena may be described. To perform these measurements, variations in the phase, intensity, polarization, wavelength, frequency, timing, or modal content of the optical fiber field are detected. Interferometric optical fiber methods, which interrogate changes in the phase of a single propagating optical field component, are the most sensitive of these and provide a baseline for sensor comparison. They may be implemented in several configurations and used for the detection of ultrasonic wave fields, as described below.

19.1.3. Optical Fiber Sensors for Ultrasonic Wave Analysis

Most conventional sensors for the detection of ultrasonic waves use piezoelectric elements that are attached to the host material directly or through a mechanical support housing assembly (8). Oscillatory particle displacements associated with wave motion strain the transducer element, producing an output voltage that may be interpreted in terms of displacement amplitudes. Major uses of such sensors include the nondestructive analysis of materials, the detection of acoustic emission events in materials, the detection of ultrasonic stress waves associated with impact events, and in instrumentation that uses ultrasonic waves for signal processing or the environmental monitoring of physical phenomena. Optical fiber sensors may be advantageous in comparison to such transducers, particularly in applications requiring immunity to electromagnetic interference, operation at temperatures above the Curie temperature of the piezoelectrics, or a low-profile geometry.

Particle motion associated with a surface acoustic wave is elliptical retrograde with respect to the direction of propagation and decays as a function of distance away from the surface (9). The wave motion may be modeled as being composed of instantaneous orthogonal displacement components oriented parallel and normal to the surface of the material. The shape of the ellipse and the decay depend on the acoustic frequency and the density and moduli of the material. Thus the measurement of one of the displacement components allows evaluation of the entire surface acoustic wave field.

Although the ultrasonic stress wave fields generated by phenomena such as acoustic emission events in materials or impact events caused by projectiles may be very complicated, a Heaviside step function displacement source on the surface of or buried within an infinite half-space has been used to model analytically the acoustic impulse response of the material (10, 11). Approximations to such a source on the surface of a material may be obtained in a repeatable way experimentally using small piezoelectric elements, electric arc es, or the break of a pencil lead (12). The latter method has been used) demonstrate the performance of optical fiber sensors with respect to nventional input acoustic signal conditions.

Several types of optical fiber sensors have been studied to measure propagating ultrasonic stress waves on or in materials, and they fit all of the categories indicated above (13). First, the end of a probe fiber may be arranged so that light exiting the fiber reflects off a surface supporting a surface elastic wave, and the detected light may be interrogated to determine the amplitude of the wave. This extrinsic fiber method is limited by the detection geometry, which requires external access to the surface supporting the wave and that the surface be a good optical reflector at the wavelength of the optical source. Alternatively, both direct and indirect in-line methods may be used to detect ultrasonic waves which perturb light propagating in a fiber or fiber sensor assembly attached to or embedded in a material that supports an elastic wave. Several conventional interferometric configurations in particular may be used to detect ultrasonic waves with gauge lengths that are generally large with respect to the acoustic wavelength, and methods that measure wave-induced polarization or modal modulation have been demonstrated similarly. Short-gauge-length interferometric-based detection methods, however, offer performance comparable in their compact and low-profile surface geometry, and superior in minimum-detectable displacement, to piezoelectric element–based ultrasonic wave transducers. All of these methods are described below.

19.2. OPTICAL FIBER INTERFEROMETRY

Optical interferometers measure the changes in the phase difference between light in different optical paths. The phase difference is observed as an intensity modulation dependent on constructive and destructive interference. They are thus intrinsically differential methods that rely on the existence of two or more optical paths in the measurement geometry. Several different geometries may be implemented, depending on the arrangement of these paths. As shown in Fig. 19.2, Michelson, Mach–Zehnder, and Fabry–Perot two-path geometries may be implemented using bulk optical components.

Michelson, Mach–Zehnder, and Fabry–Perot optical fiber interferometers may also be constructed using optical fiber waveguides and in-line optical fiber components as shown in Fig. 19.3, instead of bulk optical components. Because the operation of each interferometer configuration depends on the measurement of the phase of a single signal field with respect to the constant phase of one reference field, the optical fiber used in these configurations must be single mode rather than multimode. Alternative methods that use few-mode and multimode fiber are described below.

19.2.1. Optical Fiber Interferometry Theory

The theory of operation for optical fiber interferometers is essentially the same regardless of geometry. The delay in the optical phase measured in radians of

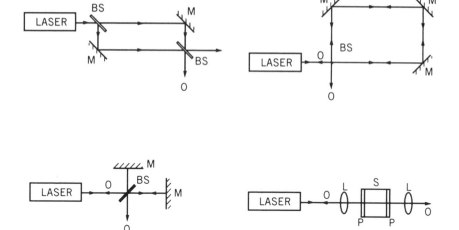

Figure 19.2.   Bulk optical interferometers: Mach–Zehnder, Michelson, Sagnac, and Fabry–Perot.

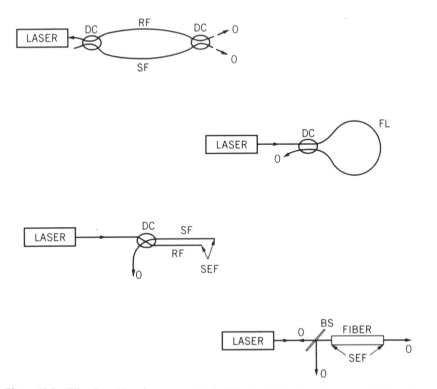

Figure 19.3.   Fiber-based interferometers: Mach–Zehnder, Michelson, Sagnac, and Fabry–Perot.

light traveling through a single-mode optical fiber is

$$\phi = nkL$$

where n is the index of refraction of the core of the fiber, k the optical wavenumber in vacuum, and L the length of the fiber over which the phase delay is measured. A small change in this phase delay which may occur at some position along the length of the fiber may be expressed to first order as

$$\frac{\partial \phi}{\phi} = \frac{\partial L}{L} + \frac{\partial n}{n} + \frac{\partial k}{k}$$

Such a change may be associated with effects internal to the fiber for intrinsic sensors, in which light remains within the fiber waveguide all along its length, or external to the fiber for extrinsic sensors, in which light exits the fiber at some point along its length, interacts with the local environment, then reenters the waveguide and is transmitted to detection circuitry. Several types of extrinsic and intrinsic fiber sensors have been used to monitor ultrasonic wave motion on materials, as described in following sections. It should be noted that the gauge length of the sensor for intrinsic sensors is a length of fiber, while for extrinsic sensors is typically air. For the measurement of the local high-frequency stress wave field associated with ultrasonic wave propagation, for example, cross terms such as

$$\frac{\partial n}{n \, \partial L} \quad \text{and} \quad \frac{\partial k}{k \, \partial L}$$

would appear in the equation above in the case of intrinsic sensor operation due to the modulation of the index of refraction of the glass in the core of the fiber due to the photoelastic effect that occurs when the fiber is strained, and the change that occurs in the propagation constants due to the strain-induced deformation of the waveguide. Extrinsic sensors avoid such first-order cross-sensitivity problems for small changes in strain, or similarly in temperature, due to the relative insensitivity of air to external modulation effects.

19.2.2. Fiber Interferometer Stabilization

The light intensity at the interferometer output is read as an instantaneous voltage from a photodetector accessing the fringe pattern. The time-dependent part of this intensity can be written as

$$dI = 2I_0 \alpha \cos[(p_s \sin \omega t) + p_d]$$

where I_0 is the intensity of each interfering beam, α the fringe contrast, p_s the amplitude and ω the frequency of the signal-induced phase shift, and p_d represents phase drift. For small phase shifts, $p_s \ll 1$, and the equation above may be simplified to

$$dI = 2I_0\alpha(\sin p_d)p_s \sin \omega t$$

Thus the detected signal is maximum at $p_d = (2m + 1)\pi/2$, called the quadrature point, and vanishes when $p_d = m\pi$, where m is an integer, as shown in Fig. 19.4. This results in signal fading and p_d drifts due to fluctuations in ambient temperature and pressure. Further, distortion occurs from changes in the polarization states of the two beams, and source intensity and frequency fluctuations. The elimination of these problems has been the objective of several phase detection schemes, as discussed below.

The simplest in concept is heterodyne detection, wherein the optical phase appears as the phase of the intensity modulation at the beat frequency produced by a Bragg cell in one arm of the interferometer. On integration of the output of an FM discriminator tuned to the beat frequency, a signal linear in p_s is obtained. As the Bragg cell introduces alignment complications, a synthetic heterodyne signal could be produced with a piezoelectric stretcher. More attractive than these schemes are the active and passive homodyne techniques. In the active approach, the photodetector output is integrated and fed back to a phase modulator in one arm of the interferometer. The interferometer is thus stabilized and constantly maintained at the quadrature point. In the low-gain mode, drift alone is tracked and the signal is present at the detector, whereas in the high gain mode, both the signal and the drift are tracked, with the signal present at the integrator. The latter has the advantage that the output signal is insensitive to changes in both system attenuation and fringe contrast. However, using these active feedback methods, electronic reset glitch occurs, due to the limited dynamic range of the feedback element.

By contrast, the passive approach does not involve feedback. The objective using this method is to synthesize two signals shifted differentially in optical

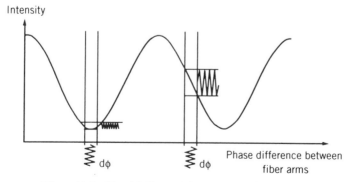

Figure 19.4. Signal-fading problem in optical fiber interferometers.

phase by 90°; fading is avoided in this case because the signal cannot vanish simultaneously from both outputs. The resulting two quadrature outputs are then manipulated to obtain the signal. The objective is met with the use of a 3×3 coupler or two detectors placed 90° apart in the fringe pattern or of a phase-generated carrier. A novel way of achieving this has been devised and is described later. The homodyne approach has a high dynamic range of about 230 dB and a minimum detectable phase shift of about 1 μrad. The corresponding minimum detectable strain would be approximately 10^{-14}.

19.3. EXTRINSIC OPTICAL FIBER SURFACE ACOUSTIC-WAVE SENSORS

Extrinsic interferometric methods may be used to directly observe the time-varying normal surface displacements associated with surface wave motions. Figure 19.5 shows one possible extrinsic geometry (14). Here one arm of the Michelson interferometer is arranged so that light which exits the end of the single-mode "signal" fiber reflects off the SAW-supporting surface and back into the fiber, or into a nearby second fiber, while light from the reference fiber reflects off a reflecting material placed at its end. These two reflected optical fields are combined in a 2×2 single-mode fused biconical tapered coupler as shown, and this combined output signal is detected by an optical detector.

From the preceding equation, the output signal from this sensor is proportional to the instantaneous phase difference between the two signals, or

$$ d\phi = \phi_1 - \phi_2 = \left(2A + \frac{\partial \phi_1}{\partial L_1} \Delta L_1 + \frac{\partial \phi_1}{\partial n_1} \Delta n_1 + \frac{\partial \phi_1}{\partial k_1} \Delta k_1 \right) $$
$$ - \left(\frac{\partial \phi_2}{\partial L_2} \Delta L_2 + \frac{\partial \phi_2}{\partial n_2} \Delta n_2 + \frac{\partial \phi_2}{\partial k_2} \Delta k_2 \right) $$

where the subscripts 1 and 2 refer to the signal and reference fibers, respectively, and A is the local surface displacement of the ultrasonic surface wave, which

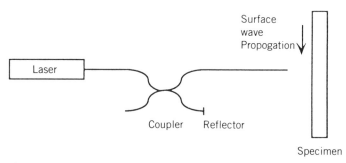

Figure 19.5. Extrinsic Michelson interferometric configuration for detection of surface acoustic waves.

is normal to the surface and parallel to the long axis of the signal fiber. Here it is A alone that we typically want to measure, so extraneous and unwanted phase variations that occur along the length of the fiber may interfere with this measurement. Practically, such variations occur due to temperature differences between the two arms of the fiber interferometer, which in turn cause changes in local fiber length and radius due to thermal expansion, in index of refraction due to photoelastic effects, and in the quasi-static displacement of the surface supporting the ultrasonic surface wave.

For these reasons, two different similar extrinsic approaches have been developed to reduce such interference effects. The first, shown in Fig. 19.6, combines the signal and reference arms in a single differential interferometric configuration (14). Here the two fiber arms are maintained close together until they reach the sensor head near the surface of the material in order to minimize differential thermal or strain effects. More important, however, light from both fibers reflects off the surface and interferes, to produce an output optical intensity signal proportional to the instantaneous difference in the local normal displacements of the material surface. As described in detail below, if the spacing between the two local focus points is an odd integer multiple of half the acoustic wavelength, the amplitude of this differential output signal is maximized. Further, this differential signal is immune to the type of quasi-static displacement interference as noted above, and the output signal may be analyzed to yield a calibrated quantitative measure of the acoustic field.

The second extrinsic approach that has been demonstrated for the measurement of ultrasonic surface waves, shown in Fig. 19.7, effectively does eliminate spurious temperature and strain fluctuation noise, because the sensing and reference arms have been reduced to short lengths. The TIMI (temperature-insensitive Michelson interferometer) diagrammed is fabricated by cleaving a 2×2 fused biconical tapered coupler immediately after the coupling region, then depositing a reflector over the end of one of the exposed fiber cores

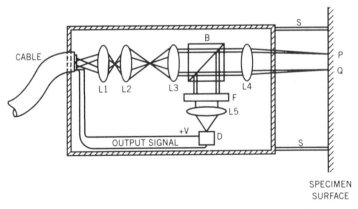

Figure 19.6. Remote detection head of extrinsic differential optical fiber interferometer for surface-wave detection.

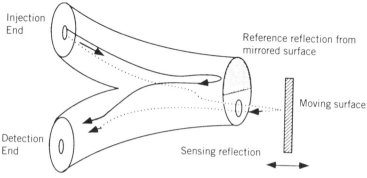

Figure 19.7. Temperature-insensitive Michelson interferometer (TIM) for surface acoustic wave detection.

(15, 16). Light input into the TIMI is partially reflected by this reflecting surface, and that reflected signal serves as the reference arm signal for the interferometer. Light from the exposed fiber end exits the coupler, reflects off the surface of the material to be examined, and reenters the fiber end. The resulting interference between this signal and the previous reference signal may be interpreted to determine surface-wave amplitude and may be calibrated absolutely.

19.4. INTRINSIC MICHELSON AND MACH–ZEHNDER INTERFEROMETER DESIGNS

Michelson and Mach–Zehnder optical fiber interferometer geometries may also be used to detect surface elastic waves, and if the fibers are embedded within a material may be used to detect components of bulk acoustic waves as well. In both of these cases, as shown in Fig. 19.8, the length of optical fiber along which the ultrasonic wave modulates the phase of the fiber by varying its geometry and index is long with respect to an acoustic wavelength, typically several meters in length from source to detector optics and electronics (17). Such a long gauge length requires that the fiber be positioned in such a way that all of the fiber be located at a plane of constant phase of the propagating wave. If the fiber, instead, cuts across many acoustic wavefronts, the total modulation of the output signal is proportional to the integral of the phase change induced along the entire fiber length. In some cases the experimental geometry may be controlled so that the plane of constant phase assumption is approximately met. In the specific case of very low frequency acoustic waves, where the acoustic wavelength may be on the order of meters or longer, geometrical arrays of fibers may be designed and arranged to intercept known and approximately plane-wave acoustic fields. This method eliminates the wave integration problem by taking direct advantage of it to effectively perform one- or two-dimensional field signal processing (18). For cases in which

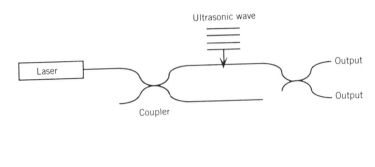

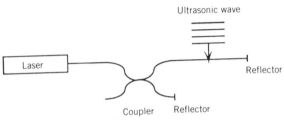

Figure 19.8. Michelson and Mach–Zehnder configurations for detection of ultrasonic waves.

long-gauge-length interferometric systems have been used for ultrasonic wave detection, the quantitative and calibrated analysis of wave amplitude fields has not been reported, due to these effects. Such methods may, however, be used when only an indication of the existence of such a wave is of interest.

19.5. MODAL AND POLARIMETRIC FIBER SENSOR IMPLEMENTATION

Modal domain and polarimetric optical fiber sensor methods have also been demonstrated for the qualitative detection of ultrasonic waves in and on materials. Each may be considered comparable to an intrinsic interferometric method. The most effective modal domain optical fiber sensors depend on the difference in the propagation constants of the LP_{01} and LP_{11}^{even} modes in two-mode elliptical-core fibers to allow effective interferometric operation (19). Strain or temperature applied to the fiber results in a change in both the geometry of the fiber and the index of the fiber core and cladding materials. Both of these effects affect the two propagating modes differently because they have slightly different propagation constants. This differential modal modulation effectively contributes to a change in the modal field distribution at the output of the fiber, and such a modal field modulation may be analyzed to determine the environmental field perturbation along the fiber length.

Modal domain fiber interferometer measurements of bulk ultrasonic waves been demonstrated for more than 10 years using fibers embedded in material specimens (20). Although the integrating effect of the sensor due to its relatively long gauge length does not allow simple quantitative analysis of the wave, such sensors may be useful for simple event detection. The major limitation of modal sensors for such applications may be their relative intensitivity to strain compared to that of other interferometric schemes, and the small strains produced by ultrasonic stress waves.

Similarly, polarimetric fiber sensor methods may be used, with similar results and limitations. Polarimetric optical fiber sensors operate on the principle that the different polarization components of optical fields propagating in a fiber are modulated differently by external fields. By controlling the polarization of the input light into a polarization-maintaining fiber and subsequently observing the output when the fiber is insonified, an integrated ultrasonic wave effect can be observed. The limitations of this method are similar to those of the modal domain fiber sensor above; the gauge length is long, so integrating over several acoustic wavelengths is likely, and as a result, little quantitative ultrasonic wave information is recoverable unless the gauge length is limited to a distance on the order of the acoustic wavelength (21).

19.6.   ULTRASONIC WAVE DETECTION USING FABRY–PEROT OPTICAL FIBER INTERFEROMETRY

Perhaps the most effective optical fiber method developed to date for the quantitative analysis of ultrasonic wave fields is by the use of extrinsic Fabry–Perot interferometry. As shown in Fig. 19.9, the interferometer consists of a small cavity along the length of a single-mode optical fiber (22–24). The

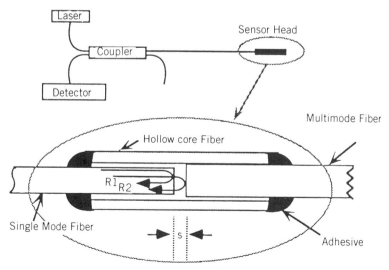

Figure 19.9.   Fabry–Perot sensor system with detail of sensor.

cavity is formed by the region separating the end of the input single-mode fiber and the output multimode fiber, which is used only as an optical reflector. Both of the fibers are held inside a short length of small-diameter hollow-core tube to maintain end alignment. The sensor is extrinsic, in that part of the input field exits the input fiber and reflects off the target fiber, although the sensor is in-line and of low profile. The gauge length of the sensor element is determined by the distance between the points where the input and output fibers are connected to the surrounding hollow-core support structure and the host-material surface or volume. Typically, this is approximately the length of the hollow-core tube, because epoxy may be applied at either end of the tube Alternatively, the epoxy may be wicked into the tube to make the effective gauge length of the sensor as small as the separation distance between the ends of the two fibers in the tube. Instead of depending on the wicking properties of the epoxy, the glass fiber elements may be fusion spliced into place to define the sensor gauge length.

The operation of this sensor for ultrasonic wave detection is straightforward and similar to that of the differential extrinsic interferometer described above. If the sensor is positioned on the surface of a material that supports a surface elastic wave, the wave may be described as a superposition of time-varying tangential in-plane and normal out-of-plane displacement components, and each component may be observed to interact differently with the sensor head element. The behavior of these components as a function of distance into the material is well known and depends on the density and moduli of the material.

If the sensor is attached to the surface of the material in such a way that the long axis of the fiber is parallel to the propagation vector of an idealized plane wave SAW, the component of motion of the wave tangential to the surface modulates the cavity length of the sensor (25). This, in turn, causes an intensity modulation at the output of the sensor. In many cases, surface-wave amplitudes are far smaller than the approximately 1-μm wavelength of the optical signal. Thus the interferometer operates well within one fringe for a full peak-to-peak variation in field amplitude, and the types of methods described above may be used to analyze displacement field components within the linear response range of the interferometer. If, instead, very large surface-wave displacements occur, the number of optical fringes that occur at the output must be counted to give an indication of that amplitude. Figure 19.10 shows a pulsed signal detected using EFPI sensor elements on the surface of an aluminum plate and a corresponding signal detected by a piezoelectric sensor element (26); The calibrated amplitude of the signal is approximately 0.6 Å. Here the ultrasonic waves that resulted in both signals were created by breaking pencil leads.

The relative locations of surface attachment determine the response of the sensor to the wave and the frequency response of the sensor system. For an EFPI sensor head that is attached to the host material at points spaced apart

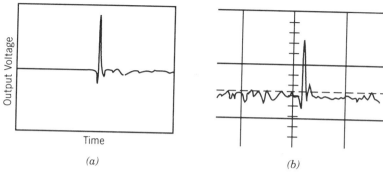

Figure 19.10. Piezoelectric sensor data (*a*) and fiber EFFI sensor data (*b*) for pencil lead break-generated simulated acoustic emission events on aluminum substrate.

along the length of the fiber by

$$D = (n + 1) \frac{\Lambda}{2}$$

where n is a nonnegative integer and Λ is the acoustic wavelength, the response of the sensor is maximized. Such multiple relative maxima in response occur because under such conditions the displacement at one attachment point is maximized in one axial direction on the fiber, while at the other attachment point the displacement is maximized in the opposite axial direction. The 3-dB frequency response bandwidth of the sensor is thus approximately two octaves around the center frequency, and this maximum response is periodic in frequency from the equation above. For a practical detection system, the bandwidth of the receiver electronics may be controlled to limit the response of the system to that associated with a single maximum and to shape the response in such a way that the two-octave bandwidth is increased (27).

These sensor elements may also be embedded within materials to yield calibrated quantitative measures of the amplitudes of bulk ultrasonic stress wave field components at different positions within the host materials and different depths below the surface for SAW. To minimize the effect of the presence of the fibers on the measurement involved, silica fibers having outer diameters as small as 35 μm, for example, have been demonstrated (28). Here the size of the fiber is relatively small compared to the approximate 1-mm acoustic wavelength, so the fiber may to first order be ignored despite the large difference between the acoustic impedance of silica and other materials.

19.7. CONSIDERATIONS FOR HARSH ENVIRONMENTS

Fabry–Perot interferometric optical fiber sensors have also been demonstrated for the detection of ultrasonic waves under practical harsh environmental conditions. For example, reference (29) describes their use for the detection of thermally induced strain as well as ultrasonic wave transients in mullite and corderite ceramic cross-flow filter (CXF) elements at temperatures exceeding 940°C, and in a flowing and corrosive coal fly ash environment. The fiber sensors used in these tests were implemented using silica fiber and high-temperature coatings and housing materials.

For operation at still higher temperatures, in-line sapphire fiber–based Fabry–Perot sensor elements also have been implemented (30). Although their operation is similar to that of their silica fiber–based counterparts, the 2040°C melting temperature of c-axis single-crystal sapphire allows sensor operation up to much higher temperatures. Major consideration at those temperatures must be given to the possible mismatch in the coefficients of thermal expansion between the fiber and the host material.

19.8. SUMMARY

Extrinsic and intrinsic optical fiber methods may be used to characterize propagating ultrasonic wave fields in and on materials. Interferometric methods have the best small-signal detection performance, being capable of sub-angstrom wave displacement measurements for practical optical and electronic system component specifications. Several interferometric geometries which allow the wave to insonify a long length of fiber and give an output signal proportional to the integral of the incremental strains in the fiber caused by the wave along the gauge length yield qualitative measurements of field presence but are subject to field shading and phase cancellation effects. In-line extrinsic Fabry–Perot methods appear to be advantageous for several reasons, including the possibility of a short and controlled gauge length, and thus quantative and calibrated output data, as well as the possibility of operation at temperatures above 1700°C. Further, low-profile fiber sensor devices of this design have been demonstrated for the analysis of propagating fields inside materials, with minimal perturbation of the internal properties of the material.

REFERENCES

1. G. Keiser, *Optical Fiber Communications*, 2nd ed. McGraw-Hill, New York, 1991.

2. A. Buckman, *Guided Wave Photonics*. Saunders, New York, 1992.

3. J. N. Blake, S. Y. Huang, B. Y. Kim, and H. J. Shaw, *Opt. Lett.* **12**, 732 (1987).

4. A. Vengsarkar, B. Fogg, W. Miller, K. Murphy, and R. Claus, Elliptical-Core Two-Mode Optical Fibre Sensors as Vibration Filters. *Electron. Lett.* **27**, 931 (1991).

5. A. Wang, A. Vensarkar, and R. Claus, Two-Mode Elliptical Core Fiber Sensor for Measurement of Strain and Temperature. *Proc. SPIE — Int. Soc. Opt. Eng.* **1584**, pp. 294–303 (1991).

6. A. Kuske and G. Robertson, *Photoelastic Stress Analysis*. Wiley, New York, 1974.

7. E. Udd, ed., *Fiber Optic Sensors: An Introduction for Engineers and Scientists*. Wiley (Interscience), New York, 1991.

8. H. Uberall, Surface Waves in Acoustics. In *Physical Acoustics* (W. P. Mason and R. N. Thurston, eds.), Vol. 10, pp. 1–60. Academic Press, New York, 1973.

9. H. Kolsky, *Stress Waves in Solids*. Dover, New York, 1974.

10. C. L. Pekeris, The Seismic Surface Pulse. *Proc. Natl. Acad. Sci. U.S.A.* **41**, pp. 469–480 (1955).

11. L. Knopoff, Surface Motions of a Thick Plate. *J. Appl. Phys.* **29**(4), 661–670 (1958).

12. F. R. Breckenridge, C. E. Tschiegg, and M. Greenspan, *J. Acoust. Soc. Am.* **57**(3), 626–631 (1975).

13. R. O. Claws, R. Dhauran, M. F. Gunther and K. A. Murphy, Quantitative measurement of in-plane acoustic field components using surface-mounted fiber sensors, *Proc. SPIE* **1798**, pp. 144–152 (1992).

14. A. O. Garg and R. O. Claus, Application of Optical Fibers to Wideband Differential Interferometry. *Mater. Eval.* **41**, 106–109 (1983).

15. K. Murphy, W. V. Miller, T. A. Tran, A. M. Vengsarkar, and R. O. Claus, Miniaturized Fiber-Optic Michelson-Type Interferometric Sensors. *Appl. Opt.* **30**(34), 5063–5067 (1991).

16. H. Hosokawa, J. Takagi, and T. Yamashita, Integrated Optic Microdisplacement Sensor Using a Y-Junction and a Polarization-Maintaining Fiber. *Proc. Optical Fiber Sensors Conference*, Vol. 2 of OSA 1922 Tech. Dig. Ser., p. 137. Opt. Soc. Am., Washington, DC, 1988.

17. M. Ohn, A. Davis, K. Liu, and R. Measures, Embedded Fiber Optic Detection of Ultrasound Detection and Application to Cure Monitoring. In *Smart Materials and Structures Workshop*. R. O. Claus, Ed., pp. 175–184. IOP Publishing, Blacksburg, VA, (1992).

18. N. Lagakos, P. Ehrenfeuchter, T. R. Hickman, A. Tveten, and J. A. Bucaro, Planar Flexible Fiber-Optic Interferometric Acoustic Sensor. *Opt. Lett.* **13**, 788 (1988).

19. K. Shaw, A. M. Vengsarkar, and R. O. Claus, Direct Numerical Analysis of Dual-Mode Elliptical-Core Optical Fibers. *Opt. Lett.* **16**, 135 (1991).

20. R. Claus and J. Cantrell, DC Calibration of the Strain Sensitivity of a Single Mode Optical Fiber Interferometer. *Proc. IEEE Reg. 3 Conf.* (1981).

21. J. R. Dunphy and G. Melts, Fiber Optic Sensor Development for High Speed Material Diagnostics. *Proc. Soc. Exp. Mech. Conf. Opt. Methods Compos., 1986.*

22. K. Murphy, M. Gunther, A. Vengsarkar, and R. O. Claus, Quadrature Phase-Shifted, Extrinsic Fabry–Perot Optical Fiber Sensors. *Opt. Lett.* **16**(4), 173–275 (1991).

23. K. Murphy, M. Gunther, A. Vengsarkar, and R. Claus, Fabry–Perot Fiber Optic Sensors in Full Scale Fatigue Testing on an F-15 Aircraft. *Proc. SPIE — Int. Soc. Opt. Eng.* **1588**, pp 134–142 (1992).

24. K. Murphy, R. Claus, et al., Sapphire Fiber Interferometer for Microdisplacement Measurements at High Temperature. *Proc. SPIE — Int. Soc. Opt. Eng.* **1588**, pp. 117–124 (1992).

25. R. Dhawan, M. F. Gunther, nd R. O. Claus, Detection of In-Plane Displacements of Acoustic Wave Fields Using Extrinsic Fizeau Fiber Interferometric Sensors. In *Smart Materials and Structures Workshop.* R. O. Claus, Ed., pp. 137–140. IOP Publishing, Blacksburg, Va., Apr. 1992.

26. W. Sachse, Cornell University, private communication.

27. C. H. Palmer, R. O. Claus, and S. E. Fick, Ultrasonic Wave Measurement by Differential Interferometry. *Appl. Opt.* **16**(7), 1849–1856 (1977).

28. A. Vengsarkar, K. Murphy, M. Gunther, A. Plante, and R. Claus, Low-Profile Fibers for Embedded Smart Structure Applications. *Proc. SPIE — Int. Soc. Opt. Eng.* **1588**, pp. 2–13 (1992).

29. K. A. Murphy, C. E. Koob, A. J. Plante, M. F. Gunther, A. M. Vengsankar and R. O. Claus, High Temperature Fabry–Perot Based Strain Sensor for Ceramic Cross Flow Filters, *Rev. Prog. Quant. Nondestr. Eval.,* Brunswick, ME *1991*, pp. 1121–1128 (1991).

30. A. Wang, S. Gollapudi, K. Murphy, R. May, and R. Claus, Sapphire-Fiber-Based Intrinsic Fabry–Perot Interferometer. *Opt. Lett.* **17**(14), 1021–1023 (1992).

20

Fiber Optic Damage Assessment

Michel Le Blanc and Raymond M. Measures
Institute for Aerospace Studies
University of Toronto
Toronto, Ontario, Canada

20.1. INTRODUCTION

Technical advances in the areas of nondestructive evaluation (NDE) and nondestructive inspection (NDI) in the past three decades can be credited for having permitted the increasing use of the *fail-safe* and *damage-tolerant* design philosophies in aircraft structural engineering. The older *safe life* approach (developed in the years 1945–1955, but still in use in many situations) requires that no significant fatigue damage in a component (or structure) should occur during its entire design life and that replacement of the component, or if that is not economical, retirement of the entire aircraft, be warranted when its conservatively predicted safe use limit is exceeded. The strict implementation of this approach necessarily means that any aircraft will have to be retired much before its expected lifetime as determined by premanufacturing fatigue tests. This is because the scatter in test data can be quite large but the safe limit must be set below the lowest of the obtained values. In addition, it is known that conditions unforeseen at the time of testing (such as effects from the environment or change in function and or use patterns from the ones originally planned) can cause unexpected premature failures.

With the damage-tolerant approach, fatigue crack initiation and growth in the structure is permitted, but it is expected that this damage will be detected early enough by a program of periodic inspections and be monitored to allow the timely repair or replacement of the component: that is, before critical flaw sizes are reached. Not only has the use of this approach allowed longer total aircraft life cycles and higher-performance designs, but safety has been improved by the continued monitoring of the aircraft's structural integrity. This philosophy permits a retirement-for-cause approach to fleet management rather than a fixed-life one. Needless to say, these inspections rely heavily on NDE methods.

Fiber Optic Smart Structures, Edited by Eric Udd.
ISBN 0-471-55448-0 © 1995 John Wiley & Sons, Inc.

20.1.1. Need for Built-in Integrity Monitoring/Damage Assessment Systems

From the discussion above, we can easily understand that each improvement in the resolution or field usability of NDE techniques can make a difference in the design of aircraft and their lifetime. Whereas NDE-NDI techniques and their reliability have been improving continuously over the years, there are still (and probably always will be) room for improvement (3). Several limitations in current NDE methods need to be considered. One of them is that, so far, very little automation of NDE exists, and this results in limitations on inspection reliability. A related concern is the lack of quantitative information provided by most NDE techniques: Such quantitative information is desired to permit accurate assessments (9). Another very important problem is the fact that damage-tolerant design can only be applied to structures that are available for inspection: Many structural components are hidden, say, underneath the skin of aircraft wings and therefore must be designed on a safe-life basis. Furthermore, new pressures are added on NDE by the introduction of new materials for which current NDE methods are not sufficient. Advanced composite materials are, in this respect, of particular importance; we still have only limited NDE capability for these materials [especially for in-the-field inspection; Cooper (9)]. However, the relative novelty of these materials warrants an even closer monitoring. Finally, depending on the rigor of the inspection program, significant costs can be incurred in carrying out the inspections: The more thorough the inspection, the higher the cost (reliability has a price). These costs can be greatly increased if there is a need to remove the structure from service temporarily for inspection.

From these considerations we can see that there is a considerable incentive to produce structures with a built-in integrity monitoring/damage assessment (IM/DA) capability. Progress in microprocessor-based computing, data analysis, and sensors now makes this feasible. The underlying principle is that with such a self-monitoring capability, a conventional inspection would only be performed when necessary (inspection-for-cause concept). In this chapter we look at the role optical fibers can play in such damage assessment systems. In the next section we provide an overview of the various ways that optical fibers can be used as the basis of an integrity monitoring/damage assessment system. A review of past and current sensor developments for the purpose of IM/DA follows. An illustrative example of a built-in fiber optic damage assessment system for impact damage in composites is then discussed in more detail. The choice of this particular concept demonstration work is not at all fortuitous, as both authors have been actively involved in this work at the University of Toronto Institute for Aerospace Studies (UTIAS). We conclude this chapter with our thoughts on research priorities and future prospects for this technology.

20.1.2. Fiber Optic Approaches to Structural Integrity Monitoring and Damage Detection

20.1.2.1. Load Monitoring Systems. As described elsewhere in this book, optical fiber strain gauges can be used as the basic sensing elements of a continuous load monitoring system. With a network of such sensors strategically located and with appropriate data analysis and recording equipment, the load history of all the major structural elements could be known. This load history is very useful, as a much truer evaluation of the structure's real use can be made compared with, for example, having data only on the number of hours flown. With this load history information, one can establish a flexible maintenance and structural health inspection schedule that depends more effectively on the structure's use.

Technically, however, load monitoring has limited capability for damage assessment. It can provide only an indirect indication of damage: either from a predictive model on fatigue or from abnormal strain response under known inputs (strain anomalies). Instead, the value of load monitoring lies more in the improved capability of inferring possible damage states than in detecting damage and assessing it. This is because the sensor is sensing strain, not damage.

20.1.2.2. Damage Monitoring Systems. For structural integrity monitoring, load monitoring may not be sufficient. First, having good predictive models for fatigue damage can be hard to achieve (especially for new materials), and this affects the quality and reliability of the inference process. Also, damage may be caused in unexpected ways, such as through secondary load channels or due to external impact events. For this type of damage, relying on strain anomalies may be inadequate because damage may be present but not affect the strain response at the sensor locations.

A form of monitoring that can detect damage directly is the acoustic emission (AE) method. This technique relies on the detection of stress waves generated in the structure by the energy release that occurs at the rupture zone of created or growing damage. The technique can be applied as a continuous monitoring system or be used periodically by subjecting the structure to preestablished loads. Collecting and analyzing the AE data obtained from a network of sensors can provide information on the severity and extent of damage and its localization. This method has been investigated extensively for metallic structures using piezoelectric transducers. However, the technique has not yet reached the point of general acceptance and utilization for monitoring of in-service aerospace structures (9). The problems lie mainly in the low signal amplitudes, which require that the sensors be close to the region of damage, and also in the small quantity of signals that are detected during the safe extension of cracks (the period where AE monitoring would be most useful).

Nonetheless, much work is under way in this field and enthusiasm is still present among the proponents of this method.

There is potential for extending the AE technique to composite materials, but there the situation is even more complex. As we describe in Section 20.2.4, some success has been achieved in developing fiber optic sensors that can detect acoustic emission from within composite structures. However, the fiber optic sensors developed so far are much less sensitive than those based on conventional sensors. Nonetheless, optical fiber sensors can offer other advantages and new possibilities, such as long sensing lengths and real in situ measurements, and therefore their use needs to be investigated further.

20.1.2.3. Damage Assessment Versus Continuous Monitoring. The function of damage assessment is somewhat independent of monitoring. A damage assessment system is supposed to provide answers to the questions: Is there damage? If so, where? How extensive is it? Answers to these questions can be obtained from built-in sensors without the need for continuous monitoring. The problem with monitoring is that it has to be done over extended periods of time (or in some cases, all the time). This generates a lot of data that have to be processed and presented in a simplified way; all this requires software and hardware and increases the cost significantly. This seriously limits the domain of application of continuous monitoring techniques based on economic considerations alone. The systems described next do not require continuous monitoring.

Threshold Detection and Memorizing Strain Sensor. One way to reduce the amount of data is to limit the load information to the maximum loads observed. It is even preferable if the sensor itself is built to hold this information. An optical fiber designed to fail at a given strain is such a sensor. As we will see in Section 20.2, chemical treatment can be used to tailor the failure strain of optical fibers. We also discuss a sensor proposed by Sirkis and Dasgupta (42) that makes use of the properties of the fiber coating to memorize the maximum load imposed on the optical fiber without failure of the fiber.

Built-in System as Aid to Periodic Inspections. Another way of limiting cost and hardware requirements of built-in damage assessment systems is to keep the built-in sensing capability but limit the data collection process to periodic inspections. The interrogation and acquisition hardware would be part of the inspection system that is used periodically for the inspection of different components with similar sensing systems. It is assumed that restriction to the built-in damage-sensing capability and not the continuous monitoring aspect can still provide some advantages, such as faster inspection times, more reliable inspections, and inspection of hard-to-access components. The authors have been working on such a system for impact damage detection in composite materials. The basic sensor for this system is a fiber that fails in the presence of damage, so it also combines elements of the preceding paragraph. This system is described in detail in Section 20.3.

20.1.2.4. Engineering Trade-offs. The technology we are covering in this chapter, fiber optic damage assessment, is only at the early stage of its development. However, as the reader will notice, there are many ways in which an eventual system could work. This variety in potential systems requires one to consider carefully the pros and cons of each approach before investing a lot of effort (and dollars) into a particular one. In the paragraphs above, we have mentioned that one can eliminate the need for costly, bulky, and potentially unrealiable continuous monitoring electronic equipment by designing sensors that are damage specific and can be interrogated periodically by equipment on the ground. But there are some problems associated with this approach: (1) since the sensors being used are tailored to be damage specific, they are less versatile than strain-gauge sensors*; (b) if the sensor relies on a permanent change in the sensor (memory), such as a crack detector based on fiber failure, its sensing capability may be lost after the triggering event and may be not usable after repair of the structure or component; and (c) manufacture and installation of these sensors may be more difficult, especially if the approach relies on a large number of sensors.

As for most new technology, system cost will be a major consideration in its implementation. A nonexclusive list of factors to consider when determining total cost is the following:

- Sensor development
- Sensor manufacturing
- Installation on structure and modification needed in structure manufacturing
- Monitoring/interrogation system
- Level of user training and skill required
- Inspection frequency and duration
- Structure downtime and need for disassembly

Engineering trade-off studies will be required to determine the best approach to be chosen for a given application. Of course, cost considerations will need to be counterweighted with a fundamental criterion: the required performance of the system. So far, however, the area of fiber optic damage assessment is new. Research has concentrated primarily on sensor development and less on applications. Experiments with fiber optic–based systems have been limited to basic demonstrations. One can guess that there are still a wide variety of approaches that have not been tried. At this stage, one of the difficulties to overcome is the lack of communication between sensor developers and potential users; good communication needs to be established. We hope that this chapter can help in bridging the gap that exists.

---

*As mentioned in Chapter 8, it should be a goal of fiber optic sensor development for smart structures to make sensors that are sensitive to multiple parameters and can be used in many ways: such as using the same sensor for cure monitoring, acoustic emission detection, and strain sensing.

In the following section we review and assess the status of research in this field. We describe in more detail the systems that have been developed the furthest but also look ahead and comment on those systems which, in our opinion, show the most promise. We focus particularly on composite materials application. The reason for this is twofold: (1) we are more experienced with the development of sensing systems for composite materials, and (2) it is with composite materials that the specific advantages of optical fibers over other types of sensors can be most utilized (particularly, the ability to embed optical fibers in composites).

20.2. SENSING MECHANISMS AND SENSOR DEVELOPMENTS

20.2.1. Crack Sensing in Metals

The use of optical fibers as crack sensors was first described by Hale et al. (25). This sensor resulted from a program of research at the National Maritime Institute (Hythe, U.K.) on structural integrity monitoring of offshore struc- tures; it has been improved over the years (22–24) and its commercial exploitation and development has been handed over to RFJ Associates (Hythe, U.K.). The principle of operation of the sensor is very simple and is similar to that of metallic foil crack propagation sensors, commonly used in testing structural materials. As shown in Fig. 20.1, the sensor consists of a set of optical fibers embedded in an epoxy backing material that can be bonded to the structure. When a crack develops in the metal underneath the sensor, the opening stretches the optical fiber to the point where it breaks. This failure can be detected from the loss of light transmission at the output of the optical fiber

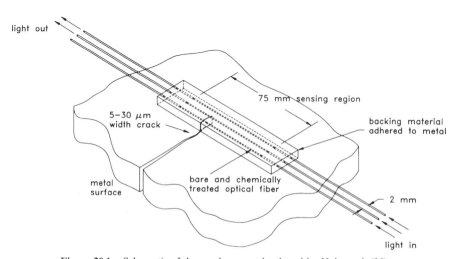

Figure 20.1. Schematic of the crack sensor developed by Hale et al. (25).

or from light bleeding at the point of failure. (The latter method of detection requires that a transparent or translucent embedding material be used and that the transmitted light be in the visible spectrum, although an infrared viewer could also be used.) A typical NMI/RFJ sensor has a sensing region of 75 mm and consists of three 100-μm core/140-μm cladding optical fibers with a 2-mm space between them. The sensing portions of the optical fibers have been stripped of their coating before being embedded, to ensure good strain transfer.

As fused silica glass optical fibers typically have a very large failure strain ($\sim 5\%$), the sensitivity of these sensors can be increased by chemical surface treatment. Long-duration acid etching or the use of silane solutions reduces the effective strength of the fiber. (The actual composition of the eching solutions used by Hale et al. was not disclosed.) It was shown (22) that the ultimate strain of the treated fibers can be reduced to $\sim 0.5\%$. The sensitivity is also affected by the shear modulus of the backing material and the distance between the optical fibers and the metallic surface. According to Hale (23), crack openings in the range 5 to 10 μm are sufficient to break the treated optical fibers.

Based on their results, Hale and Boyle (24) and Hale (23) conclude that the sensors can be made to withstand and work properly under severe environmental conditions (-196 to $+120°C$ at normal humidity, and after several months of underwater test conditions on steel and aluminum).

20.2.2. Impact Damage Sensing in Composites

Fiber-reinforced plastic composite materials are sensitive to large transverse loads such as can occur during impact with foreign objects. Of particular concern is low-velocity (i.e., subperforation) impact, which causes internal delamination and matrix cracking but leaves no, or very little, sign of damage on the exterior (impact side) surface. In effect, this seriously limits the usefulness of a visual inspection for such structures. The terms *barely visible impact damage* (BVID) and *sub-BVID* are used to describe such impact damage for which a simple visual inspection cannot be relied upon for its detection. The possibility of growth of this undetected damage is also a concern.

In a 1982 paper, Crane et al. (10) reported on the use of optical fibers embedded between plies of graphite–epoxy undirectional tape and also in glass–epoxy fabric laminates. They suggested that a grid of optical fibers could be used to cover large sections of composites and form the basis of a built-in damage assessment system. The example of a glass-reinforced plastic (GRP) submarine sonar dome was cited as a structural component that would very much benefit from such a system. They tested this concept in 6 × 6 in. (15 × 15 cm) and 3 × 6 in (8 × 15 cm) flat specimens of each material. The larger samples were supported with clamped boundary conditions on all four sides (4 in. square), and the narrower ones were clamped on only two sides. The embedded optical fibers were Corning Glass Works multimode 50-μm core/125-μm cladding and had a 250-μm-outer-diameter polymer coating. These optical fibers were shown to fail in impacted specimens at higher impact energies than were required to cause delamination in the material. However, only fibers in the smaller specimens were caused to fail. It was suggested that the test samples were not free to bend in the 4-in. (10-cm) square clamped boundary conditions used for the larger specimens. For both types of composites, fibers near the tensile surface (directly opposite the impact) were much more sensitive. However, substantial damage had to be incurred before the optical fibers failed. [As we will see later when describing the work of Glossop et al. (21), removal of the coating and chemical treatment of the fiber is required to reduce the strength of glass fiber if detection of BVID is required.]

The usefulness of optical fibers as crack detectors in many aircraft structural applications was demonstrated by Hofer (27), who reported using long optical fibers bonded to metallic aircraft components in a way similar to Hale et al. Tests were done on aluminum aircraft frames and stringers and near rivet holes of the skin. The concept was also tested on large steel rotor blades of a wind power plant.

Hofer also embedded optical fibers crack sensors in GRP and carbon fiber–reinforced plastic (CFRP) composite panels, and even in the complex GRP structure of a train undercarriage under development at the time. In the latter example, a simple visual approach to detecting light continuity was suggested as the eventual method of using this system. With this method, the optical fibers ends, forming a row on one side of the structure, are illuminated while the other side is monitored visually for continuity of the light spots.

Among the many potential uses suggested by Hofer, of significance was the suggestion that a number of these systems could be implemented in an Airbus aircraft and be monitored by the central fault display system (CFDS) in the cockpit as is done with the regular aircraft built-in test equipment (BITE) — so far, CFDS systems do not include any form of structural integrity monitoring. He and a co-worker were issued several patents on the subject (28, 36, 37).

Research on the treatment of optical fibers for damage detection in composites has also been done by Waite and Sage (49), who measured Weibull statistical coefficients describing the failure of silane solution–treated (SST) optical fibers which were either embedded in GRP composites or loaded individually. From their results they determined that for SST fibers the failure stresses are the same whether the fibers are embedded or not, suggesting that the failure mechanisms in the two environments are the same.

A somewhat different damage sensitizing approach was conceived by Lymer, Glossop, Measures, and others at the University of Toronto Institute for Aerospace Studies (19, 21, 34, 38). Their research resulted in a patented etching process that produces very-damage-sensitive optical fibers (35). This process consists of etching only periodic sections of the glass fiber, as shown in Fig. 20.2. The result of this process is alternating 5-mm sections of larger and smaller diameters with a rough transition region. It is believed that the strength reduction is the result of the compounded effect of the stress concentration, due

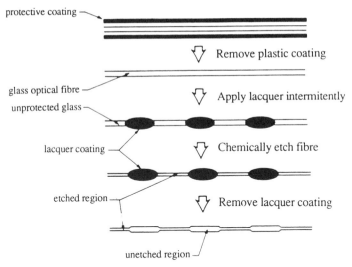

Figure 20.2. UTIAS optical fiber treatment process.

to the geometrical tapering and the surface roughness created in the transition regions between the etched and unetched sections. Extensive testing of this sensor for impact damage sensing was done on laminates of Kevlar–epoxy unidirectional plies to determine optimum sensor location and orientation within the laminate as well as optimum etching duration. The choice of this translucent material was useful, as it allowed mapping of the damage by image-enhanced backlighting, which is a very quick method of obtaining an image of internal delamination (20). The translucence of the material was also useful for finding the location of the fiber breaks by the bleeding of red He–Ne laser light launched from the fiber ends.

The result of this testing showed that:

- Optimal damage sensitivity is obtained when the fiber is embedded between collinear plies and oriented perpendicular to the direction of the adjacent composite material fibers.
- In thin laminates, the best location for impact detection and damage extent assessment is in the tensile region (opposite the impact surface).
- In thicker laminates, the optical fibers near the front surface are more sensitive to impact damage.

Following this ground-laying work, developments were made to permit the demonstration of this concept on a full-scale aircraft leading edge. This development is described in Section 20.3. In the following section we explain in greater detail the mechanism for impact damage detection in composites.

20.2.2.1. Impact Damage Formation and Its Detection

Damage Structure. When subject to low-velocity impact, a laminate made of oriented plies of uniaxial fiber-reinforced plastics will fail in very characteristic ways. The damage consists of matrix cracking (parallel to the reinforcing fibers direction) and delamination. Several plies can be delaminated. Each delamination (always involving two plies of different orientation) has a characteristic peanut shape. The long axis of the peanut is oriented parallel with the direction of the reinforcing fibers of the lower ply. The reason for this is explained graphically in Fig. 20.3 using a model by Clark (7). Two categories of matrix cracks are observed: (1) larger ones that tend to bound the lateral dimensions of the peanut-shaped delamination, and (2) smaller cracks that are distributed along both faces of the delamination. The larger cracks are formed due to the large shearing and tensile forces resulting from the large transverse forces imposed by the impactor. The occurrence of these through-the-thickness cracks triggers the formation of delamination of the interfaces above and below the cracks. This results in a top-hat geometry which forms the unit cell of impact

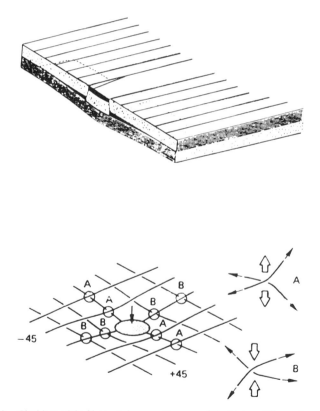

Figure 20.3. Clark's model of impact damage. Because of the higher stiffness along the reinforcing fiber's direction, the transverse load is transferred so as to promote delamination along the fiber direction of the lower ply (points A) and prevent it in the fiber direction of the upper ply (points B). [After Clark (7), by permission of the publishers, Butterworth Heinemann Ltd. ©.]

damage. These cracks are also called *cricical cracks* because they cause delamination to occur at smaller impact-energy values than would otherwise be the case.

As they form, delaminations also produce short matrix cracks (or microcracks), shown in Fig. 20.4. These cracks occur because of the large shear forces along the propagating crack front. The three-dimensional envelope of the entire damage zone has a conical structure, with the smaller-diameter delamination region being toward where the impact occurs. Depending on whether it is bending effects or contact stresses that dominate (i.e., whether we are dealing with a thin or a thick laminate), most of the damage occurs either near the far face (for thin laminates) or near the impact face (for thick laminates).

Optical Fiber Detection of Impact Damage. The understanding of how optical fibers fail when embedded inside a composite material which itself is undergoing some form of failure is a difficult problem. The interaction between the surrounding material and the optical fiber and the nature of the optical fiber–matrix interface determines how a particular type of composite material damage will affect the embedded optical fiber. Through-the-thickness cracks in

Impact Damage Growth Mechanism

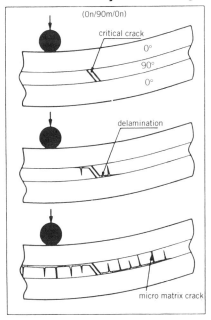 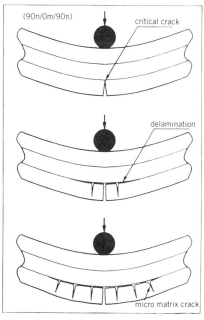

Figure 20.4. Initiation of damage and propagation due to impact. A critical crack initiates the delamination of the interface above and below the broken ply. Microcracks are formed due to shear along the delamination propagation. [After Choi et al. (5), by permission of the publishers, Technomic Publishing Co., Inc. ©.]

the composite are caused either by large tensile stresses or by large shear stresses. Hence the optical fiber located in the path of such a crack is submitted either to large tensile or transverse shear forces. In the tensile situation, the fiber–matrix interface can fail, leading to slippage of the fiber. This will reduce the intensity of the stress inside the fiber and the fiber may not fail even if the opening of the crack is significant. A shear crack, on the other hand, will also result in a complex interfacial failure of the fiber–matrix system near the location of the crack. Depending on the debonding lengths involved, the optical fiber will either see a large tensile force (long debonding lengths), strong bending over a tight radius (medium lengths), or pure shear (very short debonding). If one adds the effects of irregularities (locations of stress concentration) along the optical fiber and how this affects the optical fiber–matrix adhesion, we can see that having a detailed explanation of how the optical fibers fail in the presence of material damage is difficult to achieve.

If one is interested primarily in detecting delamination resulting from impact damage, we can see that it is desirable that the optical fiber be rather insensitive to tensile cracking in the material and very sensitive to shear cracking. This is because tensile matrix cracks can occur and develop without delamination being present — that is, as microcracks or through-the-thickness cracks that do not degenerate into or result from delamination. Therefore, the tensile mode of failure of the optical fiber is not selective to delamination. On the other hand, transverse shear failure in the material can practically result only from an impact, To detect these shear cracks, the optical fiber needs to be located either across collinear plies where critical cracks have developed (the through-the-thickness cracks that trigger a delamination), or be in the direct path of a delamination front that causes microcracks in a shear fashion. Failure of the optical fiber resulting from a transverse shear crack in the material is also more likely to cause a full misalignment of the optical fiber broken faces; this is a desirable feature.

Having mentioned these considerations, we have to say that the development of optical fiber damage sensing systems based on fracture has so far been done without considering the exact failure mode of the optical fibers, nor, for that matter, a sophisticated a priori knowledge of where damage will be located inside the composite for a range of impact loads of interest. Work has relied on the experimental assessment of the best working configurations without full knowledge of how the optical fibers fail. The best fiber locations and orientations as obtained from experiments, however, do reflect the nature of impact damage and its formation characteristics: the difference between thin and thick samples, for example, and the orthogonal placement between collinear plies for ultimate sensitivity to critical cracks.

20.2.3. Detection of Other Types of Damage in Composites

20.2.3.1. Fatigue Damage. Waite (47) embedded silane solution–treated (SST) optical fibers (Corning, 140-μm outer diameter, multimode) in laminates of weaved GRP. It was predicted that the fatigue life of such a treated fiber would be lower than that of the GRP composite material and its failure could be used to signal fatigue damage. The results were not very conclusive for laminates with optical fibers oriented along the load direction in laminates of the type $[w0]_{ns}$ (w indicating weaved material); the optical fibers tended not to fail as early as expected. But in a configuration involving plies at $\pm 45°$,* it was found that the fiber failed directly in the presence of matrix cracking along the $\pm 45°$ direction much before the optical fiber itself was expected to fail.

In another study by the same author (48), more conclusive results were obtained using Ensign Bickford hard clad silica (HCS) optical fibers for the detection of significant fatigue damage. These fibers have a polymer cladding material (as opposed to a polymer coating covering a silica cladding) and no other coating. It was shown that in the presence of fatigue damage the cladding of these optical fibers becomes damaged, resulting in the attenuation of light. It was mentioned that for about 70% of the optical fibers, this attenuation was progressive with the extent of damage as the samples were being fatigued. Fracture of the embedded optical fibers was correlated to accumulated fatigue damage, causing a laminate strength reduction between 22 and 42%. Thus, it was suggested, optical fiber sensors such as this could be used to indicate significant fatigue damage in GRP composites.

20.2.3.2. Growth of Impact Damage. It would be very useful if the same sensors used to detect impact damage could be used to map the growth of the created damage under subsequent service loads. Tsaw et al. (46) reported on experiments to measure the growth of damage due to static loading. The basic sensor was the same as that developed by Lymer et al. (35). Their results showed that the optical fibers were able to map the extent of delamination in the same manner as was obtained in impact testing. However, this is not very surprising because the static loading was done with a plunger that applied a transverse load to the samples. Thus the loading forces were very similar to those caused by an impact. [Such experiments are more characteristic of a slowed-down impact, as was concluded by Tsaw (45).]

20.2.3.3. Interfacial Properties and Water Absorption. In a 1974 paper, Leach and Ashbee (31) investigated the use of light transmission in glass fibers to study interfacial phenomena in polyester resin–glass fiber composites. This is of interest for the development of embedded optical fiber sensors based on similar phenomena. The fibers in Leach and Ashbee's experiments were the reinforcing fibers themselves. The index of refraction of the resin being below that of the 15-μm glass fibers, these fibers could be used as light-pipes. He–Ne laser light (at 6328 Å) was launched at 15° to the polished fiber ends. This angle is very close to the total internal reflection critical angle; it can therefore be

---

*The full stacking sequence was $[0_2, \{90\}/+45_2, \{90\}/-45_2/0_2/-45_2/+45_2/0_2]$, where { } indicates the location and orientation of the embedded optical fibers.

assumed that any change in the condition of the interface will affect the transmission. (The small dimension of the fibers ensures that many reflections of the light occur at the interfaces.) The results showed that water absorption in the composite resulted in reduction of light transmission (see Fig. 20.5). It was suggested that the deterioration of the glass surface increases dispersion of light at each reflection and that this explained the reduction of light transmission. It was also shown that stress-induced debonding at the glass–resin interface contributes to the reduction of light transmission (for both the case of torsion applied to composite bars and for simulated fiber pullout tests).

20.2.4. Fiber Optic Acoustic Emission and Ultrasonic Sensors

The detection of acoustic emission for the purpose of structural integrity monitoring and material evaluation has been the subject of many research efforts over the last 45 years. Acoustic emissions are stress waves generated by the rapid local redistribution of stress that accompanies many damage mechanisms inside materials. There are companies, such as Physical Acoustics Corporation, that specialize in the installation and use of AE systems for

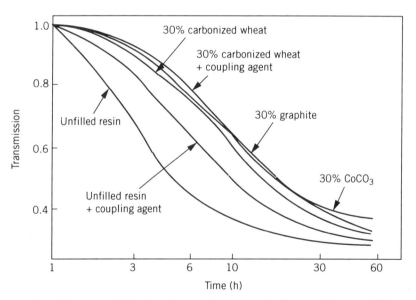

Figure 20.5. Light transmission in polyester–glass composites immersed in water. The many curves compare the effect of using various hydrophobic fillers in the resin. From the results it appears that these fillers appear to have little influence on the total degradation of the resin–glass interface, only on the rate of deterioration. [After Leach and Ashbee (31), by permission of the publishers, Butterworth Heinemann Ltd. ©.]

in-service metallic structures using piezoelectric transducers. Many efforts have been made to develop optical fiber sensors for the detection of ultrasound in water for sonar and hydrophone applications (4, 13, 14, 18, 30, 41a). Similar sensors have been proposed for the detection of acoustic emission and, in general, ultrasonic waves in solids.

In a 1980 paper, Claus and Cantrell (8) described an experiment by which they were able to detect ultrasonic pulses with a coiled optical fiber embedded in a solidified plastic resin. The pulses were generated with a spring-loaded punch striking the surface of a 5.1 cm × 1.4 cm plastic disk. The sensor configuration used was a fiber Mach–Zehnder interferometer. Cielo and Lapierre (6) described two types of optical fiber sensors to be used in a way similar to the surface-contact piezeoelectric ultrasonic transducers. One sensor relied on the photoelastic effect of the applied stresses on the birefringence of the optical fiber (polarimetric sensor). The other also relied on the photoelastic effect, but in a different way. It used a plastic-clad silica core fiber. Pressure applied on the fiber increases the index of refraction of the plastic cladding more than in the silica core. This effect is responsible for a decrease in the light transmission due to the loss of wave guidance. This method gave a somewhat lower sensitivity than the polarimetric approach, but it was suggested that it could be improved with different materials selection.

In 1986, Bennett et al. (2) reported, for the first time, the detection of acoustic emission generated in a graphite–epoxy laminate with an embedded optical fiber. The sensor was a single optical fiber used as a modal-domain sensor. Their results showed detection of AE caused by matrix cracking and fiber failure (Fig. 20.6). Comparison with a piezoelectric transducer showed that the response of the optical fiber to the acoustic pulses was not slower than that of the piezoelectric detector.

In another set of experiments, Meltz and Dunphy (39) used a high-birefringence optical fiber embedded or adhered to various materials. A CO_2 laser was used to generate pulses on the surface of these materials. In some cases a fine layer of oil was deposited on the surface; such a layer ablates quickly and generates a very sharp pulse in the material when hit by a laser pulse. It was shown that such a sensor has a very broad frequency response (better than 10 MHz). Measurements of the induced pressure was comparable to those of other researchers who used other sensing techniques. Localized sensitivity was achieved by removal of the coating on the fiber over a specific region of the optical fiber.

Liu et al. (33) embedded optical fiber Michelson interferometers in Kevlar–epoxy (Fig. 20.7). They were able to detect AE signals resulting from delamination induced by a transverse load applied to 10 cm × 10 cm specimens. Compared with those of piezoelectric transducers, the signals from the Michelson interferometer contained much more noise (Fig. 20.8). This was due to laser phase noise (induced by temperature variations and inhomogeneities in the laser cavity) and the inherent lower sensitivity of the optical fiber sensor compared with piezoelectric devices.

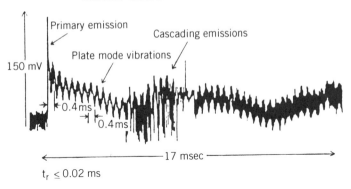

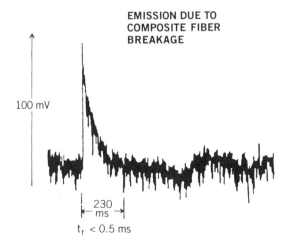

Figure 20.6. Acoustic emission detection using optical fiber modal-domain sensor embedded in graphite–epoxy. [From Bennett et al. (2), by permission of the publishers, Plenum Press ©.]

Ferguson (17) used the foregoing method and demodulating system on several Kevlar–epoxy laminates to further explore its use for the detection of damage. As is often reported in AE studies in composites, her results did not clearly distinguish the specific type of damage that was responsible for a given type of AE signal. However, she did notice a shift in the spectral domain of the power of emitted signals, from dominance in the region 300 to 600 kHz before initiation of delamination to the lower-frequency region of 100 to 300 kHz after delamination. This trend, however, was not very clear and more work is required to explore this effect. Another set of experiments showed that the count of AE events has a charcteristic knee when plotted on a graph with

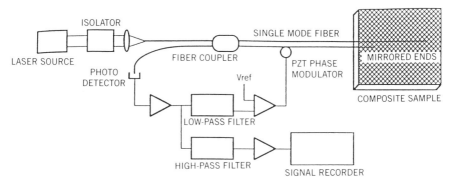

Figure 20.7. Configuration of the acoustic emission monitoring system based on an embedded fiber Michelson interferometer used by Liu et al. (33). Ultrasonic frequencies in the range 100 to 1000 kHz are detected. The signal of frequencies lower than this (separated by the low-pass filter) is used by the feedback circuit to actuate a piezoelectric cylinder attached to one of the sensor arms; this keeps the interferometer signal at the quadrature (most sensitive) level.

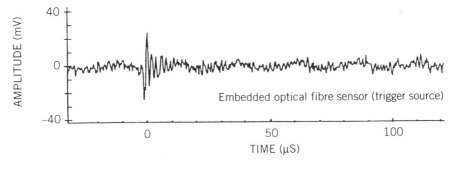

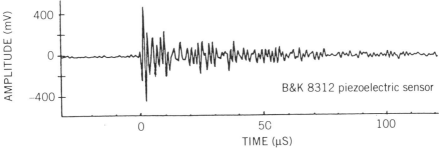

Figure 20.8. Comparison between the response of an embedded fiber Michelson interferometer and a B&K 8312 piezoelectric sensor. [From Liu et al. (33).] Reprinted with permission.

increasing deflection on the x-axis (Fig. 20.9); this is similar to the results of many others [as reported by Williams and Lee (50)].

Alcoz et al. (1) reported on the use of an embedded localized fiber Fabry–Perot (FFP) sensor for detection of ultrasound. This sensor is based on a set of two reflecting surfaces created along the fiber length. Each reflector is

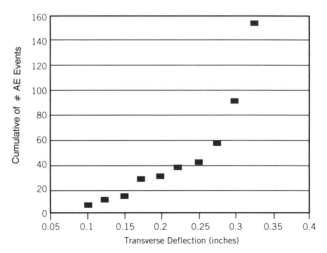

Figure 20.9. Number of AE events versus increasing deflection (and damage) in transversely loaded Kevlar–epoxy laminate. The AE is detected using an embedded Michelson sensor with a 48 mm gauge length. [After Ferguson (17).]

created by a thin coating of TiO_2 deposited on the cleaved end surface of one of the fiber ends and then fusion-spliced with an uncoated fiber portion (see Fig. 20.10). The advantage of this type of sensor over both the Michelson and Mach–Zehnder interferometers is that only one fiber is required and that clear localization of the sensing region is obtained. Quadrature control for the sensor was obtained by altering the dc bias current in the laser diode. However, this particular type of quadrature control method to maintain constant sensitivity may be difficult to achieve in a practical system working continuously for long periods of time. The FFP represents one of the most promising sensors for AE monitoring with optical fibers.

Use of the AE method for damage detection and assessment in composites is difficult, even with piezoelectric transducers. The major limitation of this modality is the difficulty of distinguishing between the various failure modes: matrix cracking, delamination, fiber failure, fiber disbond, and pull-out. As Hamstad (26) points out, it is much easier to obtain AE data than to interpret them. There is also the combined problem of much larger attenuation of signals in the composites compared with metals (especially at high frequencies) and also the fact that composites are inherently dispersive materials. If optical fibers are to be used, the sensitivity needs to be improved. However, the sensitivity of embedded optical fiber sensors themselves cannot be increased. Contrary to hydrophone work, coiling cannot be used, and contrary to hydrostatic static pressure measurements, the sensitivity cannot be increased by the use of soft coatings. Any improvement in the sensitivity must be done outside the optical fiber, by improving the signal-to-noise ratio of the source and detection electronics.

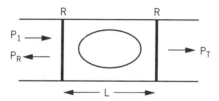

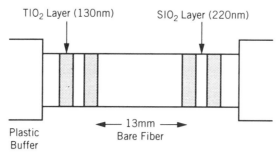

Figure 20.10. Schematic of the fiber Fabry–Perot sensor used by Alcoz et al. (1), by permission of the publishers, © 1990 IEEE.

Despite these challenges, optical fibers offer many new capabilities that warrant further continued research into their use as ultrasonic sensors. Because they can be embedded within composites, they provide the unique ability to monitor locations previously unaccessible. Furthermore, optical fibers sensors can provide the capability, also unique, of long sensing lengths. In composite materials, this means more chance that the sensor will be closer to the location of the AE source. Because damage in composites can be distributed over the entire material, it may not be possible, or relevant, to use the AE method to determine the location of the AE source. Detection of all AE events and their specific characteristics, such as amplitude and pulse duration, may be more important. But even if source location is desired, having long sensing lengths in no way reduces the capability of localization; it just places a different set of requirements on the sensing architecture than for localized sensors. If AE source localization in composite is difficult (because of the anisotropic and highly attenuated wave propagation), it will be just as difficult (if not more) for localized sensors as for long sensors. This is because AE source location relies on the relative time of arrival of the AE pulses on the detectors. The detected time of arrival on a long sensor corresponds to the initial contact between the AE wavefront and the sensor. The location of this initial contact can be derived from the signal given by the other sensors if the (anisotropic) group velocities of the AE waves inside the material are known and a suitable sensor architecture is used.

Nonetheless, there may be practical aspects that will limit the sensing length of AE optical fiber sensors. One of them has to do with the near-dc and normal structural (or acoustic) vibrations in the structure, which, in interferometric sensors, can result in high-frequency signals at the detector end. That is, dynamic strains occurring uniformly over long lengths of the optical fiber can result in high-frequency optical signals at the interferometer output that overlaps a portion of the useful AE range. For example, if the structure is subjected to a global (or *macro*) dynamic strain $\varepsilon_m(t)$ of frequency f_m that affects the fiber over its entire gauge length L [i.e., $\varepsilon_m = \varepsilon_o \cos(2\pi f_m t)$ uniformly over the gauge length L], the detected signal spectrum will contain frequencies as high as $f_d^{max} = 2f_m \varepsilon_o GL$, where G is the phase-strain sensitivity. A typical value for G is $0.65°/\mu\varepsilon \cdot \text{mm}$ (or $11 \text{ rad}/\mu\varepsilon \cdot \text{m}$). If, for example, $\varepsilon_o = 1000\mu\varepsilon$, then $f_d = 22{,}000\text{m}^{-1}f_m L$. Thus we can see that in this case, if L is on the order of a few meters, a 10-Hz macro load frequency results in a signal at the detector already in the hundreds of kilohertz. Since the low-pass quadrature control system needs to keep track and compensate for this shift, the monitored AE signal would be restricted to the >1 MHz range in this example. This effect (of clipping the low-frequency portion of the AE signal) is much less of a problem in polarimetric sensors. This is because G, in a polarimetric optical fiber sensor, is about 100 times smaller than for interferometric sensors. However, the sensitivity to the AE signal is also about 100 times smaller for a polarization sensor, and this means that the signal-to-noise ratio will probably be too low for polarimetric sensors to be used. So far, the use of optical fibers with long sensing lengths for the detection of acoustic emission in the way described above has not been reported.

20.2.5. Other Fiber Optic Damage Detection Methods

It is preferable for an optical fiber damage assessment system to be able to detect and assess damage without relying on fracture of the optical fibers. Such a system would, ideally, be able to preserve its sensing function after damage and repair of the structure. If such a system could be based on fiber optic strain gauges, a definite advantage would be gained: the same sensor could be used to monitor the loads on the structure and be used to indicate its state of damage. (In Section 20.1 we made the case that load monitoring and damage assessment are distinct approaches to structural integrity monitoring, but in principle, nothing should prevent the same sensor to be used for the two functions.) The FFP acoustic emission sensor discussed above is an example of a damage sensor that can also be used as a strain gauge: as an AE sensor, the FFP can be viewed as a strain gauge with a very fast response.

There may be other ways of detecting damage without relying on optical fiber fracture. Many authors have reported observing effects on the optical signals of embedded optical fibers during an impact. For example, Tay et al. (43) showed that light in single-mode fibers embedded in CFRP is attenuated during the time of the impact load. (This can be explained from the increased microbend losses during the loading.) However, so far, no one can realistically claim a direct and unmistakable detection scheme based on such effects.

An interesting device has been proposed by Sirkis and Dasgupta (42, 42a). They suggested that an optical fiber used as a strain gauge (such as a fiber Fabry–Perot sensor) but coated with an elastic–plastic coating would retain some "memory" of the maximum load to which the fiber was subjected. For instance, a metallic coating with a given yield strength would alter the response of the sensor when the load applied to the metallic coating would have exceeded the yield strength of the metal. A similar effect would be observed in the case of a thermal loading. Thus by properly selecting the properties of the coating, a load threshold could be defined. The important parameters and material properties that can be selected to tailor the sensor are the coefficient of thermal expansion (CTE) of the coating material, the ratio of coating and glass fiber radii, and the coating yield strength. The great advantage with this approach would be that the sensor would not lose its ability to measure strain. As Sirkis and Dasgupta's paper was strictly theoretical, many unknowns remain with respect to the practicability of such a device.

More conventional methods may be used to interrogate the structure and monitor its response using an embedded strain gauge network. These approaches would be similar to attempts that have been made with ordinary strain gauges or with accelerometers for the detection and assessment of damage in structures. An example of such an attempt is that of Jones and Goldman (29), who used conventional accelerometers to monitor the vibration characteristics of graphite–epoxy panels containing a delamination. Unfortunately, only very small changes in the natural frequencies of impact damaged specimens compared with undamaged specimens were observed. Changes of this small magnitude in a real structure could be due to many other parameters, such as thermal effects, water absorption, deterioration of resin due to exposure to light, and so on. This indicates the probable impracticability of the structural frequency analysis method of damage assessment for composite materials. However, should there be special types of structures or materials for which a structural frequency analysis would be useful, embedded or adhered optical fiber strain gauges might prove very useful, as their frequency response is usually much greater than their electrical counterparts. This means that both the static and dynamic structural response could be monitored with the same sensor network.

20.3. DEVELOPMENT OF AN EXPERIMENTAL FIBER OPTIC IMPACT DAMAGE ASSESSMENT SYSTEM FOR AN AIRCRAFT COMPOSITE LEADING EDGE

In 1988 through 1990, the Fiber Optic Smart Structures Laboratory of the University of Toronto Institute for Aerospace Studies was involved in a project in collaboration with aircraft manufacturer de Havilland Inc., then a division of Boeing of Canada Ltd. The goal of this project was to demonstrate a fiber optic damage assessment system (FODAS) for a leading-edge section of the

Dash 8 aircraft based on the sensor developed by Lymer et al. (35) (discussed in Section 20.3.2). From UTIAS's point of view, an attempt to implement this sensing approach in a "real" and relatively complex structure would permit a better understanding of the difficulties that would face the development of a working (commercial) system. De Havilland was interested in the possibility of such a FODAS for the leading edge because the design of that component called for a rubber deicing boot cobonded with the Kevlar–epoxy–Nomex structure during the cure. This rubber surface prevents the use of ordinary inspection procedures without removal and necessary discarding (even when the component turns out to be undamaged) of the rubber deicing boot.

20.3.1. Tests on Small Coupons

An extensive series of impact tests were performed on 10 cm × 10 cm coupons by Dubois (15) to find the best sensing configuration for the two representative stacking sequences shown in Fig. 20.11. Tests were made on samples with and without honeycomb core (the core did affect the response of the sensors) and with and without a rubber surface to simulate the deicing boot. It was found that the rubber surface only increased the amount of energy required to produce the same damage, but did not affect the ability of the sensors to detect this damage. More important, it was quickly realized that the sensitized optical fibers could not sustain a cure under pressure with the honeycomb core. To avoid this premature failure of the optical fibers, the manufacturing process was modified such that the outer skin (containing the optical fibers) was first cured separately. The combined results of the tests for the two main stacking sequences is shown in Fig. 20.12. For each type of coupon, the two best optical fiber orientations were selected. These optimum orientations, at two different depths inside the material, were used to form a two-dimensional grid of optical

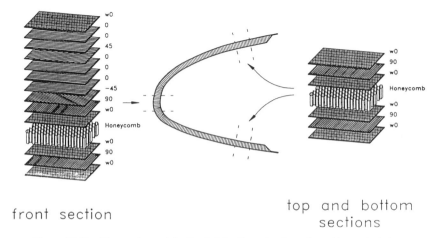

front section

top and bottom
sections

Figure 20.11. Two regions of the Dash 8 leading edge investigated by Dubois (15).

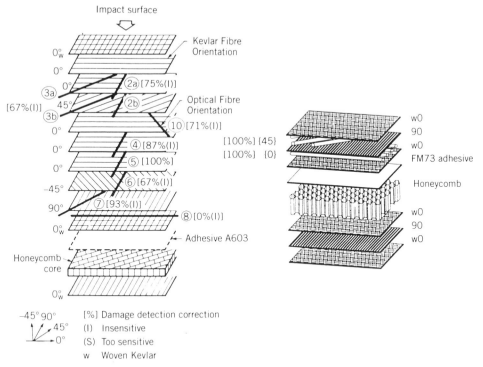

Figure 20.12. Results of the testing on small samples done by Dubois (15). The best two orientations for each region were used for the design of the final sensor configuration.

fibers. Treated optical fibers cannot cross each other between the same two plies for the following reasons: to prevent bends in the fiber, which can induce fiber failure (or, at the very least, large microbend-induced optical signal loss), and to minimize the negative impact of the fiber's presence on the interlaminar adhesion of the plies. A set of two fiber orientations can be used to pinpoint, in a cartesian fashion, the location of damage.

20.3.2. Sensing Layers in the Leading Edge

To embed a large number of optical fibers between two layers of the leading edge (done in a female mold), a manufacturing procedure using an epoxy layer to hold the optical fibers in place was adopted. A representative configuration for one layer is shown in Fig. 20.13. The epoxy material of this layer is the same as the matrix material in the Kevlar–epoxy unidirectional prepregs. On small samples it was found that most of the extra epoxy flows out, being absorbed by the bleeder cloth, and the remaining epoxy becomes evenly distributed in the laminate (adding about 1% extra weight per epoxy layer in a 10-ply laminate).

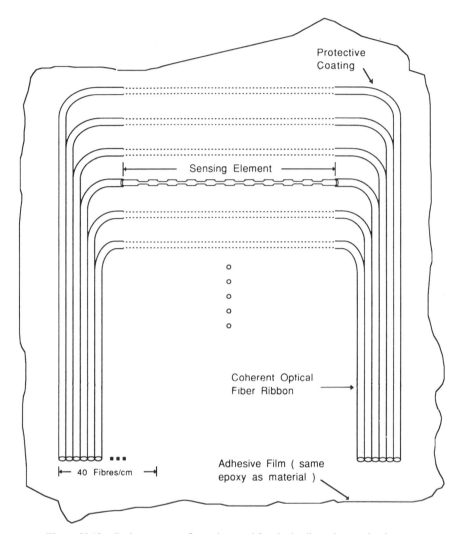

Figure 20.13. Basic sensor configuration used for the leading-edge sensing layers.

Only the sensing regions of the optical fibers were chemically treated. The lead-in and lead-out portions were left with their polymer coating on. This effectively renders these portions insensitive to low-velocity impacts (significant indentation is required to break embedded coated fibers) but allows them to survive cure in the vicinity of steps in the layup. The core temperature of 120°C is somewhat above the maximum permissible temperature for this coating (41); thus a different coating would need to be used in an eventual real system. The final configuration adopted for the leading edge is shown in Fig. 20.14. A total of 250 optical fibers were used in three layers.

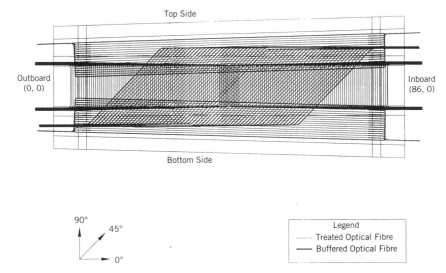

Figure 20.14. Final configuration of the sensing layers embedded in the leading edge.

The optical fibers were prepared by hand, which was a very strenuous and tedious task. The laying-up on the sensing layers was facilitated by the temporary use of guiding pegs punched through the epoxy layers. The actual positioning of the sensing layer on top of the underlying material ply in the female mold was the most delicate task of all. A lot of optical fibers failed during this step. Ripples and a hardening of the resin before embedment were responsible for the many failures when the plastic backing of the sensing layer was removed. The optical fibers went through the two cures and the trimming process with no further disruption.

Fiber ribbon input–output ports were devised to test an experimental light transmission monitoring system. This system was based on a EG&G Reticon wide-area linear photodiode array (LPA). This 1024-element LPA equipped with a fiber optic faceplate could be connected to the ports directly without the need of focusing optics. (Each element of the LPA is 12.5 μm wide, 2.5 mm high, and there is a 12.5-μm-wide gap between each element.) In each sensing layer, the outermost optical fibers (one at each end) were left coated over their entire length and were therefore insensitive to damage. This was done in order to use the output of these fibers as position references on the LPA. This way, the interpreting software could tell the location of each fiber contained between the two outermost fibers on the LPA from the position of the outer peaks on the signal output and the total number of fibers in the port. The ribbons are designed such that the fibers are in intimate contact with each other in a plane. The optical signal from the core of each optical fiber affects about five diodes, which leaves "valleys" of five signal free diodes (as seen in the output of the LPA) between the fibers. Illumination of the input port was done using a microscope illuminator projecting light on a diffuser (opalescent glass plate) attached to the port. We should mention that these ports were also very useful for launching red He–Ne laser light into the optical fibers; this made it possible to see the location of the individual fibers and the bleed spots associated with fiber failures through the Kevlar material.

20.3.3. Testing Results

As mentioned above, about half of the optical fibers failed because of the difficulty with laying the sensing layers in the female mold and during the peeling of the backing material. This did not result in the complete loss of these fibers for testing: each fiber, including the broken ones, could be illuminated at either end and their location and failure points monitored by the visual observation of the laser light bleeding through the Kevlar–epoxy material. Thus all areas of the leading edge could be tested. The impacts were inflicted by a specially built apparatus consisting of a pendulum with a height-adjustable impacting head and two types of holding fixtures: one to hold the leading edge for impacts on the front section and the other for impacts along the top and bottom surfaces. Impact energies up to 72 J could be delivered with this system; however, the testing only involved energies of less than 20 J.

A key observation in the results is the identical response of the damage sensors on the full-scale leading edge to what was obtained on small samples. On all 65 impacts inflicted on the leading edge, fiber fracture occurred only when the fibers were in the vicinity of material damage. The failure points were always inside the damaged zones as determined by image-enhanced backlighting. (A sample impact result is shown in Fig. 20.15.) The optical fibers tended to be less sensitive on the outer boundary of the impact damage, such that for the front section, it was evaluated that the {45} layer mapped on average, from the composite of all tests involving this layer, 79% of the delamination extent, while the {90} layer mapped 76%. (In essence, the fiber optic system tended to underestimate the extent of the delamination by 20 to 25%.) The lower coverage of the {90} is understandable from the damage detection mechanism discussed in Section 20.2.2.1: damage is initiated and the delamination extent is larger near the lower plies. Figure 20.16 shows the damage detection threshold for the fibers in the front section of the leading edge. The sensing layers also provided adequate detection in the tapered regions on the top and bottom of the leading edge (where the number of plies in the outer skin decreases from 10 plies to 3 plies).

20.3.4. Prospective on the Use of This Method

The conclusions gained from this work on the experimental fiber optic damage assessment system concerning the potential manufacturability of this approach can be summarized by the following: bringing this approach into a practical

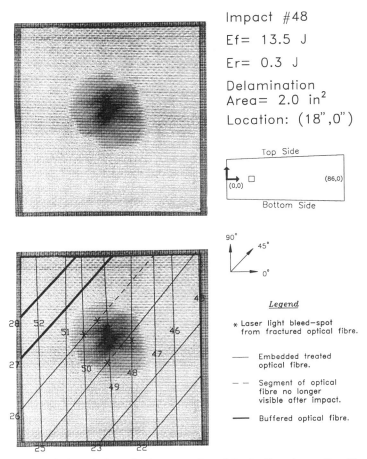

Figure 20.15. Typical result from the impact testing of the leading-edge section. The impact damage, as mapped by image-enhanced backlighting, can be seen to have caused multiple failures in the optical fibers passing through the regions of damage. [From LeBlanc (32).]

system would be difficult but not impossible. The priorities for further development of this FODAS approach would lie in the following tasks:

- Finding the right coating material and a suitable procedure for the processing of optical fibers. Speed and low cost of the process are key requirements. The current method is much too labor intensive.
- Smaller-diameter optical fibers should be used.
- Another supporting medium to permit the inclusion of the optical fibers should be selected. The new supporting medium would not need to disappear after the cure as was aimed here with the use of the same matrix material as contained in the unidirectional prepregs. Adhesive sheets that

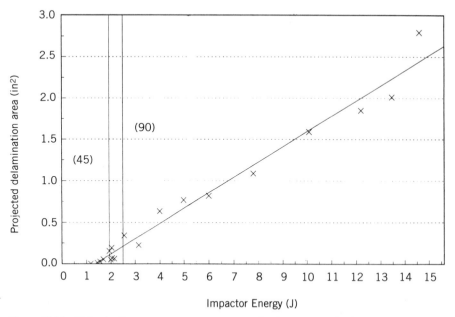

Figure 20.16. Delamination area versus energy curve for impacts on the front section of the leading edge. The thresholds for failure of the optical fibers in the two layers covering this region are indicated by the vertical lines.

can be used as toughening interlayers in laminated composites would probably be a better choice. Not only would such a layer permit the inclusion of the optical fiber sensors and increase the impact resistance of the composite, but it would serve to reduce any negative effects of the presence of the optical fiber on the properties of the host material (especially the compressive properties) by reducing stress concentration effects and virtually eliminating any risk of internal damage being initiated due to the presence of the optical fiber sensors.

• Every effort should be made to minimize the number and size of the input–output ports. We believe that structurally compatible connectors (ideally, concealed within the structure) are possible for such a sensing system.

20.4. CONCLUSIONS

We have reviewed the application of embedded optical fiber sensors for the detection and assessment of damage in structures. Such systems basically fall in one of the following categories:

1. Sensors that rely on the permanent alteration of some physical properties of the sensor with excessive strain or damage:
   - Fracture of the optical fiber
   - Modified nature of the sensor–host interface
   - Plastic deformation or other change in the optical fiber cladding or coating
2. Strain sensors
3. Acoustic emission sensors, which may or may not also be capable of strain sensing

We have argued that the monitoring of strain, although very useful as a means of knowing the use and history of instrumented structures, can only provide, at best, an indirect means of damage assessment. For strain anomaly to be used as an indication of damage, the strain gauge would need to be close to the damaged region; otherwise, it would not be affected. This proximity of the sensor may be possible for critical structures where the high-risk potential failure initiation locations are known and where the load characteristics and structural response are well understood.

On the other hand, sensors that are specifically tailored to detect damage offer better promise for forming the base of a built-in damage assessment system. For instance, long embedded optical fiber sensors that are sensitive to a damage-specific measurand along their entire lengths have more chances of (1) being near the damage and (2) detecting this damage. Such is the case for fibers treated to fail in the proximity of delamination within composites and also for AE sensors with long sensing lengths. Also, as discussed in Section 20.2, many sensing approaches are based on mechanisms where the optical fiber characteristics are permanently affected by the presence of damage and hence require very simple interrogation systems and no continuous monitoring.

Simple optical fiber sensors that are chemically treated to fail in the presence of specific levels of damage have been developed and are understood to the extent that they could be implemented in the short to medium term for specific applications. There is a necessity to improve the treatment procedure and the method of embedding these sensors within composites. Such improvements, we believe, can only be achieved with the drive and commitment that comes from the engineering requirements of a well-defined and specific application (as demonstrated, to some extent, by the composite aircraft leading-edge example). Because the signal recovery is very simple (light continuity in multimode fibers), inexpensive monitoring systems and straightforward inspection methods can be used; this permits the use of many sensors to cover large areas of a structure.

The use of embedded fiber optic sensors for the detection of acoustic emission based on the same demodulation approach as fiber optic strain gauges (although in a different frequency regime) has been demonstrated. What is needed for this approach is very sensitive "strain" detection; this sensitivity has to come from the high signal-to-noise ratio of the optoelectronic source and demodulation components and systems (such as laser source, detectors, and preamplifiers). Very little can be done to increase the sensitivity of the optical fiber sensors themselves. Much improvement in these systems is required for fiber optic AE detection to be feasible in a practical system. But also, more fundamental understanding of the acoustic emissions inside damaged composites, how they can be related to specific modes of failure, and how the signal is affected by its propagation inside the material is required.

In connection with the need for more direct measurements of damage (i.e., having damage itself as the measurand), sensors that make use of effects on the cladding or the coating of optical fibers make a great deal of sense. The pioneering work of Leach and Ashbee (31) studying interfacial phenomena on reinforcing glass fibers used as light-pipes in a resin of higher refractive index suggests that novel optical fiber sensors could be developed on a similar basis. Because novel optical fiber types and coatings are required for these kinds of approaches, it is understandable that very little research has been done on this type of system. It is hoped, however, that the current momentum to develop specialized and novel optical fibers will lead to interesting devices in the near future.

There are other applications of optical fibers to the nondestructive evaluation of materials and structures that were not discussed here. These include the use of optical fibers for the delivery of optical energy for laser-generated ultrasonic pulses or the detection of surface displacements. With the exception of the use of embedded optical fibers for generation of ultrasonic pulses (12, 44), these applications are for external inspection systems and are not specific to smart structure technology. The reader is referred to Culshaw (11), Monchalin (40), and Duddebar et al. (16) for reviews of these applications.

Structural integrity monitoring and damage assessment are very important prospective applications of smart structure technology. We hope to have conveyed to the reader the impression that such a capability will not easily be achieved. Nonetheless, optical fiber sensors offer very important new possibilities for built-in IM/DA systems and we are confident that some of the approaches reviewed here, or new ones yet to be tried, will eventually form the basis of such systems. Eventually, we think that this technology will provide the means to establish a new methodology in engineering practice whereby the engineers will routinely include built-in IM/DA systems in their structural designs.

REFERENCES

1. J. J. Alcoz, C. E. Lee, and H. F. Taylor, Embedded Fibre-Optic Fabry–Perot Ultrasound Sensor. *IEEE Trans. Ultrason. Ferroelectr., Freq. Control* **UFFC-37**(4), 302–306 (1990).

2. K. D. Bennett, R. O. Claus, and M. J. Pindera, Internal Monitoring of Acoustic Emission in Graphite–Epoxy Composites Using Imbedded Optical Fibre Sensors.

In *Review of Progress in Quantitative Nondestructive Evaluation* (D. O. Thompson and D. E. Chimenti, eds.), Vol. 6. pp. 331–335. Plenum, New York, 1986.

3. L. D. Bond, Impact of NDE-NDI Methods on Aircraft Design, Manufacture and Maintenance, from the Fundamental Point of View. *AGARD Conf. Proc.* **462**, R2.1–R2.5 (1990).

4. K. S. Chiang, H. L. W. Chan, and J. L. Gardner, Detection of High-Frequency Ultrasound with a Polarization-Maintaining Fibre. *IEEE J. Lightwave Technol.* **LT-8**(6), 1221–1227 (1990).

5. H. Y. Choi, H. T. Wu, and F. K. Chang, A New Approach Toward Understanding Damage Mechanisms and Mechanics of Laminated Composites Due to Low-Velocity Impact. Part II. Analysis. *J. Compos. Mater.* **25**, 1012–1038 (1991).

6. P. Cielo and J. Lapierre, Fibre-optic Ultrasound Sensing for the Evaluation of Materials. *Appl. Opt.* **21**(4), 572–575 (1982).

7. G. Clark, Modelling of Impact Damage in Composite Laminates. *Composites* **20**(3), 209–214 (1989).

8. R. O. Claus and J. H. Cantrell, Detection of Ultrasonic Waves in Solid by an Optical Fibre Interferometer. *IEEE Ultrason. Symp., 1980*, pp. 719–721 (1980).

9. T. D. Cooper, Specialist Comments: Applied Point of View. *AGARD Conf. Proc.* **462**, R1.1–R.6 (1990).

10. R. M. Crane, A. B. Macander, and J. Gagoric, Fibre Optics for a Damage Assessment System for Fibre Reinforced Plastic Composite Structures. In *Review of Progress in Quantitative Nondestructive Evaluation* (D. O. Thompson and D. E. Chimenti, eds.), Vol. 2B, pp. 1419–1430. Plenum, New York, 1982.

11. B. Culshaw, Optical Fibres in NDT: A Brief Review of Applications. *NDT Int.*, **18**(5), 265–273 (1985).

12. A. Davis, M. M. Ohn, K. Liu, and R. M. Measures, An Opto-ultrasonic approach to Cure Monitoring. *Proc. SPIE — Int. Soc. Opt. Eng.* **1588**, 264–274 (1992).

13. R. P. De Paula, L. Flax, J. H. Cole, and J. A. Bucaro, Single-Mode Fibre Ultrasonic Sensor. *J. Quantum Electron.* **QE-18**(4), 680–683 (1982).

14. R. P. De Paula, J. H. Cole, and J. A. Bucaro, Broad-Band Ultrasonic Sensor Based on Induced Optical Phase-Shifts in Single-Mode Fibers. *IEEE J. Lightwave Technol.* **LT-1**(2), 390–393 (1983).

15. J. M. S. Dubois, Research Towards the Development of a Structurally Integrated Optical Fibre Sensor System for Impact Detection in Aircraft Composite Leading Edge. M.A.Sc. Thesis, University of Toronto (1989).

16. T. D. Duddebar, B. R. Peters, and J. A. Gilbert, Fibre Optic Sensors Systems for Ultrasonic NDE: The State-of-the-Art and Future Potential. *IEEE Ultrason. Symp., 1989* (1989).

17. S. M. Ferguson, Fibre-Optic Acoustic Emission Detection. M.A.Sc. Thesis, University of Toronto (1990).

18. L. Flax, J. H. Cole, R. P. De Paula, and J. A. Bucaro, Acoustically Induced Birefringence in Optical Fibers. *J. Opt. Soc. Am.* **72**(9), 1159–1162 (1982).

19. N. D. W. Glossop, An Embedded Fibre Optic Sensor for Impact Damage Detection in Composite Materials. Ph.D. thesis, University of Toronto (1989).

20. N. D. W. Glossop, W. Tsaw, R. M. Measures, and R. C. Tennyson, Image-Enhanced Backlighting: A New Method of NDE for Translucent Composites. *J. Nondestr. Eval.* **8**(3), 181–193 (1989).

21. N. D. W. Glossop, S. Dubois, W. Tsaw, M. LeBlanc, J. Lymer, R. M. Measures, and R. C. Tennyson, Optical fibre damage detection for an aircraft composite leading edge. *Composites* **21**(1), 71–80 (1990).

22. K. F. Hale, The Application of Optical Fibres to Structural Integrity Monitoring. *IEEE Oceans*, pp. 344–348 (1982).

23. K. F. Hale, An Optical-Fibre Fatigue Crack Detection and Monitoring System. *Proc. SPIE—Int. Soc. Opt. Eng.* **1777**, 147–150 (1992).

24. K. F. Hale and H. B. Boyle, Use of Optical Fibre for Crack Detection. *Proc. ERA-COMRAD Semin. Cond. Monit. Hostile Environ., 1985*, pp. 2.2.1–2.2.10 (1985).

25. K. F. Hale, B. S. Hockenbull, and G. Christodoulou, The Application of Optical Fibres for Witness Devices for the Detection of Elastic Strain and Cracking, Rep. NMI R-2, OT-R-8006, Natl. Maritime Inst., Feltham, England (1980).

26. M. A. Hamstad, A Review: Acoustic Emission, a Tool for Composite Materials Studies. *Exp. Mech.*, **26**(1), pp. 7–12 (1986).

27. B. Hofer, Fibre Optic Damage Detection in Composite Structures. *Composites* **18**(4), 309–316 (1987).

28. B. Hofer and S. Malek, Crack Detection Arrangement Utilizing Optical Fibres as Reinforcement Fibres. U.S. Pat. 4,772,092 (1988).

29. R. Jones and A. Goldman, *Evaluation of a Vibration Technique for Detection of Barely Visible Impact Damage in Composites*, Struct. Tech. Memor. 370. Aeronautical Research Laboratories, Department of Defence, Australia (1983).

30. N. Lagakos, J. H. Cole, and J. A. Bucaro, Ultrasonic Sensitivity of Coated Fibers. *IEEE J. Lightwave Technol.* **LT-1**(3), 495–497 (1983).

31. P. Leach and K. H. G. Ashbee, Development of a Light-Pipe Technique for Investigating Interfacial Phenomena in Fibre-Reinforced Composites. *Composites* **5**(2), 67–72 (1974).

32. M. Le Blanc, A Prototype Fibre Optic Damage Assessment System for an Aircraft Composite Leading Edge. M.A.Sc. Thesis, University of Toronto (1990).

33. K. Liu, S. M. Ferguson, and R. M. Measures, Damage Detection in Composites with Embedded Fibre Optic Interferometric Sensors. *Proc. SPIE—Int. Soc. Opt. Eng.* **1170**, 205–210 (1990).

34. J. D. Lymer, The Characterization of Low Velocity Impact in Composite Materials Using an Embedded Optical Fibre Assessment System, M.A.Sc. Thesis, University of Toronto (1988).

35. J. D. Lymer, N. D. Glossop, W. D. Hogg, R. M. Measures, and R. C. Tennyson, Damage Evaluation System and Method Using Optical Fibers. U.S. Pat. 4,936,649 (1990).

36. S. Malek and B. Hofer, Determination of the Integrity of Part of Structural Materials. U.S. Pat. 4,603,252 (1986).

37. S. Malek and B. Hofer, Measuring Device for Determining Cracks. U.S. Pat. 4,629,318 (1986).

38. R. M. Measures, N. D. W. Glossop, J. Lymer, J. West, S. Dubois, W. Tsaw, and R. C. Tennyson, Fibre Optic Impact Damage Detection of Composite Materials. *Proc. SPIE—Int. Soc. Opt. Eng.* **949**, Paper 32 (1988).

39. G. Meltz and J. R. Dunphy, Optical Fibre Stress Wave Sensor. *Proc. SPIE—Int. Soc. Opt. Eng.* **838**, 69–77 (1987).

40. J. P. Monchalin, Laser-Ultrasonics for Industrial Applications. In *Review of Progress in Quantitative Nondestructive Evaluation* (D. O. Thompson and D. E. Chimenti, eds.) Vol. 7B, pp. 1607–1614. Plenum, New York, 1988.

41. G. Orcel, Optical Fiber Coatings for Sensing/Smart Skins Applications. In *Fiber Optic Sensor-Based Smart Materials and Structures*, papers presented at the Fifth Annual Smart Materials and Structures Workshop, Blacksburg, Virginia, (Edited by R. O. Claus), pp. 7–12. Institute of Physics Publishing, Bristol and Philadelphia (1992).

41a. S. C. Rashleigh, Acoustic Sensing with a Coiled Monomode Fiber. *Optics Letters* **5**(9), pp. 392–394 (1980).

42. J. S. Sirkis and A. Dasgupta, Thermal Plastic Metal Coatings on Optical Fibre Sensors. *Proc. SPIE — Int. Soc. Opt. Eng.* **1588**, 88–99 (1992).

42a. J. S. Sirkis and A. Dasgupta, Analysis of a Damage Sensor Based on Elastic-Plastic Metal Coatings on Optical Fibers. *IEEE J. Lightwave Technol.* **11**(8), 1385–1393 (1993).

43. A. Tay, D. A. Wilson, A. Demirdogen, and J. R. Houghton, Microdamage and Optical Signal Analysis of Impact Induced Fracture in Smart Structures. *Proc. SPIE — Int. Soc. Opt. Eng.* **1370**, 328–343 (1991).

44. C. O. Thompson, W. V. Miller, III, and R. O. Claus, Generation and Detection of Ultrasonic Stress Waves in Materials Using Embedded Optical Fibers. *Proc. SPIE — Int. Soc. Opt. Eng.* **1370**, 324–327 (1991).

45. W. Tsaw, An Embedded Fibre Optic for Damage Growth Monitoring in Kevlar Composites. M.A.Sc. Thesis, University of Toronto (1992).

46. W. Tsaw, M. LeBlanc, and R. M. Measures, Growth of Damage within Composites Determined by Image Enhanced Backlighting and Embedded Optical Fibres. In *Review of Progress in Quantitative Nondestructive Evaluation* (D. O. Thompson and D. E. Chimenti, eds.), Vol. 9, pp. 1207–1212. Plenum, New York, 1990.

47. S. R. Waite, Use of Embedded Optical Fibre for Early Fatigue Damage Detection in Composite Materials. *Composites* **21**(2), 148–154 (1990).

48. S. R. Waite, Use of Embedded Optical Fibre for Significant Fatigue Damage Detection in Composite Materials. *Composites* **21**(3), 225–231 (1990).

49. S. R. Waite and G. N. Sage, The Failure of Optical Fibres Embedded in Composite Materials. *Composites* **19**(4), 288–294 (1988).

50. J. H. Williams and S. S. Lee, Acoustic Emission Monitoring of Fibre Composite Materials and Structures. *J. Compos. Mater.* **18**, pp. 348–370 (1978).

21

Fiber Optic Smart Structures for Aircraft

HERB SMITH
McDonnell Douglas Aerospace
St. Louis, Missouri

21.1. INTRODUCTION

The military aircraft industry has been one of the major potential markets for smart structures technology. The Air Force has invested a considerable amount of money in the development of smart structures for military aircraft applications and NASA has been very active in developing applications for space structures and air vehicles, such as the National Aerospace Plane. This interest is due to the significant potential benefits that smart structure concepts offer in the areas of the survivability and supportability of aircraft. The primary focus in this chapter is on military aircraft applications, with some discussion of potential benefits to commercial aircraft. Clearly, military aircraft present the most significant challenges. We begin the chapter with a discussion of the overall concepts of supportability and survivability of military aircraft and the role that smart structures can play in implementing these concepts. Other chapters have been devoted to specific types of sensors for a variety of physical phenomena, so the focus in this chapter is on the bigger picture of what the goals and objectives of the smart structures system should be and what types of problems need to be solved by the smart structures system. This is followed by a discussion of the challenges facing production implementation of smart structures technology. Along with the discussion of challenges, we review some of the work that has been done to address these issues.

21.2. SURVIVABILITY OF MILITARY AIRCRAFT

Military aircraft are, by definition, designed to participate in combat, or hostile, activities. The aircraft will, therefore, be subject to damage from enemy fire. In the event that damage is incurred, the pilot must decide whether the aircraft

Fiber Optic Smart Structures, Edited by Eric Udd.
ISBN 0-471-55448-0 © 1995 John Wiley & Sons, Inc.

can continue to be flown, or if he must eject from the craft (a process not without risk). Smart structure concepts offer the potential to improve this situation in several ways. Detecting the occurrence of ballistic damage, locating the damage, and assessing the extent of damage are all important to improving survivability.

21.2.1. Damage Detection

Damage detection has two aspects. First, the occurrence of the impact must be recognized, and second, the location of the damage must be determined. Although it may be obvious to the pilot (at least in some cases) that ballistic damage has occurred, situations are not so obvious to the smart structure system. Transient data must be captured that contain information about the damage event in a form that can be easily recognized amid the very noisy environment found on a fighter aircraft. The location of the damaged area is of vital importance in determining how much the aerodynamics of the aircraft will be affected. Control surfaces may be damaged or lost. In addition, fuel, hydraulic, electrical, or other systems might be damaged, and the severity of the physical damage to the structure must be determined. Primary load paths may be degraded to the point where even minor maneuvers could cause catastrophic failure of the structure.

21.2.2. Extent of Damage

Detecting the occurrence of an impact and even determining the location of the impact do not yet tell how much the health of the structure has been impaired. The smart structure system must be able to quantify the severity of the impact, that is, to determine the physical extent of the damage to the structure. Composite structures may be delaminated or blown away completely. The damage may be primarily to the skin and its stiffeners, or spars may be destroyed, causing major changes in the primary load paths, strength, and stiffness of the structure.

A 23-mm high-explosive incendiary (HEI) round can cause extensive damage to an aircraft structure, particularly if the impact occurs in a fuel storage area. The hydraulic ram effects amplify the damage caused, in addition to presenting the potential for secondary explosion and fire. Figure 21.1 shows the type of damage that can be caused in typical composite structure. This significantly degrades the strength and aerodynamic performance of the structure.

Another type of damage that is of concern with composite structure is delamination caused by low-velocity impact (LVID), which can occur during ground handling and maintenance. A dropped hammer, drill, or toolbox can cause a delamination in the composite skin of an aircraft without leaving any visible evidence that it occurred. Catastrophic failure can result when the aircraft reenters service. The smart structure system must be able to perform a

Figure 21.1. Ballistic damage to composite structure can be extensive.

structural health self-test prior to beginning a mission to ensure that the structure is not impaired.

21.2.3. Response to Damage

If the smart structure system can detect, locate, and quantify the extent of damage to the structure, what is to be done with that information? In the case of LVID detected prior to flight, indications can be made to the pilot, or ground personnel, that damage exists and that further inspections or repairs must be performed.

Ballistic damage that is detected in flight must be handled differently. A number of systems are in operation on the aircraft which perform a variety of tasks essential to the operation of the aircraft or which assist the pilot as he completes a mission. One such system is the overload warning system (OWS). The OWS determines, based on flight parameters and stored information, when the aircraft is approaching the boundary of the safe flight envelope. Current systems are configured only for a completely healthy structure. The smart structure system, having data showing that damage has occurred and quantifying its effect, can interact with the OWS to allow the safe flight envelope to be reduced an amount appropriate to the extent of damage. This would minimize the potential for a pilot to overstress a damaged structure and cause catastrophic failure after ballistic impact.

The flight control system is an essential element in fly-by-wire systems and contains the control laws necessary to produce stable flight. The aerodynamics and dynamic response of the aircraft change dramatically as a result of ballistic damage, invalidating the basic control laws. Research is ongoing in reconfigurable flight controls, and significant developments have been demonstrated. These new systems can modify the flight control laws in response to known damage. This resilience in the flight control system allows the aircraft to fly more like an undamaged aircraft, thereby giving the pilot greater opportunity to complete a mission or to return safely to base.

The smart structure system will make other interactions possible, also. Fuel leaks, for example, are difficult to ascertain. The fuel sloshes in the tank, making accurate measurement difficult; thus an average must be taken over a longer period. The rate of fuel consumption by the engines must be monitored over the same period, and finally, a comparison made between the change in fuel level estimated in the tank and the amount of fuel consumed by the engines to determine if more fuel is missing than can be accounted for through normal engine use. If a leak is determined to exist, fuel can be relocated to an area free of leaks. In the meantime a large quantity of fuel can be lost. Interaction between the smart structure system and the fuel system could allow damage information to function as an early warning of potential fuel leaks. Much fuel, and valuable flight time, could be saved.

Other interactions such as these may be possible as well. The advent of the smart structures systems will enable the structural health to become a part of the total aircraft state at any point in time. The old concept of the structure being a separate entity that simply carries the pilot, avionics, and payload, will be changed forever. The smart structure system will be fully integrated into the aircraft avionics structure, and structural health will be a standard piece of information available to all other decision-making processes. The schematic of a typical aircraft avionics bus shown in Fig. 21.2 illustrates this interaction. Communication occurs over the MIL-STD 1553 bus in most of these communication schemes. In the future, data buses will rely more heavily on optical fibers. One significant challenge remaining is the definition of these interfaces between the numerous aircraft systems and the fusion of their information into a comprehensive picture of the aircraft's health.

21.3. SUPPORTABILITY OF MILITARY AIRCRAFT

Often, 50 to 75% of the total life-cycle cost of an aircraft is incurred after the aircraft enters service. Although not merely as dramatic as some of the survivability issues, the supportability of the aircraft is extremely important.

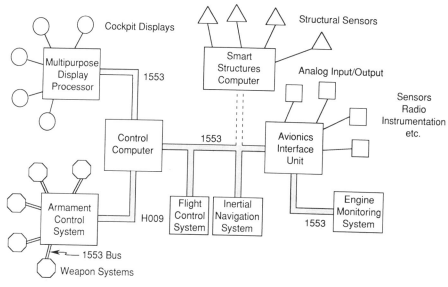

Figure 21.2. The smart structure functions must be compatible with the overall avionics architecture of the aircraft.

Smart structures technology provides several potential improvements in the support and maintenance of an aircraft.

Inspection and maintenance events on a military aircraft occur at specific times in its life. The life of an aircraft is determined by the number of flight hours and the severity of those hours. One major contribution of smart structures is the potential to monitor the structural loads accurately and enable precise determination of the life.

Current fatigue-life tracking systems rely on loads computed from flight parameters or through direct measurement of strains at a limited number of points with strain gauges. Several problems with these systems lead to a data recovery rate of only 50 to 70%. Present systems typically record data on a tape storage device that must be removed from the aircraft periodically. Failure to remove the tape in a timely manner often results in data being written over and lost. Tapes can sometimes get misplaced or lost in transit to the ground processing station. Electromagnetic interference (EMI) results in the need for extensive evaluation of the data by ground personnel to locate and remove EMI and other bad data from the flight record. Current systems are sampled-data systems operating generally in the range 10 to 20 Hz. Higher-frequency structural disturbances, such as buffet caused by turbulence of flow separation that occurs in the range 45 to 60 Hz, have been shown to be extremely important to fatigue crack growth (1). These events are completely missed by slow sampling rates.

When strain measurements are made, strain gauges are generally utilized which are bonded to the structure at strategic locations. Strain gauges will occasionally become disbonded or fail internally and thus provide erroneous data. Backup gauges are usually available; however, the response time of the current tracking system is so slow that 6 months or more may pass before the maintenance personnel get notification that a gauge is bad and that the system should be switched to the backup gauge.

All of this adds up to a fatigue life tracking system that is usually slow, inaccurate, and labor intensive. The most current fatigue life information available to the squadron commander for fleet management decisions is often 6 to 12 months old. Fiber optic smart structure systems offer the means to deal with these issues. Embedded optical fiber sensors are extremely durable and immune to electromagnetic interference. More sophisticated data acquisition systems with increased sampling speeds, or data-driven systems which capture and digitize only the peaks and valleys of the strain for fatigue tracking, are being developed. At McDonnell Douglas we have been developing software to acquire and processs strain sensor data in real time. For the purposes of fatigue tracking, this involves capturing the successive peaks and valleys of the strain, executing a rainflow cycle pairing routine (2) and then computing either the crack initiation damage increment or the crack growth increment. The basic architecture of this process is shown schematically in Fig. 21.3. The crack initiation times computed on a cycle-by-cycle basis compare well with traditional blocked spectrum analysis procedures (Table 21.1), as does the crack growth behavior (Fig. 21.4).

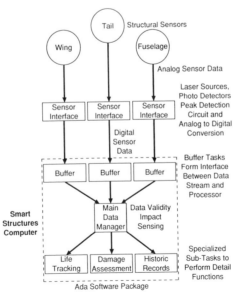

Figure 21.3. Architecture of real-time life and damage tracking software.

**Table 21.1   Comparison of Cycle-by-Cycle Crack
Initiation Software (Ada) with a Traditional
Blocked-Spectrum (CI91) Process**

Spectrum	Procedure	Life
F/A-18 wing fold	Ada	1,992
	CI91	1,884
F/A-18 wing root	Ada	4,522
	CI91	3,470
F-15 stabilator	Ada	22,001
	CI91	22,321
F-15 wing	Ada	12,080
	CI91	11,885

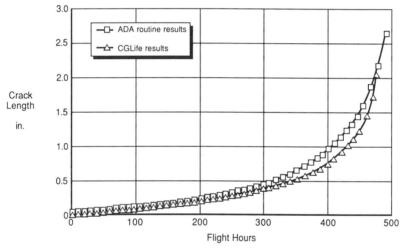

Figure 21.4. Crack growth results compare closely between blocked spectrum and cycle-by-cycle
approaches.

21.4.   CHALLENGES TO IMPLEMENTATION

Smart structure concepts offer significant potential benefits in a number of
areas. Much of the development work has occurred over the last 5 years in
research laboratories (as would be expected); therefore, some challenges remain
before smart structures becomes a technology that is a standard part of a
modern fighter aircraft. Successful transition from development to production
will depend on manufacturing methods that do not significantly increase the
complexity or cost of the composite parts, embedding methods that do not
significantly degrade the strength or durability of the parts, and implementation schemes that do not significantly increase the weight or the supportability

requirements of the aircraft. Any new system or technology must buy its way onto the aircraft; that is, the benefits of having it on the aircraft must outweigh the costs associated with having it on the aircraft. Each of the major areas where challenges remain is addressed in more detail in the sections that follow.

21.4.1. Material Design Allowables

One of the significant benefits of optical fibers has been their ability to be embedded within the laminate of a composite structure. Quantifying the effects that the embedded optical fibers have on the design allowables of those laminates is one of the challenges that remains. In addition to the optical fibers, some sort of connector, or egress fitting, must also be embedded to allow the fiber to exit the part safely and durably. Also, any electrooptic devices, such as couplers, required to implement the specific architecture of the sensor network must be embedded. The latter items are all larger than the basic optical fiber and present a significant concern. The effects of these "foreign objects" on the strength allowables of the laminate must be determined before safe designs can be produced.

Figure 21.5 shows photomicrographs of composite laminates with embedded optical fibers. In part (*a*) of the figure the optical fibers are oriented parallel to the carbon reinforcing fibers and present little or no disruption. In part (*b*) of the figure the optical fibers are oriented perpendicular to the carbon reinforcing fibers. In the latter case, a significant resin pool can be seen around the optical fiber. Also, the carbon fibers must curve around the optical fiber, creating a potential instability under compression load.

Composite materials exhibit considerable scatter in the material design allowables. As a result, the strain levels allowed are low enough that fatigue is not a design issue. The static strength properties are of considerable importance to the design process, and any degradation due to embedded foreign objects must be well characterized. Testing was conducted at McDonnell Douglas to determine the potential severity of the embedded optical fibers on the compression and tension ultimate strengths. The results of this testing are shown in Fig. 21.6 compared with the theoretical predictions, and little deviation can be seen. More extensive testing to characterize the statistical effects fully are under way. Specific situations may need to be addressed individually if design details are present that cause concern.

21.4.2. Bond Quality

A high-quality bond must be maintained between the optical fiber and the resin matrix. This will ensure proper strain transfer from the laminate to the optical fiber and enable an accurate sensor measurement to be made. The buffer material must be chosen to provide a good bond to the matrix material and must have sufficient temperature resistance to prevent breakdown during the manufacturing process (350°F for thermoset materials). Several optical fiber

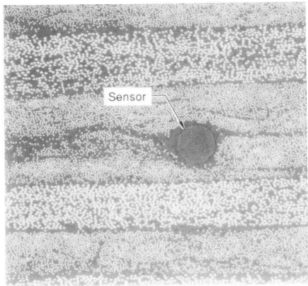

(a)   **Sensor Parallel With Surrounding Fibers
(Preferred)**

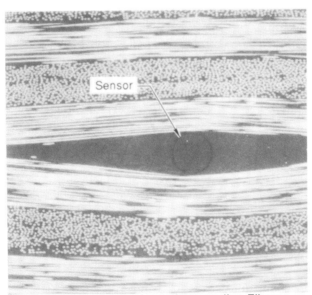

(b)   **Sensor Perpendicular to Surrounding Fibers
(Not Preferred)**

Figure 21.5. Photomicrographs of optical fibers embedded in carbon-epoxy material.

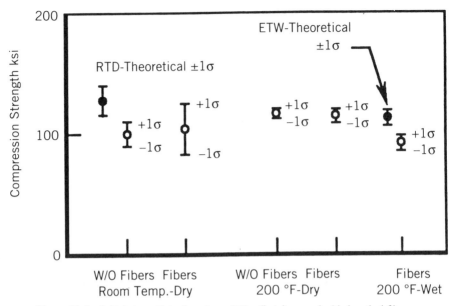

Figure 21.6. Initial strength testing shows little effect from embedded optical fibers.

manufacturers have high-temperature polyimide coatings available that perform well in the typical thermoset composite materials used in many aircraft
parts. Thermoplastic materials are gaining in popularity in the aircraft industry, but they have significantly higher processing temperatures (750°F). In
addition, the nature of the bond between the matrix and the buffer material
will be different than with thermoset materials. Bond quality should always be
assured for a specific combination of buffer material and laminate matrix
material. (A more detailed look at the integrity issues associated with embedded optical fibers is presented in Chapter 5, the coatings issue is addressed in
Chapter 3 and the optical fiber-composite matrix interface is explored in
Chapter 4.)

21.4.3. Manufacturing Compatibility

Manufacturing of composite parts is labor intensive. Individual precut plies of
material are laid up by hand on hard tools. Vacuum bagging or mating hard
tools are used to apply pressure. The tooling, along with the laid-up plies, is
autoclave processed. Optical fiber sensors embedded in the laminate must be
positioned very accurately to ensure that part-to-part consistency in sensor
location is maintained. The optical fibers themselves are fragile, and improving
their handling quality during the layup process not only increases the survivability of the sensor network, but also speeds the process.

In addition to the optical fibers, the egress connectors, or fittings, must
easily integrate with the play layup process must be easy to position accurately,

and must be compatible with the tooling or bagging used in the autoclave. (These issues are addressed in greater detail in Chapter 6.)

21.4.4. Maintenance and Logistics Support

The maintenance requirements of the smart structure architecture must be considered early in the development process. Maintenance personnel must be able to test the system and diagnose any problems. The equipment required and the trained personnel must both be available. Austere conditions can exist at forward locations, and some equipment, such as arc fusion splicers, cannot be used near a fueled aircraft. The smart structures system should incorporate as much self-test capability as possible to minimize the amount of maintenance required by ground personnel.

21.4.5. Data Validity

The goal of smart structures systems development is to have a system that operates in real time. Critical decisions about the aircraft health will be made in split seconds based on data provided by the sensors. It is essential that some means be in place to ensure that the sensor data are real and meaningful. Even the best sensor will experience occasional failures that must be detected as sensor failures by the smart structure system.

Several approaches to this problem are possible. Knowledge of the aircraft structure and the means wherein it carries load provides the ability to establish limits on specific strain readings as functions of other flight parameters (Fig. 21.7). This type of constraint is currently used in manual validation operations of tracking data and can easily be embedded in a rule base to provide this fundamental data check.

A sensor that ceases to operate will peg out at a maximum value or drop completely to zero and stay there. Correlating the time factor to the sensor gives the ability to detect a sensor that has not changed for a significant length of time. Other failure modes may exist for specific sensor types that exhibit very characteristic behaviors. The system can include rules to detect these situations as well.

The symmetry of the aircraft can be used to ascertain a sensor that is out of range. This would involve a slightly higher level of logic, and the added assurance must be traded off against the computation time penalty. Neural networks may provide some utility in this effect as well.

21.5. COMMERCIAL AIRCRAFT APPLICATIONS

The commercial aircraft industry has been shocked by a number of tragedies and near tragedies in recent years that have focused attention on several issues.

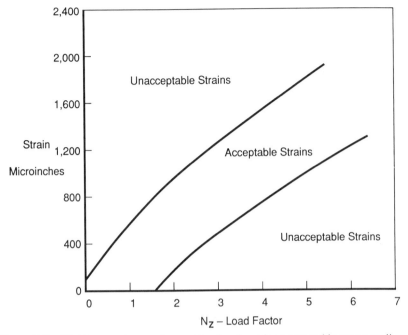

Figure 21.7. Flight parametees can be used to place bounds on acceptable sensor readings.

We will not soon forget the pictures of the remains of PAN AM 103 strewn over Lockerby, Scotland, nor the Aloha Airlines 737 with a 20-ft-long section of the passenger cabin peeled off. Two major issues: aging aircraft and terrorism. The FAA has active programs in the areas of aging aircraft and in aircraft hardening and survivability. There are certainly many design issues involved in these areas, but there is also a significant role for new sensing concepts. The sensing needs are very similar to those discussed previously in relation to combat aircraft.

Accurate assessment of the damage associated with the age of the aircraft requires an accurate knowledge of how the aircraft has been, and is, being flown. Commercial aircraft use has traditionally been assumed to be fairly uniform, with approximately one ground–air–ground (GAG) or pressuriz-ation cycle per 2 hours of flight time. In the case of the Aloha Airlines 737, the very short island hops created a higher number of GAG cycles per flight hour and aggravated the aging problems. Over-ocean use also creates a high-corrosion environment. Long-distance intercontinental flights have a very low number of GAG cycles per flight hour.

Gust loads are a higher frequency load, in the range 1 to 2 Hz, which are similar to the buffet loads experienced in combat aircraft, although at a lower frequency. They contribute a significant amount to the consumption of the fatigue life, in that cracks grow rapidly under buffet, or gust, conditions, particularly when superimposed on a steady-state load.

Strain sensors can be utilized to monitor both the pressurization cycles and the gust effects on the wing loads. Some sensing concepts can monitor corrosion (or the corrosion environment) and provide a more complete understanding of the conditions under which cracks may develop.

Explosive devices in the cargo hold (or passenger compartment) create damage similar to that experienced in battle damage scenarios. Sensor architectures designed to locate and quantify the extent of damage in these situations can provide valuable information to the pilot and greatly increase the survivability of the aircraft.

21.6. SUMMARY

Smart structures technology provides the promise of significant improvements in the supportability and survivability of modern aircraft structures for both military and commercial aircraft. The most innovative concepts are for composite structures, where the sensors can be embedded within the material; however, many of the technology elements can be utilized with sensors bonded to metallic structure.

More sophisticated and accurate tracking of loads is important to commercial and military aircraft, particularly as age extension programs push the real life of the aircraft well beyond its initial design lifetime. Sensing and quantification of structural damage is a concern to the survivability of military aircraft, and in these days of terrorism, to the survivability of commercial aircraft as well.

A number of issues remain to be resolved before smart structures technology will be ready for full implementation in production aircraft; however, there are a number of ongoing research programs that are addressing these issues. It will not be long before the technological challenges will have been met and new, "smart" aircraft become the standard of performance.

REFERENCES

1. R. Perez, S. Harbison, H. G. Smith, Jr., and C. R. Saff, *Development of Techniques for Incorporating Buffet Loads in Fatigue Design Spectra*, NADC-90071-60; Naval Air Development Center, Warminster, Pa. 1990.
2. S. D. Downing and D. F. Socie, Simple Rainflow Counting Algorithms. *Int. J. Fatigue*, **4**, p. 31, January (1982).

22

Control of Smart Space Structures

ANDREW S. BICOS
McDonnell Douglas Aerospace
Huntington Beach, California

22.1. INTRODUCTION

All vehicles, whether spacecraft, aircraft, submarines, automobiles, or others, and some civil structures, such as bridges and power plants, have control systems that use inputs from various sensors to perform their functions using various actuator outputs. Simple examples include an automobile cruise control, aircraft autopilot, and a spacecraft attitude control system. These classical systems control one or more rigid-body characteristics of the vehicle, such as speed and orientation. Similarly, we can control the shape and vibration of a structural component of these vehicles and structures. These structures have their flexible-body characteristics controlled, such as deformation. These controlled structures are not what we call smart structures, even though they possess the required components, which include sensors, actuators, and a control processor. Throughout the 1980s (1), investigations in smart structures have studied the integration of one or more of these components into the structural system, for example, by embedding within or bonding onto the structure the sensors and/or actuators. Most efforts have used a control-driven, multidisciplinary approach to the design of these actively controlled structures. The ultimate objective is to simulate a biological system in design as well as function. We can consider a human being to be made in part of a smart structure, the bones being the structural components and the muscles being the actuators. The various sensory organs are the sensors and the brain is the processor. The nervous system is composed of the processor and sensors that are connected to actuators to form the smart structure control system.

In this chapter, structural control systems that apply to aerospace systems are discussed, but the material is generally applicable to other systems as well. The emphasis of the discussions will be on control systems and what governs

Fiber Optic Smart Structures, Edited by Eric Udd.
ISBN 0-471-55448-0 © 1995 John Wiley & Sons, Inc.

their use in fiber optic smart structures. First, control of spacecraft is discussed and some basic control theory is introduced. This is used to discuss some of the problems associated with spacecraft structural control and some of the techniqus used to address these. A brief discussion of some applications of control of smart space structures follows. Finally, we discuss the types of fiber optic sensors that are useful in the control of smart space structures.

22.2. SPACE STRUCTURE CONTROL

There are two types of spacecraft control: attitude control and vibration/shape control. The traditional control of spacecraft has been attitude control or the control of the spacecraft orientation assuming that the spacecraft is a rigid body. If the spacecraft cannot be assumed a rigid body, then added to attitude control is the control of the vibration and/or shape of the spacecraft. The first deals with six rigid body modes, and the latter deals with the infinite number of flexible body modes. We first discuss the difference between these control types and why both will be needed on future spacecraft. Some basic control theory is discussed and used to describe some of the problems encountered in control of spacecraft. An overview of some control techniques for addressing these problems is also given.

22.2.1. Types of Spacecraft Control

Most space structures of the past 30 years have been small and stiff enough so that the approximation of their dynamics as rigid-body dynamics has served well in most cases of attitude controller design. In this case the control system dynamics can be separated from the structural dynamics, as shown in the top plot of Fig. 22.1. The assumption that the spacecraft is essentially rigid was valid as far as the attitude controller was concerned since the controller bandwidth was rolled off well below the frequency of the first spacecraft structural mode. There have been some glaring exceptions in the past, however, in most cases the rigid-body assumption has been adequate for control design purposes. More of today's spacecraft, as well as those being planned for future missions, are larger, more flexible, and/or require more precise pointing or shapes to be maintained. This leads to the possible interaction of the control system with the flexible modes of the spacecraft, as shown in the bottom plot of Fig. 22.1. If planned for and implemented into the design, this interaction can be very useful in reducing vibration problems and/or maintaining correct shape. However, if this interaction was not planned for or was overlooked in the design, serious consequences can occur, resulting in anything from degraded performance to loss of the spacecraft.

Both types of control will be needed on future spacecraft. This includes launch vehicles, space platforms, and interplanetary vehicles. Attitude control of some sort is required on all spacecraft; however, a considerable number of

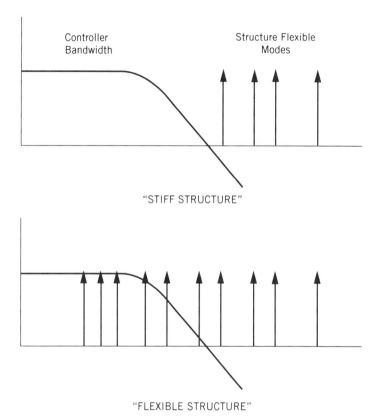

Figure 22.1. It is harder to separate the controller from the structural frequencies for a flexible structure than a stiffer or rigid structure.

future spacecraft will be larger or have more stringent pointing requirements, or both. This will necessitate that the attitude control system not interfere with shape or vibration control systems in any way that is detrimental to the performance and stability of the spacecraft and its mission. Keeping the two types of controllers from interacting may not always be possible; however, designed-in useful interaction can be beneficial and will be used.

22.2.2. Basic Control Theory

The components of a control system for a large flexible spacecraft are shown in Fig. 22.2. The plant is defined as any physical object to be controlled—in our case, a spacecraft. The spacecraft is made up of primary structure to which are attached appendages such as solar arrays, radiators, instruments, and other payloads. Included in the plant are any passive damping treatments that increase the amount of energy dissipation in the spacecraft structure. The plant is excited by disturbances that cause responses that must be controlled in order for the spacecraft to satisfy its mission performance requirements. The sensors

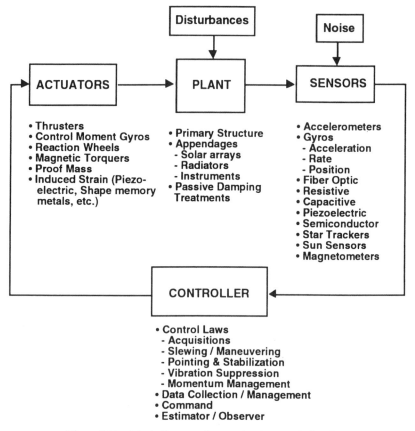

Figure 22.2. Block diagram of space structure control system.

measure this response and feed this information to the controller. The sensor's inputs to the controller are made up of the measured quantity plus any noise existing in the sensor system. The controller determines what actuator forces are required to keep the disturbance from degrading the performance of the spacecraft, yet maintaining overall system stability. The actuators provide that required force at the required locations.

22.2.3. Classical Methods

22.2.3.1. Open-Loop Response. The system open-loop response is the response of a system, a structure in our case, to an excitation. The excitation is the input and is usually a force. The response is the output and is usually given as the displacement of the structure or its derivatives, velocity and acceleration.

22.2.3.2. Closed-Loop Response. The structure closed-loop response is the response of the structure to an excitation that is made up of some external excitation plus some closed-loop control system excitation. The control system excitation is determined from some sensor measurement(s) and control law so as to effect the desired response when added to the external excitation. This is traditionally shown in a block diagram as in Fig. 22.3, which shows a single input/single output (SISO) proportional–derivative feedback (PD) control system. The transfer function for the plant or structure, $G(s)$, is the Laplace domain ratio of the open-loop response, $x(t)$, to the input, $u(t)$, obtained by taking the Laplace transform of the equation of motion.

22.2.4. Stability of Closed-Loop Systems

The transfer function of the closed-loop system from $R(s)$ to $X(s)$ is of the form

$$H(s) = \frac{X(s)}{R(s)} = \frac{G(s)K}{1 + G(s)KH}$$

The zeros of $H(s)$ are defined as the roots of

$$G(s)K = 0$$

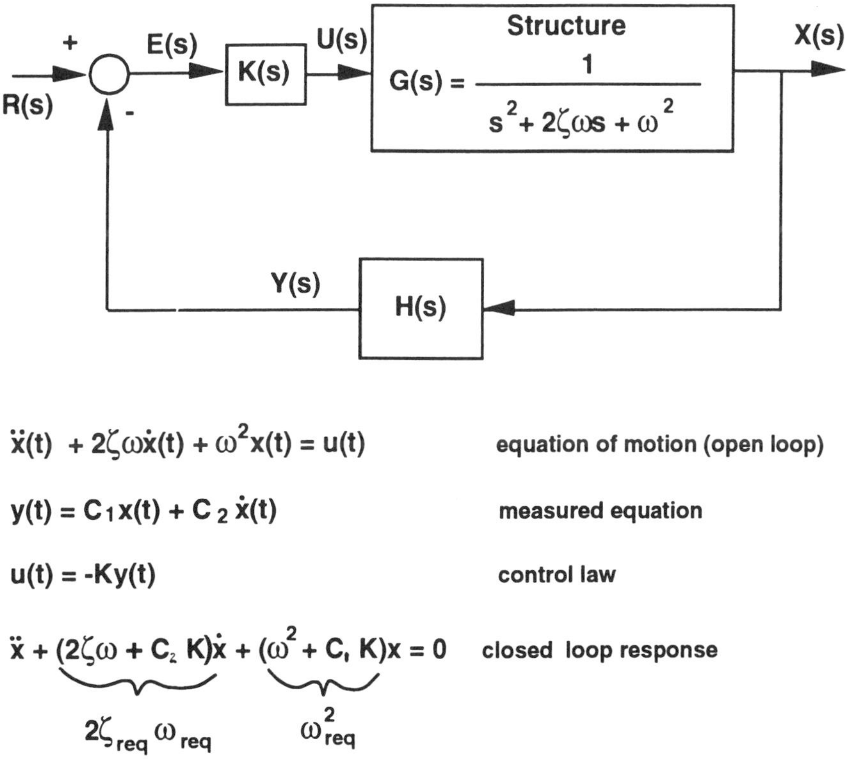

$$\ddot{x}(t) + 2\zeta\omega\dot{x}(t) + \omega^2 x(t) = u(t) \qquad \text{equation of motion (open loop)}$$

$$y(t) = C_1 x(t) + C_2 \dot{x}(t) \qquad \text{measured equation}$$

$$u(t) = -Ky(t) \qquad \text{control law}$$

$$\ddot{x} + \underbrace{(2\zeta\omega + C_2 K)}_{2\zeta_{req}\omega_{req}}\dot{x} + \underbrace{(\omega^2 + C_1 K)}_{\omega^2_{req}}x = 0 \qquad \text{closed loop response}$$

Figure 22.3. Block diagram of closed-loop control of a structure.

and the poles of $H(s)$ are defined as the roots of

$$1 + G(s)KH = 0$$

The latter equation is the characteristic equation (determinant) of the block diagram. The system is stable if the characteristic equation has no poles (roots) in the right half-plane of the s-plane. Therefore, $K(s)$ is chosen such that the system is stable. This is not, however, the only consideration for the selection of $K(s)$. $K(s)$ must also be selected so as to provide the required performance.

The several SISO system analysis techniques available fall into two categories: frequency-domain and time-domain techniques. The frequency-domain techniques fall into two subcategories: the Nyquist diagram, Nichols chart, and Bode plots, which utilize the open-loop transfer function; and the Routh array and root-locus techniques, which utilize the closed-loop transfer function. More about these techniques can be found in Ogata (2). The time-domain techniques involve looking at the transient response to determine the performance parameters of the system, such as maximum overshoot, settling time (damping), and frequency content.

For the purposes of our discussions here, we will use the root-locus technique. In the characteristic equation, $K(s)$ is a function of a variable gain, k. The root-locus technique is a graphical method in which the locus of the roots of the characteristic equation are plotted for k varying from zero to infinity. For the design of SISO control systems, the root-locus technique is used to indicate the manner in which the open-loop poles and zeros should be modified such that the response meets the system performance and stability requirements.

Phase and gain margin are two measures of the degree of stability of a system. Figure 22.4 shows the phase and gain margins in logarithmic plots for a stable system and an unstable system, respectively. Phase margin is defined as the amount of additional phase lag at the gain crossover frequency required to bring the system to instability. Gain margin is defined as the reciprocal of the magnitude $|G(j\omega)|$ at the frequency where the phase angle is $-180°$.

The performance of a control system (i.e., the ability to control the error due to the disturbance to the required level) is dependent in this example on the controller gain used. However, too much gain can cause the system to become unstable. Other stability problems occur when the control system sensors are not located correctly. Sensor placement is important for flexible structure control, both attitude and vibration control. An example is given in Fig. 22.5. In Fig. 22.5a, a rigid "spacecraft" has mass m, a sensor that measures its position, x, and an actuator that applies a force u. For this system and the PD control law, the root-locus plot shows that the system is stable for all values of gain (i.e., the poles all remain in the left half-plane for all values of control gain k). In Fig. 22.5b, however, the spacecraft is no longer rigid. If the control system sensor measures position x_2 and the control force is applied at x_1, the root-locus plot shows that for even the smallest gain, the system is unstable.

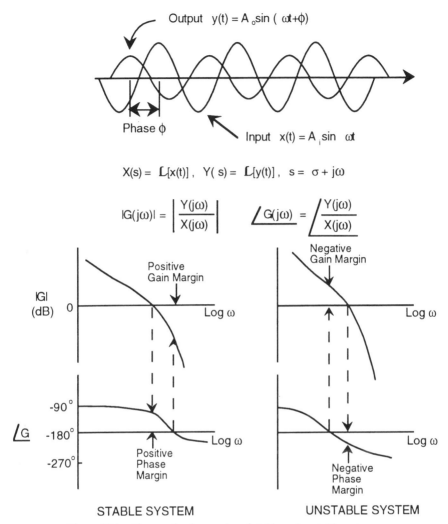

Figure 22.4.   Phase and gain margins of stable and unstable systems.

One easy way to stabilizes the system is to collocate the sensor and actuator, as shown in Fig. 22.5c. The same control law is used but now the sensor measures position x_1 and the actuator applies the control force at x_1. The root-locus plot shows that for all values of gain, the system is stable.

Figure 22.5b also illustrates the problem of spillover. This problem occurs when energy from the control system is added unintentionally to unmodeled modes or modes not included in the truncated design model. Flexible spacecraft have a hierarchy of models used. The actual spacecraft has infinite degrees of freedom that is modeled by a high-order model (e.g., a finite-element model) with thousands of degrees of freedom. The higher modes of this model are inaccurate. This model is then truncated to a medium-order model with

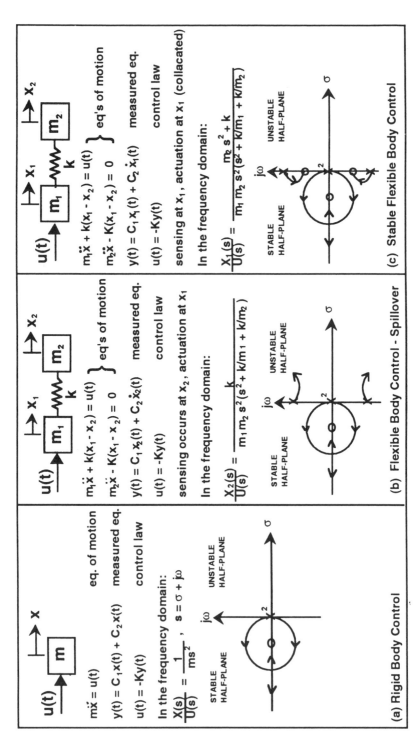

Figure 22.5. The location of the rigid-body (or attitude) control system sensors and actuators is very important to the stability of the structural system.

hundreds of degrees of freedom. All the modes of this model are relatively accurate, but this is still too many modes for use in controller design. This model is truncated further still to a low-order model with tens of degrees of freedom. These are not necessarily the lowest modes but those modes that significantly affect the performance of the spacecraft. This is the design model on which the controller is designed. The controller design is evaluated on the medium-order model (evaluation model). This approach has been used for controlling conventional spacecraft that are relatively rigid, where only the rigid-body modes are used in the design model but several of the prominent flexible modes are also included in the evaluation model. The controller will be designed for the modes in the design model but will interact with the remaining modes of the evaluation model as well. This interaction can cause degradation in performance and even instability. For the structure shown in Fig. 22.5b the control model contained only the rigid-body mode. In addition, the evaluation model also contained the flexible-body mode. If the structure were rigid, as in Fig. 22.5a, there would be no problem. If the structure were flexible, however, spillover might be a problem (depending on sensor and actuator placement).

Another, more realistic example is shown in Fig. 22.6. Here a launch vehicle and its rigid body and first two flexible bending modes are shown. The control objective is to damp the modes of the launch vehicle. Again the location of the sensor can significantly affect the stability of the system. Some ways around these stability problems are shown in Fig. 22.7. The two-mode system discussed above is used again (Fig. 22.7a).

- Adding passive damping moves the flexible mode poles into the stable half-plane (Fig. 22.7b); however, it must be enough to allow the required amount of controller gain to be used. Usually, for flexible space structures the inherent passive damping is not enough.
- Adding a notch filter decreases the magnitude of the elastic modes response and alters the phase, which favorably changes the pole departure angle on the root-locus plot (Fig. 22.7c), which allows stability to be maintained as the gain is increased.
- Adding lag to the control system also alters the phase, which favorably changes the pole departure angle on the root-locus plot (Fig. 22.7d).
- The notch is sensitive to relatively small changes in frequency, whereas the lag is not (i.e., lag is more robust).

22.2.5. Modern Linear Optimal Control Methods

22.2.5.1. State-Space Form for Structures. For a structural system of the form

$$M\ddot{q} + C\dot{q} + Kq = \beta u$$

where q is the generalized modal coordinate vector, βu the generalized force vector, and M, C, and K the mass, damping, and stiffness matrices, the

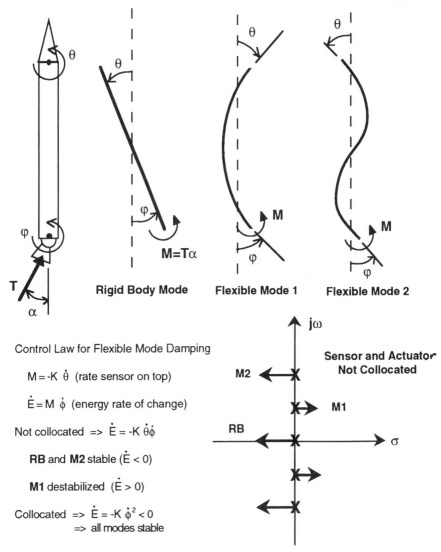

Figure 22.6. The location of the control system sensors and actuators is very important to the stability of the structural system.

finite-dimensional state-space equations are

$$\dot{x}(t) = A(t)x(t) + B(t)u(t) + w(t)$$
$$y(t) = C(t)x(t) + D(t)u(t) + v(t)$$

where $x(t)$ is the state vector $= [q \quad \dot{q}]^T$, $u(t)$ the input vector of applied control forces, $y(t)$ the measurement vector, $w(t)$ and $v(t)$ the noise vectors, and the matrices A, B, C, and D completely characterize the system dynamics.

(a) Negligibly-damped system

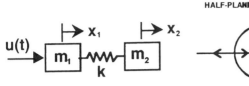

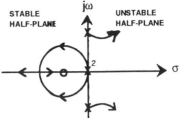

(b) Passive damping added to negligibly-damped system control

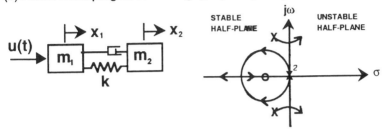

(c) Notch filter added to negligibly-damped system control

Notch filter transfer function

$$N(s) = \frac{s^2 + as + b}{(s + 1)^2}$$

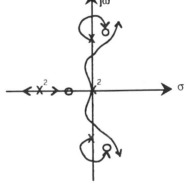

(d) Lag added to negligibly-damped system control

Lag transfer function

$$L(s) = \frac{t_o}{(s + t_o)^n}$$

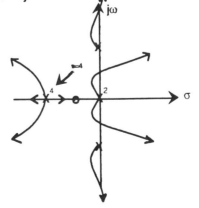

Figure 22.7. Several solutions to the stability problems.

22.2.5.2. Stability of State-Space Systems. A time-invariant state-space system (A, B, C, D are not functions of time) is stable if all the eigenvalues of A or the roots of

$$\det(sI - A) = 0$$

have their real part less than zero.

22.2.5.3. Optimal Control Systems. In many practical structural control problems, a selected performance index or cost function (e.g., a function of the control energy required and/or a function of the error signal) is to be minimized. Usually, these are quadratic functions of the form.

$$J = \int_0^\infty (x^T Q x + u^T R u)\, dt$$

Quadratic performance functions are used because they have several useful characteristics: (1) they are positive definite; (2) they define a control that yields the minimum rms response, which is a common engineering criterion; and (3) the solution yields a linear control law. By definition, the system that minimizes the selected performance index is optimal. It is important to note, however, that hardware implementation of a particular optimal control law may be very difficult and expensive.

22.2.5.4. Controllability. A system is defined to be controllable at time t_0 if it is possible by means of an unconstrained control vector to transfer the system from any initial state $x(t_0)$ to any other state in a finite time (2). The controllability is a measure of the degree to which the state of a system can be driven from any initial state to a specified state by an appropriate choice of control inputs $u(t)$. The condition for complete controllability will be given without proof [see Ogata (2) or Kwakernaak and Sivan (3) for proof]. The n-dimensional linear time-invariant system

$$\dot{x}(t) = Ax(t) + Bu(t)$$

is completely controllable if and only if the column vectors of the controllability matrix

$$P = [B|AB|A^2B| \cdots |A^{n-1}B]$$

span the n-dimensional space (i.e., the rank of P is n). The condition for complete controllability can also be stated in terms of the system transfer function. A necessary and sufficient condition for complete controllability is that no pole–zero cancellation occurs in the transfer function. If a pole–zero cancellation does occur, the system cannot be controlled in the direction of the canceled mode (2). Controllability is intimately tied to the placement and authority of the control system actuators.

22.2.5.5. Observability. A system is defined to be observable at time t_0 if with the system in state $x(t_0)$, it is possible to determine this state from the observation of the output over a finite time (5). The observability or reconstructability of a system is a measure of the degree to which the state vector can be obtained from the measurements $y(t)$. The condition for complete observability or reconstructability will be given without proof [see Ogata (2) or Kwakernaak and Sivan (3) for proof]. The n-dimensional linear time-invariant system.

$$\dot{x}(t) = Ax(t) + Bu(t)$$

$$y(t) = Cx(t)$$

is completely observable if and only if the row vectors of the observability matrix

$$C = \begin{bmatrix} C \\ CA \\ CA^2 \\ \vdots \\ CA^{n-1} \end{bmatrix}$$

span the n-dimensional space (i.e., the rank of C is n). The condition for complete observability can also be stated in terms of the system transfer function. A necessary and sufficient condition for complete observability is that no pole–zero cancellation occurs in the transfer function. If a pole–zero cancellation does occur, the canceled mode cannot be observed in the output (2). Observability is tied intimately to the placement and sensitivity of the control system sensors.

22.2.5.6. Full-State Feedback Control. The dynamics of the n-dimensional time-invariant system

$$\dot{x}(t) = Ax(t) + Bu(t)$$
$$y(t) = Cx(t)$$
(22.1)

can be altered by using a full-state feedback control law

$$u(t) = -Fx(t)$$

Implementing this control results in the closed-loop system

$$\dot{x}(t) = (A - BF)x(t) = A_{cl}x(t)$$

The poles of A_{cl} may be placed arbitrarily for the control required if and only if the system (22.1) is controllable. If the system is not controllable, it can at least be stabilized, with a degradation in performance, if the unstable subspace of A is within the controllable subspace of the system (22.1).

22.2.5.7. Linear Quadratic Regulator. The state feedback control law that minimizes the performance index

$$J = \int_0^\tau (x^T R_{xx} x + u^T R_{uu} u)\, dt$$

subject to

$$\dot{x}(t) = Ax(t) + Bu(t) \qquad x(0) \text{ specified}$$

$$R_{xx} = R_{xx}^T \geqslant 0 \qquad R_{uu} = R_{uu}^T > 0$$

in the linear quadratic regulator (LQR). The choice of R_{xx} and R_{uu} is not always obvious. If the important state is $z(t) = Hx(t)$, for example, the line-of-sight pointing error, then the choice for R_{xx} is $H^T H$. If the objective of the control is to damp the structure, R_{xx} is chosen to represent the energy in the structure.

The LQR problem can be solved as an optimization problem where the control law

$$u(t) = F(t)x(t) = [R_{uu}^{-1} B^T S(t)]x(t)$$

where $S(t)$ is found from the solution of the Riccati equation. In algebraic form, the Riccati equation is

$$0 = A^T S + SA + R_{xx} - SBR_{uu}^{-1}B^T S$$

The control law leads to a stable, closed-loop system. LQR results in very robust control that has excellent gain and phase margin. In practice, this can be fully realized only in a linear system with perfect sensors and actuators.

The only problem with the LQR is that it cannot be implemented unless the states are fully specified, and generally, the states are not fully measured. Therefore, to implement the LQR control law, an estimate of the state is required.

22.3. CONTROL OF SMART SPACE STRUCTURES

Control of structural vibrations and shape of future space systems will involve smart structures technology in both their primary and secondary structures. A smart or intelligent structure is defined as a structure that contains actuators, sensors, processors, and signal and power electronics that are integral parts of the structure, forming a well-designed system that meets the structural require- ments in a cost-effective and energy-efficient manner. The actuators and sensors may have structural functionality as well. These subsystems allow the charac- teristics of the structure to be altered through control of the various properties of the structure, such as mechanical, thermal, electromagnetic, chemical, and optical properties. There are currently no existing smart structures as defined above. The foregoing definition is only a goal at this time. The requisite technologies are currently being heavily researched in the United States, Europe, and Japan. These include structural systems called adaptive, sensory, controlled, and active. Figure 22.8 (1) shows the relationship among these technologies.

The required disciplines to be integrated for smart structures are structures, materials, sensors, actuators, controls, and communications. This integration is not just a conglomeration but a well-designed system that meets the structural requirements in a cost-effective and energy-efficient manner.

What makes smart structures possible at this time is the maturation of several technologies to the point that the required components for the control of a structure can be integrated into the structure so that in effect the structure and its control components become the same part. These technologies include:

- Laminated material systems for structural components.
- Components for exploiting the off-diagonal terms in the material consti- tutive equations.
- Advances in microelectronics.
- Advances in fiber optics.

The applications of smart structures include:

- Aeroservoelastic control and maneuver.
- Structure-borne noise reduction.

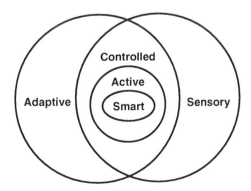

Figure 22.8. Smart structures are a subset of active and controlled structures

- Acoustic noise reduction.
- Precision assembly of components.
- Isolation of machinery noise.
- Shape control of mirrors and reflector antennas.
- Jitter suppression in optical systems.
- Load alleviation in structures.

The characteristics of the sensors used in a control system can greatly affect the performance and stability actually achieved in practice. An example of this is given in Spanos (4), where the stability of a system is compared for a perfect sensor versus a sensor with a limited bandwidth.

The primary sensor characteristics for smart structure control are sensitivity, spatial resolution and bandwidth. Those characteristics of secondary interest are transverse sensitivity, temperature sensitivity, linearity, hysteresis, EMC, and size. Ideally, for the control of space structures we would like to have sensors that have high sensitivity, can be distributed adequately for good spatial resolution, and have high bandwidth compared to the highest mode to be controlled (5).

22.4.   FIBER OPTIC SENSORS FOR CONTROL

The characteristics of fiber optic sensors make them a very good choice for smart structure control. Application of fiber optic sensors to smart structure control will allow a variety of control schemes to be implemented. The fiber optic sensor types most applicable to the structural control systems discussed above are:

1. *Fabry–Perot.* The optical observable is phase, which is related to the material observables of strain and temperature. For control purposes, the strain capability would be used with a specified sensor gauge length. Other characteristics include displacement resolution of less than 1 nm, intrinsic and extrinsic sensing, and multiplexibility.
2. *Fiber Gratings.* The optical observables are reflectivity and transmission, which are related to the material observables of strain and temperature. For control purposes, the strain capability would be used with λ-mixed spatial resolution. Other characteristics include displacement resolution of less than 1 nm, intrinsic sensing, and multiplexibility.
3. *Optical Time-Domain Reflectometry (OTDR).* The optical observable is time delay, which is related to the material observables of strain and temperature. For control purposes, the strain capability would be used with specified sensor spatial resolution.
4. *Two-Mode Elliptical-Core Fiber Gratings.* The optical observable is

phase, which is related to the material observable of vibration. For control purposes, this would be used to measure vibration over a weighted (variable-diameter fiber) gauge length. The sensitivity of these sensors is about 100 times less than the others; however, for control purposes this is still good.

The material observables can be used to obtain other control-relevant observables, such as acceleration. An example of this is the fiber optic accelerometer, which contains a microbend intensity sensor (6). In this case the intensity is related to the strain in the optical fiber, which is induced by an inertial mass. Comparison of fiber optic sensors to other types used in control are given in Table 22.1. Table 22.2 gives the quantitative ranges of strain sensors in these categories.

Table 22.1   Sensor-type Comparisons

Sensor type	Advantages	Disadvantages
Piezoelectric	Small, highly sensitive, can be weighted and shaped, well demonstrated	EMI, ground loops, multiplexing difficulties, temperature limitations
Resistive	Small, sensitive, well demonstrated	EMI, ground loops, single measurand, frequency response (low bandwidth), temperature limitations
Fiber optic	Small, highly sensitive, multiplexable, can be weighted, high-temperature capabilities, no EMI, no ground loops, multimeasurand	New technology, multiplexing not conclusively demonstrated

Table 22.2   Comparison of Strain Sensors

	Sensor Type				
	Foil[a]	Semi-conductor[a]	Fiber Optic[b]	Piezo Film[c]	Piezoceramic[c]
Sensitivity	$30\,V/\varepsilon$	$1000\,V/\varepsilon$	$10^6\,V/\varepsilon$	$10^4\,V/\varepsilon$	$2 \times 10^4\,V/\varepsilon$
Active length (in.)	0.008	0.03	~ 0.04	< 0.04	< 0.04
Bandwidth	dc–acoustic	dc–acoustic	$\sim$ dc–acoustic	~ 0.1 Hz–GHz	~ 0.1 Hz–GHz

[a]10 V excitation.
[b]0.04 in. gauge length.
[c]0.001 in. sensor thickness.

Source: Adapted from Crawley (7).

There has been much work reported in the literature using piezoelectric sensors for control of structures in the laboratory (8,9) and some work reported using shape-memory alloy sensors (10). To date, however, there has not been much published on fiber optic sensors used in control systems.

REFERENCES

1. B. Wada, J. Fanson, and E. Crawley, Adaptive Structures. *J. Intell. Mater. Syst. Struct.* **1**(2), 157–174 (1990).

2. K. Ogata, *Modern Control Engineering.* Prentice Hall, Englewood Cliffs, NJ, 1970.

3. H. Kwakernaak and R. Sivan, *Linear Optimal Control Systems.* Wiley, New York, 1972.

4. J. T. Spanos, Control-Structure Interaction in Precision Pointing Servo Loops. *J. Guidance, Control Dyn.* **12**(2), 256–263 (1989).

5. S. Hanagud, M. W. Obal, and A. J. Calise, Optimal Vibration Control by the Use of Piezoceramic Sensors and Actuators. *Proc. 27th Structures, Structural Dynamics and Materials Conf.*, San Antonio, TX, p. 177–185, *1986* (1986).

6. D. R. Miers, D. Raj, and J. W. Berthold, Design and Characterization of Fiber Optic Accelerometers. *Proc. SPIE—Int. Soc. Opt. Eng.* **838**, p. 314 (1987).

7. E. F. Crawley, SDM Lecture: Intelligent Structures. *33rd Struct. Struct. Dyn. Mat. Conf.*, Dallas, *1992* (1992).

8. C. Lee, W. L. Garrard, and T. O'Sullivan, Piezoelectric Modal Sensors and Actuators Achieving Critical Damping on a Cantilevered Plate. *Proc. 30th Structures, Structural Dynamics and Materials,* Mobile, AL, p. 2018–2026, *1989* (1989).

9. D. K. Lindner and K. M. Reichard, Weighted Distributed-Effect Sensors for Smart Structure Applications. *Proc. American Defense Preparedness Association, American Institute of Aeronantics and Astronautics, American Society of Mechanical Engineering, Society of Photographic, Instrumentation Engineers–the International Society for Optimal Engineering Conf. Act. Mater. Adap. Struct.*, Alexandria, VA, *1992*.

10. D. G. Wilson, R. Ikegami, and G. J. Julien, Active Vibration Suppression Using Nitinol Sensors and Actuators. *Proc. Damping Conf.*, West Palm Beach, FL, **1989** (1989).

23

Fiber Optic Smart Civil Structures

DRYVER R. HUSTON and PETER L. FUHR
College of Engineering
University of Vermont
Burlington, Vermont

23.1. INTRODUCTION

Civil structures such as bridges, buildings, and dams offer many unique opportunities for the use of smart structures technologies. A smart civil structure can be defined as one that can measure external loads, can assess the effects of these loads, and can respond appropriately. Such structures will be safer, will perform better such as through reduced vibration levels, and will be more economical to own and operate than will their conventional counterparts. Although the number of actual civil engineering structures that have been built using these techniques are, at present, relatively few and limited in scope, the potential number of applications is huge. It is quite probable that fiber optic sensors and systems will be an integral component of future smart civil structural systems. The ultimate extent to which civil structures become fitted with smart components will depend on the engineering, economic, and safety cost–benefits of the technology. Most probably the initial smart civil structures will be high-performance structures such as skyscrapers and bridges, and those facilities where a failure presents large safety concerns, such as nuclear waste containment vessels, bridge decks, dams, and structures under construction.

The three components of smart structures technology are (1) sensing, (2) assessing sensory data, and (3) reacting, if necessary. In civil structures, the sensing issues are complicated by the large size of the structure and the multitude of different measurands. Data processing can be performed at several levels of integration. One level is to use an onboard computer that is capable of making judgments and recommending courses of action, such as driving a set of actuators. A more modest approach is to use a computer to preprocess the data so that only the relevant facts are presented to the owners of the

Fiber Optic Smart Structures, Edited by Eric Udd.
ISBN 0-471-55448-0 © 1995 John Wiley & Sons, Inc.

structure, who then make decisions as to how to react, such as performing preventive maintenance.

The primary sensing requirements for civil structures are (1) sensing the external loads, (2) measuring the reaction of the structure to the external loads, and/or (3) determining the internal state of health of the structure. In many respects structural sensing has a very broad application. The construction and performance of virtually every structure is sensed and monitored to a certain degree. The sensing can range from simple visual observations to the use of very sophisticated modern electronic and optoelectronic systems (10). A major challenge in sensing the performance and response of civil structures is that a wide variety of physical parameters need to be measured across the span of a large structure with a lifetime that can extend for decades and even centuries. The sensing requirements are complicated by the fact that most of the measurands exhibit dynamic behaviors with time constants that vary over several orders of magnitude. An additional difficulty is that modern data acquisition hardware can assemble enormous amounts of data, yet only a small fraction of the data is usually significant.

Sensing the external loads applied to civil structures can provide very useful information to the owners, operators, and designers of the structure. Since comprehensive load measurements are often quite different to obtain, especially without the help of modern instrumentation, most of the loads that structures experience are known imprecisely. This results in the necessary making of design and rating decisions with incomplete and uncertain information. If the external loads are measured, the design and rating procedures can be improved. Active structural control systems also require real-time measurements of the external loads.

The external environmental loads that act on a structure during its lifetime are often quite diverse. These loads can include gravity, traffic, earthquakes, weather, waves, floods, fires, chemical attack, and so on. Some of the gravity loads that are worthwhile to measure are the weight of heavy trucks on bridges; the weight and placement patterns of heavy objects in a building, particularly over a 50- or 100-year lifetime; the distribution of the loads in suspension and cable-stayed bridges; and the loads experienced by falsework, shoring systems, and partially completed structures. Construction loads are very important to measure because almost all structures are very vulnerable to damage during construction. Such measurements can increase site safety and reduce construction costs. The measurement of earthquake loads has received a considerable amount of attention over the years. However, despite large expenditures, many major structures in earthquake-prone regions are not yet instrumented for earthquakes. An example is the lack of instrumentation on the San Francisco Bay bridge during the Loma Prieta earthquake. Weather-induced loads in the form of wind, snow, ice, and temperature gradients are rarely measured. Instead, load prediction methods that are based on wind tunnel, computer, and climatic simulations are used, with their attendant shortcomings. Detecting the presence of chemicals that can attack structures, such as the presence of chloride ions in reinforced concrete, can be quite useful in alerting the owner so that palliative actions can be undertaken.

Measuring structural responses in the form of accelerations, strains, deflections, and internal forces is also of great interest. The internal load distribution of most structures is difficult to predict, since most structures are statically indeterminate and slight deviations from nonideal behavior can cause large unexpected stress concentrations. This is particularly the case with structures that have time-dependent material properties, such as occurs in concrete structures in early stages of construction. The reported load and deformation measurements on these structures in the field are few. Since excessive localized internal loads can damage the structure and can be quite dangerous, it would be useful to quantify the load distribution and to develop an early-warning system that can warn of impending damage or collapse, particularly during construction.

Another quantity that is of interest to measure is change in the material state of the system. Most engineering materials, including steel and concrete, have properties that change with time and load history. The change is often manifested as a degradation in the form of fatigue, cracking, or corrosion. However, certain materials, such as concrete and consolidating soils, may actually strengthen with time. Presently, there are very few embedded sensors that are capable of measuring these changes. This is an excellent opportunity for the use of suitably configured embedded optical fiber sensors.

Fiber optic sensor systems have the potential to offer many advantages over conventional electronic systems when applied to civil structures. The overall size of most civil structures poses many problems for the sensor system, some of which can possibly be overcome through the use of fiber optic sensors. One of the major problems with conventional sensor systems is the cabling. A sensing system on a major structure may require more than 100 data channels. A conventional electronic analog sensor cable requires at least five high-quality parallel wires (two for signal transmission, two for power supply, and one for shielding) inside a weatherproofed jacket with strain relief for each transducer. These cables can be very expensive, heavy, and bulky. Many cabling problems can be reduced significantly with the use of fiber optic cables, which can carry the excitation and signal light over the same line, can be multiplexed, and are relatively light in weight. The immunity of fiber optic sensors to electromagnetic interference is also a significant advantage. The long cable lengths associated with the use of conventional electronic sensors can cause the formation of large antennas which can pick up noise, create ground loops, and are susceptible to lightning strikes. Fiber optic cables are relatively immune to these problems. Another advantage of optical fibers is that they are resistant to corrosion, which can be a serious problem for long-term monitoring systems that are exposed to the elements. Additional advantages of fiber optic sensors are due to their geometry. Fiber optic sensors can be configured to measure quantities that are distributed over a distance. Distributed sensing may be useful in many applications, due to the large size of the structure. The ability of fiber optic sensors to be embedded in tight areas can be particularly advantageous for concrete and geotechnical structures.

All of the perceived advantages of fiber optic sensor systems raises a question as to why more structures are not now configured with such sensors. The answer lies primarily in the time scales required for the testing, verification, and acceptance of new civil structural engineering technologies. The time scales of the design, fabrication, and useful life of civil structures is quite long, sometimes being measured in centuries. As a consequence, the adoption and implementation of new concepts and technologies requires longer time periods than those for most other engineering disciplines.

23.2. EXISTING APPLICATIONS

Fiber optic sensors applied to civil structural engineering components and systems have now been tested in several laboratories and have been installed in a few full-sized structures. The laboratory results have been quite promising. The majority, if not all, of the field installations are currently of the research or demonstration type. Following is a brief description of some of the laboratory and fields studies.

Reinforced concrete has been one of primary objects of fiber optic sensor research. Reinforced concrete is a ubiquitous construction material that is formed from several components, including aggregate, cement, sand, reinforcement, and various chemical additives. As a result, it exhibits many complex behaviors, most of which are initiated or occur underneath the surface. Since concrete is initially a reactive semiliquid material that is cast into forms, there are a significant number of sensing applications where the sensors are embedded into the concrete at the time of construction. The use of embedded fiber optic sensors to measure physical quantities in concrete was suggested by Mendez et al. (19), who indicated that the alkali nature of concrete will damage the silicon in glass, but that damage can be avoided by jacketing of the glass fibers with plastic buffers. A series of laboratory tests and installation procedures were proposed. These included using Rayleigh backscattering coupled with optical time-domain reflectometry (OTDR) to measure the temperature in large reinforced concrete structures, detecting cracks by intensity transmission measurements and interferometric strain measurements.

Some of the early laboratory studies of the use of fiber optic sensors in reinforced concrete were reported by Huston and Fuhr (11) and Fuhr et al. (7). These studies involved embedding optical fibers of various types inside small $0.1 \times 0.1 \times 1.0 \, \text{m}^3$ reinforced concrete beams (Fig. 23.1). The initial tests focused on whether optical fibers could survive the concrete embedding and curing process. Four multimode fibers with various jacketing treatments were placed into a beam. The ability of the fibers to transmit light was monitored with a simple device consisting of a regulated LED, a photodetector, and a multiplexor. The fiber transmissibility was monitored over a 28-day cure cycle

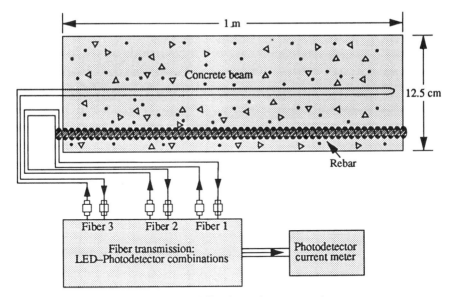

Figure 23.1. Typical fiber layout in a concrete beam.

(Fig. 23.2). Three of the fibers survived. Fiber 3 failed on day 17. A detailed analysis of the broken fiber indicated that the fiber failed at the connector and not inside the beam. Next, the beam was loaded in four-point bending until failure. The transmission through fiber 2 was monitored as a function of the applied load. At a load of approximately 55 kN, the fiber transmission began to drop off (Fig. 23.3). It is believed that the transmission loss was due to the cracking of the concrete, which caused the fiber to crack. The internal cracking of the fiber was verified by an OTDR measurement (Fig. 23.4).

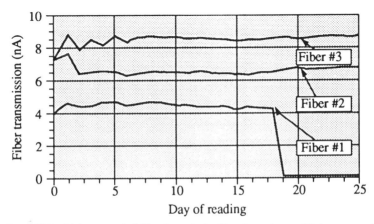

Figure 23.2. Light transmissibility through the fiber throughout a 28-day cure cycle.

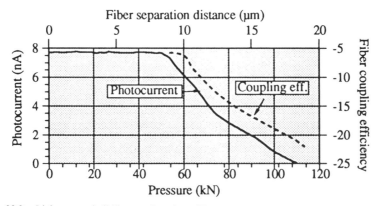

Figure 23.3. Light transmissibility as a function of four-point bending load in the concrete beam.

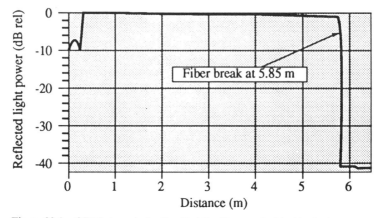

Figure 23.4. OTDR trace indicating that the fiber cracked inside the beam.

The presence of vibrations in a similar concrete beam was detected using multimode optical fibers fitted with statistical mode sensor (SMS) vibration sensors (13). The beam was mounted on supports placed at 0.22 of the length of the beam from its ends. This support condition is the same as used in a xylophone and simulates free–free boundary conditions. Vibrations were induced by shaking the beam at midspan with an electromechanical shaker (Fig. 23.5). The motion was monitored with a surface-attached accelerometer. Figure 23.6 shows spectra of both the accelerometer and the fiber optic vibration sensor when the shaker was driven at 80 Hz. The two signals were highly correlated. The SMS interferometric fiber optic vibration sensor has also been used to monitor the vibrations of a steel member on a one-third scale mode of a truss bridge (12) (Fig. 23.7). These tests indicated that the SMS vibration sensor could detect the vibrations and compared well with that of an accelerometer (Fig. 23.8).

Additional laboratory studies reported by Escobar et al. (5) used bonded

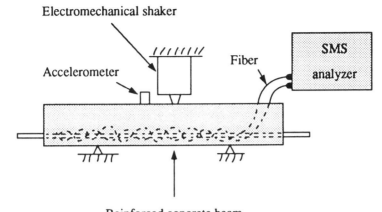

Figure 23.5. Vibration measurement setup for the concrete beam.

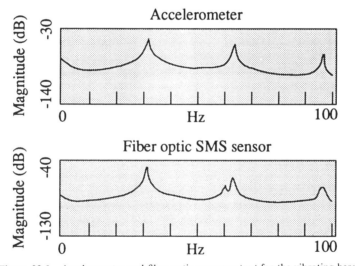

Figure 23.6. Accelerometer and fiber optic sensor output for the vibrating beam.

fiber optic strain gauges on concrete beams, and Kruschwitz et al. (15) used extrinsic Fabry–Fizeau interferometric (EFFI) strain gauge in concrete beams. Maher and Nawy (16) report the successful use of fiber optic Bragg grating strain gauges in laboratory concrete beam tests. Masri et al. (17) used an array of extrinsic Fabry–Perot interferometric (EFPI) strain gauges in a large cruciform concrete beam test.

Ansari (2) reported the development of a fiber optic probe that is capable of measuring the amount of air entrained in fresh concrete. The proper level of air entrainment makes concrete much more resilient to freeze–thaw cycles. However, it is difficult to measure the level of entrainment when the concrete is in a fresh semiliquid state. The sensor is based on measuring the difference

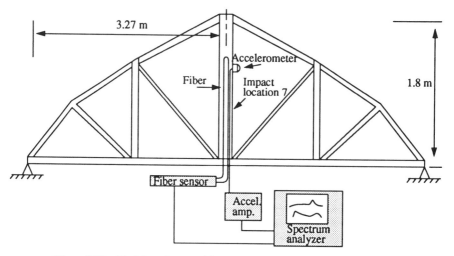

Figure 23.7. Model steel truss with bridge with fiber optic sensor attached.

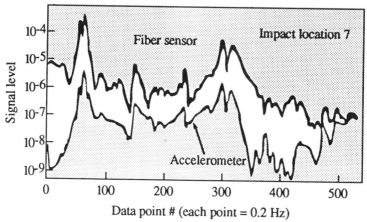

Figure 23.8. Vibration spectra from the accelerometer and fiber optic sensor attached to the model bridge.

in the index of refraction caused by air bubbles in the concrete. Air bubbles will cause light to reflect back from the end of the fiber. If the concrete does not contain any air bubbles, it will not reflect light.

Microbend technology has been used to measure the tension in a posttensioning strand of a rebar (1). The fiber is spiral wrapped around a posttensioning strand. When the strand is placed in tension, the microbend action of the fiber and the strand prevents light from traveling through the fiber. When the tension is relaxed, light can pass through the fiber and the state of relaxation can be detected. These sensors have been installed in a Paris subway and at the Marienfeld pedestrian bridge. A microbend technique was used by Ansari and Navlurkar (3) to measure the dynamic propagation of cracks in concrete beams under bending.

The microbend technique can be adapted to measure debonding between reinforcing bars and the concrete matrix. Debonding and the ability of structures to resist it are crucial to the survival of many structures in their limit states, as occurs in strong earthquakes. Unfortunately, there are very few techniques available for the measurement of debonding and for determining the internal mechanisms of debonding activity in complex structures. A simple debond detector can be constructed by running a fiber through a predrilled reinforcing bar. Debonding at the steel–concrete interface will damage the fiber. Complete debonding can cause the fiber to crack and to fail to transmit light. Simple intensity measurements on such a system as it is loaded to failure can indicate the presence of localized debonding action. The results of these simple debond tests are shown in Figs. 23.9 to 23.11.

The ability to wrap optical fiber into metal rope also opens the possibility for embedded strain detection along the length of the metal rope. May et al. (18) used EFFI strain sensors to monitor steel ropes. A similar installation in a linelike structure is reported by Harrison and Funnel (8), who monitored the overheating of power transmission lines with fiber optic sensors.

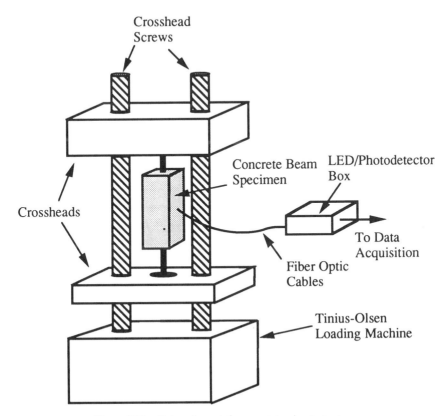

Figure 23.9. Debonding reinforcement tension test setup.

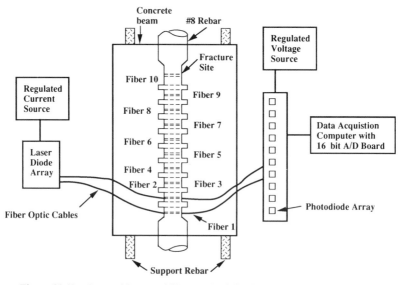

Figure 23.10. Internal layout of fibers and reinforcing bars in the debond test.

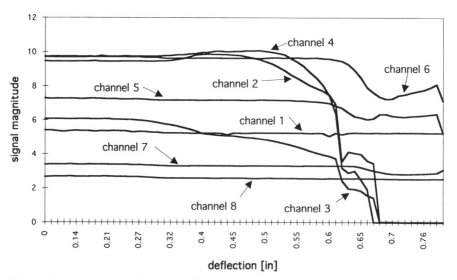

Figure 23.11. Intensity readings of the fibers versus reinforcing bar deflection and debonding.

Large-scale structures that have actually had fiber optic sensors embedded into them are still relatively few. Holst and Lessing (9) have installed fiber optic displacement gauges in dams to measure shifting between segments. Fuhr et al. (6) report the installation of fiber optic sensors into the Stafford Building at the University of Vermont (Fig. 23.12). This study describes many of the important issues that must be addressed if fiber optic sensors are to be successfully buried in a structure during construction. These include coordina-

Figure 23.12. Installation of the optical fibers in the formwork of the Stafford Buildings as it is under construction.

tion with the construction crews, placing the optical fiber along the reinforcing bars so that it will not be damaged when the concrete is poured, and developing schemes for gracefully exiting from the concrete members without damage when the forms are stripped away from the members. Figure 23.13 shows an exit detail that was quite successful. This involved placing an electric conduit box in the concrete, butted up against the form. The excess fiber was placed in the box and wrapped up with tape to prevent the concrete from entering the box. Figures 23.14 and 23.15 show the results of simple tests that verify the efficacy of the fiber optic sensor installation in a vibration isolation slab that is located in the basement of the Stafford Building. These tests consisted of impact-response measurements of elastic waves propagating through the slab. A sledgehammer fitted with a piezoelectric load cell provided the impact source of energy. The response was measured with an embedded fiber optic cable and an SMS sensor.

Similar installations have been conducted at the Winooski One dam in Winooski, Vermont (14), in the physics building at the University of Cork in Ireland (B. McCraith, personal communication, 1993), and in a railway overpass bridge in Middlebury, Vermont (14). Figure 23.16 is a layout of the Winooski One dam site on the Winooski River in Vermont. Figure 23.17 shows some of the fiber layout in the Winooski One dam. Figure 23.18 shows a comparison of accelerometer and fiber optic time histories that were collected at the Winooski One dam in a location that was close to one of the hydraulic power generation turbines. The fiber optic (SMS) sensor shows a distinct peak at 168 Hz (Fig. 23.19), which has been determined to correspond to the meshing of gears in the turbine–generator transmission.

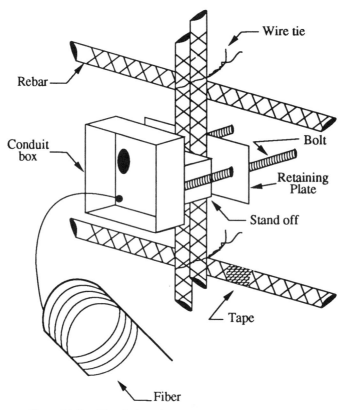

Figure 23.13.   Electrical conduit box for gracefully exiting the forms.

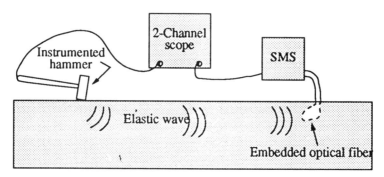

Reinforced concrete vibration isolation slab

Figure 23.14.   Impact-response measurements with fiber optic sensor in a vibration isolation slab in the Stafford Building.

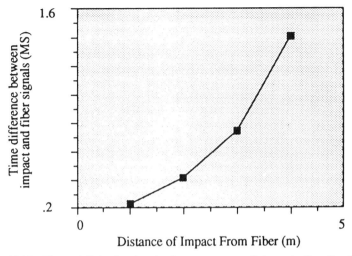

Figure 23.15. Time of flight for the elastic waves versus distance in the vibration slab impact-response tests.

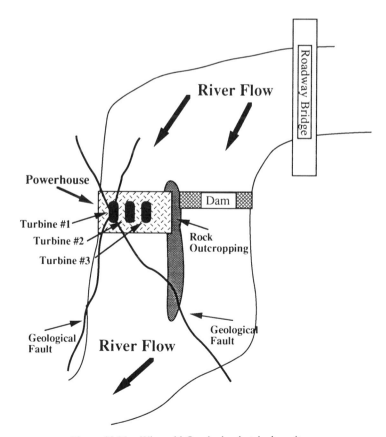

Figure 23.16. Winooski One hydroelectric dam site.

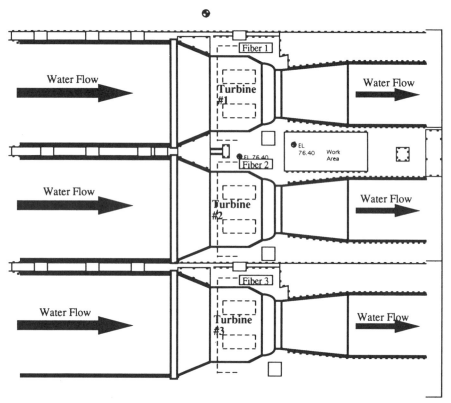

Figure 23.17. Top-down view of the Winooski One hydroelectric dam powerhouse, showing the turbine mounts and some of the fiber layout.

Other field installations have used fiber optic sensors to measure the weight of trucks as they drive over sensitive strips at highway speeds (21, 22). Similar techniques have been applied to intrusion detection systems. Another reported field stress measurement is by Caussignac et al. (4), who measured the internal stresses in neoprene bridge bearings.

23.3. FUTURE STUDIES

In this section we provide some suggestions and speculations on future areas of research for fiber optic sensing applications to civil structures. Since this is a rapidly advancing technical field, it is highly likely that several of these procedures will be undergoing development at the time of publication. One area of research is in the development of sensors for monitoring the curing process in concrete. Concrete curing is a complex chemical reaction between aggregate, lime, water, and various specialty chemicals. The amount of water or hydration in the concrete as it cures is very critical to the performance of

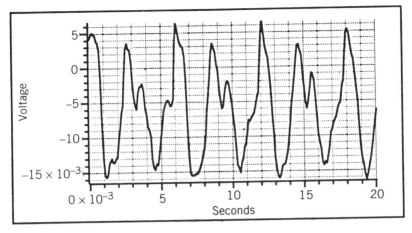

FIBER RESPONSE

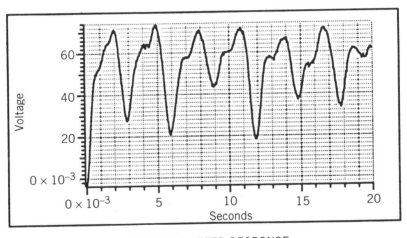

ACCELEROMETER RESPONSE

Figure 23.18. Simultaneous fiber optic and accelerometer time histories gathered from a turbine mount at the Winooski One dam.

the concrete. If there is too much water in the mix, the concrete is weaker and it is prone to shrinking and cracking during curing. If not enough water is used in the mix, the concrete is difficult to handle and there may be insufficient water available for the reaction to be completed. The hydration and curing process could be monitored through an embedded chemical type of fiber optic sensor that measured the amount of hydration in the concrete as a function of time. A measurement of the amount of curing that had occurred would be very useful during the construction process because the decision as to when to

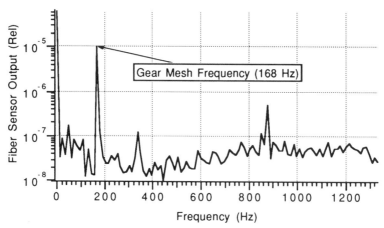

Figure 23.19. Fiber optic vibration spectrum from a turbine mount at the Winooski One dam.

remove forms and shoring is highly dependent on the state of cure. If the forms are removed too quickly, the structure is unsafe and may collapse. If the forms are removed later than necessary, the contractor loses money.

There is considerable interest worldwide in intelligent vehicle highway systems (IVHSs), which will create an intelligent and integrated traffic management and control system, primarily for congested urban environments. A key aspect of IVHS is to sense the presence of vehicles. Fiber optic vehicle presence detectors could easily be placed into pavements. Such embedded sensors could possibly be more durable, yet have less expensive cabling requirements than their conventional counterparts.

Corrosion of civil structures is a persistent problem. Structures that are prone to corrosion are often those that are subjected to deicing salts, such as bridges, pavements, parking garages, and those that lie in coastal regions that are subjected to ocean spray. Both steel and reinforced concrete structures are subject to corrosive attack. However, due to the difficulty in repairing reinforced concrete, its corrosion problems are perhaps more critical. The precise mechanism by which salt causes corrosion damage to concrete is not entirely well understood. It is generally believed that the basic phenomenon is that chloride ions from the salt penetrate the concrete through either pores or cracks and attack the reinforcement, thereby accelerating corrosion. Corroded reinforcement steel swells to a volume that is up to five times larger than that of the original steel. This swelling causes spalling and cracking of the concrete, which further exposes the reinforcement to corrosion. Most methods of corrosion reduction depend on preventing the chlorides from reaching the steel reinforcement, such as through the use of impermeable membranes, improved drainage, reduction in the use of road salt, and high-quality crack-free concrete construction practices.

At present there are very few methods of nondestructively determining whether corrosion processes are under way inside a given concrete structure, until the corrosion has progressed to where visible damage such as spalling has occurred. One possible sensing technique would be to embed a fiber optic chemical sensor into the concrete that is sensitive to chloride ions. Another possibility would be to use fiber optic sensors to detect corrosion-induced color changes that occur in reinforced concrete. A simple scheme would be to detect the brown color of rust. These sensors could then be multiplexed by time-domain resolved spectroscopy.

Corrosion of critical steel members that are not easily subjected to visual inspection are excellent candidates for embedded or surface-attached fiber optic corrosion sensors. One example is that of the steel cables in cable-stayed and suspension bridges. The steel cables usually consist of a large number of parallel strands that are wrapped in several layers of protective coatings. Severe corrosion of these cables would seriously compromise the integrity of the structure. The present method of corrosion detection requires stripping away the outer coatings and inspecting the cables visually. Since the procedure is costly, and since most portions of the cables are awkward to access, an embedded cable corrosion detector would be quite attractive.

The measurement of aerodynamic (wind) and hydraulic loadings on structures with conventional sensors is limited the number of channels that must be monitored in order to take the measurements. Since pressures are distributed over the surface of the structure, it is necessary to take a multitude of point pressure measurements or a few distributed measurements to gain an accurate picture of the loading. Fiber optic sensors offer some interesting possibilities in this regard. One possibility would be to multiplex several fiber optic pressure sensors on a single line. Another possibility would be to use distributed-type pressure sensors. At present, most fiber optic pressure transducers are not designed for wind or hydraulic loads. The requirements for accurate wind pressure measurements are quite demanding since wind pressures are often the same order of magnitude as the shift in the barometric or atmosphere pressure. The main applications of these sensors would include dams and the sides of skyscrapers, when there is still considerable uncertainty as to what the pressures are for the design of the windows and cladding.

Electric power transmission systems offer an excellent opportunity for the application of fiber optic sensing technologies. In addition to overheating, power lines can be subjected to damage by wind and ice, such as by galloping (20). The EMI resistance of the fibers enables them to be wrapped around the power lines without any degradation of the optical signals. A major problem with power lines is galloping, which occurs when ice forms on the cables, creating a noncircular aerodynamic cross section. The combination of the aerodynamic forces on the nonsymmetric cable and dynamic cable forces can cause large amplitude vibrations that are difficult to suppress. Fiber optic vibration sensors might offer an effective method of measuring and monitoring the extent of galloping problems on remote transmission lines. Fiber optic sensors can also use OTDR instrumentation to detect damage in the lines and to locate failures such as may happen after an intense storm. A further application is to use optical fiber for the control of circuit breakers along a broken live transmission line, where the EMI may cause a conventional control system to fail. One of the more interesting applications of fiber optic sensors in concrete is to use them to detect acoustic emission events. Acoustic emission occurs when permanent damage occurs in the concrete. This causes the concrete to crack, which creates high-frequency elastic waves, which can be detected. The standard technique for detecting acoustic emission signatures is to use an array of ultrasonic transducers that are attached to the surface of the concrete member. This technique has proven to be quite effective in the laboratory with small concrete members. However, using acoustic emission to monitor the health of ultrasonic transducers in the field is probably impractical, due to the rapid attenuation and scattering of the waves in concrete. Embedded fiber optic sensors might be more practical, especially when the monitoring system is configured for event counting rather than counting and location.

A final possible application of fiber optic sensors is in geotechnical systems. One of the more promising developments in recent years in geotechnical engineering has been the use of engineered geotextiles to stabilize structures. A major issue of concern is the long-term performance and efficacy of the geotextile systems. Weaving fiber optic sensors directly into the fabric of the geotextile could result in an effective measurement system.

REFERENCES

1. Anonymous. *The Pedestrian Bridge to Marienfelde Leisure Park*. Strabag Bau-AG, Berlin Branch, Bessemer Str. 42a, 1000 Berlin, 1992.

2. F. Ansari, Rapid In-Place Air Content Determination in Fresh Concrete. *ACI Concr. Int.*, January, pp. 39–43 (1991).

3. F. Ansari and R. K. Navlurkar, A Fiber Optic Sensor for the Determination of Dynamic Fracture Parameters in Fiber Reinforced Concrete. In *Applications of Fiber Optic Sensors in Engineering Mechanics* (F. Ansari, ed.), pp. 160–176. *Am. Soc. Chem. Eng.*, New York, 1993.

4. J. M. Caussignac, A. Chabert, G. Morel, P. Rogez, and J. Seantier, Bearings of a Bridge Fitted with Load Measuring Devices Based on an Optical Fiber Technology. *Proc. SPIE — Int. Soc. Opt. Eng.* **1777**, 207 (1992).

5. P. Escobar, V. Gusmeroli, and M. Martinelli, Fiber-Optic Interferometric Sensors for Concrete Structures. *Proc. SPIE — Int. Soc. Opt. Eng.* **1777**, 215 (1992).

6. P. L. Fuhr, D. R. Huston, P. J. Kajenski, and T. P. Ambrose, Performance and Health Monitoring of the Stafford Medical Building Using Embedded Sensors. *Smart Mater. Struct.* **1**, 63–68 (1992).

7. P. L. Fuhr, D. R. Huston, T. A. Ambrose, and S. Snyder, Curing and Stress Monitoring of Concrete Beams with Embedded Optical Fiber Sensors. *J. Struct. Eng. Am. Soc. Civ. Eng.* **119**(7) (1993).

8. B. J. Harrison and I. R. Funnell, Remote Temperature Measurement for Power Cables. *CIRED Conf., 1991*.

9. A. Holst and R. Lessing, Fiber-Optic Intensity-Modulated Sensors for Continuous Observation of Concrete and Rock-Fill Dams. *Proc. SPIE — Int. Soc. Opt. Eng.* **1777**, 223 (1992).

10. D. R. Huston, Smart Civil Structures—An Overview. *Proc. SPIE—Int. Soc. Opt. Eng.* **1588**, 21 (1992).

11. D. R. Huston and P. L. Fuhr, Fibr Optic Monitoring of Concrete Structures. *Proc. SPIE — Int. Soc. Opt. Eng.* **1798** (1992).

12. D. Huston, P. L. Fuhr, J.-G. Beliveau, and W. B. Spillman, Structural Member Vibration Measurements Using a Fiber Optic Sensor. *J. Sound Vib.* **149**(2), 348–353 (1991).

13. D. R. Huston, P. Fuhr, P. Kajenski, and D. Snyder, Concrete Beam Testing with Optical Fiber Sensors. *Proc. ASCE Minisymp. Nondestr. Test. Concr.*, San Antonio, TX *1992* (1992).

14. D. R. Huston, P. L. Fuhr, and T. P. Ambrose, Dynamic Testing of Concrete with Fiber Optic Sensors. In *Applications of Fiber Optic Sensors in Engineering Mechanics* (F. Ansari, ed.), pp. 134–143. Am. Soc. Civ. Eng., New York, 1993.

15. B. Kruschwitz, R. O. Claus, K. A. Murphy, R. G. May, and M. F. Gunther, Optical Fiber Sensors for the Quantitative Measurement of Strain in Concrete Structures. *Proc. SPIE — Int. Soc. Opt. Eng.* **1777**, 223 (1992).

16. M. H. Maher and E. G. Nawy, Evaluation of Fiber Optic Bragg Grating Strain Sensor in High Strength Concrete Beams. *Applications of Fiber Optic Sensors in Engineering Mechanics* (F. Ansari, ed.), pp. 120–133. Am. Soc. Civ. Eng., New York, 1993.

17. S. F. Masri, M. S. Agbabian, A. M. Abdel-Ghaffor, M. Highazy, R. O. Claus, and M. J. de Vries, *An Experimental Study of Embedded Fiber-Optic Strain Gauges in Concrete Structures,* Civ. Eng. Rep. M9213. University of Southern California, Los Angeles, 1993.

18. R. G. May, R. O. Claus, and K. A. Murphy, Preliminary Evaluation for Developing Smart Ropes Using Embedded Sensors. *Proc. SPIE — Int. Soc. Opt. Eng.* **1777**, 155 (1992).

19. A. Mendez, T. F. Morse, and F. Mendez, Applications of Embedded Optical Fiber Sensors in Reinforced Concrete Buildings and Structures. *Proc. SPIE — Int. Soc. Opt. Eng.* **1170** (1990).

20. E. Simiu and R. H. Scanlan, *Wind Effects on Structures: An Introduction to Wind Engineering*, 2nd ed. Wiley, New York, 1986.

21. S. Teral, Vehicle Weighing in Motion with Fiber Optic Sensors. *Proc. SPIE — Int. Soc. Opt. Eng.* **1777**, 139 (1992).

22. K. W. Tobin and D. Muhs, Alogorithm for a Novel Fiber-Optic Weigh-in-Motion Sensor System. *Proc. SPIE–Int. Soc. Opt. Eng.* **1589**, 12 (1992).

Index